SPOTLIGHT
THE
AGRICULTURAL
REVOLUTION

Christopher Martin

Wayland

SPOTLIGHT ON HISTORY

Spotlight on the Age of Exploration and Discovery
Spotlight on the Age of Revolution
Spotlight on the Agricultural Revolution
Spotlight on the Cold War
Spotlight on the Collapse of Empires
Spotlight on Elizabethan England
Spotlight on the English Civil War
Spotlight on the First World War
Spotlight on the Industrial Revolution
Spotlight on Industry in the Twentieth Century
Spotlight on Medieval Europe
Spotlight on Post-War Europe
Spotlight on the Reformation
Spotlight on Renaissance Europe
Spotlight on the Rise of Modern China
Spotlight on the Russian Revolution
Spotlight on the Second World War
Spotlight on the Victorians

Cover Illustration: Cyrus McCormick's reaping machine (1831).

First published in 1986 by Wayland (Publishers) Ltd,
61 Western Road, Hove, East Sussex, BN3 1JD. England

© Copyright 1986 Wayland (Publishers) Ltd

British Library Cataloguing in Publication Data
Martin, Christopher, 1939
   Spotlight on the Agricultural Revolution
   ——(Spotlight on history)
   1. Agriculture——England——History——
Juvenile literature
   I. Title   II. Series
   630′.942   S455

ISBN 0-85078-757-2

Typeset, printed and bound in the UK by The Bath Press, Avon

CONTENTS

1 Farming before the Revolution	4
2 Enclosures	15
3 Men who made the Revolution	21
4 The Labouring Poor	36
5 The Second Agricultural Revolution	48
6 Afterwards: Golden Age and Decline	60
Date Chart	72
Glossary	73
Further Reading	74
Index	75

1 FARMING BEFORE THE REVOLUTION

Farming methods have changed throughout history, as more productive ways to use the land have been devised. The Agricultural Revolution, of approximately 1730 to 1870, marks a particular phase of rapid change in this long evolution.

The triumphs of the Revolution were two-fold. Firstly, the new farming matched a huge growth in Britain's population (6 million in 1750; 21 million in 1851) with a parallel expansion in food production. It kept Britain self-sufficient, and therefore possibly helped the country to gain victory, during the Napoleonic Wars, and it fed the industrial workers who were making Britain 'the workshop of the world'. Secondly, British farming methods were exported overseas to open up vast farmlands in British colonies in Canada and Australasia, and also in the Americas. These years of change had a human cost, however: another, darker kind of revolution in the lives of the rural poor cast a shadow over the triumphs.

Factories in the industrial cities of the north of England provided alternative employment to farming.

Legumes, like sainfoin, proved valuable as fertilizer and animal feed, helping to increase agricultural production.

Prelude to Revolution

In the seventeenth century, the first beginnings of the Revolution could be seen alongside the last vestiges of the medieval order of farming.

Progressive, commercially-minded landowners and yeoman tenant or freehold farmers had discovered the secret behind the massive increase in agricultural products that marked the later Revolution. By observation, or influenced by farming practice in Holland and Flanders,

A DISCOURS OF HUSBANDRIE
USED IN
Brabant and Flanders:
SHEWING
The wonderful improvement of Land there; and serving as a pattern for our practice in this
COMMON-WEALTH.

The Second Edition, Corrected and Inlarged.

LONDON,
Printed by *William Du-Gard,* dwelling in *Suffolk-lane,* near London-stone, *Anno Dom.* 1652.

Weston's Discours *showed the value of legumes and turnips in crop rotation.*

they had found that legumes (crops like clover, sainfoin or lucerne) were beneficial both as fertilizer and as animal feed. Nitrogen from the air was 'fixed' in the roots of these plants: when ploughed in, they helped to replenish depleted soil. They also provided a good animal feed, and this allowed the keeping of more livestock and the production of more manure. Arable land could therefore be doubly enriched, and so produced more corn. The practice of reserving one weed-infested fallow (or resting) field could be avoided and the land was always growing something useful. The fairly widespread adoption of legumes is shown by the substantial sale of clover seed in London in the 1620s.

The Surrey landowner, Sir Richard Weston, confirmed and took this discovery a step further. Exiled as a Royalist in Flanders during the Civil War, he studied local farming and was impressed by the Flemish use, not only of clover but also of turnips in their field rotations. These could provide cattle with food throughout the winter: large numbers therefore need not be slaughtered in autumn, as was the custom. Weston saw that the new crops would dramatically increase the production of corn and would revolutionize the keeping of cattle. He set down his ideas in his important book, *Discours of the Husbandry used in Brabant and Flanders* (1651). Over a century later, the farming writer, Arthur Young, looked back to see Weston as 'a greater benefactor than Newton'. 'He did indeed offer bread and meat to millions,' commented the historian, Lord Ernle.

Open Fields
The old order was seen in the survival of the open field system. About half the arable land in England was still cultivated in this way in 1700. The open fields traced their origins back to the Anglo-Saxon settlement of Britain. Three huge fields beside a village were divided into strips.

Medieval farmers plough the field strips with oxen, while other labourers smash clods of earth with hammers.

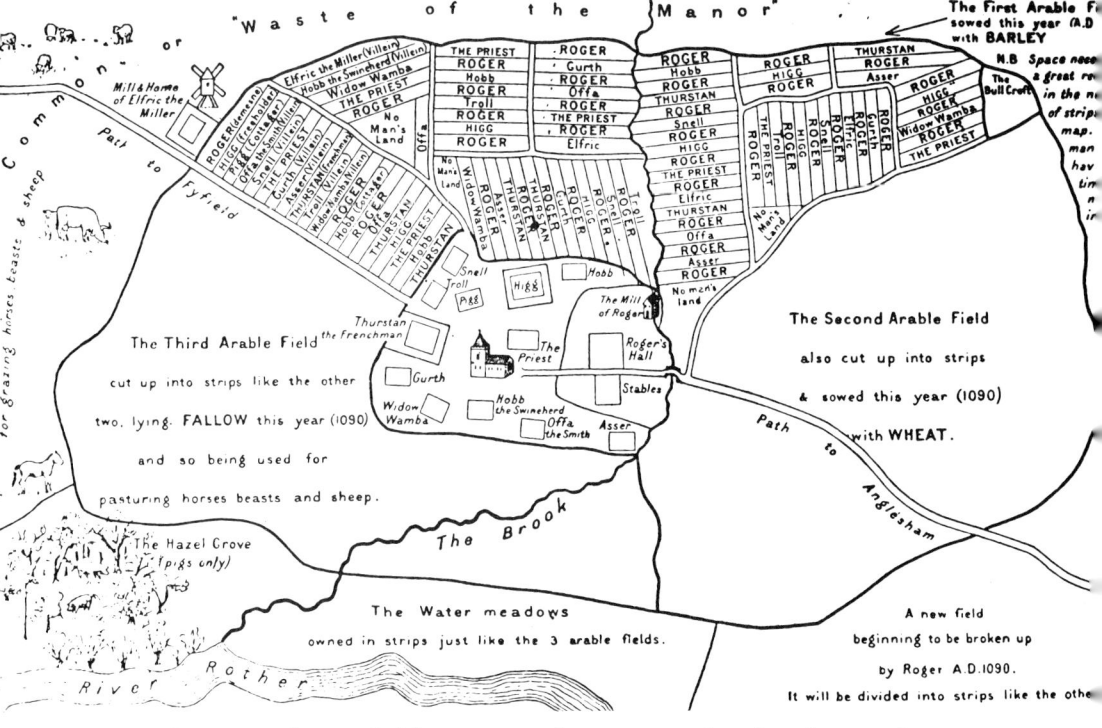

A diagram of open fields, waste and common, showing the rotation of crops. Each owner would have had five times the number of strips that are shown here.

Village farmers owned a selection of these strips. Men joined together to provide labour for harvest or animals for ploughing. In some places the ancient three-course rotation of crops was still followed, one field lying fallow each year, while the other two were sown with cereals. Beside the fields was the common meadow-land, where each strip owner had hay and grazing rights. Then there was the wild common or 'waste', rough ground or woodland, where villagers had traditional cherished rights to gather food or fuel.

There were advantages to this type of community. Almost everyone occupied a useful place. There were chances for modest ambition. A labourer might rent a cottage with common rights, and save up to buy strips of land. His wife would often be employed in domestic industry, such as straw-plaiting in Bedfordshire, blanket-making at Witney, or stocking weaving in Nottinghamshire. Above all, the cottager was self-sufficient: the strips grew his food; the common fed his cow and pig; its turf provided fuel for the fire which smoked his bacon and baked his bread.

Although the physical labour was very hard, and the threat of disease and early death ever present, there was a harmony about country life under this old order. We can glimpse this in a letter of Dorothy Osborne, written to a friend in 1653: 'You ask me how I pass my time here ... I walk out into a common that lies hard by the house, where a great many young wenches keep sheep and cows and sit in the shade singing of ballads. I talk to them and find they want nothing to make them the happiest people in the world, but the knowledge that they are so ...' The simplicity and beauty of rural life were not just a poet's invention.

Beginnings of enclosure

Propagandists of the Agricultural Revolution stressed the defects of the open fields. Ground was wasted in the grass balks ridges that divided the strips. There were continual arguments about encroachments. Time and money were lost in working the scattered strips. The common rights encouraged a 'lazy, thieving sort of people'. Weeds and animal diseases spread quickly. Livestock kept on the commons were underfed and observers spoke of 'greyhound-like sheep, pitiful, half-starved animals' or cattle like 'living skeletons'.

Grass balks divided village farmers' strips in open fields. In this print of Cambridgeshire, you can see the divisions.

Such propagandists ignored the improvements that were already changing the open fields: the use of legumes on the fallow in some places, or the cutting off of hedged closes for animal rearing (at Laxton in Nottinghamshire, where open fields still exist today, nearly half the land lay in such closes by 1691). Instead they supported enclosure, that is the amalgamation of strips into compact farms with hedged fields, which would give the efficient farmer more control, better productivity and more profit.

Enclosures were not a new phenomenon. Some parts of England had been enclosed already, or had never adopted the open field system when the land had originally been cleared from the forests. In medieval,

At Laxton, in Nottinghamshire, open fields survive and are still worked today.

The practice of sheepfolding. Villages were often deserted as the introduction of farming under enclosure prevented labourers from making a proper living off the land.

Tudor and Stuart times, many enclosures had been made by private agreement. Some, especially during the years of the booming wool trade in the fifteenth and sixteenth centuries, had been made by force or fraud. Greedy speculators had enclosed land as sheep runs, sometimes depopulating whole villages in the process. Sir Thomas More, the statesman and writer, commented bitterly: 'Your sheep that were wont to be so meek and tame ... they consume, destroy and devour whole fields, houses and cities.' On the whole, such enclosure was seen as a social menace and was resisted by Tudor and early Stuart monarchs and Parliaments.

It was perceived, however, that enclosure might be useful in the 'wastes', the areas of England still untouched by cultivation. In the late seventeenth century it was estimated that a third of the landscape was still waste: marsh, forest, heath or mountain. These areas were, it was felt, a national disgrace—'filthy blotches on the face of the country'. Sherwood Forest, for example, still covered part of Nottinghamshire; expanses of Derbyshire were 'black regions of ling'; and acres of broom spread across Northumberland. Near to London, Hounslow Heath was described as 'fit only for Cherokees and savages', while Epping Forest was infested with robbers.

The drainage of the Fens was carried on from the seventeenth to the nineteenth century.

The most dramatic advance into the 'waste' had been made in East Anglia, where the Fenlands, today one of the most productive districts of England, were 700,000 acres of marsh and bog. A group of local landowners invited the Dutch engineer, Vermuyden, to drain this tract of country, a task which continued from 1630 to the 1650s. The work was only partially successful as the local Fenmen resented the creation of farmland from their swamps, where they had survived by fishing and fowling. Anxious to return the land to 'Captain Flood', they broke the embankments and opened the sluices. The work was not fully completed until the nineteenth century.

A decisive change of climate in favour of enclosure came during the Commonwealth era after the period of Civil War (1642–1653). Extremist groups, known as Diggers and Levellers, argued that the land belonged to the people, that 'the common people ought to dig, plough, plant and dwell upon the commons, without hiring them or paying rent to any'. When the Diggers put their theory into practice and occupied St George's Hill in Surrey in 1649, they were promptly driven off by Cromwell's troops. Later, in 1656, Parliament threw out an anti-enclosure Bill. This was a significant shift of opinion in favour

Bustling market scenes like this one at Smithfield featured frequently in Daniel Defoe's description of England.

of enclosure, and marked the real beginning of the great enclosure movement.

Another foundation stone of the Agricultural Revolution was the spirit of scientific inquiry fostered by the Royal Society, which began in 1662. 'The new philosophy of Science', wrote the Society's historian, Thomas Sprat, 'shall enrich us with all the benefits of fruitfulness and plenty'. At this time, botanical gardens were established to encourage the study of growing and improving plants, and an intelligent, progressive farming literature began.

Scottish and Welsh drovers moved thousands of cattle and sheep to London markets.

The commerical climate of British farming on the eve of the Revolution is vividly described by the journalist-writer, Daniel Defoe, in his *Tour of Great Britain*, published in the 1720s. He presents a picture of bustling activitiy and flourishing markets, dominated by those of the mighty capital. He gives fascinating glimpses: of Scottish and Welsh drovers moving thousands of cattle across England to be fattened in the home counties before sale at Smithfield Market; of hundreds of Norfolk turkeys being laboriously walked in flocks to London; of five hundred thousand sheep crowding the great fair at Weyhill, Andover; of the many specialist products (like cheeses or cider) produced by the various regions and carried to London by barge or coastal ship to avoid the 'deep, dirty roads' of the time.

It was in this vigorous atmosphere that the Agricultural Revolution had its beginnings.

2 ENCLOSURES

The enclosure movement of the eighteenth and nineteenth centuries was different from previous practice in these respects: it was supported by Parliament; it took in large tracts of hitherto unexploited country; it swallowed much of the old village common; and, with astonishing speed, it changed the English landscape.

The enclosure process
There were only eight Enclosure acts put through Parliament before 1714, eighteen under George I (1714–27) and 229 in George II's reign (1727–60). Yet in the eighty years after 1760 there were 2,500 Acts, incorporating some four million acres of open fields. Besides these, 1,800 Acts enclosed nearly two million acres of 'waste', mostly in the north and west, between 1760 and the end of the nineteenth century. The rage for enclosure reached a high point during the Revolutionary and Napoleonic Wars (1793–1815), when even parts of Dartmoor and Exmoor were ploughed to grow corn.

The enclosure movement mainly affected the great tract of land running across the centre of England from Dorset to Yorkshire. Here the open fields, with their patchwork of strips, gave way to a chequer board of

These late seventeenth-century views show enclosed fields on hills near Oxford, while open fields still existed near Cambridge.

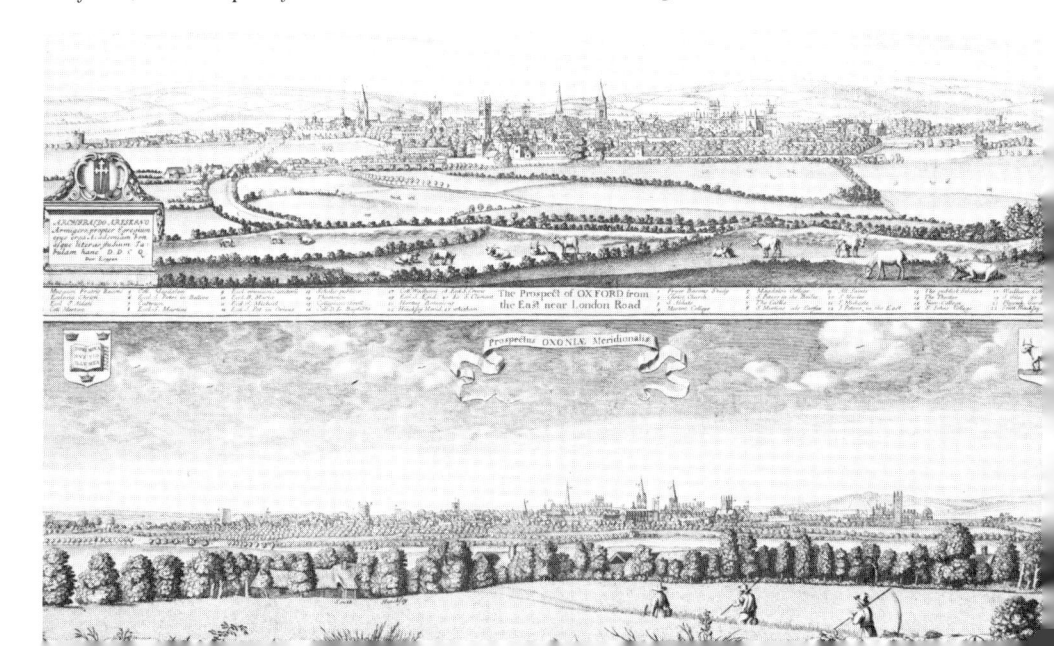

A wealthy landlord tours his estate. Such men were influenced by Townshend and Coke, and began to farm their land more productively.

small hedged fields, often with new roads, trees and farmhouses. Half of Northamptonshire was altered in this way. By 1820, only five per cent of farmland was worked as open fields.

Large landowners and the larger freeholders or tenant farmers benefited most from enclosure, the former because of larger rents and the ease of dealing with fewer tenants, the latter because enclosed land was easier to run and yielded better profits than any set of strips. It was men like these who initiated Parliamentary Acts of Enclosure. Although it was expensive, the Parliamentary process was soon preferred to private local agreements between interested parties. Large areas of land could be redistributed in one block, and the owners of the largest part of the village land could force their scheme onto the lesser men.

Procedures

An Act began when a local landowner sent a petition to Parliament, setting out the advantages of the proposed change. A Bill supporting the petition would be read twice and referred to a Commons Committee, which considered evidence against it. Then the Bill would receive Royal Assent. At each stage of the procedure, the large landowner was supported by men of similar interests. This may seem unfair to us, but that was an undemocratic era when property rights were dominant. Public discussion of the plan was minimal. Proposals were displayed briefly in notices pinned to the church door of the affected district. The General Enclosure Acts of 1801, 1836 and 1845 made the process easier and cheaper, again favouring the larger landowner.

Commissioners—usually country gentlemen, farmers or land agents— were appointed to listen to claims and to divide the land. The size of the allotments were made in proportion to legal rights established by claims. Some historians have doubted the justice of the commissioners' work, but recent studies have shown that they were, in fact, very fair and painstaking. It was true, however, that the smallholder often had to sell the land he was given because he could not afford the legal fees or the fencing costs of enclosure.

Sometimes (but not often) there was violent opposition to a scheme. Otmoor in Oxfordshire was commonland, used by cottagers to rear large flocks of geese. In 1814, a proposal to drain and cultivate the

Commissioners' maps like this one from East Heslerton in Yorkshire show how awards were made after enclosure.

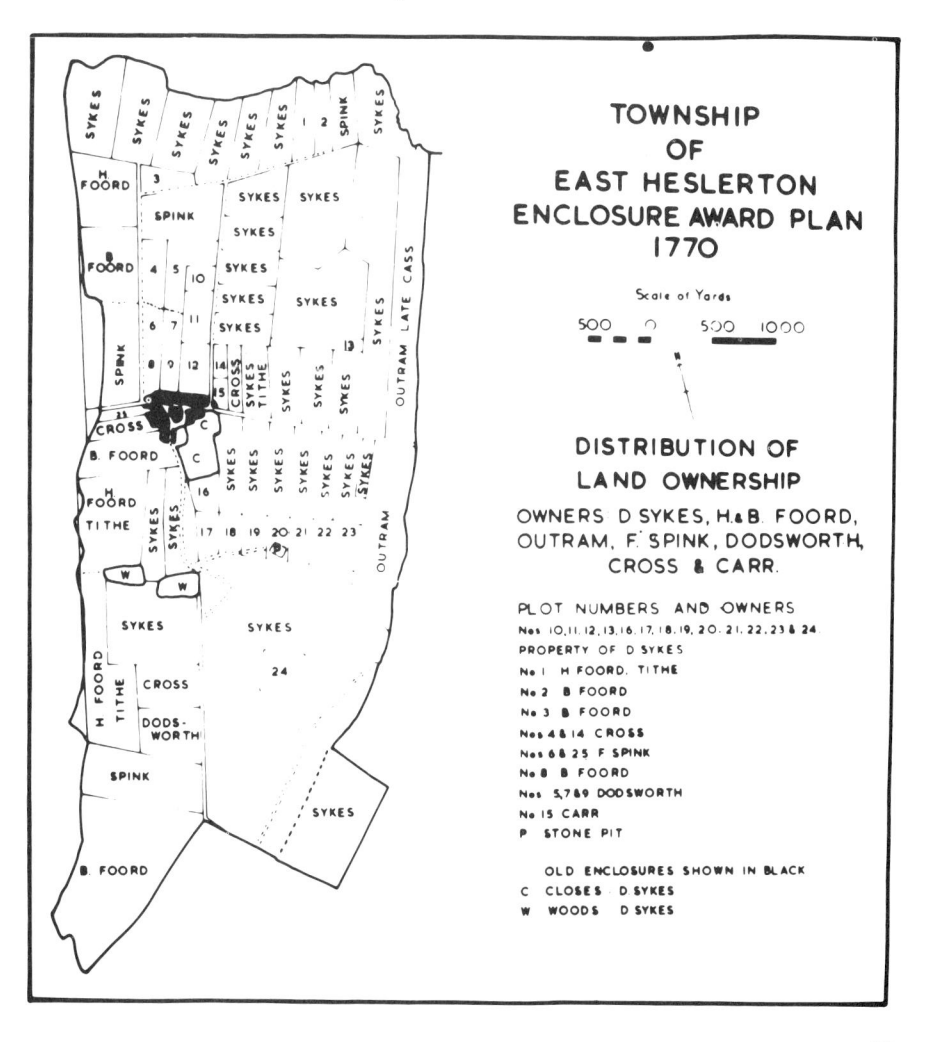

TOWNSHIP OF EAST HESLERTON ENCLOSURE AWARD PLAN 1770

Scale of Yards
500   0   500   1000

DISTRIBUTION OF LAND OWNERSHIP

OWNERS D SYKES, H.&B. FOORD, OUTRAM, F. SPINK, DODSWORTH, CROSS & CARR.

PLOT NUMBERS AND OWNERS
Nos 10,11,12,13,16,17,18,19,20,21,22,23 & 24
PROPERTY OF D SYKES
No 1 H FOORD, TITHE
No 2 B FOORD
No 3 B FOORD
Nos 4 & 14 CROSS
Nos 6 & 25 F SPINK
No 8 B FOORD
Nos 5,7 & 9 DODSWORTH
No 15 CARR
P STONE PIT

OLD ENCLOSURES SHOWN IN BLACK
C   CLOSES   D SYKES
W   WOODS   D SYKES

common was torn from the church door. One of the scheme's advocates criticized local poor: 'In looking after a brood of goslings, a few rotten sheep, a skeleton of a cow ... they acquired habits of idleness and dissipation and dislike to honest labour'. When the enclosure was made, resentment turned to defiance: a thousand men, women and children smashed the new hedges. Troops were called and arrests made. For several years the 'war' continued, moonlit nights being a time for breaking down the fences.

The new landscape
Once allotments had been made, successful men would fence off their land. Many of the commissioners' maps of the new landscapes survive: they show how bold straight, wide-verged roads were cut, or how streams were diverted. The new fields could be handsome: William Cobbett, the journalist and social reformer on one of his *Rural Rides* in 1821, described some near Royston in Hertfordshire: 'The fields on the left seem to have been enclosed by Act of Parliament and they certainly are the most beautiful fields that ever I saw ...'

The hedges were quickthorn, which was fastest to grow. At first the seedlings were protected by a double wooden fence. This meant that the first effect of enclosure was to create a bleaker landscape because the hundreds of miles of fencing meant the felling of thousands

A typical pattern made by enclosed fields.

John Clare, the Northamptonshire poet, who protested against enclosure.

of trees. After some years, the hedges were laid and became another glory of the English countryside, especially in spring. Many years later, the writer, Katherine Mansfield, looked from a train at these hedges: 'The country was simply bowed down with beauty—heavy, weighted down with treasure. Shelley's moonlight may glittered everywhere.' Elm trees were planted in the hedges: elm was valuable because of its use in ship-building. Isolated farm houses were constructed on some of the new holdings.

John Clare at Helpston
Enclosed fields introduced more efficient agriculture. Many writers praised the improved order but an exception was the 'peasant poet', John Clare (1793–1864). In childhood he had loved to explore Helpston Heath near his Northamptonshire home. The enclosure of Helpston in 1809 devastated Clare's world:

> But now, alas! your hawthorn bowers
> All desolate we see!
> The spoiler's axe their shade devours,
> And cuts down every tree . . .

Petticoat Lane in London in the eighteenth century. As towns became larger and more congested, the townspeople demanded that the land around them should be left as commons.

His poem, *The Moors*, is his elegy for a lost Eden.

> Now the sweet vision of my boyish hours
> Free as spring clouds and wild as summer flowers
> Is faded all . . .
> Enclosure came and trampled on the grave
> Of Labour's rights and left the poor a slave.

Clare was a dreamer, set apart from the economic order by his imagination. Yet he expressed the threat to poor countrymen like himself of 'the rage of blundering plough' that made enclosure 'a curse upon the land'.

The steady advance of enclosure was only halted in the end by articulate townsmen. From the 1840s onwards, protests grew against the destruction of the remaining commons near congested towns and cities. The Commons Preservation Society supported the city-dweller's plea for open spaces that would offer 'air and exercise'. Enclosure ended with the saving of Berkhamsted Common and Epping Forest in the 1880s: the tide of opinion had finally turned.

3 MEN WHO MADE THE REVOLUTION

Certain individuals stand out in the history of the Agricultural Revolution. Modern historians see them less as innovators than as well-publicised representatives of the new farming. They were popularizers of improvements already put into practice by a host of other farmers whose names and achievements are lost to history.

Jethro Tull (1674–1741)
Tull alone was a real innovator. His seed drill of 1701 was the first effective agricultural machine, offspring of the spirit of science and interest in mechanism that were to be typical of the coming age, and

Jethro Tull, the gentleman farmer and inventor of the seed drill.

This picture shows seed being sown broadcast in the foreground. This wasteful method continued until seed drills became common in the nineteenth century.

true parent of the multitude of complex devices at the disposal of the modern farmer. Brand's iron plough of 1780, or the Rotherham plough with its curved mould-board may be considered as more important for farming but Tull's machine is in another line of development.

Son of a wealthy landowner, Tull was born in Berkshire. When he left Oxford University, poor health forced him to seek an open air life instead of his planned legal career, and he began to farm one of his father's properties at Crowmarsh, Oxfordshire. Isolated by the dreadful roads of the era, Tull tried to work in the spirit of the 'new science', using a microscope and home-made apparatus, and carefully recording results of experiments in plant growth that he conducted in his garden. He decided to plant his farm with sainfoin to improve the soil. He found that a thin sowing in a shallow channel produced the best crop.

Sainfoin seed was expensive. Tull's men were used to sowing broadcast, that is by scattering the seed as the sower walked the field. To plant whole fields as Tull wanted would have been a difficult task.

In any case his workers despised him as a 'Gentleman farmer' with 'too many bees in his bonnet case'. Tull, in turn, was exasperated with his workers: 'It were more easy to teach the beast of the field than to drive the ploughman out of his way', he wrote later. He was forced to invent his machine, 'to contrive an engine to plant sainfoin more faithfully than hands would do'.

With the help of local craftsmen, he set to work. A large seed container ran on wheels. As they turned, the seed ran down hollow tubes or 'coulters'. The front of these cut the soil, making furrows, into which the seed ran. The great merit of Tull's design lay in its control of the flow of seed. To do this, Tull, a keen musician, borrowed an idea from the mechanism of a church organ: 'I examined and compared all the mechanical ideas that ever had entered my imagination, and at last pitched upon a groove, tongue and spring in the sound box of an organ'. The whole machine was light enough to be pulled by a boy. A harrow, dragged behind, covered the drilled seeds.

The benefits of the drill soon became apparent. A quarter of the seed used in the old broadcast method now yielded a crop three times as big per acre. The traditional field was infested with weeds. The straight lines of the drilled field were easier to weed, especially when Tull devised a horse-pulled hoeing machine, which cleared the rows quickly and effectively.

A rear view of the seed-drill. The three coulters are visible between the back wheels

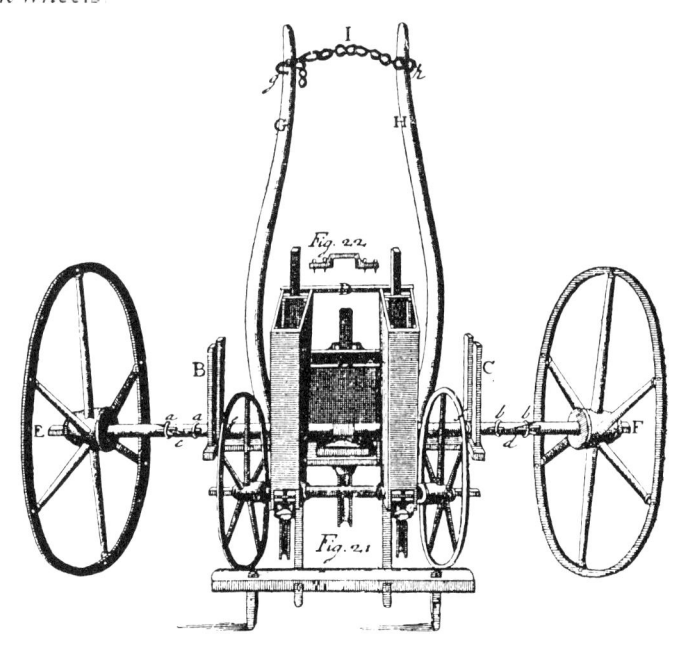

Charles Townshend, statesman and influential landowner.

Tull set down his many ideas on soil cultivation in *Horse Hoeing Husbandry* (1733), a book which was famous enough to be read and discussed at court by King George II. He remained confident that his ideas 'would one day become the general husbandry of England'. Tull's machines were years ahead of their time and only came to be used widely in the nineteenth century when more precise building techniques had been established.

Charles Townshend (1674–1738)
Townshend was a rich and powerful Norfolk landowner. After Eton, Cambridge and the 'Grand Tour' of Europe, he devoted himself to politics. He rose to become Chief Minister to King George I. When his brother-in-law, Robert Walpole, finally defeated him in a struggle for power in 1730, Townshend retired to his Norfolk estate, never visiting London again.

Townshend's land at Raynham looked unpromising. Rushy marshes lay beside a treeless sandy heath where 'two rabbits struggled for every blade of grass.' There was no proper road to the Hall, only a mile-wide tract of ground which was scarred by carriage tracks.

The statesman who had helped to sway the destiny of Europe had little difficulty in altering his estate. To improve the unproductive soil, Townshend used the old practice of marling. Marl was a mixture of lime and clay, rich in minerals, which was dug from beneath the sandy topsoil. Secondly he cultivated the turnip as a field crop, as Sir Richard Weston had suggested, and as Townshend himself had observed in Holland and Germany. Although it was already grown in the eastern counties, his enthusiasm for it won him the nickname 'Turnip', given in ridicule, but now seen as a title of honour.

Townshend slotted the new crop into the four-course Norfolk rotation: barley, clover or rye-grass, wheat, turnips. This pattern enabled him to keep more livestock. The animals in turn produced more manure to increase the fertility of the soil.

The Norfolk four-course rotation, adopted by Townshend, was more productive and made better use of land than previous practices.

LORD "TURNIP" TOWNSHEND
1674 — 1738
AND THE NORFOLK FOUR COURSE ROTATION

1. TURNIPS
2. BARLEY
3. RED CLOVER PEAS BEANS
4. WHEAT

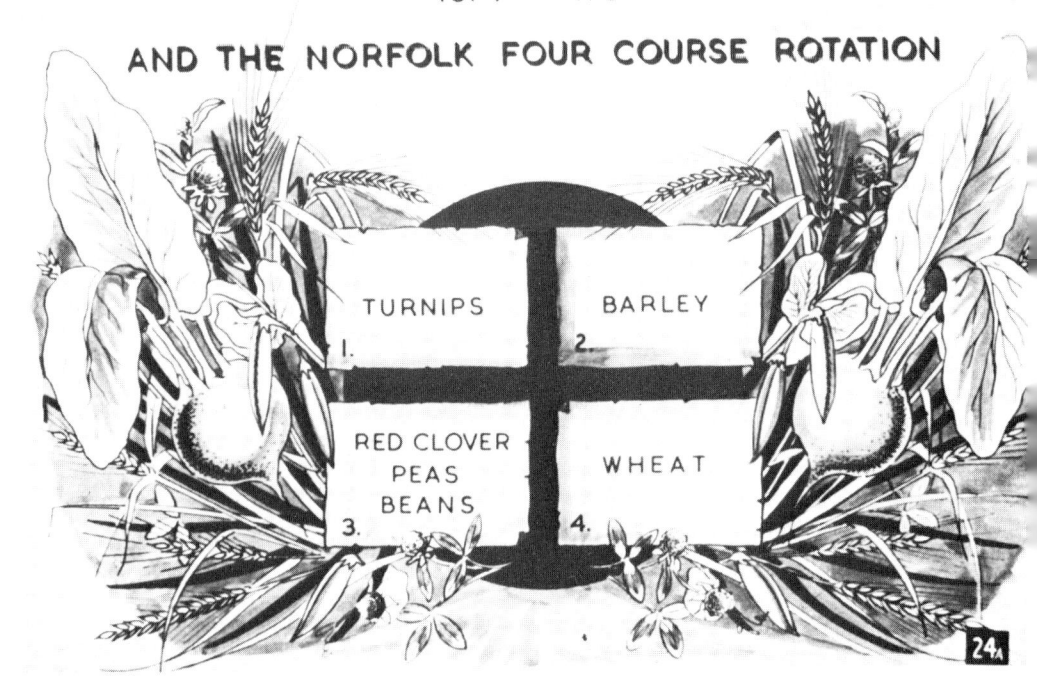

When the famous writer and agricultural journalist Arthur Young visited the estate in 1760, he found a transformation: thirty years before, the land 'yielded nothing but sheep feed, whereas these same tracts of lands are now covered with as fine barley and rye-grass as any in the world and great quantities of wheat besides.'

Townshend was not the first to use the Norfolk rotation. His importance lay in the example he set to other landowners, showing how income from estates could be increased by the application of soundly-based ideas.

Robert Bakewell (1726–95)
Changes in the appearance of livestock was a striking aspect of the Revolution. If the grotesque size and shapes in the fashionable animal portraits of the late eighteenth century were an exaggeration, the truth is given by figures for average weights of animals sold at Smithfield Market in London:

1710:	cattle 370 pounds	sheep 28 pounds	lambs 18 pounds
1795:	800 pounds	80 pounds	50 pounds

Thanks to the publicity given to him by Arthur Young, Bakewell was the best-known of several figures who established the stock-breeder's art. Contemporaries compared his work to that of famous inventors such as Arkwright or Watt: 'Like the immortal Newton, what his genius has conceived, he happily possessed the patient industry to execute'.

Bakewell was born at Dishley, Leicestershire in 1726. This district of fine grazing was already known for its stock-breeders. When he succeeded to management of his father's farm in 1760, he began his experiments. An affable, John Bull-like figure, Bakewell kept generous open house, as there was no inn in the village. British lords, French and German Dukes, Russian Princes all stayed with him, although he never altered his bachelor routine: 'At half past ten, let who would be there, he knocked out his last pipe.'

Arthur Young visited him in 1770 and was impressed by his neatness and ingenuity in the management of his farm. His livestock was treated kindly. Even rams, bulls and stallions were docile: a massive prize bull was led in for display, guided only by a cotton thread held by a child of seven. Young commented on 'the amazing gentleness in which he brings up these animals. All his bulls stand still in the field to be examined. The way of driving them from one field to another is by a little swish ... All his cattle are as fat as bears, yet his land is no better than his neighbours.'

Some of his animals were famous: his bull 'Twopenny', his ram 'Two pounder' and his black stallion. In his house he kept a museum of

Robert Bakewell, who conducted many successful experiments in stockbreeding. He was famous for the kindly treatment of his animals, who thrived and achieved fantastic sizes.

Smithfield Market in London, where the average weight of animals sold increased dramatically as a result of the work of men like Bakewell.

skeletons of his best animals. 'Old Comely', the heifer that lived to 26, was preserved complete in pickle, as were carcasses or joints that illustrated certain points (one sirloin had fat four inches thick!)

    Stockbreeding before Bakewell had been a casual affair. Sheer size had been the standard of merit: 'Nothing would please', said one of Bakewell's fellow-breeders, 'but Elephants and Giants'. In the horse or ox, used for pulling loads, large bones and long legs were valued; in cows, milking capacity; in sheep, length of wool. Bakewell perceived the rising demand for meat, and therefore valued flesh. He wanted well-covered animals, 'small in size and great in value'. He travelled the country, patiently collecting specimens of animals from which to breed. Carefully-selected animals, which possessed the features he wished to reproduce, were brought together, and selections made again from their offspring.

Bakewell's great success was with sheep. He produced his New Leicester breed, which was compact but quick to fatten, giving two pounds of meat where only one was produced before. Bakewell was secretive about his methods; he sold few animals, and hired out his rams at high prices, and, in 1789, received 3,000 guineas for their hire. Other less-known contemporaries also worked with sheep. John Ellman, from Sussex, produced his famous Southdown, ultimately more adaptable than Bakewell's New Leicesters.

The fat, longhorn cattle that Bakewell produced were not widely successful because of their inferior milking capacity. The Darlington herd of shorthorns led the way: a Darlington bull 'Hubbach' was the ancestor of the faultless Durham Ox, which was carried around England from 1801 to 1810 on a special display vehicle. Devons and Herefords were other favourites. This time saw the beginning of the important pedigree stock industry, and the formation of prize herds became a particular interest of rich landowners.

The Durham Ox, a faultless shorthorn bull, was exhibited throughout England from 1801 to 1810 and did much to popularize the breed.

Bakewell also transformed the heavy black Midland horse into a smaller, more powerful animal, two of which could plough four acres a day.

'Nowhere have I seen works that do their author greater honour,' wrote Young. 'Let me exhort the Farmers of England to take Mr Bakewell as a pattern'. Although he died a bankrupt, Bakewell spread his influence across England.

Arthur Young (1741–1820)
Incompetent as a farmer himself, Young nevertheless had a prominent place in the Revolution as a brilliant and prolific writer who produced some 250 books, most of which were concerned with promoting the spirit of progress in farming. If modern historians see his rival, William Marshall, as more expert, Young remains the typical voice of his time.

Arthur Young, a pioneer in agricultural journalism.

George III was a keen farmer and wrote articles for Arthur Young's magazine under the name 'Ralph Robinson'.

Born in London in 1741, Young began work in the wine trade, but soon began writing—so that he could buy the latest styles in clothes! Before he was nineteen, he had published four novels. Then he tried farming in Suffolk, losing three farms in succession.

However, he continued to write. By the time he was thirty-five, he had earned thousands of pounds with his pen. He was the pioneer of a new trade—agricultural journalist. He travelled incessantly, observed sharply and reported crisply, spreading news of sound innovations and denouncing practices that hindered progress. In 1767, he began a series of 'Tours' of England and Wales that, along with Cobbett's later *Rural Rides*, have become the most vivid descriptions of England during the Agricultural Revolution. His *Farmer's Letters* were carried everywhere by that keen farmer, George III. In 1784, he began the influential journal, *Annals of Agriculture*, which continued monthly until 1809.

Young himself wrote a quarter of the material in its forty-six volumes. George III, known as 'Farmer George', contributed articles under the name 'Ralph Robinson', borrowed from one of his shepherds at Windsor.

So well-known did Young become that, when, in 1793, Prime Minister Pitt formed the Board of Agriculture, he was the obvious choice to be Secretary. The Government-supported Board was to collect information and stimulate discussion. Surveys of each county were undertaken, six of them by Young himself. His report on Oxfordshire (1809) was his last piece of writing, yet even when blindness hindered him as he grew older, he still dictated his ideas.

Young supported large, enclosed farms, new rotations and crops, improved stock breeds, good irrigation and drainage. He hated waste and commonland, absentee landlords, bad roads, old-fashioned cultivation, outdated implements and inefficient use of land. It was he, perhaps, who invented the idea of an Agricultural Revolution. He knew he was living through a period of important change. To him, the farmers of the enclosures were 'quite new men, in point of knowledge and ideas ... A vast amelioration has been wrought, and is working. The Goths and Vandals of open fields touch the civilization of enclosures ...'

Thomas Coke of Holkham (1754–1842)

Arthur Young admired the large, efficient landowner. 'Where,' he asked, 'is the little farmer to be found who will cover his whole farm with marl at the rate of 100 or 150 tons per acre? Who, to improve the breed of his sheep, will give 1,000 guineas for the use of a single ram for a single season? Who will send across the kingdom to distant provinces for new implements, and for men to use them?'

Thomas Coke, of Holkham, Norfolk, might have been the man Young was describing. In the 'rage for agriculture' that seized the aristocracy in the late eighteenth century, Coke led the way. Handsome, charming and vital, Coke had inherited his estate at the age of twenty two. Although he was Member of Parliament for his county for half a century, he was never an active politician. As a Whig, opposed to George III, he had little chance of wielding power. Instead he devoted himself to his lands.

Coke led his tenants by personal example. When two of his estate farms fell vacant, he worked them himself, sometimes wearing a labourer's smock frock to go into the fields. He demonstrated new methods personally. He encouraged tenants with long leases, helping them financially so that they could follow the clauses in their contracts which demanded that they cultivate their land in modern style. Good farmers from all over England competed to be Coke's tenants, and the total rents from his estates rose from £2,200 in 1776 to £20,000 in 1816.

Coke began by enriching his soil with marl, the newly discovered

Thomas Coke of Holkham, Norfolk.

Coke's 'Clippings' at Holkham brought together farmers and agricultural experts from all over Europe.

bone fertilizer, and clover crops. Poor grass lands were ploughed and resown. Certain grass seeds were known to produce the sweetest hay. He showed estate children the seeds and sent them out to scour the fields for samples. Novelties like the swede (hardier than the turnip) and the potato were grown. He kept large herds of Southdown sheep and Devon cattle. The bare Norfolk landscape was transformed by the planting of fifty acres of trees each year. As an old man, Coke enjoyed sailing in a ship made from oaks grown from the acorns he himself had planted!

Coke's neighbours were suspicious of his methods and his Southdowns were said to be 'Whiggish sheep'! He calculated that his improvements spread across Norfolk at the rate of a mile a year. Coke had far more influence outside his own district. In 1778, he began his celebrated 'sheep shearing' meetings (or 'Coke's Clippings'). Farmers were invited to stay at the Great Hall to discuss agriculture. As the years passed, the meetings became grander and more spectacular: in 1818, the 'Clipping' lasted a full week, gathering hundreds from Britain, Europe and America. The Emperor of Russia even sent a representative. At the gatherings, politics were set aside, following the toast, 'Live and let live'. All were equal: the small farmer sat beside the Prince. At the final meeting in 1821, 7,000 people were present. Coke's work as a publicist for the new farming was perhaps his finest achievement.

Coke lived on until 1842. By his agency, two million acres of waste were taken into cultivation. Even during the black years of post-war depression after 1815, no Holkham worker lost his place, and the estate continued to prosper. When Coke died, his tenants erected a monument to his memory. At its four corners stood models of a Devon ox, a Southdown sheep, a plough and a drill. 'He was,' wrote the historian, C. J. Hall, 'witness to all England and indeed to Europe of what might be done by an enlightened landlord.'

The New Leicester sheep gave more meat and better wool than previous breeds.

4 THE LABOURING POOR

Parallel to the Agricultural Revolution ran another more sombre pattern of change in the condition of the lives of the rural poor. The one was only partly to blame for the other. Machines, for example, did not replace men. Mechanization was a mid-Victorian phenomenon,

Poverty and starvation forced many poor farm workers' families into the workhouse.

The labourer's cottage was often overcrowded. In parishes which adopted the Speenhamland system, men were spurred on to marry and have large families by the need to qualify for bread allowance.

introduced as a result of labour shortage in farming. Nor did enclosure create mass unemployment. For a time, the new fences, hedges, roads, and the thousands of acres of waste taken into cultivation needed more labourers, not fewer. The most serious effect of enclosure—'this scythe of desolation' as the country poet, Robert Bloomfield called it—was to take away the cottager's common rights. This loss, claimed George Sturt, in his book *Change in the Village* (1912) ruined the balance of the old system: it was 'like knocking the keystone out of an arch. The keystone is not the arch, but, once it is gone, gradually the whole structure falls down'. Whatever the actual effect of the loss, it remained as a grievance in the popular imagination.

Causes of rural poverty
Various other factors contributed to the rise of poverty in the countryside. The first was sheer greed. The agricultural population was made up of these elements: the landlords (and in the nineteenth century about 4,000 people owned half the country); tenant and freehold farmers, who actually worked the land; and smallholders, country tradesmen and labourers.

Poverty arose when the first two groups, encouraged by the boom in enclosures and by the high prices of the French wars era, took too much of the farming product and left the third group deprived. The second was the rapid growth of the rural population (rising from 1.5 million in 1751 to 2.9 million in 1831). A surplus of labour was created, especially in the southern counties: here, unlike the north with its factory towns, there were no other industries to provide work. Trapped by the working of the Settlement and Poor Laws, and by their own fear and ignorance, families could not easily move to look for work elsewhere. There were changes, too, in the pattern of employment at farms. In the north, 'farm servants' lived in and ate with the family. Part of a worker's payment came in food. In the south, this tradition had been dying. A man was taken on as a short-term labourer. He had no security. When it rained and work was impossible, he was not paid. If he complained, he was dismissed. There were plenty of men ready to take his place. His cash wage was a doubtful blessing. He, and not the farmer, now bore the cost of any increase in prices. 'Why do not farmers now feed and lodge their work people, as they did formerly?' wrote the great journalist, William Cobbett. 'Because they cannot keep them upon so little as they gave them in wages . . .'

William Cobbett, an eloquent spokesman for the rural poor.

An early nineteenth-century cartoon by James Gilray, who attacked the social aspirations of the rich farmer.

The farmer, however, during the 'good times' of the late eighteenth and early nineteenth centuries, began to grow rich. To the disgust of observers like Arthur Young, he began to assume the trappings of a gentleman: 'I see sometimes a pianoforte in a farmer's parlour, which I always wish was burnt; a livery servant is sometimes found and a postchaise to carry their daughters to assemblies. These ladies are sometimes educated at expensive boarding schools and the sons often go to the University to be made Parsons ... Let all these things and all the folly, frippery, foppery, expense and anxiety that belong to them remain among gentlemen.' Meanwhile, the farming labourers grew poorer.

The first trouble came in 1795. The price of wheat had risen steadily; labourers' wages much less in proportion. Many were soon too poor to buy bread, which was especially dear because it was taxed. Disturbances broke out all over England. Rioters seized food stocks and sold them to the poor at what they considered fair prices. To the rich, these riots were a danger signal.

Food riots, about the cost of bread, broke out after 1795.

The Speenhamland system
A solution was the adoption of a minimum wage, fixed in relation to the price of bread-corn. In May 1795, Berkshire magistrates gathered at the 'Pelican' inn at Speenhamland, near Newbury to discuss the idea. Instead, they chose an alternative: a dole from the poor rate in each parish would supplement a man's wages, the amount depending on the price of bread and the number of children in the family. This 'Speenhamland system' passed rapidly from county to county, and was supported by Parliament. The large farmer benefited because he did not have to raise wages. The small farmer, who did not employ many labourers, was hurt because his poor rates increased. The labourer, as the eighteenth century economist Thomas Malthus pointed out, was encouraged to marry early and produce a large family: single men could not draw the dole, and were the first to be laid off by a farmer. Above all, the Speenhamland system demoralized the worker. He was treated as a pauper, even when he was in full-time work.

The farming boom of the Napoleonic war years disguised the problem of rural poverty for a time. Peace in 1815 brought a depression. Land cultivated in wartime now became uneconomic. Farms fell vacant as bankruptcies spread. The rural labour market was swamped by the demobilization of 250,000 men from the Army and Navy. The Corn Law of 1815 attempted to help the farmer by prohibiting corn imports until home prices reached a certain level. But the Law made bread dear and the poor suffered.

Cruel man-traps were laid to catch poachers.

Rick burning, a powerful form of protest, spread across southern England during the Labourer's Revolt of 1830.

Riots began taking place in 1816. In Suffolk, mobs marched under the banner, 'Bread or Blood!', determined to reduce prices. Ricks and barns were fired. Soldiers were sent to quell the rioters, who threatened to march on London. At Littleport, near Ely, there was a serious battle in which two rioters were killed.

Swollen crime figures were another indication of rural distress. Poaching was a characteristic crime. A succession of Game Laws, each harsher than the last, was passed in the early nineteenth century, as game preservation became fashionable. Armed gamekeepers patrolled preserves; spring guns, which could kill, and brutal man-traps, which could cripple a poacher for life, were set to catch the unwary. Nevertheless, ferocious night battles between gamekeepers and poaching gangs continued to be waged in the woodlands. The penalties for those caught were harsh: hard labour, transportation, even hanging. The poor stubbornly persisted: 'It is better to be hanged than to starve,' a young man told Cobbett.

THE HOME OF THE RICK-BURNER.

This cartoon from the magazine 'Punch' recognizes the plight of the rick burner.

Armed gamekeepers patrolled estates in search of poachers. If they found them, there would often be violent confrontations.

By the 1820s, the Speenhamland system was inadequate to support life. In 1821, the writer William Cobbett, always an eloquent crusader against social injustice, began his horseback *Rural Rides* to assess the state of England. If there was much to admire, there was more to deplore. Near Leicester he observed 'hovels made of mud and straw, bits of glass, or of old cast off windows merely stuck in the mud wall. Enter them and look at the bits of chairs or stools, the wretched boards tacked together to serve for a table, the floor of pebble, broken brick or of bare ground; look at the thing called a bed, and survey the rags on the backs of the wretched inhabitants.' He contrasted the abundance of food produced with the condition of those who worked to provide it: 'What injustice, what a hellish system it must be, to make those who raise it skin and bone and nakedness, while the food and drink are almost all carried away to be heaped on swarms of tax-eaters.'

The labourers' revolt

In spring, 1830, Cobbett could sense the approaching revolt. Near Lincoln he met 'three poor fellows digging stone for the road, who told me that they never had anything but bread to eat, and water to wash it down ... Just such was the state of things in France at the eve of the Revolution! Precisely such!' In his paper, *The Political Register*, he angrily reported how four harvesters had been found dead of starvation behind a hedge 'with nothing but sour sorrel in their famished bellies.'

Rioting began in Kent in June 1830 and spread across the southern counties, where poverty was worst, and then into the Midlands and East Anglia. The threshing machine, invented by Andrew Meikle as far back as 1786, had become quite widely adopted by farmers, its use depriving the poor of winter barn work with the hand flail, the 'poverty stick' as it was called by labourers. The machines were the centre of agitation. Threatening letters, signed by a mysterious 'Captain Swing', announced the mobs' intentions of smashing them. Despite the initial sympathy of some farmers and magistrates, hundreds of ricks were burned. Sympathy faded. Troops were again sent to troubled areas, and magistrates and landlords organized armed groups of 'specials' to oppose the rioters. The revolt collapsed. Soon 1,900 men crowded prisons in twenty counties, waiting to be tried by Special Commissions. Six men were hanged, and four hundred transported. Of the Winchester trial, The 'Times' reporter wrote: 'There is scarcely a hamlet in the county into which anguish and tribulation have not entered. Wives, sisters, mothers, children beset the gates daily ... I cast my eyes down into the felons' yard, and saw many of the convicts weeping bitterly, some burying their faces in their smock frocks'.

Transportation to penal settlements in Australia was often the penalty imposed for poaching or rioting.

The Swing Riots of 1830 concentrated on the destruction of threshing machines, which deprived the poor of winter work. Tithes—land taxes paid to the Church—were also resented.

The Tolpuddle Martyrs were transported to Australia in 1834, after swearing an oath of loyalty to each other—illegal at the time.

The satirical poet, Ebenezer Elliott, described the changes in the village brought by these dark years:

> One farm prospers now, where prospered five.
> Ah, where are they?—wives, husbands, children, where?
> Two died in gaol, and one is dying there;
> One broken-hearted fills a rural grave
> And one still lives, a pauper and a slave.
> Where are their children? Some beyond the main,
> Convicts for crime; some here, in hopeless pain,
> Poor wanderers, blue with want; and some are dead,
> And some in towns, earn deathily their bread.

The last protest of the rural labourer came from Dorset. Trade Unions had been made legal in 1825, although their activities were severely limited. When, in 1834, workers in the village of Tolpuddle in Dorset formed a Friendly Society of Agricultural Labourers, they were prosecuted—not for belonging to a Union but because they had sworn an oath of loyalty, such oaths being illegal. Six men were transported. An outcry against this harsh treatment of the 'Tolpuddle Martyrs' caused them to be pardoned two years later, yet the sentence deterred other such Union activity until the 1870s.

The poor were broken. In the 1840s, an American observer, Henry Colman, described English labourers as 'in a very low condition, ignorant and servile. I cannot help thinking that the condition is a hard one in a country where the accumulations of wealth in some hands, growing out of this same labour are enormous.'

5 THE SECOND AGRICULTURAL REVOLUTION

The end of a long, dark period of post-war depression in farming coincided roughly with the accession of Queen Victoria in 1837. The first decades of her reign were marked by striking and widespread innovation: it was a second Agricultural Revolution.

In 1837, many parts of England remained unaffected by changes in farming practice. 'A hundred farmers plodded along the Elizabethan road', said the historian, Lord Ernle, 'while a solitary neighbour

Queen Victoria's accession in 1837 appeared to mark the beginning of a farming boom, as new machinery was invented.

This picture shows the old-fashioned farming implements which were often still in use after machinery had been introduced.

marched in the track of the twentieth century.' Some open fields still survived. Livestock was housed in draughty, decaying sheds, surrounding a filthy stockyard. Little machinery was used. Heavy wooden ploughs, pulled by as many as twelve oxen, were still seen. Seeded fields might be rolled by a tree trunk or large stone. Corn was cut by hand, and elaborately 'dressed' (threshed and sorted) in the barn.

Increasing prosperity

Yet there were signs of promise. Banks, more stable after two uneasy decades, gave the farmer financial support. New means of transport stimulated trade: canals, better roads and above all, railways. Spreading rapidly in the 1830s and 1840s, they gave a tremendous impulse to every kind of economic activity. The farmer could now send cattle and farm products to market swiftly: the age of the drover was at an end. In this business climate, meat and cereals sold briskly and farmers began to do well.

Canals and railways allowed farmers to send products to market swiftly, avoiding the bad roads of the early nineteenth century. **Above.** *A typical canal scene.* **Below.** *Travelling on the Liverpool and Manchester railway in 1831.*

Sir J. B. Lawes, the pioneer agricultural scientist.

The second Agricultural Revolution aimed to produce greater efficiency and a higher yield per acre. Industrialists and scientists began to feed more of their discoveries into the work of the farm.

New institutions led the way. The old Board of Agriculture had lapsed in 1822 during the depression. The Royal Agricultural Society was established in 1838, under the motto 'Practice with science'. Its first Show at Oxford in 1839 attracted 20,000 visitors. In 1866, the Society began the 'Agricultural Returns', the first reliable farming statistics.

The Royal Agricultural College at Cirencester was opened in 1845 to advance technical education in the field. More influential was the Rothamsted Experimental Station in Hertfordshire, started in 1843 by J. B. Lawes on his family estate. He had studied chemistry at the new University College, London and when he inherited his land, he began the Station, working with his lifelong colleague, J. H. Gilbert. Lawes' first work was with fertilizers, with which he made a fortune. He lived to see fifty-seven successive crops grown on his Broadbalk field without natural manure. Farming owes the station an immense debt. The two men also set down the scientific basis of rotation, studied the exact effects of various feeds on animals, and initiated investigations into botany, chemistry, meteorology, and plant and animal physiology.

Sir James Smith, who made important advances in the improvement of drainage.

Drainage by machine was one of the triumphs of Victorian farming.

Innovations

If the use of roots and legumes had transformed the working of lighter soils, districts where soil was heavy, wet clay desperately needed efficient drainage to make farm holdings more profitable. Current methods were clumsy: deep furrows or trenches were filled with stones, brush-wood, thorns or twisted straw and then re-covered with earth. In 1831, James Smith of Deanston Perthshire, published his *Remarks on Thorough Drainage*. His system combined deep ploughing with a network of parallel underground drains running downhill to a main drain. Visitors flocked to see how he had transformed a waterlogged marsh into a garden. Twelve years later a gardener, John Reade, invented a cyclindrical claypipe. By the 1850s, a machine had been devised to make 20,000 of these claypipes a day on site, while a 'mole plough' laid and buried them. Such inventions were some of the wonders of the 1851 Great Exhibition: pipes were made 'by machinery, which squeezes out clay from a box through circular holes, exactly as macaroni is made at Naples'. The cost of drainage was reduced dramatically. After 1846, government and private loans were available to help farmers drain land. Within twenty years millions of acres had been drained in this way.

The influence of the chemist on agriculture began in 1803, with the Board of Agriculture lectures by the brilliant scientist, Humphrey Davy. Davy described the relationship between plant growth and the composition of the soil. The result of his researches was his *Elements of Agricultural Chemistry*, published in 1813. His work was carried further by the German, Justus von Liebig, who produced *Organic Chemistry in its applications to Agriculture* in 1840. Liebig saw how

Sir Humphrey Davy, whose lectures on the relationship between plant growth and the composition of the soil reflected the influence of the chemist on agricultural methods.

As the mechanization of agriculture progressed, threshing by steam engine became common.

the application of certain chemical substances to a plant's roots could stimulate its growth. Lawes and Gilbert (who had worked with Liebig) gave his theory a practical basis by creating artificial fertilizers. Over the centuries, farmers had learned by observation the power of certain materials to revive the soil. Eighteenth-century farmers had claimed that 'there was nothing like muck' but some also tried wood-ash, woollen rags, soot, or fish and seaweed. The use of bone meal had become so widespread that merchants were accused of digging up Europe's battlefields to supply the demand! In 1835, guano (bird droppings) was first imported from South America; by 1847, 300,000 tons a year were used. To make his artificial fertilizer, Lawes treated bones (and later mineral phosphates) with sulphuric acid to produce superphosphate of lime. He patented the process and set up a factory in London. By the 1870s, half a million tons were sold each year.

New machines
The most spectacular innovation was the farm machine. The period 1830 to 1880 was the first major phase in the mechanization of agriculture. It was itself part of the second, more sophisticated phase of the Industrial Revolution, marked by improved design, more accurate

machine tools and more experience of production. Firms like Ransomes and Garretts grew into international companies, triumphantly displaying their achievement at the Great Exhibition. 'The present age,' noted Henry Stephens in his *Book of the Farm* (1844), 'is perhaps the most remarkable that time has produced, for the perfection of almost every kind of machine and tool required.'

Steam provided the motive power. The threshing machines, so hated in the 1820s, had been powered by horses or water. After 1840, the work was done by steam engines, either fixed near the barn, or taken round by contractors.

After 1850, machines also winnowed, divided, and sacked the corn. Steam devices powered feed mills and root slicers. In the field, steam was less valuable. Heavy, self-propelled engines only sank into the soil. For a time steam ploughing seemed promising: stationary engines at the field edge pulled ploughs on steel cables, cutting over eight furrows at a time. But costs were high and use was limited to flat ground. The farmer had to wait until the twentieth century for the tractor to replace his horse teams.

The American McCormick harvester, with its patent knotting device to bind the sheaves.

The Great Exhibition of 1851 in Hyde Park, London, where many new agricultural machines were displayed.

The reaping machine was perfected by the American, McCormick, for use on the prairies, where fields were huge and labour short. British inventors may have provided him with inspiration. In 1812, John Common, a Northumberland millwright, devised a machine that he had to test by moonlight in order to avoid his neighbours' hostility. Models were sent to Canada and one of them may have reached McCormick. The Scotsman, Patrick Bell, invented a reaper pushed by two horses. He tried the machine secretly in a barn, cutting corn stalks placed in cracks in the floor. Indifferent workmanship made the machine a failure. However, rebuilt by an engineering firm as the 'Beverley Reaper', it proved a success alongside the American machines at the Great Exhibition. These early reapers still needed teams of men to gather the cut corn. The self-binding reaper, with a patent string knotter, dates from 1873. The ultimate 'combined-harvester', horse-drawn, was first seen in California in 1887.

The Leviathans

Some remarkable farmers, 'Leviathans' as they were nick-named, tried to run their farms on industrial lines, using all the new technology. J. J. Mechi is an example: with his fortune made from his patent 'Magic Razor Strop', he bought some poor, wet heathland at Tiptree, Essex. He drained the soil with ninety miles of pipe, and divided it into squares, each served by metalled roads. Liquid manure from his animal yards was pumped into the fields by a steam engine that also threshed his corn and worked slicers, millstones and irrigation pumps. A steam plough worked his soil and American reaping machines cut his harvest.

Ploughing by stationary steam engine and cable was only successful on flat fields like this one.

J. J. Mechi, one of the farming 'Leviathans'. His revolutionary innovations ultimately met with failure.

For a time he made a profit, until the bad weather of the 1870s defeated him. In his essay, *Man of Progress*, Richard Jefferies described a farmer called Middleditch, another 'Leviathan' from Swindon. His land was one huge field, across which ran a light railway carrying men and materials. Everything on the farm was done by machine: 'The beat of the engines never seemed to cease'. 'In no other way but by science, by steam, by machines and intelligence, can we compete with the world,' commented Middleditch. To the relief, perhaps, of their old-fashioned neighbours, both men failed in the end.

6 AFTERWARDS: GOLDEN AGE AND DECLINE

In 1846, following a series of bad harvests and the disastrous Irish potato famine, the Corn Laws were repealed. The consequent throwing open of English markets to grain exports from abroad seemed to mark the beginning of bad times for farmers. Wheat prices fell; farms were abandoned; arable farmers turned to pasture. Such was the depressed state of farming in 1850 that the *Times* newspaper sent the Scottish farmer-journalist, James Caird, to tour England to study conditions. He found uneven practice but was impressed by the capacity of the industry to feed the nation. He recommended what he called 'High Farming' as the way forward: that is, high investment and the application of the discoveries of the second Revolution—drainage, fertilizers, superior livestock and new implements. Capital spent in this way would yield high profits. 'In this country,' said Caird, 'the agricultural improver cannot stand still.'

The Great Irish Potato Famine of 1846 wiped out many families and brought about a period of mass emigration.

Victorian novelists often depicted farming life as idyllic, but the truth was very different.

High farming

For the next twenty years, Caird seemed to be right. Farming prosperity increased rapidly, as gold discoveries in Australia and California improved trade. Most agricultural prices rose by thirty per cent in this quarter century. The expanding railway system allowed remote areas access to the ever-growing market of London, and the cities of the Midlands and the North. Even milk was carried into town by train, creating a prosperous new industry. Foreign competitors were hampered by war: the American Civil War reduced corn imports to a trickle. In any case the railway had not yet reached the prairies of the Mid-West. The English wheat area expanded to four million acres. The seasons were favourable and there were ten good harvests in succession. These, then, were the years of the farming 'Golden Age', when the Agricultural Revolution came to affect farming practice most widely. During these years it was said of some farmers that money was not so much made as brought home in buckets.

The Golden Age decades were reflected in the exquisite pastoral images found in the works of Victorian novelists, like George Eliot in her *Adam Bede* (1859) or Thomas Hardy in his *Far from the Madding Crowd* (1874)

The Victorian novelist, George Eliot.

or *Tess of the D'Urbervilles* (written 1891 about the 1860s or 1870s). In the diaries of the clergyman, Francis Kilvert, who wrote in the 1870s, we can see pictures of a countryside improved by the farming revolution, and as yet unspoilt by depression or the motor car: 'I stopped to listen to the rustle and solemn night whisper of the wheat . . . The corn seemed to be praising God and whispering its evening prayer . . . Laughing voices were wafted from farms and hayfields out of the wide dusk . . .'

As the golden years saw land become a sound investment, much building took place, especially after the tax on bricks was removed in 1850. It was a time for embellishment of mansions, restoration of churches and the creation of village elementary schools. The worker on a large estate might find himself moved to a handsome brick and slate 'model' cottage with three bedrooms. A Frenchman, Hypolyte Taine, admired the latest design of farm when he visited England in the 1860s: 'We stopped at a model farm. No central farmyard. The farm is a collection of fifteen or twenty low buildings in brick, economically designed and built. Bullocks, pigs, sheep, each in a well-aired, well-cleaned stall . . .

Steam engines for all the work of the arable land. A narrow gauge railway to carry the food to the animals. Farming in these terms is a complicated industry, based on theory and experiment ... The most remarkable thing is, the place makes money'.

The labourer's life

The condition of farm workers at this time varied from place to place. In the north wages were higher; elsewhere, wages and conditions were often poor, and a relentless drift from the countryside continued, even in the 'Golden' years. Between 1851 and 1871, the number of workers fell by 250,000 or nearly a quarter.

Housing was a major problem. The country cottage was often a rural slum. Victorian artists like Myles Birket-Foster made a good income from paintings of idyllic thatched cottages with children happily playing. The reality was a two-room hovel, with leaking roof and earth floor, with no sanitation and polluted water supplies. Families were crammed together. The Rev. James Frazer, in an 1867 report to Parliament, described how 'in one small chamber ... two and sometimes

By 1873, cattle trucks had been equipped to provide the animals with water during the journey.

In this picture, the Victorian painter, Myles Birkett-Foster, portrayed the life of poor country people as quaint and attractive.

THE COTTAGE.

Mr. Punch (to Landlord). "YOUR STABLE ARRANGEMENTS ARE EXCELLENT! SUPPOSE YOU TRY SOMETHING OF THE SORT HERE! EH?"

A 'Punch' cartoon attacking farmers who spent heavily on their farm buildings, while allowing their workers to live in slum cottages.

three generations are herded promiscuously ... Human nature is degraded into something below the level of the swine. It is a piteous picture; and the picture is drawn from life'.

Photographs and paintings of farm labour, especially those showing harvest work, made it look attractive. 'The sight of the men,' wrote the novelist, Rider Haggard, 'as they cut down the ripe corn with wide sweeps of the scythe, made a fine picture of effort strenuous and combined'. The actuality was different: 'The hell of a life was the harvest,' remembered a fen-man, W. H. Barrett. Until compulsory schooling was introduced after 1870, children joined farm work almost as soon as they could walk. They might be set to watch a gate, whiling away the hours by carving labyrinth patterns in the turf. The Trade Union leader, Joseph Arch, recalled a twelve hour day bird scaring, constantly in fear of 'a smart taste of the farmer's stick'. Women had always been drudges in

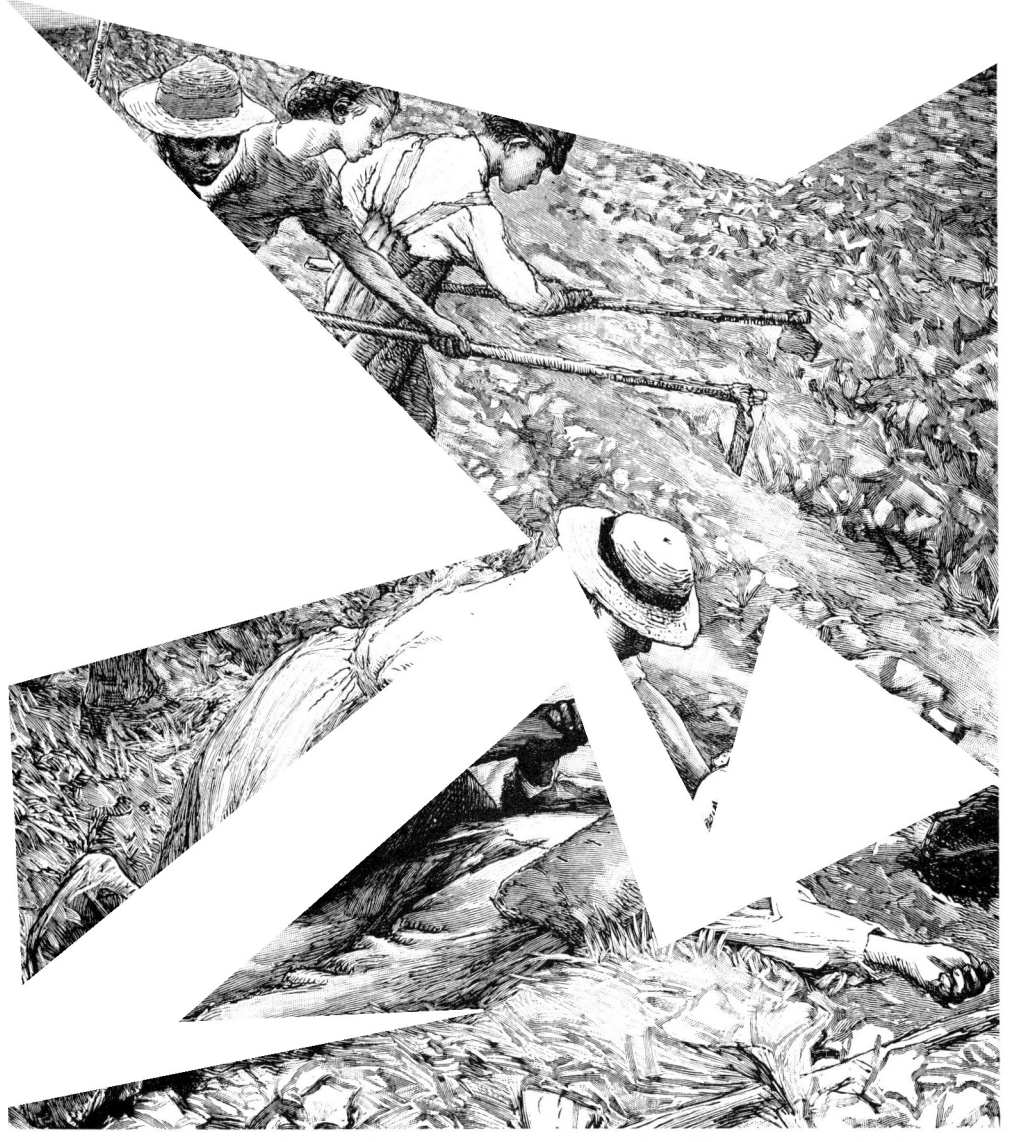

The scandal of exploitation of women and children in East Anglian agricultural gangs was ended by the Gangs Act of 1867.

the fields. Cobbett recorded their hard life in his picture of some Wiltshire girls 'ragged as colts and pale as ashes ... their blue arms and lips would have made any heart ache'. Not until the 1860s was real concern shown for women in agriculture. An 1867 Report noted how the work 'could almost unsex a woman in dress, gait, manners, character, making her rough, coarse, clumsy, masculine'.

The gang system was a particular scandal. In the eastern counties, where labour was scarce, men, women and children were recruited to work for farmers, in lifting potatoes, digging weeds or picking up stones. In the late 1860s, an outcry arose against the gangs—for the degrading work, the immoral behaviour, the illiteracy of the children—and the Gangs Act of 1867 curbed the worst excesses of the system, forbidding employment of children under eight.

The drift from the land accelerated as education made people more aware of conditions and as railways made escape easier. Some moved to the cities. A man who left Norfolk to work in Sheffield in 1870 wrote to his friend: 'We are getting on right well for instead of working for 13 shillings [65p] a week, we get 22/6 [£1.13p], and instead of working with bread and cheese, we get a thumping bit of beef ... Tell our poor fellowmen not to stop there and be starved and ruined by Mr Farmer'. Those with real initiative emigrated to America, Canada or Australasia where farming skills were highly valued. There were sometimes village emigration societies whose members paid into a fund that would provide one family with a new life.

The loss of labour forced farmers to improve conditions: wages, for example, rose by nearly forty per cent between 1850 and 1870. In this climate, Unionism once more emerged. In 1872, Joseph Arch, a talented, self-educated leader and speaker, addressed a meeting of many hundreds in Warwickshire. In his *Autobiography*, he remembered: 'I mounted an old pig-stool, and in the flickering light of the lanterns I saw the earnest upturned faces of those poor brothers of mine—faces gaunt with hunger and pinched with want—all looking towards me'. By 1873, the Union

A cartoon by the magazine 'Punch', showing the squalid lives of the British at home, in sharp contrast to the opportunities to make a new life abroad.

Joseph Arch addressing farm workers in Warwickshire in 1872.

Grain from America unloaded at Hull in 1882. Heavy imports like these brought depression to English farming after 1875.

was a national movement with 150,000 members. Strikes forced up wages but farmers responded with lock-outs and evictions from tied cottages. The agricultural work force was too scattered to group effectively, and the depression of 1875 onwards reduced the need for labour. By 1879, the Union was, as Arch had feared, 'falling to pieces like a badly-made box'. It had, however, some influence in winning the farm labourer the vote in 1883.

The farming slump
The sudden collapse of agriculture after 1875 was the long-delayed result of the repeal of the Corn Laws. In the late 1870s, a series of bad harvests culminated in the disaster of 1879, the worst summer of the century. But cereal prices failed to rise as a massive inflow of American wheat flooded the market. Production in the USA trebled between 1860 and 1880. Extended railways, steamships and the elevator system of handling grain in bulk cut costs dramatically: the expense of carrying corn from Chicago to Liverpool fell by three quarters in the last decades of the century. Cheap American cheese, and dairy products from Holland, France and Denmark provided fierce competition for the English farmer. By the 1880s it was also possible to bring frozen meat from Australia, New Zealand and South America to this country. The British government could do little: cheap food was a benefit to the industrial cities, whose products were the foundation of British exports. Farming was allowed

The invention of the steam plough was one of the great triumphs of the second Agricultural Revolution.

As a depression set into English farming, families often chose to seek a new life abroad. Emigration scenes such as this were common, as people set out for America or Australia, hoping for a better life.

to fall into decline. In 1857, it employed a fifth of the work force that produced a fifth of the national product. In 1900, it employed less than a tenth of the working population, producing a fifteenth of the nation's income. Agriculture ceased to be the centre of national life and became just one of many industries, and one troubled by serious problems.

The year 1875 marks the end of the Agricultural Revolution. The machines whirring their way across the fields of the 1860s, where fine crops grew and handsome animals grazed were its positive achievements. The rural depopulation —deplored by social critics like C. F. G. Masterman, who described Essex in 1909 as 'becoming one vast wilderness, a retreat for foxes and a shelter for conies, with the houses tumbling into decay (and) apathy settling down like a grey cloud over all'—was its other, more sombre result.

DATE CHART

1651	Sir Richard Weston: *Discours of the Husbandry used in Brabant and Flanders*
1701	Jethro Tull's seed drill
1730–38	'Turnip' Townshend transformed his Norfolk estate
1733	Jethro Tull: *Horse Hoeing Husbandry*
1760	Beginning of great wave of Parliamentary enclosure (to 1850) Robert Bakewell began livestock breeding experiments at Dishley, Leicester
1776	Thomas Coke took over Holkham estate, Norfolk
1778	First Holkham 'sheep shearing' meeting
1784	Arthur Young began *Annals of Agriculture* (until 1809) Andrew Meikle's threshing machine
1795	Speenhamland system of poor relief
1813	Humphrey Davy: *Elements of Agricultural Chemistry*
1815	Corn Laws to protect farmers from threat of imported grain
1826	Patrick Bell's mechanical reaper
1830	'Swing' riots
1831	James Smith: *Remarks on thorough draining*
1834	New Poor Law Tolpuddle Martyrs transported
1838	Royal Agricultural Society founded
1843	J. B. Lawes: Rothamsted Experimental Station first artificial manures
1846	Repeal of the Corn Laws
1851	The Great Exhibition: British and American farm machines displayed
1850 (to 1875)	High Farming era: the Golden Age
1867	Agricultural Gangs Act
1872	Joseph Arch and the National Farm Labourers Union
1875 (to 1914)	Agricultural depression

GLOSSARY

Arable Land used for growing crops.
Balk Grass barrier between strips in the open field.
Broadcast To sow by scattering seeds freely by hand as the sower walks the field.
Corn Laws Laws passed to protect the British farmer from foreign competition. A duty, or tax, decided by the government, was imposed on all imported grain, until British grain reached a certain price per quarter of a ton.
Drill To plant seeds in rows using a dibber or a machine.
Enclosure Amalgamation of land into a compact farms with hedged fields.
Fallow Land that is rested to allow it to recover fertility after successive crops.
Guinea One pound and one shilling (£1.05).
Husbandry Another word for farming, in the sense of a skill or a science.
Legumes Crops like clover, sainfoin and lucerne, that absorb nitrogen from the air and revitalise the soil.
Leviathons A great or powerful figure, so called after a monster in the Bible.
Marl A mixture of lime and clay, used to enrich light, sandy soil.
Open Field System Three large fields divided into strips.
Poor Laws Laws passed from 1600 onwards to try to combat poverty in Britain. Minimal relief was to be provided for the poverty-stricken through a poor rate contributed to by all householders of a parish. The New Poor Law of 1834 condemned the poor to the harsh conditions of the workhouse if they failed to find employment.
Root crop Turnips, swedes, mangolds, potatoes etc.
Rotation The growing of different crops in a regular order to avoid soil exhaustion.
Tenant farmer One who rents his land from a landlord.
Threshing Beating out or separating the grain from cut corn.
Transportation Criminal punishment whereby an offender was sent to a penal settlement in Australia.
Waste Wild, uncultivated land.

FURTHER READING

Addy, J. *The Agrarian Revolution* Longmans, 1964.
Crowther, J. G. *The Story of Agriculture* Hamish Hamilton, 1958.
Ernle, Lord *English Farming, Past and Present* Heinemann, revised, 1961.
Hall, C. J. *A Short History of English Agriculture and Rural Life* A. C. Black, 1924.
Hammond, J. R. *The Village Labourer 1760–1832* Longmans, 1911.
Hobsbawm, E. J. and Rude G. *Captain Swing* Lawrence and Wishart, 1969.
Hoskins, W. G. *The Making of the English Landscape* Hodder/Penguin, 1955.
Huggett, F. E. *A Day in the Life of a Victorian Farm Worker* Allen and Unwin, 1972.
Mingay, G. E. *Rural Life in Victorian England* Heinemann, 1977.
Trevelyan, G. M. *Illustrated English Social History* Longmans, 1942.

Selections from contemporary writings
Cobbett, William *Rural Rides* (1830) Penguin Classic, 1985.
Defoe, Daniel *A Tour through the whole Island of Great Britain* (1724–6) Penguin, 1971.
Eliot, George *Adam Bede* (1859) Penguin, 1980.
Hardy, Thomas *Far From the Madding Crowd* (1874) Penguin, 1973.
Hardy, Thomas *Tess of the D'Urbervilles* (1891) Penguin, 1978.

PICTURE ACKNOWLEDGEMENTS

The illustrations were supplied by: Ann Ronan Picture Library *front cover*, 42, 43, 50 (below), 63, 67, 71; BBC Hulton Picture Library 7, 8, 14, 28, 35, 64; by permission of the Syndics of Cambridge University Library 9, 15, 39; Institute of Agricultural History and Museum of English Rural Life University, of Reading 5, 6, 10, 11, 17, 18, 25, 29, 34, 57, 65, 70; Mansell Picture Collection 19, 21, 22, 27, 30, 37, 38, 40, 46, 47, 50, 53, 54, 56, 58, 68; Mary Evans Picture Library 12, 13, 16, 23, 24, 33, 36, 41, 44, 45, 49, 50 (above), 52, 55, 59, 60, 61, 69; the remaining pictures are from the Wayland Picture Library.

INDEX

America 4, 35, 47, 57, 61, 70
Arch, Joseph 65, 67
Australasia 4, 61, 67

Bakewell, Robert 26, 28–30
Bell, Patrick 57
Board of Agriculture 32, 53
Bread riots 39, 42
Building 18, 62, 63

California 57, 61
Canada 4, 57, 67
Captain Swing 45
Cattle
   Durham Ox 29
   pedigree 9, 26, 29, 34, 49
Chemistry 53, 55
   Davy, Sir Humphrey 53
   Gilbert, J. H. 55
   Lawes, J. B. 53, 55
Chicago 70
Child labour 65, 66
Clare, John 19, 20
Cobbett, William 18, 38, 42, 44, 65
Coke, Thomas 32, 34–35
   'Coke's clippings' 34
Combined harvester 57
Commissioners 16, 18
Commons Preservation Society 20
Corn Laws 40, 60, 70

Defoe, Daniel 14

Diggers and Levellers 12
Domestic industries 8
Drainage 12, 32, 53, 58, 60
Drovers 14, 49

Elliot, Ebenezer 47
Emigration 67
Enclosure 10, 11, 12–16, 20, 32, 37–38

Farmer 5, 8, 10, 16, 32, 37–40, 49, 53, 60, 65–67
Farming 4, 5, 37, 40, 60, 71
Fenlands, East Anglia 12
Fertilizer 6, 55, 60
Friendly Society of Agricultural Labourers 47

Game Laws 42
Gangs Act, 1867 67
Gang system 66
George III, King 15, 31, 32
Great Exhibition, 1851 53, 56, 57

Industrial Revolution 4, 55
Irish Potato Famine, 1846 60

Labourer 8, 37–40, 47, 63, 67

75

Legumes 6, 10, 22, 23, 53
'Leviathans' 58
   Mechi, J. J. 58
   Middleditch 59
Liebig, Justus von 53
Littleport, Ely
   riots 42
Livestock 6, 9, 25–26, 49, 60, 62

Machinery 36, 45, 49, 55–57, 59–60
Man-trap 42
Markets 14
   Smithfield 14, 26
McCormick, Cyrus 56, 57
   reaper 56, 57

Napoleonic Wars 4, 15, 40
Norfolk crop rotation 26

Open field system 7, 9–11, 16, 49
   Laxton, Nottinghamshire 10
Otmoor riots, 1814 17

Poaching 42
Poor laws 38
Poor rate 40

Railways 49, 61, 63, 70
Reaping machine 56–58
Riots
   1795 39
   1816 42
   1830 44, 45

Royal Agricultural Society
Rural poverty 4, 36–38, 42, 44–45, 63, 65

Settlement 38
Sheep
   pedigree 9, 26, 29, 34
Smith, James 53
Speenhamland System 40, 44
Sprat, Thomas 13
Steam 56, 58, 63
Stockbreeding 26, 28, 29
Strips 7–10, 16

Taine, Hypolyte 62
Threshing machine 45, 56
Tolpuddle Martyrs 47
Townshend, Charles 'Turnip' 24, 26
Tull, Jethro 21, 22
   seed drill 23

Unionism 47, 67

Vermuyden 12
Victorian novelists 61, 62
Village depopulation 11

Weston, Sir Richard 7, 25
Winchester Trial 45
Winnowing machine 56

Young, Arthur 7, 26, 30–32, 39

Contents

Preface	*page* vii
How to use this book	ix
List of processes covered	xi
PART ONE A–Z of treatment and testing processes	1
PART TWO Major groups of processes	327
Blasting	329
Electroplating	339
Hardness testing	367
Heat treatment	383
Mechanical testing	439
Non-destructive testing	459
Painting	481
Welding	497
PART THREE Tables for identification and comparison of processes	521
Tables	524
1 Adhesion testing processes	524
2 Alteration to properties – casting, forging etc.	525
3 Blasting processes	525
4 Brazing and soldering processes	526
5 Build-up or additive processes	527
6 Cleaning treatments	528
7 Corrosion resistance processes	529
8 Procedures for evaluating corrosion resistance	530
9 Decorative treatments	530
10 Fatigue resistance improvement processes	532
11 Flaw or crack detection	533
12 Friction characteristics improvement processes	534
13 Hardening treatments – increase in tensile strength	535

14	Heat treatments	537
15	Inspection processes	538
16	Joining techniques	540
17	Metal removal processes	541
18	Metal identification methods	542
19	Painting processes	542
20	Electroplating processes	543
21	Softening treatments – reduction in tensile strength	544
22	Strength testing methods	545
23	Welding processes	546

PART FOUR Appendices: useful information		547
1	Elements with their symbols and atomic numbers	549
2	Metallic elements with their melting points and specific gravities	551
3	Temperature conversion table	552
4	Strength conversion table	556
5	Hardness and tensile values approximate conversion table	559
6	Impact data conversion table	562
7	Specific gravity conversion table	563
8	Thickness conversion table	565
9	Companies and organizations involved in metal treatments and testing	566

Preface

For this second edition of the *Handbook of Metal Treatments and Testing* the opportunity has been taken to divide the main body of the book into two parts, and to include about 10 per cent additional entries.

Part One is an alphabetical A–Z 'encyclopaedia' of methods of treating and testing metals. Part Two is a fuller exposition of eight major processes, which are further subdivided alphabetically into individual techniques.

The purpose of the book remains to provide practising engineers with details of as many methods of treating and testing metals as I could find. While it may be of interest to a specialist, the book is not intended as a metallurgical or chemical treatise.

From the comments obtained on the first edition it appears that the majority of processes and treatments used with metals have been included.

There are many occasions where my opinion is given when assessing the comparative values of processes or techniques. These judgements are my personal opinion based on experience. There will be occasions where, for various reasons, the information will not be agreed by all concerned. This may be because of local industrial or geographical conditions where a different interpretation of the same name applies.

It is hoped that this second edition will prove useful to the reader. Comments will always be welcome.

Robert B. Ross
1988

How to use this book

There are three basic reasons for referring to this book apart from those occasions when the casual reader uses it for general information. These are:

1. Where some process or technique used with metals has been found then it should appear in Part One of this book in the alphabetical A–Z section, and there will be a brief reference defining the technique. This will then also supply further information on the technique or refer the reader to Part Two where additional information will be available.

2. The second main section, Part Two, of the book brings together eight major groups of techniques under separate headings. The headings are Blasting, Electroplating, Hardness Testing, Heat Treatment, Mechanical Testing, Non-destructive Testing, Painting and Welding. This means that the reader can evaluate and have a better understanding of the complete range of processes.

3. The third purpose of the book is to allow the reader to compare the various techniques. This is achieved by two means, with each of the individual items attempting to give as much information as possible on techniques which will achieve the same result.

(a) Wherever possible in the A–Z section of the book, the technique described is compared with other processes which might achieve the same result.

(b) Part Three consists of a series of Tables which help the reader to isolate those treatments which will enable him or her to achieve a particular aim, together with those which are relevant for the particular metal. Thus there is a list of treatments which are useful for 'corrosion protection' and another list for 'increase in hardness'.

A reader needing to find out how to improve the surface hardness of a specific metal for example, could refer to Part Three, list the processes given under 'Hardening treatments' for the metal concerned, and then refer to Parts One and Two for detailed information on the various techniques indicated.

Finally, the Appendix section, Part Four, provides useful information such as the names and addresses of companies and organizations involved with metals (Appendix 9). Wherever a proprietary name has been given in the A–Z section the address will be found in this appendix.

List of processes covered

Listed below are all the metal treatment and testing processes covered in this book. The major sections which cover whole 'families' of processes are listed in capital letters, and appear in Part Two.

Abrasive Blasting
Abrasive Testing
Accelerated Corrosion Testing
Acid Descaling
Acoustic Testing
Activation
Adhesion Testing
Aerocase Process
Age Hardening
Ageing
Air Blasting
Air Hardening
Air-pressure Testing
Airless Blast
Airomatic Welding
Alachrome
Albo
Aldip Process
Al-fin Process
Alkaline Descaling
Alloy Plating
Alloying
Almen Test (Almen Number)
Alochrom
Alodine
Alplate Process
Alrak Process
Aludip Process
Alumilite Process
Aluminium Chromium Duplex Coating
Aluminium Plating

Aluminium Soldering
Aluminizing
Aluminothermic Process
Alzak Process
Amalgamating
Ammonia Carburizing
Angus Smith Process
Annealing
Anodic Etching
Anodic Oxidation
Anodic Protection
Anodizing
Anolok Process
Antifouling
Aquablast
Arc Welding
Argon Arc Welding
Argonaut Welding
Argonox
Argonshield
Artificial Ageing
Atomic Arc Welding
Atrament Process
Auruna Process
Austempering
Austenizing
Autofrettage
Autogenous Welding
Autronex Process
Back-etching
Bailey's Creep Test
Balling
Banox Process
Bar-drawing

Barffing
Barr–Bardgett Creep Test
Barrel Plating
Barrelling
Bauer–Vogel Process
Bend Test
Bengough Stuart Process
Berketeck
Bethanizing
Bi-Ni
Bit Soldering
Black Anncaling
Black Anodizing
Black Nickel
Black Oxide
Blackening

BLASTING
    Glass-bead Blasting
    Grit Blasting
    Plumstone Blasting
    Shot Blasting
    Shot Peening
    Vacu-Blast
    Vapour Blasting

Blue Annealing
BNF Test
Bonderizing
Bond Film
Borax Treatment
Boriding
Bower Barff Process
Box Annealing
Brass Colouring

xii / LIST OF PROCESSES COVERED

Brassing
Brass Plating
Brass Welding
Brazing
Bright Annealing
Bright Chrome Plating
Brinell Hardness Test
British Non-ferrous Jet
 Test (BNF Jet Test)
Bronze Plating
Bronze Welding
Bronzing
Browning Process
Brunofix
Brunorizing
Brutonizing
Brytal Process
Buffing
Bullard Dunn Process
Burnishing
Butt Welding
Buttering
Buzzard Process
Cadmium Plating
Calorizing
Carbo-nitride
Carbon Arc Welding
Carbon Dioxide
 Welding
Carburizing
Case Hardening
CASS Test (Copper
 Accelerated Acetic
 Acid Salt-spray)
Castenite Process
Casting
Cathodic Etching
Cathodic Protection
Cavitation Erosion
Ceconstant
Cementation
Centrifugal Casting
Chapmanizing
Charpy Test
Chasing
Chemag Process
Chemical Deburring
Chemical Machining

Chemical Polishing
Chemical Vapour
 Deposition
Chesterfield Process
Chisel Test
Chromate Coating
Chromating
Chromic Acid Anodize
Chromium Aluminium
 Duplex Coating
Chromium Plating
Chromizing
Cinera Process
Cladding
Classical Anneal
Cleaning
Close Annealing
Cloud-bursting
Cobalt Plating
COD
Coining
Cold Drawing
Cold Galvanizing
Cold Rolling
Cold Welding
Cold Working
Colour Anodize
Compression Test
Composite Coatings
Contact Tin Plating
Controlled Anneal
Controlled-atmosphere
 Furnace Brazing
Conversion Tin Plating
Copper Plating
Corrodekote Test
Corronizing
Corrosion Protection
Coslettizing
Crack Opening
 Displacement
Crack Testing
Creep Test
Cromalin
Cromcote
Cromodizing
Cromylite
Cronak

Cupping Test
Cuprobond
Cuprodine
CVD
Cyanide Hardening
Cyclic Anneal
Dalic Plating
Damascening Process
Deburring
Decarburizing
Deep Anodize
Deep-drawing
Degreasing
Demagnetization
Deoxidine
Depletion Gilding
Descaling
Dew-point Control
Diabor
Diamond Pyramid
 Hardness Test
Die-casting
Die Welding
Differential Heating
Diffusion Bonding
Dinanderie
Dip Brazing
Dip Moulding
Dip Soldering
Dispersion Hardening
Dot Welding
Double Refining
Double Temper
Drawing
Drawing-back Process
Drifting Test
Dri-loc
Drop Forging
Drop Stamping
Drop Test
Dry Blasting
Dry Cyaniding
Dry-drawing
Dry-film Lubrication
Dry film Painting
Drying
Ductility Measurements
Durboride

LIST OF PROCESSES COVERED / xiii

Durferrit
Durionizing
Durometer Test
Dye-penetrant Crack
    Testing
Eddy Current Sorting
Eddy Current Test
Edgewick Hardness Test
Efco-Udylite Process
Elasticity Test
Elcoat CRC
Elcoat MZ
Elcoat TC
Elcoat 37
Elcoat 101
Elcoat 240
Elcoat 360
Electric Arc Welding
Electric Cleaning
Electrochemical
    Machining
Electrocleaning
Electrocolour Process
Electrodeposition
Electroforming
Electrogalvanizing
Electrogranodizing
Electrography
Electroless Nickel Plating
Electroless Plating
Electrolytic Etch
Electrolytic Polishing
Electron Beam Welding
Electro-osmosis
Electropercussion
    Welding
Electrophoresis

ELECTROPLATING
    Alloy Plating
    Aluminium Plating
    Barrel Plating
    Brass Plating
    Bronze Plating
    Cadmium Plating
    Chromium Plating
    Cobalt Plating
    Composite Coatings

    Copper Plating
    Gold Plating
    Indium Plating
    Iron Plating
    Lead Plating
    Lead–Tin Plating
    Levelling
    Nickel Plating
    Palladium Plating
    Platinum Plating
    Rhodanizing
    Rhodium Plating
    Ruthenium Plating
    Silver Plating
    Solder Plating
    Speculum Plating
    Tin Plating
    Tin–Copper Plating
    Tin–Nickel Plating
    Tin–Zinc Plating
    Zinc Plating

Electroslag Welding
Elongation
Eloxal Process
Elphal Process
Embossing
Emulsion Cleaning
Enamel Plating
Enamelling
Endurance Process
Endurance Testing
Endurion
Equi-tip
Erichsen Test
Etching
Eutectrol Process
Evaporation Process
Explosive Riveting
Explosive Welding
Extrusion
Fadgenizing
Fafcote
Falling Weight Test
Fasbond
Fatigue Test
Feroxil Testing
Ferrite Testing

Ferrostan Process
Ferrous Blasting
Fescolizing/Fescol
    Process
File Test
Fillet Welding
Fingerprint Testing
Fink Process
Fire Cracker Welding
Fire Gilt Process
Firth Hardometer
    Hardness Test
Flame Annealing
Flame Cleaning
Flame Cutting
Flame Descaling/Flame
    Cleaning
Flame Hardening
Flame Plating
Flame Scaling
Flash Butt Welding
Flattening Test
Flaw Detection
Flex Testing
Flow Soldering
Fluidized Bed Heating
Fluxing
Footner Process
Foppl Test
Forge Weld
Forging
Forward Welding
Foslube
Fracture Test
Fremont Hardness Test
Fremont Impact Test
Friction Welding
Frosting
Fuess Testing
Full Anneal
Furnace Brazing
Fusion Welding
Galvanic Protection
Galvanizing
Galvannealing
Gas Carburizing
Gas Welding
Gator–card

xiv / LIST OF PROCESSES COVERED

Gilding
Glass-bead Blasting
Glueing
Gold Plating/Gilt Plating
Goldschmidt Process
Graduated Hardening
Grain Refining
Granodizing
Granosealing
Graphitizing
Green Gold
Grit Blasting
Haig Prism Test
Hammer Welding
Hand Brazing
Hanson–Van Winkle
    Munning Process
Hard Chrome Plating
Hard-drawing
Hard Facing
Hard Plating
Hard Soldering
Hardas Process
Hardening

HARDNESS TESTING
  Brinell Test
  Diamond Hardness Test
  Edgewick Test
  Equi-Tip Test
  File Test
  Firth Hardometer Test
  Foppl Test
  Fremont Test
  Haig Prism Test
  Herbert Test
  Jagger Test
  Keeps Test
  Kirsch Test
  Knoop Test
  Ludwig Test
  Microhardness Test
  Mohs Test
  Monotron Test
  Muschenbrock Test
  Pellin's Test
  Poldi Test
  Pusey Test

  Pyramid Diamond Test
  Rockwell Test
  Scleroscope Test
  Scratch Test
  Shore Test
  Tukon Test
  Turner's Sclerometer Test
  Vickers Test
  Warmens Penetrascope
    Test

Harperizing
Hausner Process
Haworth Test
Heat Tinting

HEAT TREATMENT
  Age Hardening
  Annealing
  Austempering
  Austenizing
  Boriding
  Bright Annealing
  Carbo-nitride
  Carburizing
  Classical Anneal
  Controlled Anneal
  Cyanide Hardening
  Cyclic Anneal
  Double Temper
  Flame Hardening
  Full Anneal
  Gas Carburizing
  Graphitizing
  Hardening
  Homogenizing
  Hydrogen Annealing
  Induction Hardening
  Isothermal Annealing
  Magnetic Anneal
  Maraging
  Marquenching
  Martempering
  Nitriding
  Normalizing
  Pack Anneal
  Pack Carburize
  Plasma Nitride

  Postheating
  Precipitation Hardening
  Preheating
  Quench and Temper
  Quench Tempering
  Refine
  Secondary Hardening
  Selective Annealing
  Selective Carburizing
  Selective Hardening
  Self-annealing
  Self-hardening
  Skin Annealing
  Softening
  Solution Treatment
  Spheroidal Annealing/
    Spheroidization
  Stabilizing
  Stabilizing Anneal
  Step Anneal
  Stress-equalization
    Anneal
  Stress-relieving/Stress-
    releasing
  Subcritical Anneal
  Subzero Treatment
  Tempering

HEEF
Heliarc Welding
Herbert Hardness Test
High-frequency Induction
    Welding
Holiday Test
Homocarb Process
Homogenizing
Hot Quenching
Hot Working
Hot-dip Coating
Huey Test
Hull Cell Test
Humidity Testing
Hydraulic Pressure Test
Hydrocarb process
Hydrogen Annealing
Hydrogen Brazing
Hydrogen Embrittlement
Hydrogen Welding

LIST OF PROCESSES COVERED / xv

Hynac
Ihrigizing
Immersion Coating
Immersion Plating
Impact Test
Impregnation
Impressed-current
    Corrosion Protection
Imprest Process
Inchrome Process
Indium Plating
Induction Brazing
Induction Hardening
Induction Heating
Inert-atmosphere
    Furnace Brazing
Inert-gas Shielded Metal
    Arc Welding
Inertia Welding
Integral Welding
Interrupted Ageing
Interrupted Quench
Inverse Annealing
Ion Nitriding
Iron Plating
Isothermal Annealing
Izod Test
Jacquet's Method
Jagger Test
Japanning
Jet Test
Jetal
Jominy Test
K4
Kanigen Plating
Karat Clad
Kayem Process
Keeps Hardness Test
Keller's Spark Test
Kenmore Process
Kephos
Kern's Process/Kern's
    Test
Kirsch Test
Knoop Hardness Test
Koldweld
Kolene
Korel Method

Kuftwork
Kynar
Lacquering
Lap Welding
Laser Welding
Laxal Process
Lead Annealing
Lead Patenting
Lead Plating
Lead–Tin Plating
Levelling
Lime Coating
Linde Plating
Liquid Honing
Liquor Finishing
Lithoform
Loctite
Lost Wax Process
Ludwig Test
Luminous Painting
Macroetch Test
Madsenell Process
Magna Flux
Magnet Test
Magnetic Anneal
Magnetic Particle
    Inspection (MPI)/
Magnetic Crack Test
    (MCT)
Magnetic Particle
    Examination (MPE)
Malcomizing
Malleabilizing
Maraging
Marquenching
Martempering
MBV Process
McQuaid Ehn Test
MCT
Mechanical Alloying
Mechanical Refining

MECHANICAL
    TESTING
  Bend Test
    (Full Bend Test
    Repeated Bend Test
    Reverse Bend Test)

  Compression Test
  Crack Opening
    Displacement
  Crack Tip Opening
    Displacement
  Creep Test
  Elasticity Test
  Elongation Test
  Fatigue Test
  Fracture Test
  Impact Test
  Nick Break Test
  Nicked Fracture Test
  Notch Bar Test
  Proof Stress/Proof
    Strength
  Proof Test
  Shear Test
  Tensile Test
  Torsion Test
  Young's Modulus of
    Elasticity

Mechanical Working
Mellozing
Melonite
Mercast Process
Merilizing
Mesnager Test
Metallic Arc Welding
Metallic Cementation
Metallic Coating
Metallic Painting
Metallization
Metal-Lock
Metalock Process
Metal-sorting
Metal-spraying
Metascope Testing
Metcolizing
Method X
Microcharacter
Micro-Chem
Microhardness Test
MIG Welding
Modified Bauer Vogel
    Process
Mohs Hardness Test

Mollerizing
Molybond
Molykote
Monotron Test
MPI
Muschinbrock Hardness Test
Natural Ageing
NDT
Needle Descaling
Negative Hardening
Negative Quenching
Nertalic Process
Nervstar
Ni-Carbing
Nichem
Nick Break Test
Nickel Ball Test
Nickelex
Nickel Plating
Niflor
Nigrax
Nirin
Nisol
Nitralizing
Nitrarding
Nitration
Nitriding
Nitrogen Hardening
Nitrox
Nitrox P
Nivo

NON-DESTRUCTIVE TESTING (NDT)
  Acoustic Testing
  Air-pressure Test/Gas-pressure Test
  Crack Detection Testing
  Dye-penetrant Crack Test
  Eddy Current Test (for defects)
  Etching
  Gas Pressure Test
  Holiday Test
  Hydraulic Test/ Hydraulic Pressure Test
  Magnetic Crack Test/ Magnetic Particle Inspection/Magnetic Flaw Detection
  Porosity Test
  Pressure Test
  Radiographic Test
  Sonic Testing
  Spark Test (Porosity Testing)
  Strain Gauging
  Thickness Testing
  Ultrasonic Flaw Detection
  Vibration Testing
  Visual Inspection

Normalizing
Noskuff
Notch Bar Test
Nubrite
Oil Hardening
Olsen Test
Onera Process
Osronaut
Oven Soldering
Overageing
Over-drawing
Oxalic Acid Anodize
Oxyacetylene Welding
Pack Anneal
Pack Carburize

PAINTING
  Paint Preparation
  Brush Painting
  Curtain Painting
  Dip painting
  Electrophoresis/Electropainting
  Electrostatic Painting
  Enamelling
  Lacquering
  Powder Painting
  Roller Coating
  Spray Painting
  Stove Enamelling

Palladium Nickel Plating
Palladium Plating
Pallnic Plating
Paraffin Test
Parkerizing
Passivation
Patenting
Peel Test
Peen Plating
Peening
Pellin's Test
Penetrol Black Process
Penybron Plating
Percussion Welding
Pfanhauser's Plating
Phosphating
Physical Testing
Physical Vapour Deposition (PVD)
Pickling
Pillet Plating
Planishing
Plasma Cutting
Plasma Heating
Plasma Nitride
Plasma Plating
Plasma Welding
Plating
Platinizing
Platinum Plating
Plug Weld
Plumstone Blasting
Poldi Hardness
Polishing
Porcelain Enamelling
Porosity Testing
Postheating
Pot Annealing
Pot Quenching
Powder Painting
Precipitation Hardening
Preece Test
Preheating
Press Forging
Pressure Test

LIST OF PROCESSES COVERED / xvii

Pressure Welding
Primer Painting
Process Annealing
Progrega
Progressive Ageing
Projection Welding
Proof Stress/Proof Strength
Proof Testing
Protal Process
Protolac Process
Pull-off Test
Punching
Pusey Hardness
Pylumin Method
Pyramid Hardness Testing
Pyro Black
QPQ
Quench Ageing
Quench Hardening
Quench Tempering
Quicking
Radiography
Ransburg Process
Recarburization
Red Gilding
Redux Process
Refine
Reflectogage Testing
Reflowing
Refrigerated Anodizing
Regenerative Quench
Reheating
Repeated Bend Test
Repousse Process
Resistance Soldering
Resistance Welding
Reverse Bend Test
Rhodanizing
Rhodium Plating
Rivet Test
Rivet Weld
Rockwell Hardness Test
Rolled Gold
Roller Spot Welding
Roller-tinning
Root Bend Test

Rose Gilding
Rumbling
Rustproofing
Ruthenium Plating
Sacrificial Protection
Salt mist/Salt-spray Testing
Sand Blasting
Sand Ford Process
Sandberg Treatment
Sanding
Sankey Test
Satin Finish
Satin Kote Treatment
Satin Nickel Plating
Satylite Nickel Plating
Sawdust Drying
Schnadt Test
Schoop Process
Schori Process
Scleroscope Hardness Test
Scouring
Scragging
Scratch Brushing
Scratch Test
Sealing
Seam Welding
Seasoning
Secondary Hardening
Selective Annealing
Selective Carburizing
Selective Hardening
Selective Plating
Selenious Acid Treatment
Self-annealing
Self-hardening
Sendzimir Process
Sensitizing
Sermetel
Sermetriding
Servarizing Process
Shallow Hardening
Shape-strength Test
Shear Test
Shell Moulding
Shepherd Process

Shepherd Test
Sherardizing
Shielded Arc Welding
Shimer process
Shock Test
Shore Hardness
Short-cycle Annealing
Shorter Process
Shot Blasting
Shot Peening
Shrink Fitting
Sieve Test
Siliconizing
Silk-screen Printing
Silver Ball Test
Silver Plating
Silver Soldering
Sintering
Skin Annealing
Skin Pass
Skip Weld
Smithing
Snarl Test
Snead Process
Soaking
Sodium Hydride
Soft Facing
Softening
Solder Plating
Solderability Testing
Soldering
Solid Carburizing
Solution Treatment
Solvent Cleaning
Solvent Degreasing
Sonic Testing
Spark Erosion
Spark Testing
Speculum Plating
Spheroidal Anneal/ Spheroidization
Spin Hardening
Spinning
Spot Welding
Spra-bond
Spread Test
Spring-back Test
Sputtering

xviii / LIST OF PROCESSES COVERED

Stabilizing
Stabilizing Anneal
Steam Blueing
Steelascope Testing
Step Anneal
Step Quenching
Stitch Weld
Stop-off
Stove Enamelling
Straightening
Strain or Stress Ageing
Strain Gauging
Strain Hardening
Strauss Test
Stress-equalization
　　Anneal
Stress-relieving/Stress-re-
　　leasing
Stress-rupture Test
Stretcher Straining
Stretching
Stromeyer Test
Stud Welding
Subcritical Anneal
Submerged Arc Welding
Subzero Treatment
Sulphinuz
Sulphuric Acid Anodize
Sulphur Printing
Super Gleamax
Superficial Hardness Test
Supersonic Testing
Surface Hardening
Sursulf
Swageing
Sweating
Swilling
Taber Abraser
Tack Weld
Taylor-White Process
Techrotherm Rokos
　　Process
Temper Blueing
Temper Hardening
Temper Rolling
Tempering
Tensile Test
TFD

Thermal Treatment
Thermit Welding
Thickness Testing
Thuriting
Tiduran
TIG Welding
Tin Plating
Tin–Copper Plating
Tin–Lead Plating
Tin–Nickel Plating
Tin–Zinc Plating
Tinning
Tocco Process
Tool Weld Process
Torch Brazing
Torch Hardening
Touch Welding
Toyota Diffusion Process
Tribomet
Tri-Ni
Trisec Drying
Tufftride Process
Tufftride QPQ
Tukon Hardness Test
Tumbling
Turner's Sclerometer
Twisting Test
Ultrasonic Cleaning
Ultrasonic Crack
　　Detection
Ultrasonic Soldering
Unionmelt Welding
Unshielded Metal Arc
　　Welding
Upset Welding
Vacu-Blasting
Vacuum Brazing
Vacuum Coating
Vacuum-coating
　　Deposition or
　　Plating
Vacuum Metallization/
　　Vacuum
　　Evaporation
Vapour Blasting
Vapour Degreasing
Vapour-phase Inhibitors
　　(VPI)

Vaqua Blast
Vibration Testing
Vickers Hardness Test
Vitreous Enamel
Walnut Blasting
Walter Black
Walterizing
Warman Penetrascope
Water Gilding
Water Hardening
Watt Nickel Process
Wave Soldering
Weathering
Weibel Process
Weld Deposition
Weld-decay Testing

WELDING
　Argon Arc Welding
　Argonox Welding
　Argonshield Welding
　Atomic Arc Welding
　Carbon Dioxide (CO_2)
　　Welding
　Electric Arc Welding
　Electron Beam
　　Welding
　Electropercussion
　　Welding
　Electroslag Welding
　Explosive Welding
　Flash Butt Welding
　Forge Welding
　Friction Welding
　Fusion Welding
　Heliarc Welding
　High-frequency
　　Induction Welding
　Hydrogen Welding
　Inert-gas Shielded
　　Metal Arc Welding/
　　Inert-gas Shielded
　　Welding/Inert Gas
　　Welding
　Metal Inert Gas
　　Welding (MIG
　　Welding)
　Percussion Welding

Plasma Welding
Pressure Welding
Projection Welding
Resistance Welding
Roller Spot Welding
Seam Welding
Spot Welding
Stitch Welding
Submerged Arc Welding
Tack Welding
Thermit Welding

Tungsten Inert Gas Welding (TIG Welding)

Wet Nitriding
Wiped Joint
Wire Brushing
Wohler Test
Work Hardening
Wrapping Test

X-ray
Xeroradiography

Yield
Young's Modulus

Zartan
Zerener Process
Zinc Coating
Zinc Phosphatizing
Zinc Plating
Zincate Treatment
Zincing
Zincote Process
Zinkostar
Zyphos

PART ONE

A–Z of treatment and testing processes

Abrasive Blasting
A *blasting* technique designed to cut and remove metal, to be distinguished from the alternative blasting techniques where shot impinges.

Application range ALL METALS

Shot blasting will efficiently remove brittle material from metallic surfaces such as millscale oxide, remains of paint, etc. The *shot blasting* technique will not generally, however, remove metal and may – as in the case of *shot peening* – be designed to produce a surface which has special characteristics. Abrasive blasting, on the other hand, will cut. The process is more generally referred to as *grit blasting*.

This uses a variety of materials in the form of sharp abrasive grit. This can be tightly controlled regarding analysis and particle size, or left-over slag from other operations such as mining or metal refining. Sand as such, that is silicon dioxide SiO_2, is a forbidden substance for abrasive blasting in the United Kingdom but substances such as fractured chilled iron, alumina (aluminium oxide) blast-furnace slags and similar materials are commonly used. Described fully in Part Two **Blasting**.

Abrasive Testing
A destructive testing technique designed to measure the resistance of a surface to abrasion.

Application range ALL METALS

The surface being tested is subjected to repeated movements either against the materials used for testing, or by being rotated or reciprocated under load. Two specific tests exist, '*Haworth*' and '*Taber*', but very often the abrasive testing is not carried out to scientific levels, but is intended simply to give some indication of the capacity to resist abrasion.

Abrasion resistance is a function of the hardness of the surface, and thus *hardness testing* itself may give results which preclude the necessity to carry out specific tests to measure abrasion resistance. Therefore, the abrasive test is more often required when the hardness of the abrading material or surface is unknown. It is a fact that no softer material will cause a cutting action on a harder material. Thus, if the hardness of the two components being abraded is accurately known, there is no necessity for any other abrasive testing.

Accelerated Corrosion Testing

Techniques used to ascertain whether materials or coatings give satisfactory corrosion protection.

Application range ALL METALS

The purpose is to produce a test which in itself is reproducible and also gives results in a short period of time which reproduce the actual corrosion potential over a number of years. There is now being built up a body of experience to relate the results of testing carried out over a period of days, or even hours, which clearly shows that the information gained can be used to predict the corrosion resistance over a period of years.

There are a number of accelerated corrosion tests. At their simplest, these can take the form of *thickness testing*, or in more sophisticated form there is *salt-mist testing*, where the salt droplets are controlled for size, chemical composition, temperature, etc. There are simpler tests: the relatively simple *humidity test* can also be a form of accelerated corrosion testing.

In general, however, it must be stressed that corrosion testing is a method of ensuring that the corrosion-protection system or metal chosen is within specification, rather than any long-term guarantee that the system being inspected will withstand a specific number of years under any condition. The reaction may range from an industrial-type marine atmosphere with high humidity and temperature, where corrosion will be very severe, to an inland rural atmosphere, where humidity may be variable but will not be high, and where corrosion conditions will exist but probably at only one-tenth or one-twentieth the severity of the previous case. There is, finally, the condition of zero humidity and controlled, normally very low, temperature, where no appreciable corrosion on even unprotected mild steel will occur. It will be seen, therefore, that no single corrosion test can supply meaningful results which will always apply in this wide range of conditions.

There will be found in this book a number of tests, such as the *salt-mist test*, the *weld-decay test* and the *copper, acetic acid accelerated (CASS) tests*. In addition there is available from other sources information from the many published charts and books which list various metals and their resistance to corrosion.

There can be no doubt that the life of a protective coating is not only to some extent related to its thickness, but also to the technique of application and the lack of porosity. It has now been shown that control of the variable parameters which exist in the various coating treatments can result in coatings which are homogeneous, pore-free and give long-life corrosion resistance.

Coatings can be inert, which when correctly applied, will have an indefinite life, but may fail rapidly when damaged, or they can be 'active'. These 'active' coatings will be constantly used up to protect the base metal, and thus will have a finite life. The length of this life will be related to the corrosion of the environment, the type of coating and its thickness.

To summarize, accelerated corrosion testing is carried out in order to evaluate in hours the life of the coating or metal. This evaluation should be used more to ensure that the surface treatment is within specification, rather than to guarantee any specific life. This is often used in conjunction with *adhesion testing, abrasive testing* and solvent resistant tests.

Acid Descaling

An alternative name for *pickling*, a process using acid to dissolve oxide and scale.

Acoustic Testing

A form of *non-destructive testing*. Described fully in Part Two **Non-destructive Testing**.

Activation

A term occasionally used in the *electroplating* process to denote *etching*. This process removes the last trace of any oxide on the metal surface, and by removing a thin layer of the metal itself, ensures that the metal subjected to the plating process is in an active condition. (Further details are given under the heading *Etching*.) Described fully in Part Two **Electroplating**.

Adhesion Testing

A destructive test to assess adhesion.

Application range ALL METALS

There are four types of adhesion testing, as follows:

1. *Scratch test* In this test attempts are made to cut the coating to assess the degree of adhesion by the ease or difficulty with which the coating can be peeled from the substrate. The usual method is to scratch a rough triangle and to judge the ease of failure at the internal apex. A cross-hatch pattern can also be produced and the failure rate evaluated.

2. *Pull-off test* In this test an adhesive tape is stuck to the surface of the film and the difficulty of removing it is assessed. This technique generally requires that the film is cut in order to give a positive evaluation. With paint, this technique can be quite sophisticated, supplying an exact measure of adhesion

of the film to the substrate. This is sometimes referred to as the 'Peel test'. It takes several forms, the simplest probably being the cross-hatch type: the paint film is scored with a number of parallel lines (usually 5) each approximately 0.15 in apart (4 mm). A second series of parallel lines are then scribed at right-angles to the first. This results in a number of squares each 0.15 in^2 (4 mm^2). If failure of the paint film occurs during the scoring operation, some idea of the adhesion can be obtained by noting when failure occurred. Thus, if failure takes place during the first scoring, adhesion is very poor. If failure is of only a very few squares on completion of the cross-hatch, this indicates much better adhesion. If no failure can be seen, then the adhesion of the paint to the substrate must be reasonable.

The cross-hatch is then subjected to a 'pull-off' test by taking adhesive tape and placing this in intimate contact with the prepared cross hatch. It is important that all air bubbles are removed. The adhesive tape is then pulled in a manner such that the direction of pull applied to the paint film is 45 degrees to the vertical. The surface of the adhesive can then be examined and will give a quantitative measurement of the adhesion. Where all the squares are removed adhesion is relatively poor, where no squares are removed adhesion is excellent.

Peel testing can also be applied as a destructive quality-assessment test by using some form of *stop-off* to prevent adhesion of the paint film locally. After any curing etc., this local poor-adhesion film is carefully cut and removed to give a strip of paint which still remains connected to the remainder of the paint film. By using weights or a *tensile test* machine on this strip, it is possible to evaluate the adhesion of the film to the substrate. This test takes various forms some of which require that the test strip be cut. Accuracy will depend on a number of factors and duplicate tests should always be carried out.

3. *Visual examination test* This test should always be carried out to examine for blisters or areas where adhesion is seen to be below standard.

4. *Blistering test* This test is confined to *electrodeposits* of ductile material. The plated surface is rubbed, using a soft smooth metal, generally copper (a standard method is to use a copper coin with a worn edge). The purpose of this test is to cause expansion of the plated film by rubbing. Where adhesion is of a high standard, this will not be possible, whereas any poor adhesion will be shown by the formation of blisters.

Aerocase Process

A form of *cyanide hardening*, in which the normal sodium potassium cyanide salt has additions of sodium and calcium chloride.

The result of this process is a case which is closer to a *carburizing* than *nitriding* case, in that there are less nitrides in the final case. This means that the case obtained will be less brittle than with the normal *cyanide hardening* technique. Apart from this slight advantage, this process suffers all the disadvantages of the basic technique, details of which will be found in Part Two **Heat Treatment**.

It is now largely replaced by carbonitriding.

Age Hardening
An alternative term for *precipitation hardening, precipitation treatment* or *ageing*.

This process can be applied to a number of specific alloys of several metals and must always be preceded by some form of *solution treatment*.

The combination of solution treat and *ageing* results in an increase in mechanical properties. Described fully under *Precipitation Hardening* in Part Two **Heat Treatment**.

Ageing
See *Age Hardening*.

Air Blasting
The method by which shot or grit is given the necessary velocity to carry out blasting. The term air blasting is used more to differentiate between *airless blasting* and *vapour blasting* than as a specific form.

The term may also apply to the use of compressed air for drying etc.

Air Hardening
The term used when a steel can be quenched by cooling in air. Described under *Hardening* in Part Two **Heat Treatment**.

Air-pressure Testing
A form of *non-destructive testing*. Described fully in Part Two **Non-destructive Testing**.

Airless Blast
A method where the velocity of the shot or grit used results from a centrifugal impeller.

Application range ALL METALS

The blast material is fed at a controlled rate to the impeller which has vanes to throw the blast material at right angles to the direction of rotation. Using abrasive resistance materials for the guides, the shot or grit can be directed to the area being blasted.

In general, this technique is confined to blasting inside cabinets. These cabinets can be of the walk-in type, the automatic type, where the components are rotated through the blast cabinet, or the hand-operated type.

The term airless blast refers to the method of imparting velocity to the blast material and, thus, distinguishes it from *air* and *vapour blasting*; it is not, therefore, a *blasting* method.

Airomatic Welding

An alternative name for *electric arc welding*.

Alachrome

Alternative spelling for Alochrome. See *Chromate Coating*.

'Albo'

This is the proprietary name for a nickel plating solution. It is claimed that this has a simple formula, is easy to control and has few maintenance problems. It is therefore of use where low productive output is required in small jobbing type plants.

Further information can be obtained from Messrs Cannings.

'Aldip' Process

A process for coating ferrous materials with aluminium and its oxide for protection and heat-resistant purposes (see *Corrosion Protection*).

Application range STEEL ONLY

After grease and dirt have been removed by an alkali cleaner, the parts are *pickled* in acid, rinsed and furnace-dried. They are placed in a preheating salt bath at 690–760°C, for 4 minutes, then into a molten aluminium bath, covered with flux, for about 45 seconds, and returned to the preliminary salt bath, slowly removed and air-cooled. An aluminium and aluminium oxide surface remains. Any excessive roughness or unwanted metal can be removed by *shot blasting*.

This process is also called *mollerizing*. It is one of the forms of *aluminizing*; more details are given under *Aluminizing*.

'Al-fin' Process
The formation of an aluminium coating.

Application range STEEL AND CAST IRON

The steel or cast iron component is cleaned, fluxed, then dipped in molten aluminium.

The process is used prior to casting aluminium on to steel or cast iron. *Casting* follows immediately after the hot-dipping when the surface aluminium is still molten, thus ensuring a good intermetallic bond. It is commonly used on internal-combustion engine components where aluminium cooling fins are an efficient means of dissipating the unwanted heat.

This is a form of *aluminizing*, where an adherent surface layer of the intermetallic compound of iron and aluminium is used as the bond between the steel or cast iron and the aluminium of the cast fins.

The Al-fin process is never used as an isolated method, but only as a pretreatment to the casting of aluminium on steel. Good bonding is essential as conduction of heat across the bond is necessary. This is a specialist type of process where alternatives are only possible by varying one or more of the parameters involved.

Alkaline Descaling
A chemical process for removing scale.

Application range ALL METALS

When certain materials such as aluminium are descaled additives are necessary to prevent excessive attack. The term alkaline descaling generally means the use of straightforward chemical action, but under some circumstances electrolytic action is also used, and the *sodium hydride process* has been described as a form of alkaline descaling.

The usual alkaline descaling solution uses caustic soda as a base, sometimes with additives. At the correct strength and temperature, an alkali attack on steel, particularly iron oxide is possible, but it is quite slow. With aluminium, the attack on the base metal would not stop and would be rapid. Thus, additions must be made to the solution to act as inhibitors.

The procedure is that the components to be descaled are immersed in the alkaline solution. With steel and many other metals, the time of immersion is not critical, but with metals such as aluminium and to a lesser extent certain copper alloys, magnesium and zinc, excessive time may result in a local undesirable pitting and overetching.

The alkaline solutions, which have considerable variations, will attack the metal, and the attack will vary depending on the solution used and the metal involved. The reader is advised to obtain specialist advice before using this technique as the process is generally more expensive than *pickling* and with some materials can lead to local attack.

The advantage of alkaline descaling is that there is no danger from *hydrogen embrittlement*. This means that components with a relatively high tensile strength in stressed condition, can be descaled. Very often components such as springs cannot be *pickled* using acid without serious embrittlement problems, sometimes resulting in fracture during the pickling operation. A second advantage is that the action is much more gentle and selective, and this can be useful under some circumstances.

The process almost invariably will be more expensive than acid *pickling*. Alternative methods would include *blasting* to remove scale, though this can be difficult where complicated components are involved, as for example pipe bores which require *descaling*. Also of use could be the *sodium hydride process*, which is a more sophisticated, expensive process.

Alloy Plating

The *electrodeposition* of more than one metal at the same time. This includes *brass, bronze* and *speculum plating*. Described fully in Part Two **Electroplating**.

Alloying

The addition to a metal of another metal or non-metal or combination of metals.

As a rule, pure metals have limited applications outside their ability to withstand corrosion. In general, metal alloys are the most useful materials found in engineering.

The most common alloy used in engineering is steel. Steel is an alloy of iron and iron carbide. To this basic alloy may be added all sorts of other materials, such as nickel, chromium, silicon, etc. Alloys can also be made by the addition of materials such as tungsten carbide, which are themselves compounds.

With non-ferrous materials, the same conditions apply. Aluminium forms alloys with magnesium, manganese, silicon and copper, and also forms intermetallic compounds with copper, zinc, etc.

Thus, it will be seen that alloying materials either can be other metals and non-metals, or a combination of a metal and non-metal or combinations of the metals themselves, known as intermetallics.

It is not usually possible to predict the properties of an alloy by extrapolating from the properties of the alloying materials. Thus, magnesium has poor seawater corrosion properties, but when added to aluminium, it produces an alloy with marine corrosion resistance superior to other aluminium alloys. Many other similar examples exist. Reference should, therefore, always be made to the specification of an alloy before assuming its properties. In case of difficulty

ALOCHROM / 11

the various metal associations can be contacted, for example the Aluminium Federation; Copper Development Association; etc., which are listed in Appendix Two.

Alloying is a basic manufacturing technique and thus does not come within the description of treatment or testing of metals.

Almen Test (Almen Number)

A test piece used in conjunction with the *shot-peening* operation to evaluate the degree of *work hardening* which has been given.

Application range STEEL AND SOME ALUMINIUM ALLOYS

In essence this is a technique to increase the compressive strength on the surface of the component being treated, and in this way increase the fatigue strength. In order to evaluate the amount of peening carried out, it is necessary strictly to specify the variable parameters involved and then to ensure by use of the Almen test that these achieve the correct effect. This is carried out using a strip of metal of thin gauge and approximately 2 in (50 mm) long by $\frac{1}{2}$ in (12 mm) wide. This is held rigidly in contact with a flat, hardened surface in such a manner that it cannot be readily moved in any direction.

This test strip is then subjected to the *shot-peening process* exacctly as laid down by the design requirements. This means that one surface only of the material is peened, and as the material involved is thin and soft and the test strip is restricted from moving, the surface peened will stretch. This results in the strip curving upwards. The height of the curve is measured before and after the peening and the increase in height measured in thousandths of an inch is termed the Almen number.

A standard Almen number for *shot peening* of steel components would be 15–25, indicating that the increase in height or the degree of bending of the Almen test piece was 0.015–0.025 in (0.37–0.5 mm).

Briefly, therefore, the Almen test is a means of measuring the amount of *cold work* which is applied by *shot peening*. The test requires that the strip of metal used is in a standard material, generally fully softened mild steel or aluminium.

As it is a means of reproducing the shot-peen process there is no obvious alternative to this control technique. Described fully in Part Two **Blasting**.

'Alochrom'

A proprietary process to improve corrosion resistance and to act as a key to paint.

Application range ALUMINIUM AND ALLOYS

There are several types available, which can be applied by brush, spray or dip, and are identified by a number following the 'Alochrom' name. The process results in the formation of an oxide film into which is absorbed some of the chromate. There are various spellings of the name, and to some extent 'Alochrom' is the common name applied to the *chromate* conversion treatment of aluminium. The process requires, first, thorough cleaning and removal of any existing corrosion products. For best results, it is recommended that the surface of the aluminium should be lightly etched either with the cleaning solution itself, or if this is not desirable or possible, with a separate *etching* solution of phosphoric acid. The surface must then be thoroughly washed and the 'Alochrom' solution applied to this clean, active surface by dipping, brushing or spraying. The ideal treatment is that the material should then be thoroughly washed, using a dipping or spraying process with copious application of running cold water. If this is not possible, then provided the material is warmed to ensure complete drying, a satisfactory oxide can be produced. Care must then be taken to ensure that no pockets of liquid are present, and that the solution has not contacted other materials, such as paint or plastic which will be attacked by the strongly oxidizing chromate solution. The chemicals used in the process are very aggressive and must be handled with great care.

The purpose of the 'Alochrom' treatment is to produce an adherent aluminium oxide which will absorb some chromate solution. This seals the surface and adds to the corrosion resistance.

While this is not as satisfactory as the anodizing process, it is much cheaper to apply, and, where subsequent painting is correctly applied, will result in a high standard of corrosion resistance. Paint applied directly to aluminium with no prior treatment will have poor adhesion. Details of 'Alochrom' can be obtained from ICI Ltd.

'Alodine'

The name given to the the *'Alochrom'* process in the USA.

Information can be obtained from the American Chemical Co. Ltd. Further details of the process, which is a chromate conversion treatment to improve the corrosion resistance and paint adhesion of aluminium, are given under *Chromate Treatment*.

'Alplate' Process

A proprietary process of the *aluminizing* type.

In general this process is only used on steel but it can be used on other materials, such as nickel alloys, when necessary.

The surfaces to be treated are heated in a hydrogen atmosphere at 1000°C. The parts are immersed from this atmosphere into a bath of molten aluminium at about 700°C. This technique prevents the formation of surface oxides and, thus, allows the correct alloying of aluminium to the steel or nickel alloy.

With the advent of better control of standard *aluminizing* techniques, this method is no longer commonly used. Further information on alternative methods will be found under *Aluminizing*.

'Alrak' Process
A chromate treatment.

Application range CAST IRON, STEEL, NICKEL ALLOYS

A method of applying an oxide conversion coating to enhance corrosion resistance and improve paint adhesion, and to some extent, friction characteristics.

The process consists of thorough cleaning, removal of the natural oxide film in an aqueous phosphate solution and *activation* of the surface with dilute acid. The coating of sodium carbonate–potassium dichromate is normally applied by immersion for 20 minutes at a temperature of 90°C. The parts are rinsed and dried usually above 70°C, which increases the hardness and abrasion resistance, giving a grey–green coating which is suitable for painting and has better corrosion resistance than the natural aluminium oxide.

This process is applicable to all aluminium alloys and is convenient in that the solution may be applied by immersion, spraying, brushing or any wetting method. It is similar to the '*Alochrom*' process.

'Aludip' Process
An alternative name for *aluminizing*, using the immersion process.

'Alumilite' Process
An alternative term for *anodizing*, using the sulphuric acid process.

Aluminium Chromium Duplex Coating
This is a process where aluminium and chromium are diffused into the base surface for corrosion and oxidation resistance.

Application range NICKEL ALLOYS, STAINLESS STEELS

This is a two-stage process where the first stage is chromizing to produce a chromium-enriched layer followed by a second stage which is aluminizing where aluminium is diffused into the chromium layer and through the chromium layer into the base metal.

The reader is referred to *Aluminizing* and *Chromizing* for information on these two processes.

This duplex coating produced with this technique has been shown to give excellent protection against sulphur atmospheres and also hot oxidation at quite high temperatures.

This process needs specialized equipment using high temperatures and is expensive and is only economically justified on high technology components. The surface is readily damaged and thus has limited application for normal commercial purposes.

Aluminium Plating
A recently developed technique, not yet applied as a general production process as it uses molten salt electrolytes. Described fully in Part Two **Electroplating**.

Aluminium Soldering
Joining of aluminium using a low-melting-point alloy (see *Soldering*).

Application range ALUMINIUM

To solder aluminium, it is necessary to remove the adherent oxide. Two methods are used:

1. With active flux and specially alloyed solders, generally tin–zinc, tin–cadmium or tin–zinc aluminium. There are also more recently developed, more highly alloyed solders. These result in excellent solder joints, provided the manufacturers' instructions are followed. Care must be taken to remove all trace of flux as these will be corrosive and can result in joint failure during service. The fluxes used dissolve the aluminium oxide which prevents the wetting of the surface, and allows the special solder to alloy with the base material.

2. With ultrasonic equipment. This uses a less active flux and ultrasonic energy, to break down the adherent oxide and allow the molten aluminium solder to alloy with the base material.

The joint achieved by both methods is comparable and technically identical to tin–lead soldered joints. Method (1) requires simple treatment, and scrupulous care to remove all trace of flux. Method (2) requires more sophisticated equipment, but uses less active flux.

It is also possible to achieve a joint by mechanically cleaning and using active flux. With this method, the skill of the operator is in reverse proportion to the activity of the flux for satisfactory joints. Again, however, thorough cleaning is essential after soldering to remove all traces of flux. The difficulty of cleaning will be directly proportional to the activity of flux used.

Soldering aluminium using these techniques is now a satisfactory production process, but because of the activity of aluminium regarding oxidation, and the stability of the oxide formed, it will always be more difficult than conventional soldering of steel or copper alloys. Alternative joining methods include *riveting, brazing* and *welding*; also crimping and gluing.

Aluminizing

A process in which the surface of the material is impregnated with aluminium and its oxide to improve oxidation resistance in the medium temperature range of 400°C–500°C.

Application range CAST IRON, STEEL, NICKEL ALLOYS

1. *The molten-bath process* The parts are first cleaned and dried. This can be achieved by the *grit blasting* process, which ensures the complete removal of all scale and produces a satisfactory standard of cleanliness. Acid *pickling* after alkaline cleaning is also satisfactory, but it is important that the parts are dried. The cleaned and dried components are immersed in a molten bath of aluminium, which may contain additives in certain proprietary methods.

2. *Metal-spraying* The objects are first *grit blasted*, and then sprayed with aluminium metal. The freshly-sprayed components are then heated to a high enough temperature to ensure that the aluminium surface is converted to the oxide. The efficiency of the process relies on excellent preparation, good quality metal spraying and careful control at the oxidation stage.

3. *The sealed-box process* The components are placed in a heat-resistant box, separated by a proprietary powder. The components must be thoroughly cleaned, and generally lightly *blasted*, to produce an active surface, and handled with extreme care. Each component is placed such that it is separated from its neighbour, and is then covered with a least 1 inch (25 mm) of powder and the process repeated until the box is filled. Normally components are relatively small and a complete batch will seldom weigh more than 200 kg.

This box is then sealed either by welding, or by some other mechanical means, and is purged with argon or nitrogen and inserted into a furnace and brought to operating temperature. At the end of the appropriate time, which will be several hours, the box is removed from the furnace and, after cooling, the seal is broken and the components removed. A thin film of aluminium will be present, which will have an adherent oxide at the surface.

The first two aluminizing processes are used on mild and low-alloy steels. They increase the resistance to hot oxidation and give some protection against corrosion. This can give them a useful life at 600°C–800°C for considerable periods, provided the conditions are freely oxidizing. This allows the use of mild or low-alloy steel for articles such as furnace furniture, etc. when otherwise stainless steel would be required. The components cannot withstand rough handling without damage as the oxidation resistance is thus destroyed. The third process is generally applied to nickel alloys to increase the hot-corrosion resistance where free oxidizing conditions do not exist or where there is a sulphur-bearing atmosphere. The surface obtained is comparatively fragile and must be handled with considerable care.

A variety of trade and descriptive names are applied to this process; these appear in this book under their appropriate headings, and include: "*Aldip*', "*Al-fin*', "*Aludip*', "*Alphate*', "*Calorizing*', "*Elphal*', "*Fink*', "*Metcolizing*', "*Mollerizing*' and "*Servarizing*'.

Alternatives to aluminizing will depend on the specific conditions, but will often use stainless steel instead of mild or low-carbon steel. Under some conditions *nickel plating*, particularly *electroless nickel*, will give superior results economically. There are also now available high-temperature paints, some of which use fusible glass frits, which might be suitable and modern vitreous enamels may also be competitive.

Aluminothermic Process

An alternative name for the *thermit welding* process. Described fully in Part Two **Welding**.

'Alzak' Process

A proprietary electrochemical brightening process applied to aluminium to achieve a surface of high reflectivity, the process being frequently followed by *anodizing* for decorative purposes, such as jewellery. Colour *anodizing* can also be applied.

The bath contains 2–5 per cent fluoboric acid at a temperature of 30°C and uses 15 V with a direct current density of 1–2 A/ft^2 (1.1–2.2A/dm^2). The non-adherent film of oxide present on removal from the bath can be removed by a dip in caustic soda followed by rinsing.

This process is a form of *electrolytic polishing*

Amalgamating

The name given to the process by which alloys are formed with mercury and many metals such as gold, silver, iron, copper and aluminium.

When mercury is in contact with any of these metals in the finely powdered condition, an alloy is formed. When the mercury is predominant, the alloy will be liquid, and when the mercury content is reduced, the alloy becomes solid. It is possible, by using special filter materials such as chamois, to remove excess mercury, leaving a solid or plastic material.

Amalgams were at one time commonly used for a variety of purposes. With knowledge of the highly toxic nature of mercury and most of its compounds, there is much less use of the technique today. Amalgamating has now largely been replaced either by *electroplating* of silver or gold, or by employing metal alloys or plastics.

Ammonia Carburizing
A specialist form of *surface hardening*.

By this process the carburizing atmosphere is enriched with ammonia gas with the result that the atmosphere contains the carburizing potential plus the nitriding atmosphere.

The process is now more commonly called *Carbo-nitride*. Described fully under this heading in Part Two **Heat Treatment.**

Angus Smith Process
A process for the prevention of corrosion, applied to sanitary ironwork.

Application range CAST IRON

The metal is heated to 600°C after casting, then plunged into a hot mixture of coal-tar and paranaphthalene. This results in the formation of an adherent oxide which is immediately sealed and impregnated with bituminous compounds. The corrosion resistance relies on the retention of this barrier to prevent the formation of corrosion products.

The processed articles have little eye-appeal; corrosion resistance is comparable to high-quality painting. This process is seldom applied to modern equipment which will either be stainless steel or vitreous enamel, if it is not ceramic.

Annealing
The name used to indicate a heat treatment process resulting in softening.

There are several forms of annealing, but in general it can be stated that to 'anneal' is to 'soften'. The different forms of annealing are variations on the *softening* process but cover a range of processing, achieving a variety of purposes, all of which include a reduction in hardness.

There are four processes covered by the general term annealing: (1) full anneal; (2) subcritical anneal; (3) cycle anneal; and (4) stress-relieving anneal. To some extent the meaning of the term used will vary between different districts, different industries and even within the same industry. Specialized names are often applied to the annealing of different forms of metals, and variations of each of the above four basic operations also exist. These names appear under their own heading in this book. The various methods of heat treating metals, including the four basic forms of *annealing*, are described fully in Part Two **Heat Treatment.**

Anodic Etching
A specific form of *electrolytic etching* where the component being etched is anodic in the electrolytic circuit.

Application range ALL METALS

Metal will be removed from the surface, and oxygen formed at the surface. This may result in reducing the size of the component and *electrolytic polishing* is generally a form of anodic etching. Further information will be found under *Etching*.

Anodic Oxidation
See *Anodizing*.

Anodic Protection
This term is not explicit but is generally used to describe a method of corrosion protection.

Application range ALUMINIUM, MAGNESIUM, POSSIBLY STEEL, TITANIUM

Such methods rely on the production of an oxide film which is produced by making the component the anode in an electrical circuit. This results in the production of atomic oxygen and the active material forms an oxide which is adherent on the surface of the metal. There appears some possibility that anodic protection could also be applied to titanium. (Further details appear under *Anodizing*.)

The term is occasionally applied when sacrificial inert anodes are used to protect steel structures or pipework. Further information is given under *Corrosion Protection*.

Anodizing
A process which results in the formation of an adherent oxide film produced by an electrolytic process.

Application range ALUMINIUM, MAGNESIUM AND TITANIUM

The purpose is to improve the corrosion resistance, to act as a key for paint, and in some circumstances, to act as a form of *crack detection*. It can also be used to increase the surface hardness of aluminium and its alloys.

The process makes use of the fact that, when an electric current is passed through an electrolyte, hydrogen is produced at the cathode and oxygen at the anode. The oxygen is in the atomic state and is, thus, extremely active and will immediately attempt to combine with any material with which it is brought into contact. The component to be treated, therefore, is made the anode in the electrical circuit and is immersed in a suitable electrolyte. This results in the production of atomic oxygen, which reacts with the clean, active surface metal anode to form an adherent metal oxide.

As the process continues the current applied to the cell will be seen to drop. This is because the oxide which forms is a non-conductor and, thus, the objects being treated gradually become insulated. When it is seen that no further current is flowing, the process has been completed.

A number of different processes exist, using different solutions and application of the current, but only two basic procedures are in common use for aluminium, together with a number of specialized processes based on these two methods.

First and by far the most common method of anodizing aluminium is the *sulphuric acid process*. It uses sulphuric acid approximately 10 per cent strength with a voltage of 10–20 V. This results in a surface where the anodic film depth varies depending on the alloy used and the temperature of the sulphuric acid. With a temperature in excess of 30°C, little or no anodizing will take place, since the sulphuric acid dissolves the oxide at approximately the same rate or faster than it is produced. As the temperature is reduced to about 20°C, an oxide film in the region of 0.0005 in (0.01 mm) is produced. It is important to realize that the temperature involved is at the liquid–solid interface, and it is at this interface that heating is produced as part of the process. It is, thus, essential to ensure adequate agitation of the solution to dissipate the locally formed heat as efficiently as possible.

As the temperature is lowered to an optimum of 3°C, measured at the liquid–solid interface, a very hard anodic film is produced, a process made use of in the *hard anodizing, black anodizing* or *refrigerated anodizing* process. With this, a film of up to 0.0075 in (0.15 mm) can be produced, which in turn produces a dark oxide and, with certain aluminium alloys, will appear almost black, hence one of its names.

A surface hardness in excess of Rockwell C50 can be achieved with this process. It is once again important to realize, however, that the hardness on the surface will depend on the aluminium alloy used.

The term *black anodize* more generally will be applied to the process where the normal sulphuric acid anodized surface is dyed black. This will not be hard. See *Black Anodizing*.

The second most important form of anodizing of aluminium uses chromic acid. This is at a strength of approximately 35 g/l. The temperature in this case is held at approximately 20°C, and it is not normal to use refrigerant to produce *hard anodizing* with chromic acid. With this process, it is usual for the voltage applied to start at about 8 or 10 V, and as the current falls to almost zero, for the voltage to be increased in small steps to a maximum of approximately 60 V. This means that the chromic acid anodizing produces an oxide film capable of withstanding a 60 V potential as compared with an oxide film capable of withstanding 10 V in the *sulphuric acid process*.

One reason for using the chromic acid process is that chromic acid, being yellow in colour, acts as a dye, and during the anodic process it will penetrate any surface imperfections such as porosity, cracks or other defects which might exist. On removal from the chromic acid and washing in cold water, most of the chromic acid is removed. When, however, the components are left standing, any chromic acid which has been absorbed into a defect comes to the surface causing staining, rendering the defect easily indentifiable. Advances in the technique of *dye-penetrant crack detection* have reduced the importance of this aspect, as it is now possible to detect minute surface defects using this system.

Oxalic acid is sometimes used for anodizing aluminium but to a much lesser extent than those described above.

Anodizing of aluminium also acts as a mordant for dyeing and *colour anodizing*, described under that related process, consists of removing the anodizing solution, usually sulphuric acid, by washing in cold water and immersing the components in the appropriate dye. After removal from the dye bath, the component should be immersed in cold water and then sealed in boiling water. It is then impossible to remove the colour from the aluminium surface without mechanically removing the metal.

Anodized aluminium has the best corrosion resistance of all aluminium surface treatments. It is superior to the *chromate* process ('Alochrom') and also acts as an excellent key for painting, but there are relatively few instances when painting will improve the *corrosion protection* of the anodic film. Anodizing will improve the wear resistance of the aluminium surface and even normal room temperature anodizing will reduce the tendency of galling to occur.

The use of *hard anodizing* results in surfaces as hard as Rockwell C50 and, thus, aluminium can be used in place of steel or other metals of similar surface hardness; it can, under some circumstances, be used in lieu of *hard chromium plating* or similar hard surface treatments.

The corrosion resistance of aluminium is better in the low pH range, that is where acid conditions are known to exist. This is the range where steel has accelerated corrosion. As the pH rises, that is becomes alkaline, aluminium tends to corrode more rapidly while steel will be protected. Anodizing helps to protect aluminium from alkali attack, but such resistance depends on the quality of the process and the aggressive nature of the alkaline corrosion agent.

Anodized aluminium is an electrical insulator. This property is sometimes used on electrical devices where local insulation is achieved by stopping off to give a conductive surface adjacent to an insulating surface.

By the use of correctly anodized aluminium, corrosion resistance at least comparable with plated or painted steel can be achieved. The surface wear is also improved to a marked extent and this can be enhanced further by the use of *hard anodizing*. The various anodizing processes appear under their own headings in this book, and are listed here: *'Alumilite', 'Anolok', 'Bengough Stuart', Black anodizing, 'Brytal', 'Buzzard', Chromic Acid Anodize, Deep Anodize, 'Eloxal', Enamel Plating, 'Hardas', 'Modified Bauer–Vogel', Oxalic Acid, Refrigerated Anodizing, 'Sand Ford', 'Shepherd'* and *Sulphuric Acid Anodize.*

'Anolok' Process
A proprietary *sulphuric anodizing* process; further information can be supplied by Alcan Ltd.

Antifouling
A process used in conditions where underwater marine growth causes drag on ships' hulls and incrustation of woodwork.

Application range STEEL

Earlier antifouling methods utilized copper, and in the days of wooden ships, hulls were copper-sheathed as it was discovered that marine growths did not form on copper or copper alloys.

Modern techniques use paint or compounds containing copper salts which are insoluble, or specially formulated paints which do not themselves dry as long as they are below the surface of the water. Compounds of arsenic and mercury are also used.

These paints inhibit growths, by allowing the toxic material to leach out slowly, thus preventing the formation of organic growth. Readers are advised to contact one of the companies specializing in the production of antifouling compounds for further information, as this is an area where rapid advances are being made.

90/10 copper/nickel alloy as a thin clad sheet is now being used. Care is necessary to prevent galvanic corrosion of steel in the area of the copper nickel.

Aquablast
An alternative name for the process of *vapour blasting*. Described fully in Part Two **Blasting**.

This uses a jet of water to impart velocity to the grit used.

Arc Welding
An alternative name for *electric arc welding*. The energy to produce fusion is supplied by the electric discharge arc, generally with air being ionized. It is, however, possible to use other gases such as *inert-gas shielded electric arc, argon arc,* or *carbon dioxide welding*. These are, however, variations on the arc welding process and are described in more detail under their appropriate headings. Described fully in Part Two **Welding**.

Argon Arc Welding
See *Inert-gas Shielded Metal Arc Welding*. This is a form of *arc welding* where the arc is struck in argon, which also shields the weld area and prevents oxide forming. Described fully in Part Two **Welding**.

'Argonaut' Welding
An alternative term for automatic *argon arc welding*. Described fully under *Inert-gas Shielded Metal Arc Welding* in Part Two **Welding**.

'Argonox'
The proprietary name of a mixture of gases, oxygen and argon used for *shielded arc welding*. Described fully in Part Two **Welding**.

'Argonshield'
A form of *shielded arc welding* where a percentage of argon is added to air. Described fully in Part Two **Welding**.

Artificial Ageing
A technique of *through hardening* which can be applied to specific alloys having certain well-defined metallurgical characteristics. This process requires reasonably low temperatures in the range of 100–700°C for a specified time, generally longer than four hours and must always be preceded by some form of *solution treatment*. The term in the main is used to differentiate alloys which must be heated to cause age hardening, from those which age at room temperature. Described fully under *Precipitation Hardening* in Part Two **Heat Treatment**.

Atomic Arc Welding
A heat source used for *fusion welding*, where electrical discharge energy is used to ionize various gases. It is now seldom used. Described fully in Part Two **Welding**.

'Atrament' Process
This is a corrosion-resistance process.

Application range STEEL, ZINC

The steel or zinc surface is converted to a complex manganese phosphate by immersion in a boiling or hot solution. This is a form of *phosphate treatment*.

'Auruna' Process
A proprietary name for a specific type of *gold-plating*, using a cyanide-free solution. It is used for decorative purposes, and is a form of white gold. Further information can be obtained from Sel-Rex Ltd.

Austempering
A form of delayed *quench hardening*.
It is very similar to *martempering* and *marquenching*. Discussed fully in Part Two **Heat Treatment**.

Austenizing
A term used when steel components are heated to above their upper critical limit. Discussed fully in Part Two **Heat Treatment**.

Autofrettage
A *work hardening* method for strengthening steel tubes, which is generally only carried out on gun barrels and similar objects.

Application range STEEL TUBE

A mandrel or ball is forced through the tube or gun barrel, such that the surface of the bore is compressed. In addition to giving considerable increase in tensile strength to the tube, the process can be used to impart a high-quality surface finish.

Autofrettage is different from *cold-drawn* tube in that only the internal surface is worked. Generally the wall thickness of the tube involved precludes a normal *cold-drawing* operation. High pressure to take the surface above the yield strengh can also be used.

Electroless nickel plating may be a suitable alternative. Under certain conditions modern *high-frequency hardening* might also be considered. Shot peening to produce a compressed surface layer can also be used.

Autogenous Welding
A name given to *fusion welding*, where the components are designed in such a way that they have an integral filler.

This integral filler can also be used for location purposes. During the *welding* process which uses an inert electrode such as tungsten, or in some cases gas heating, the filler melts and acts in the same way as the filler rod with the conventional technique. It is generally confined to high-quality, high-production fusion welds, where weld integrity is essential.

'Autronex' Process
A proprietary name of a gold *electroplating* process for particular use in the electronics industry. Information can be obtained from Sel-Rex Ltd.

Back-etching
A specialized etch which is applied only to *electrodeposition* of chromium.

Application range STEEL

Chromium plating is carried out using chromic acid as the electrotype. This acid has excellent *cleaning* and *etching* characteristics, and there is the practice under certain conditions of placing the clean components in the plating solution and reversing the polarity so making the component to be plated the anode when first placed in the solution. Thus, by *anodic etching*, the surface of the steel will be *activated* or *etched* in the normal manner. Immediately after completion of the specified time, the polarity is reversed and plating commences. This means that the 'ideal' situation can exist in that delay between the etch and plating is non-existent.

The problems with back-etching are that the solution used for plating becomes contaminated with iron from the surface of the component used and any other contaminant which might be on the surface, and also the ratio of anode to cathode will generally be undesirable for good control. This results in the build up of oxide of iron and chromium in the plating solution which affects the life of the solution.

Bailey's Creep Test

A specific type of *creep test* which attempts to evaluate the life period of components over 100 000 h.

Application range ALL METALS

With this, creep determinations are carried out with a series of loads at various temperatures. By plotting each stress, information is obtained on the possible creep strength of the material over this period (100 000 h). This test is, therefore, an attempt to predict the creep strength of the material. Described fully under *Creep Test* in Part Two **Mechanical Testing**.

Balling

A term used for a specific type of *annealing*. In this process the steel is treated in such a manner that the carbides, in whatever form they exist prior to treatment, are converted into spheroids, generally on a pearlitic or pearlitic–ferritic matrix.

Further details of this process are given under the heading *Spheroidal Anneal*, and the process of *Cyclic Annealing* will generally produce this type of structure. Described fully in Part Two **Heat Treatment**.

'Banox' Process

The name given to a *phosphate process* used to coat wire prior to final drawing.

Application range STEEL

The wire is treated immediately after semi-finish drawing, as described under *Phosphating*, but will not normally be chromate sealed. The resultant phosphate film protects the freshly-drawn wire from surface corrosion for a short period, but the prime purpose is to act as a *dry-film lubricant* for the final drawing. This prevents the die tearing the surface and, thus, produces a finished drawn wire free of surface blemishes.

It is superior to most liquid lubricants as they can be difficult to apply to ensure complete coverage. There are other methods of treating the surface with dry lubricants but they also tend to be difficult to apply. The use of high-pressure metallic soaps applied to the wire immediately prior to the die gives comparable results. There is also a treatment where the wire is dipped in a borax solution which leaves a thin coating on the wire surface. This is cheaper but not as effective as *phosphating*.

Bar-drawing

A manufacturing technique, which does not come within the remit of this book. It is a technique for the production of narrow-bore stainless tubing, where the tube is drawn across a bar, and can be used for seamless tubes with a very narrow bore and tightly held tolerances.

'Barffing'

A method of controlled oxidation to produce an adherent oxide.

Application range STEEL

The parts are *cleaned* to remove oil, grease and loose solid matter, then heated to 500°C–600°C in a closed container into which steam is injected for a short period. This ensures that the surface layer is oxidized to Fe_3O_4, magnetic oxide. This is the most stable oxide and, if the process is correctly controlled, is reasonably adherent. Immediately after cooling and removal from the reaction vessel, the parts can be oiled or waxed, and this will be absorbed into, and retained by, the oxide film, thus considerably enhancing the corrosion resistance.

The process is similar to and comparable with *steam blueing* and *black oxide*, and can also be compared technically to *phosphating*. None of these processes results in high-quality corrosion-resistant films as the layer is porous, thus the substrate of unoxidized steel is capable of corroding unless a completely airtight surface layer of oil, grease or paint is present.

'Barffing' is generally used where the component has a light oil film during its service life, and will not usually give a corrosive-free service life when used in the dry state.

Barr–Bardgett Creep Test

A form of *creep test* which attempts to predict the 'creep strength' of a material using a shorter-term test.

Application range ALL METALS

This is a relatively complicated test and the reader is advised to obtain further information regarding the test itself and the interpretation of the test results in specific cases. (See also *Creep Testing* in Part Two **Mechanical Testing**.)

Barrel Plating

This is the *electroplating* process carried out in a rotating barrel. There is also a barrel peening process which results in the surface of the component having a soft metal such as zinc or cadmium battered into the surface. This uses steel balls and powdered metal zinc or cadmium.

Barrelling

A name given to the *barrel plating process* and also to *barrel tumbling*. *Barrel plating* is described under **Electroplating**, and only *barrel tumbling* is described here.

Application range ALL METALS

The process essentially is the tumbling action of components in any material, including non-metallic, in contact with chips of inert materials, fine abrasive compounds such as sand, and liquids, usually water with detergent or wetting agents added, but also paraffin, etc. The purpose is to *deburr* or to produce a smooth finish.

The process relies on the inert chip, lubricated by the liquid, vibrating or moving relative to the component with the abrasive material carried on the inert chip. It is, thus, the abrasive material which removes metal from the component, but it is the shape of the inert chip which largely decides the part of the component which will be abraded. It will be seen that the first areas to be abraded are the edges, with flat areas or convex areas of large radius next. Concave and re-entrant angles are the least affected, unless the chips are carefully chosen.

The process has been used in one form or another for many years, and initially consisted of circular barrels, of the wooden beer-barrel type, rotating at an angle of approximately 45°. Components were tumbled in contact only with each other, using either no lubricant, or water or paraffin. This process was lengthy – up to 5 days – and while giving successful *deburring* of simple shapes could not produce high-quality finishes and was largely uncontrollable.

The biggest advance in *barrel deburring* was the use of horizontal hexagonal barrels with inert chips to separate the components. Again this was a lengthy process, but higher-quality finishes could be achieved. In most cases use was made of natural material, and some abrasive action was obtained by the breakdown of this material. It was not, however, until separate abrasive compounds were added under controlled conditions that the modern rapid-*deburring*, high-quality finishes were obtained.

It is now known that, in the horizontal, hexagonal barrel, *deburring* is confined to the period when the components are falling down the slope against the direction of the barrel movement. As long as the component is buried in the chips, there is little or no relative movement between part and chips, thus no *deburring* takes place. It will thus be seen that a barrel completely filled will have little effect regarding de-burring.

Using this knowledge, vibrating barrels were produced. These bear little resemblance to the original 'beer barrel', but are in the form of bathtubs. These are filled with approximately 25 per cent component, 50 per cent chips and the remainder abrasive matter and water. The 'bathtub' is spring-mounted with a motor-driven eccentric which imparts a vibrating motion. This can be controlled and is designed to give a corkscrew action to the components which are in constant vibrating contact with chips and abrasive material. The *deburring* takes 1–3 h for most components, compared with 4–8 h with horizontal barrels.

The fact that with modern barrelling techniques the components never contact each other, means that the possibility of damage is negligible. By the correct choice of chip shape and size, the correct abrasive compound and the control of wetting agents, it is possible to *deburr* any type of component, and to produce a variety of surface finishes, to fine microinch limits. Modern chips are preformed ceramics in a large variety of shapes and sizes. Using jigs to hold and rotate components, selective *deburring* can be successfully achieved.

Production barrel *deburring* has largely supplanted hand-trimming and, in some cases hand-*polishing*. It is unfortunate that the possibilities of controlled barrel *deburring* are not yet fully appreciated, and there are many examples of inefficient barrels, producing substandard work. There is also the problem that handling components at loading and unloading is often uncontrolled and causes damage.

Chemical polishing and *electropolishing* can be used for *deburring*, but it is more difficult to control and generally more expensive. *Dry blasting* and *vapour blasting* can also be used for this purpose, but again are more specialist and expensive.

The "*Harperizing process*' is an advanced form of barrelling, where the barrel is rotated through an arc, while rotating on its own axis. This gives high centrifugal forces, resulting in very rapid *deburring*, and also the formation of a controlled radius in areas of difficult access.

Bauer-Vogel Process

A form of *chromate treatment* for aluminium to increase its corrosion resistance and paint adhesion. The procedure is described under *Chromate Coating*. See *Modified Bauer-Vogel Process*.

Bend Test

A form of *mechanical testing* where the test specimen varies considerably, and the type of bending applied has several variations. It is a measure of the ductility of a local area. Described fully in Part Two **Mechanical Testing**.

'Bengough Stuart' Process

A chromic acid *anodizing* technique, using an approximately 3 per cent (30g/1) chromic acid solution.

Application range CERTAIN ALUMINIUM ALLOYS

Unless the correct alloy is used, incorrect *anodizing* will occur and the appearance of the finished component will be patchy. The process starts with a low voltage which is progressively increased to about 50–60 V over a period of about 1 h. For further details, see *Anodizing*.

'Berketeck'

This is a proprietary name given to a series of powder coatings. These fuse at particular temperatures giving protection to the underlying surface.

Application range ALL METALS

The technique is to use either a small range of powders which fuse at varying temperatures if the temperatures involved are relatively low, or when the temperatures to be used increase in value, then a larger range of powders are used with those at the lower range burning off and disappearing as the higher temperatures are reached. This allows the surface to be protected over the necessary range of temperatures.

These powders are used for a variety of purposes but commonly to protect components which are being subjected to a heat treatment process to prevent surface oxidation. The powders compete therefore, with controlled atmospheres at heat treatment or electrodeposition of metals.

Further information can be obtained from Steetley Berk Ltd.

Bethanizing

An acid *zinc plating* process.

Application range STEEL WIRE

The difference between bethanizing and electrolytic *sulphuric acid zinc plating* is that, instead of zinc anodes, inert mild-steel anodes are used, and the electrolyte itself is manufactured from zinc oxide dross which is dissolved in sulphuric acid, filtered and pH adjusted.

The deposit will not normally be of a standard as high as that obtained using zinc cyanide solution, or a controlled acid zinc sulphate solution, both of which are discussed under *Zinc Plating*. The hot-dip *galvanizing* process, with continuous wiping of the wire as it leaves the molten zinc, will probably be as economical, and is certainly technically superior.

The use of paints, under certain conditions, should also be considered as an alternative. These would probably be of the zinc-rich etch primer type, but many other forms of paint film might be satisfactory depending on the service use.

'Bi-Ni'

A name given to multi-layer nickel chrome plating, where two types of nickel, dull or semi-bright are deposited prior to chromium plating.

Application range STEEL

The process gives superior corrosion resistance to normal nickel chrome plating.

This is a proprietary name for an Oxy Metal Finishers process; further details can be obtained from the company.

Where appearance and corrosion resistance are involved there is no obvious alternative, other than the use of corrosion resistant base metal or the application of a thicker layer of bright nickel below the chrome.

Bit Soldering

This is a soldering technique; it appears under *Soldering*.

A solid piece or 'bit' is used to conduct heat to the joint being soldered, thus this is a technique for heating at soldering.

Black Annealing

A process usually applied to strip or thin sheet metal, but also applied to any form in the oxide-free or bright condition.

Application range STEEL

The process is that the steel surface is given an adherent oxide, formed by means of controlled oxidation by heating the material inside a closed box so that free oxidizing conditions are not present. With this, a thin adherent layer of iron oxide is produced.

The process is not normally used for *corrosion protection* purposes, but under some circumstances, the end-product may have sufficient adherent oxide for reasonable corrosion resistance. This will not, however, be of a high standard.

The colour produced is not black but varies from a very dark blue to light grey, depending on the surface condition and cleanliness of the raw material and the amount of oxide present during annealing. The time at temperature can also affect the colour, which is a function of the thickness of the oxide film.

'*Annealing*' and '*Pack Annealing*', described in Section Two **Heat Treatment** are alternative techniques. Further information may be found under *Heat Tinting*. The application of a tinted lacquer could be an alternative.

Black Anodizing

An oxidation, corrosion resistance process with a dark finish.

Application range ALUMINIUM AND ITS ALLOYS

The solution used is sulphuric acid at a temperature of not more than 3°C. It is essential that this low temperature is maintained at the surface of the material during the *anodizing* process, and unless this temperature is maintained, then correct black anodizing will not occur.

The process results in the formation of aluminium oxide in an adherent form. With pure aluminium and certain of the low alloys, the colour obtained is almost black. The colour becomes a lighter grey with more alloying content, and certain materials will have a colour very little different from that with normal *sulphuric anodizing*.

From the process a very hard film of measurable depth results, up to 0.025 mm (0.001 in) and a hardness up to Rockwell C50. For further details, see *Anodizing*.

This is also known as *hard anodizing, deep anodizing* or *refrigerated anodizing*.

The term is often used where components are *anodized* and dyed black, but technically my understanding is that this is *colour anodizing* and not black anodizing, where the colour is produced as part of the *anodizing* process.

Black Nickel

A form of *nickel plating* where, by the use of additives and careful control of current density and temperature, a dark form of nickel can be produced.

Application range STEEL, COPPER AND ITS ALLOYS AND ALUMINIUM

The process is difficult to control in order to obtain consistent results, but is sometimes used where a black, non-reflective surface is required on steel together with good corrosion resistance. It can also be used on copper and its alloys and, with the correct pre-treatment, on aluminium, but in general its use is confined to steel components. It will give superior corrosion resistance to *painting* treatments, and is superior to *phosphating* or '*Black Oxide*'. It is also used for its decorative appearance.

Nickel plating is described fully in Part Two **Electroplating**. The reader is advised to contact a reputable plating supply house for further information.

'Black Oxide'

Proprietary processes resulting in a surface which not only has some corrosion resistance in its own right, but also the added ability to absorb and retain oil, thus considerably enhancing the corrosion resistance.

Application range STEEL

The process uses a solution containing chromates and dichromates generally as the sodium or potassium salts. The solution is extremely alkaline and is used at a temperature close to or above 100°C. The 'Black Oxide' proprietary processes all use strong caustic solution with variation in the makeup of the added chromates and dichromates. The process forms an adherent oxide film on all surfaces. Provided this is kept oiled and undamaged, no further oxidation (rusting) will occur, under normal moist atmospheric conditions.

The 'Black Oxide' solution is used close to or above the boiling point of water, thus any moisture trapped in the components before immersion will result in the immediate formation of steam, which can cause explosive emptying of the 'Black Oxide' container. Some of the proprietary solutions are two-stage, the components being immersed in two separate solutions. The first stage will be below boiling water temperature, acting as a surface activator to ensure that when the parts are transferred to the bath, no moisture is carried over. On removal from the 'Black Oxide' solution, the components should be washed in cold and hot water, dried either in hot oil or an oven, and immediately immersed in oil.

The process has been used for several hundred years in the gunsmith industry, where gun barrels have been blued or blackened using this technique. Some of the Spanish processes were extremely complex in nature, but the result was the formation of an adherent dark iron oxide which was capable of absorbing oil. The modern processes result in an identical surface finish at a much reduced cost.

While it is possible for the solutions to be made up from basic chemicals, it is generally found that the use of proprietary solutions result in an end-product of superior quality, with easier technical control.

Provided the parts are oiled and the surface kept dry, or dried and reoiled after use, many years of corrosion resistance can be expected. For toolroom-type operations, where components are used within the factory and not liable to become damp the use of 'Black Oxide' is becoming a common method of ensuring corrosion-free tools.

On completion of machining, the components are oxidized and oiled, and during service this oil film is maintained by wiping with oily rags or use of oil-coolant at the machining operations.

The 'Black Oxide' finish is not comparable to any form of plating and will not give results as good as those obtained by *zinc* or *nickel plating*. 'Black' is to some extent misleading as the finish is very seldom true black, but varies from very dark blue to brownish or even light grey in some instances. It is not as good as a correctly applied *paint film*, but has the advantage that it does not interfere with the size of the component. 'Black Oxide' is technically superior to *phosphating*, *temper blueing* and *chromating* and has a better visual appearance.

Alternative and proprietary names for 'Black Oxide' include: '*Brunofix*', '*Chemag*', '*Jetal*' and '*Penetrol Black*'.

Blackening

One of the terms commonly found with surface treatments of metal.

Application range ALL METALS

It can mean different processes under different circumstances. In essence blackening is any process which results in the metal becoming black or dark in colour. It covers the '*Black Oxide*', *black annealing* and *black anodizing* processes, and also other forms of blackening produced by dyeing, painting or various chemical treatments. It is not confined to a single metal, but covers steel, aluminium, magnesium, copper alloys, etc.

Blasting

The term used to cover a number of processes where shot or grit is impinged onto a metal surface. Described fully in Part Two **Blasting**.

Blue Annealing

A method of tempering which results in an adherent oxide being formed, which is blue in colour, and imparts some corrosion resistance.

Application range STEEL

A process used where spring quality is involved. The term is applied to the *tempering* operation carried out after *hardening* and results in a blue-black finish. This term, however, is technically incorrect as the operation is a *temper* to increase ductility after *hardening*. The term 'annealing' should properly be reserved for operations where an increase in softness is the prime consideration.

Blue annealing is similar in all respects, therefore, to the process *temper blueing*; and is a specific form of spring tempering. The processes of *Tempering* and *Annealing* are fully described in Part Two **Heat Treatment**; and the reader is also referred to *Heat Tinting*.

BNF Test

An inspection process to measure the thickness of *electroplated* deposits. Described fully under *British Non-ferrous Jet Test*. (See also *Thickness Testing* in Part Two **Non-destructive Testing**).

'Bonderizing'

A proprietary name given to paint pretreatment which provides a high standard of *corrosion protection*.

Application range STEEL AND ALUMINIUM

Components may range from small pieces to car bodies. The treatment also provides a mechanical key for all types of organic finishes.

The term is most commonly applied to the *phosphate* coating of steel, but may be used for *chromate* treatment of steel and aluminium, or similar metal pretreatment processes. Caution is, therefore, necessary when interpreting this term.

Further information may be obtained from Pyrene Chemical Services Ltd.

'Bond Film'

The proprietary name given to a combined pretreatment and organic finish for corrosion protection.

Application range STEEL

The first stage is a phosphate system which is immediately followed by immersion in quick drying lacquer. This lacquer will be absorbed into the phosphate film, supplying the necessary protection. Subsequent treatment can include normal air drying or stoving enamel.

The process can be used where there will be known delays between components being phosphated and the application of the final corrosion resisting paint film. Further information will be found in this book under *Phosphating* and Part Two **Painting**.

Details of the process can be obtained from Pyrene Chemical Services Ltd.

Borax Treatment
A dry lubricant technique used for cold forming operations.

Application range STEEL

The material to be treated is cleaned, and very often *acid pickled*, to remove scale or corrosion. Immediately after the subsequent swill, the material is immersed in a hot boiling solution of borax, allowed to come to the solution temperature, then removed and allowed to air dry. Where necessary, hot air or oven drying can be used.

The adherent thin film achieved will be sufficient to supply some lubrication for subsequent light drawing operations on wire or straightening or light forming. The coating is alkaline in nature and will supply some *corrosion protection*, but this will not be of a high standard.

The method provides a cheaper, but less effective, alternative to *phosphating*, or other dry film lubrications.

Boriding
A form of surface hardening.

Application range LOW CARBON AND ALLOY STEEL

This uses the fact that given the correct conditions and temperatures, iron will combine with boron to form iron boride, which is extremely hard in its own right and has the ability to diffuse into the surface of the component. The process is carried out by packing the clean and preferably blasted components in the activated material. The components should be placed in such a manner that the boron rich material supports them and has a least 2 in (50 mm) of material between any part of the components. The treatment is carried out at up to 950°C but some effect can be achieved as low as 850°C. The higher the temperature, the more rapid the treatment but the greater the grain growth. The time of treatment will vary depending on the case depth required, being approximately two hours for a case of 0.005 in (0.1 mm) and eight hours for 0.010 in (0.25 mm). It will thus be seen that the process is relatively expensive.

Like *nitriding*, boriding produces an intrinsically hard surface. If the components can be slowly cooled, removed from the compound, and cleaned they will have an extremely hard surface. If desired, components can be removed and quenched directly from the boriding compound, but this will result in a relatively coarse grain structure depending on the time and temperature used. It is possible to reheat the component to obtain grain refining, and then to quench. This should be carried out under neutral or slightly reducing conditions, such as are obtained in a correctly controlled salt bath. Care must also be taken to prevent the very hard case cracking because of too rapid heating or because the design of the component includes stress raisers.

Where desired local boriding can be achieved by copper plating areas required to be soft. Boriding is a relatively modern technique which is expensive but will produce a surface harder than any other commonly used technique, such as *carburizing* or *nitriding*. It is difficult or impossible to machine the surface achieved as the hardness is claimed to be greater than 1400 DPN. Abrasive polishing can be achieved but grinding will invariably result in cracking.

Because boriding requires no subsequent quenching operation it can be produced without severe distortion. Some growth will be found but provided the components are designed to have a minimum of stress raisers, they have no rapid changes in section and are relatively homogeneous throughout, reasonably stable results will be achieved. This necessitates the correct condition prior to boriding and it is recommended that components should be normalised prior to the boriding operation and final dimensions should then be produced using only light stressing such as is achieved with grinding.

Boriding is, therefore, a form of surface hardening which should be considered only as an alternative to *nitriding* or forms of additive surface hardening such as *chromium plating* or *metal spraying* using tungsten carbide.

Bower Barff Process

An additive film produced by controlled heating to give improved corrosion resistance.

Application range STEEL

The parts are heated in a sealed container, and when they have reached approximately 900°C and the available oxygen is exhausted, steam at the same temperature is injected. This results in the formation of two oxides of iron, Fe_2O_3 and Fe_3O_4. This steam cycle is followed by carbon monoxide which converts all the Fe_2O_3 into Fe_3O_4. These cycles are repeated until the desired thickness of adherent oxide has been produced. The film of magnetic iron oxide, Fe_3O_4 gives excellent corrosion resistance, provided the film is not broken, and is adherent.

The corrosion resistance is equal to, or better than, a good-quality *paint* film, but lacks the galvanic protection qualities of zinc and cadmium plating and cannot have any colour variation. The colour will generally be a matt dark grey, but this will vary with the conditions. The advantage is that complicated steel components can have a homogeneous *corrosion protection* film applied, difficult with conventional *electroplating* or *painting* of complicated shapes.

The process requires specialized equipment and, thus, will only be economical when the above requirements are mandatory and sufficient components are required. It cannot be used with standard *heat treatment* equipment, and should not be attempted without technical control.

Box Annealing

A method identical to *black annealing*, and covering all *annealing* processes carried out in a closed box.

The process will reduce the amount of oxygen in contact with the components. Components with an adherent, or semi-adherent oxide invariably result.

It is most commonly applied to strip- or sheet-metal components, but there is no reason why both term and process cannot be applied to any form of annealing carried out as described on any metal or alloy.

For further discussion of *annealing*, see Part Two **Heat Treatment**.

Brass Colouring

A term used to denote that components have been treated to give a brass colour.

Application range BRASS

The term is sometimes applied to components which have been brass plated but this should be properly termed *brassing* and is described under that heading.

Brass colouring is the term used when a brass object is treated to alter the natural colour for any reason. Brass objects are sometimes treated to simulate bronze, see under *Bronzing*. A darker colour, black or steel grey, may be required to match a brass component to its steel neighbours on an assembly.

It should be noted that brass colouring will seldom improve the corrosion resistance of brass to any degree, and may under some circumstances actually reduce protection. In addition, the coating only affects the surface and has no abrasion resistance, thus any *cleaning* or *polishing* operations will remove the colour, exposing the underlying brass. The process is now seldom employed for these reasons, together with the expense of copper alloys. Alternatives include *electroplated* or *painted* steel, plastic, and *anodized* aluminium. There are, however, some ornamental components, cast or fabricated in brass, which are treated to alter surface colour.

The process requires that the components are thoroughly *cleaned* using any conventional method, but wiping with a cloth is not sufficient. The subsequent operation will depend on the colour required, and some trials and considerable experience are necessary to produce high-quality work.

When a dark grey or black colour is required, a copper oxide is produced by immersing the parts in copper sulphate with sufficient ammonia added to redissolve the green precipitate which forms when insufficient ammonia is added. The solution is used at approximately 60°C. The part should be removed after approximately half a minute, and after washing immersed in hot caustic solution, then finally washed and dried.

For the production of a steel grey colour, the cleaned component is immersed in a solution of arsenious oxide, ammonium chloride and hydrochloric acid. This solution is used cold, and takes several minutes, with best results obtained when the parts are brushed with a soft brush during immersion; followed by washing and drying.

A blue-black colour can be obtained using sodium thiosulphate and lead acetate at 80°C. Again brushing with a soft brush during immersion, gives best results. The colour obtained can be varied by altering the solution, the temperature and brushing technique.

Brassing
A term given to the production of a brass appearance.

Application range STEEL

The process requires that the components are cleaned, and where necessary fluxed, then immersed in molten brass. As with all molten immersion treatments, the temperature of the molten metal and time of immersion are critical in achieving adhesion and in controlling the thickness of the coating. The adhesion is seldom satisfactory and the process has been confined to low-cost articles. The use of brass plating has largely replaced this hot dip process.

Brass Plating
The *electrodeposition* of the steel with the alloy of copper and zinc, carried out for decorative reasons. Described fully in Part Two **Electroplating**.

Brass Welding
A term sometimes applied to *brazing*, and used for a specific type of *welding* where, for example, the temperature used is lower than that normally applied with steel. In general, however, the term can be assumed to imply *brazing*, but it is recommended that the definition of the term is queried.

Brazing
A form of joining, where the two metals or components do not themselves fuse but the joining material is melted and, thus, acts as a form of metallic glue, holding the two components together, when the braze metal freezes.

Application range ALL METALS

It is commonly considered that brazing is the process where copper alloys of zinc and tin are used to join cast iron or steel. This is the historic form of brazing, but technically the process can be carried out at any temperature and, as stated above, is the process of joining two metals which need not be similar, by the use of a third metal, and it is only this latter material which is melted in the process. The same definition can be applied to the process known as *soldering*, but here it is also vague as traditionally tin and lead alloys were used as the joining medium. With the advent of lower melting-point braze materials, and the addition of silver to copper alloys, the brazing process became known as *silver soldering*, thus linking traditional copper alloy brazing to the tin and lead *soldering* operation.

For good-quality brazing two factors are of vital importance. First, the clearance between the mating components should be such that the molten braze metal will rise by capillary action. Successful brazing requires that the amount of braze metal is not more than approximately 0.001 in (0.025 mm) thick. With this ideal thickness, the strongest possible joint is obtained. With an increase in the thickness of braze metal, the joint strength is reduced until, with excessively thick braze metal, the strength of the joint is equal to the tensile strength of the braze metal itself, and this is generally of a relatively low order. It is essential that the design and planning department are aware of this fact and that the correct dimensions are applied to the components being produced. For repair brazing, the same criteria exist, and it must be appreciated that if good results are to be achieved, tight control must be maintained on any components repaired by this method.

The second factor controlling good-quality brazing is cleanliness. It is essential that all dirt and grease, and as much oxide as possible, are removed from the surfaces prior to them being brought together. At the stage when the components are separate, it should be relatively easy to ensure a high standard of cleanliness, and the better the *cleaning*, the easier it will be for the operator to ensure a good-quality braze. With repair brazing, *cleaning* is more difficult, but every effort must be made to ensure that the surfaces which require to be wetted by the molten braze metal are perfectly clean. There is no doubt whatever that any dirt on the surfaces to be brazed will either prevent wetting completely, or will not allow complete wetting.

The use of fluxes is general when brazing is carried out. The purpose is to remove the last trace of oxide which cannot be removed by conventional *cleaning* processes, or by chemical or mechanical means. The flux required will be a function of the metals being joined and the braze itself. The activity of flux necessary depends on several factors, including the standard of cleanliness achieved and the skill of the operator and the type of metal being joined. The cleaner the components, and the simpler the metal (for example, mild steel or copper alloys rather than cast iron or aluminium), the less active the flux need be, and the less need for the operator's skills. The more active the flux, then the lower the skill required or the dirtier the metal or the more difficult the metal which can be brazed. Unfortunately, the active fluxes are extremely corrosive and, unless considerable care is taken subsequent to brazing, then corrosion will be encountered in service.

The brazing process itself consists of *cleaning* the components, *fluxing* the surface to be brazed, applying the heat and feeding the braze metal. The simplest form, where the operator uses a hand-held torch and feeds the braze with a feeder rod, is that the clean components have flux painted on both faces prior to fitting, and are heated gently to the required temperature and, ideally, the braze metal is fed such that the joint is filled from the bottom upwards by capillary action.

Using this technique, a self-inspecting process can be carried out. Unless braze metal appears at the upper surface all round the joint, a successful braze has not been achieved. The braze filler rod is often dipped in flux during the process, thus keeping the surface covered with flux. Where braze metal does not appear at the upper surface, this may be caused by excessive clearance. It could also be that the components are not clean enough, and thus the surfaces will not wet, or that excessive or insufficient heat has been applied. It is appreciated that in many circumstances the use of capillary feeding will not be possible, and under these circumstances a tighter control of all the parameters noted above will be necessary for high-quality brazing.

A variation on this simple method uses a preformed filler material. This can be applied as a ring or washer, or as a special shape. The components are *cleaned* and *fluxed* (or *pickled* to remove oxide), then dried and assembled with the braze metal in the desired location. This should be such that, on the melting of the filler metal, the area being joined is fed by capillary action. The method of heating will be appropriate to the components and the braze metal. It may be by *torch*, *furnace* or *induction*, or any other suitable method, but the temperature chosen should not be more than approximately 100°C–150°C maximum above the melting point of the braze filler metal. Use can also be made of brazing paste. This will be the chosen braze metal in powder form, converted into a paste or paint. This vehicle can, if desired, incorporate flux, but the technique is more often used in controlled-atmosphere *furnace* or *vacuum brazing*. Neither of these heating methods are suitable for use with active fluxes, since gases are given off which would contaminate the furnace atmosphere. Where the braze metal chosen can be *electroplated*, it is possible to use this technique to deposit selectively the desired thickness of braze metal in the area of the braze. The components can then be assembled, using the plated deposit to locate the pieces and hold them in position prior to heating. This method is confined to materials which can be readily deposited by electroplating, generally pure metals such as copper, gold and silver, but certain alloys such as bronze (copper and tin) can be used. The base metal must be capable of being electroplated.

Heating methods often give the title to the type of brazing. Thus, *torch, furnace, controlled atmosphere, hydrogen, vacuum, salt-bath, induction* and *resistance brazing*.

Filler metals will largely depend on the use of the finished assembly. Where steel and cast iron are used, the filler will generally be a copper alloy, with zinc or tin, but there are a large number of different alloys in this range, including the silver braze metals often called 'silver solder', having a lower melting range than conventional braze metals.

As stated above, the strength of a good braze joint is more a function of the design and fit of the assembly than the strength of the braze metal. Thus, the choice of braze metal will depend on the temperature at which the assembly will be operated and on the corrosive environment as much as, if not more than, on the mechanical properties of the braze metal. The designer has a choice of filler metals with a melting range below 100°C to above 1200°C, including metals such as pure gold which is virtually non-corrodible.

The temperature and technique used to supply the heat must be carefully chosen and controlled. This will depend on the filler metal, but should not be more than 100–150°C above the top end of the melting range. Thus, the use of *oxyacetylene* as a source of heat, with a possible temperature in excess of 3000°C, should not be used to braze copper alloys, which have a melting range between 500°C and 1000°C.

Brazing is a method of joining metals which has been employed for very many years, and a wide range of techniques and filler metals are available for specialized purposes. To some extent the process is being supplanted by *welding* as this technique improves, but this can never cover the full range of capabilities open to brazing. Modern glue is also competing with braze jointing, particularly for room temperature use, but again brazing is the more flexible process.

For further information, see under *Soldering*, *Glueing* and Part Two **Welding**. Companies such as Englehard Industries and Johnson Matthey, who supply filler materials, are available for technical advice.

Bright Annealing

A process which softens without scaling.

Application range GENERALLY STEEL BUT COULD BE ANY METAL

Described fully in Part Two **Heat Treatment**.

Bright Chrome Plating

The decorative deposition of chromium and nickel.

Application range ALL METALS

Described fully under *Chromium Plating* in Part Two **Electroplating**.

Brinell Hardness Test

A standard method of *hardness testing*, using a hardened ball. The principle is that a known load is applied and the size of the impression formed is measured. Described fully in Part Two **Hardness Testing**.

British Non-ferrous Jet Test (BNF Jet Test)
An inspection process used to measure the thickness of *electroplated* deposits. This is locally destructive.

Application range ALL METALS PLATED

The test works on the principle that a standard chemical will take a specific time to dissolve a certain thickness of metal. It is carried out by the impingement of a jet of a specified liquid on the plated surface being tested, and measuring the time which this solution takes to remove the *electroplated* deposit. The end of the reaction is noted by a colour change which occurs on the surface.

The apparatus consists of a laboratory stand with a calibrated glass column into which the appropriate liquid is placed. The height between the object being tested and the apparatus is controlled. Each metal deposited and base metal will have a specified test liquid, and the temperature of the test liquid must be controlled.

It should be realized that this test is selective, in that only the area tested will be measured for thickness. It is, thus, essential that careful consideration is given to the test spot, and that a number of these spots are tested in the case of important components with peculiar shapes. Graphs are used to convert the time taken to dissolve the coating into thickness of metal deposited.

The test, in the correct hands, has a high degree of accuracy which is easily reproduced. It is, therefore, a thickness test which is specified for important articles and used by national organizations. The different methods of measuring coating deposits will be found under *Thickness Testing* in Part Two **Non-destructive Testing**. Further information on this test can be obtained from the British Non-ferrous Metals Association.

Bronze Plating
The *electrodeposition* of the tin–copper alloy, bronze.

Application range GENERALLY STEEL – but could be any metal which can be plated

It is used on a limited scale for decorative purposes, and in engineering for *stopping off* at *nitriding*; it is also used to produce an interference fit on assemblies. Described fully in Part Two **Electroplating**.

Bronze Welding
An alternative name for *brazing*. With *welding*, the materials being joined and the filler metal are all involved in a 'massive mingling', to ensure that adequate *fusion* or *forging* occurs. One definition of welding is that the materials involved are joined by the use of heat to cause fusion (melting) and/or pressure. It is possible to weld with heat and pressure without fusion melting. In *brazing*,

the braze metal melts, but not the metals being joined, and thus, acts as a metallic glue. It is thus different from welding.

In bronze welding the filler metal will generally be a copper–tin alloy, but often with modern techniques it is brass, i.e. a copper–zinc alloy.

On the rare occasions when bronze components are *welded* together, using any *fusion* or *forging* process, this is correctly termed bronze welding, that is the welding of bronze.

Bronzing

A chemical process to give a bronze colour.

Application range STEEL, COPPER ALLOYS SUCH AS BRASS, AND ZINC

The process is a chemical conversion, whereby the surface of the metal is treated in such a manner that the resultant finish has the appearance of bronze. Bronze is a relatively expensive copper–tin alloy which has excellent corrosion resistance and weathers well. It has an attractive appearance and is commonly used for name-plates, furniture, architectural articles, etc. It is, therefore, quite common that cheaper materials are treated to have the appearance of bronze. The process of bronzing can be divided into two systems, one of which is applied to mild steel and the other to copper alloys such as brass.

The method of treating steel is that thorough *cleaning* of the surface is carried out to remove all evidence of oxide and to ensure that the surface has a homogeneous finish. Either *scratch polishing* or *blasting* with a fine grit will give satisfactory results. Immediately after this preparatory process, the components should be immersed in the appropriate bronze solution. The process can also be carried out by swabbing or wiping, and for many articles, superior results are obtained using the swabbing method.

It must be appreciated that, with steel particularly, the finish obtained has a bronze appearance, while having none of the corrosion-resistant characteristics of bronze metal. This, then, is not a suitable process for outside architectural purposes, but is rather for producing a satisfactory bronze-type finish on articles to be used indoors in low humidity conditions.

The solution used for bronzing steel is antimony chloride in hydrochloric acid with a subsequent treatment containing ammonium chloride in dilute acetic acid to give alternative colouring.

The treatment of copper alloys such as brass results in a more satisfactory type of finish, having superior corrosion-resistant characteristics over bronzed steel. Again, however, it must be appreciated that bronzing is a superficial finish, and any abrasive cleaning or corrosive conditions, such as marine atmospheres, will attack the surface and either remove or discolour the bronzing.

The bronzing solutions which can be applied require careful application and will not generally result in a homogeneous finish. The solutions vary and include hydrochloric acid with potassium permanganate. By heating the solution, or by controlled heating of the treated brass, variations of the bronze colour can be obtained. Other solutions are lead acetate and sodium thiosulphate.

Zinc-metal and zinc-alloy die castings can be given a bronze colour by treatment with ammonium chloride, and potassium oxalate in dilute acetic acid.

These chemical bronzing processes are a cheap method of producing an attractive colour, but have few technical advantages, and unless carefully controlled can actually result in local corrosion.

With modern *paints* and *lacquers*, the use of plastics and *colour anodized* aluminium, the same attractive visual appearance can be achieved, very often at low cost, and certainly with considerable technical advantage.

Most of the solutions noted above are toxic and all are corrosive.

Bronze plating will result in a true bronze surface, superior to bronzing, but since this is only a surface film, it will be subject to failure if abrasion occurs.

Browning Process

A treatment which results in a brown colour.

Application range STEEL

Like many terms, the method covers a wide variety of treatments which result in a brown finish. The composition of the mixture can vary considerably, but is generally based on a ferric chloride – nitric acid mixture with or without alcohol and the chlorides of antimony, bismuth, copper or mercury and other materials. A common treatment consists of cleaning the components then treating them in the browning solution at a temperature of approximately 40°C to 60°C.

The components are washed after removal from the solution, and left in a humidity cabinet where corrosion will occur. After a period, measured in hours, the corroded components are dried and then washed in clean boiling water. This complicated process is then repeated with occasional wire brushing to remove any loose adherent rust which forms.

The procedure can be compared with some of the '*Black Oxide*' treatments where the conditions are designed to give full oxidation of steel, resulting in a black finish. The browning process is essentially a method of controlled oxidation with attempts made to produce an adherent lox-oxide film. As with '*Black Oxide*' the corrosion resistance is relatively low, but by ensuring that components are oiled at all times, an attractive visual appearance can be obtained.

With modern materials and treatments, the browning process is seldom or never used. Comparable corrosion resistance can be obtained with high-quality *phosphating*, and with modern *painting* techniques a full range of colours on steel can be achieved with excellent corrosion resistance.

Colour anodizing of aluminium and *electroplating* of various metals are also aesthetic alternatives, which give superior corrosion resistance.

'Brunofix'
A proprietary name for one of the '*Black Oxide*' treatments for steel, which makes use of a hot, strong alkaline solution with an oxidizing agent.

'Brunorizing'
A proprietary name for a specialist *heat treatment* which is applied to rails. These are heated to above the critical temperature required for *hardening* and held at a lower controlled temperature prior to being reheated to above the *hardening* temperature. This is followed by cooling in air, with the ends of the rails being quenched with jets of compressed air, resulting in the structure being '*spheroidized*' and the ends being in a slightly harder condition.

With the advent of continuously welded rails, this treatment is no longer applied; the reader is referred to *Spheroidal Anneal* in Part Two **Heat Treatment**.

'Brutonizing'
A term applied to the specialized process of hot-dip *galvanizing* (zinc coating) of wire which is followed by a drawing operation.

Application range STEEL

This has a dual purpose of overcoming the problem of tears on the zinc, which can result in an uneven surface finish and unattractive appearance, and also the technical problem which arises when the thicker coating sometimes cracks, placing an undue stress on the adhesion, and allowing local attack on the substrate.

In addition to eliminating these drawbacks, and giving a wire with an even coating of zinc with an attractive finish, there is no doubt that the subsequent *cold-drawing* operation will find any defect in the adhesion of the zinc, resulting in rejection at this stage.

'Brytal' Process
A proprietary dual process for brightening and *anodizing*.

Application range ALUMINIUM

It is generally confined to pure aluminium or special-purpose, low alloys which have been developed for the specific purpose of *electropolishing* and *anodizing*.

The parts for the 'Brytal' process are *anodically etched* in a solution of sodium phosphate and carbonate, which chemically polishes the raw material, provided it has been supplied in reasonably good condition. After washing, the parts are subsequently *anodized* to protect the chemically polished surface.

There are now available a range of proprietary chemical polishing solutions based on phosphoric acid. These are used in conjunction with the *sulphuric anodizing* process and produce comparable results. It is also possible, with modern techniques, to produce a bright aluminium surface on steel comparable to the above by means of *vacuum deposition* of aluminium. This will require that the parts are used out of contact with moist air, or are *lacquered* to prevent the vacuum-deposited aluminium from dulling by oxidation.

Buffing

A specific type of *polishing*, using a high-speed disc, made from layers of cloth, leather or plastic.

Application range ALL METALS

The disc is impregnated with an abrasive in liquid slurry or solid form, and the metal article being polished, or buffed, is pressed against the disc.

Buffing should be a free cutting operation, where the surface metal is removed by the action of the abrasive carried on the revolving disc. No *cold working* or smearing of the metallic surface should occur with correct buffing.

The term is sometimes applied to the local *grinding* of *welding* flash, etc.

Bullard Dunn Process

A *descaling* process generally only applied to sheet material; it is not normally found with modern production processes.

Application range STEEL

The material is first *cleaned*, and then electrolytically treated in a bath of sulphuric acid with approximately 1 g/l of tin. The solution is used warm, at approximately 60°C–70°C. The sheet steel is made the cathode, when hydrogen forms at the interface. This reduces or removes any oxidized scale and, at the same time as the basic steel is freed of oxide, a coating of tin is deposited electrolytically. While this tin will have little or no adhesion, its presence on the fresh surface of steel will reduce the corrosion potential.

On removal from the solution, the steel must be thoroughly washed and dried, and in many circumstances the tin will be removed prior to final processing by reversing the electrolytic action to make the steel anodic, thus removing the tin and taking this back into solution in the sulphuric acid.

However, the presence of a minute quantity of tin, or tin compounds, helps to reduce the tendency of freshly *descaled* steel to rapidly corrode.

The process is of limited application with modern technology, where hot sulphuric acid with an inhibitor is used with success. Alternatively, the acid treatment is followed by immersion in a solution such as borax, or some other alkali, as a method of ensuring that the corrosion potential of the freshly *descaled* steel is reduced. However, other methods of scale removal such as *blasting, alkaline descaling, sodium hydride* treatment, *wire-brushing*, etc., will now be more commonly used. The Bullard Dunn process does not result in any positive corrosion resistance, and is a relatively expensive method of scale removal.

Burnishing

This is a form of metal finishing, where the surface is treated mechanically in such a manner that no appreciable metal is removed but the surface is smoothed.

Application range ALL METALS

The term burnishing is often misused for a form of *polishing*, but technically the operation can be applied either with a rotary or a reciprocating motion, using hardened polished steel tools, which do not themselves cut into the surface of the metal but produce a smooth finish. This finish has the advantage that it will be free of stress raisers, caused by the normal cutting or machining operations and at the same time it will have a thin surface layer of material which has been *cold worked*, thus enhancing the fatigue strength of the component, as the layer will have a higher tensile strength than the material itself. The process, then, is commonly used where a high-quality fatigue-resistance surface is required, with excellent visual appearance. It will be appreciated that the harder the basic material, the more difficult the burnishing operation will be, and also the less effective the result.

To some extent burnishing can be compared to *shot peening* as a means of increasing fatigue strength. Burnishing is generally more expensive on a production scale, but for a limited number of components, it can be an economic proposition. However, the method is more difficult to control, as the skill of the operator is the effective factor, and it is also a more difficult process to automate under most circumstances than *shot peening*.

Burnishing can also achieve a high-quality finish, comparable to *polishing*, where the fatigue strength is improved by the removal of all stress raisers. The method, correctly carried out, has the advantage over *polishing* that the *work hardening* layer produced will positively add to the fatigue strength.

The state of the surface prior to burnishing can be critical. If this is very rough, the operation can result in bending of the peaks to fill in the valleys, rather than a true smoothing. This will leave a series of stress raisers which will have poor fatigue characteristics, and a preliminary inspection of the surface of high-duty components is essential.

Butt Welding
A weld join where the two components or pieces are joined by butting against each other.

Application range ALL METALS

The term is generally used to differentiate the technique from *fillet welding*. The welding process itself will vary according to the technical requirements.

Buttering
The deposition of high-ductility weld metal on a surface prior to further *welding*.

Application range STEEL

Buttering is generally applied by the *electric arc* system with fluxed rods, the purpose is to increase the ductility of the weld.

In general, buttering is carried out only as a means of increasing local ductility when problems are envisaged or have been encountered. The technique of buttering is commonly applied when 'laminar' tearing has been identified. This supplies a high ductility 'buffer zone' which reduces the strain on the heat affected zone.

The term is, on occasion, used to denote a type of *welding* technique where the deposit is spread across an area rather than filling a prepared weld section.

'Buzzard' Process
A term, used in the USA, for *chromic acid anodizing*. It uses 100 g/l solution of chromic acid in water.

Cadmium Plating
A process which is similar to, but more expensive than, *zinc plating*. It is used to protect steel in contact with aluminium. Described fully in Part Two **Electroplating**.

'Calorizing'
A proprietary name for a form of *aluminizing* which coats the surface of steel with aluminium oxide, giving increased corrosion and oxidation resistance at temperatures up to 700°C. The components are treated in a rotating cylinder

filled with the calorizing mixture, which is aluminium suspended in aluminium oxide. The main use is in the coating of heat-treatment pots, furnace components, ladles, etc. Details are available from the Calorizing Corporation; see also *Aluminizing*.

Carbo-nitride
A *surface hardening* process, applied only to steel for comparatively shallow case depths – up to 0.005 in (0.01 mm). Described fully in Part Two **Heat Treatment**.

Carbon Arc Welding
A seldom-used form of *welding*, where the arc is produced between two carbon electrodes. Under some circumstances, the arc may be produced between a carbon electrode and the material being welded. The arc produced has certain characteristics, the heat being intense, the arc difficult to control and the atmosphere reducing, that is will prevent the formation of oxide. While this last factor can be useful, the advent of *inert-gas shielded metal arc welding* ('*argon arc*') has largely made the carbon arc weld redundant.

Carbon Dioxide Welding
A form of high-production *electric arc welding* where the welding arc ionizes the carbon dioxide and the weld is shrouded by carbon dioxide; it is generally only used on steel. Described fully in Part Two **Welding**.

Carburizing
Probably the most important method of *surface hardening* of steel, to prevent wear. Described fully in Part Two **Heat Treatment**.

Case Hardening
Any technique which results in a hard surface layer.

Application range STEEL, COPPER AND ALUMINIUM ALLOYS

The term may be used to indicate any process where a *surface hardness* is produced. The most common forms are those applied by the *carburizing* and *nitriding* processes and their variations. Thus, there is the *carburize* process itself, whereby the surface carbon is increased to produce a bimetal steel with high carbon at the surface and low carbon at the core. The terms 'case harden' and 'carburize', then, are often assumed to mean the same. *Nitriding* is a form using nitrogen, where iron nitrides are formed at the surface, producing a case in many respects similar to *carburizing*.

With steels, the cheapest method of *surface hardening* using this process will be the *carbo-nitride* and *cyanide* processes. With these, the surface has carbon and nitrogen introduced in the form of iron carbides and iron nitrides. With *carbo-nitride*, the process is gaseous, while *cyanide hardening* employs a liquid salt bath. Both have economical advantages over *carburizing* and *nitriding*, but neither produce technically as good a surface case.

Other forms of *surface hardening* which may be categorized as case hardening processes include *chromium plating*, where a hard deposit of chromium is *electroplated* and can be finally ground to give a hard, low-friction surface. This is an additive process, where the material is added to the surface, as distinct from the *carburizing* process, where the components remain at basically the same size before and after treatment.

Shot peening produces a surface which can be seen under micro examination to include a layer of hardened material which results from the *work hardening*. This, together with *burnishing* and *autofrettage*, is a process giving a very thin surface hardness, which is the result of a specialist process designed to improve the fatigue strength of the material. This improvement is usually local, but these methods are used when the more conventional forms of case hardening, such as those described above, cannot be carried out for geometrical or technical reasons.

Case hardening also includes those forms of *electroplating* where the deposit is hard and, thus, can be said to cause *surface hardening*. The modern process of *electroless nickel plating* will produce a hardened surface after the necessary *heat treatment*.

For comparative purposes, the following information may be useful. *Cyanide hardening* on a production basis is probably the cheapest form of case hardening where a heat-treatable case is required. With modern effluent control, however, the use of cyanide is discouraged, and where a cyanide-effluent treatment plant is required solely because of *cyanide hardening*, then the economics of the process are such that *carbo-nitriding* is probably cheaper. Both are ideal for producing a thin hardened case less than 0.005 in (0.01 mm) for mass-production components. If the equipment is available, it can also be used for 'one-off' jobs. The case produced, however, will be slightly more brittle than that produced by *carburizing*, and hardness and fatigue resistance will be less than with *nitriding*.

For general engineering purposes, where case depth in excess of 0.025 in (0.06 mm) is required, then *carburizing* is accepted as being the most economical method. *Nitriding* is the more expensive process but results in a case of unique properties, in that it has a high compressive strength, and thus considerably improves the fatigue resistance of the components. In addition, the process is superior with regard to distortion problems as there is no subsequent quenching operation. Finally, a case is achieved which can be used at temperatures up to approximately 350°C–400°C without any appreciable reduction in hardness. The *cyanide, carbonitride* and *carburizing* methods all start to lose their hardness at temperatures in excess of 200°C. *Boriding* is expensive but will produce a case harder than nitriding.

Induction hardening using induction coils for heating can produce a controlled case on medium carbon steels. Where large quantities of simple shapes are involved this can be very economical.

Chromium plating is comparable in hardness and has better low-friction characteristics than *carburizing* or *nitriding*. However, since less ductility is achieved and it is essentially more suited to 'one-off' methods, there is greater difficulty in controlling the process for high-quality continuous production. In addition, the case produced is more brittle than with either of the above processes and, thus, it is not generally regarded as an alternative at the design stage. There is, however, an argument for considering the use of thin *chromium plating* up to thicknesses of 0.001–0.002 in (0.025–0.05 mm) where the stress on the surface is low. Thus, where conditions produce more frettage than actual wear, the presence of a thin chromium deposit can have considerable advantages, since in many cases no subsequent *grinding* is required and chromium has unique characteristics of hardness and low friction. (It should be noted that *chromium plating*, together with the forms described below, can be applied to materials other than steel.)

Burnishing and *shot peening* are generally applied where it is necessary to improve the fatigue resistance locally. They are relatively cheap processes from the point of view of both capital and operating costs, but require considerable control to ensure that they are correctly carried out. From the design engineer's point of view, they do not as a rule compete with case hardening by heat treatment or with *chromium plating*.

Electroless nickel plating is now regarded as an alternative to *Chromium plating* under many circumstances, and *metal-spraying* and *weld deposition* of hard materials, including ceramics, may be classed as case hardening processes as they result in a hard surface.

The design engineer must evaluate the available processes, taking into consideration the load involved, bending stresses, life expected, corrosion conditions and economy.

CASS Test (Copper Accelerated Acetic Acid Salt-spray)

This is a form of *acelerated corrosion test* which uses the 'salt mist' method, but the solution contains cupric chloride and acetic acid.

The solution is 5 per cent sodium chloride, 0.025 per cent cupric chloride, and is acidified with acetic acid to a pH of 3.2. The test is carried out at 50°C.

This test is more corrosive than the straightforward 5 per cent sodium chloride *salt-mist test*, and therefore, the results can be expected to show more scatter. It is possible, however, to achieve results in hours rather than days, when testing good-quality plating deposits. The reader is referred to the section *Accelerated Corrosion Testing* for further information.

'Castenite' Process
A proprietary name given to the *impregnation process* used to salvage porous metal.

Casting
A general term covering the production technique whereby the metal is heated until it is molten and then poured into a mould, where it is allowed to cool and solidify. This book covers techniques for treating metals, and thus production methods such as casting are not discussed in detail.

Application range ALL METALS

In practice, considerable difficulties are encountered because of the general affinity of all molten metals for oxygen, in particular, and other atmospheric gases. There is, in addition, the problem of cooling contraction, the possibility of reaction between the molten or hot solid metal with the mould material and any fluxes which might be used. Therefore, a relatively limited range of metals can be cast to produce a useful component.

There are, however, certain metals which cannot be produced by any other technique, notably the cobalt alloys.

The best known cast metals are the range of cast irons.

Casting can be a highly economical method of producing intricate shapes, but in general a casting will not have the mechanical properties which can be obtained from the same material in the wrought condition. The reader is advised to obtain detailed information on the properties of casting from relevant technical sources.

Cathodic Etching
An electrolytic etch technique for pre-plating and inspection purposes.

Application range STEEL

The process makes the component the cathode in an electrolytic cell, usually with sulphuric acid as the electrolyte. The anode will generally be lead, but may be stainless steel or, under certain circumstances, titanium. When the component is the cathode, hydrogen will be evolved at the component surface, and this can result in the surface being reduced, that is any oxides will be removed. This technique is, therefore, sometimes applied immediately prior to *electrodeposition* when the surface of the metal will be presented to the *electroplating* solution in the ideal condition for good bonding. Other reasons can exist for preparing the surface in this manner, and there is an inspection technique whereby certain surface defects are accentuated using this procedure.

The fact that atomic hydrogen is produced at the metal surface means that hydrogen embrittlement could be a serious problem. The reader is advised to obtain advice on the metals which are prone to problems in this area.

This process is generally used on high-integrity *carburized* or fully *hardened* components which have been subjected to *grinding*. It is well known that *grinding* abuse of fully hardened steel surfaces can result in the formation of grinding cracks but prior to the actual production of a crack there may be serious metallurgical defects at the surface because of local high-temperature effects. These result in metallurgical changes which, because of their very local nature, can act as serious stress raisers, causing failure in service. The use of cathodic etching with approximately 50 per cent sulphuric acid at room temperature followed by thorough swilling in water, then oiling in hot oil, can give a surface *etch* which accentuates these defects, allowing their visual identification.

A similar, much simpler, technique is the use of *nitric acid etching*, where the components are immersed non-electrolytically in dilute nitric acid again followed by swilling and oiling.

These *etching* techniques are unique in identifying the metallurgical changes which can and do occur during *grinding* of hardened steel. These defects will not be identified by any other form of *non-destructive testing*.

Cathodic Protection

A technique whereby with the use of metals such as zinc, aluminium and magnesium, which are anodic to iron, *corrosion protection* is achieved.

Application range STEEL

It is also known as *sacrificial corrosion protection*, as the process relies on the fact that, where any cell exists between two different metals with an electrolyte, one of these metals will corrode and in the process of corrosion will protect the second metal. The metal which corrodes is the anode, and, thus, it 'sacrifices' itself, while the protected metal is the cathode.

By reference to the electrochemical series of metals, the potential voltage difference can be found, and any metal which is above or cathodic to the second metal will be protected. The higher the voltage between the metals the greater the corrosion which will occur on the anodic metal; thus, the anode will be more rapidly used up.

In its simplest form, cathodic protection has pieces of metallic zinc, aluminium or magnesium bolted in contact with the metal to be protected. This is the technique commonly applied to large structures such as ships, piers or pylons, immersed in water. Careful siting of the anode is necessary, and where, for example, a third metal is involved, as occurs with shipping where bronze propellers and shafts historically have resulted in severe corrosion of ships' sterns, then considerable care is necessary.

Variation of the simple cathodic protection is the use of *impressed current*. With this, the anode is unreactive but conductive. Commonly it is titanium, but under laboratory conditions it will be platinum or gold. The anode has a current imposed to ensure that it is, in fact, anodic to the component or structure being protected.

It is now common practice, with large expensive structures in hostile environmental conditions, that no single corrosion protection technique is used, but that the assembly is first painted or protected by *metal-spraying*, this initial protection being backed up by cathodic or *impressed current* protection. The reader is advised to obtain specialist advice on the technique used, and also to ensure that the correct material is applied under stringently controlled conditions, including monitoring during the useful life.

Following is the Electromotive Series for metals commonly used in engineering. Each metal is protected by the 'sacrificial' action of any metal which appears below it on the list:

Gold
Platinum
Silver
Mercury
Copper
Lead
Tin
Nickel
Iron
Cadmium
Chromium
Zinc
Aluminium
Magnesium
Lithium

The lower metal will always 'sacrifice' itself to protect a higher metal, and the farther apart the metals, the greater the corrosion potential.

Cavitation Erosion

This is a failure mechanism in which the surface is subjected to compression and vacuum or suction by liquids, or by the collapse of a high vacuum.

Application range ALL METALS

This is a defect which occurs in service rather than any process which is applied to metals. It is the result of rapid change from pressure applied to a metal, to vacuum or suction on the surface. It was traditionally identified on ships' propellers which rotate to drive the vessel through the water. This results in heavy pressure being applied by the liquid on to the propeller surface. At a certain point in the revolution of the propeller there is a rapid change from surface pressure to surface vacuum or suction. Where the liquid has a high vapour pressure the suction applied to the surface of the metal can apply a high tensile load resulting in failure. Where a vacuum is produced this also loads the metal surface in tension and can result in failure. With many metals this can result in plucking of the grains from the surface.

Cavitation erosion can be identified visually by the form of the pit. This will be rounded, will have vertical walls and will have a smooth surface. Erosion without cavitation will always result in some elongation of the erosion pits. Corrosion will always result in some roughening of the surface.

Cavitation erosion can be cured by engineering practices whereby, by altering the shape or speed, the pressures and pressure changes are reduced or eliminated. It is also possible by the careful selection and treatment of metals, to ensure that there is a fine homogeneous grain. This will reduce the effect of cavitation. Certain materials have characteristics which enhance their resistance to cavitation. Further specialized advice should be obtained if it is considered that cavitation erosion could be a problem.

Ceconstant

This is a proprietary name for a series of cyanide salt bath processes.

There are different types of salts identified by numbers. Further information can be obtained from Degussa.

Cementation

A process where carbon is diffused.

Application range STEEL

The term was originally applied to the process of converting wrought iron into steel by placing relatively thin sections of wrought iron in a carbon-containing material, commonly charcoal, and heating to above 900°C. The term is also occasionally used as an alternative to '*carburizing*', which is described in more detail in Part Two **Heat Treatment**. This is not a commonly used term.

Centrifugal Casting

A specialized *casting* technique, where the mould is rotated during the process.

Application range ALL METALS

Under some circumstances there will be no inner wall, thus the casting is produced by control of the metal flow and the speed of rotation.

It will be appreciated that only relatively simple shapes can be produced by this method, but as these will be hollow, considerable saving in machining costs are possible. Continuous centrifugal casting of pipes and tubes is now carried out and, with certain materials, relatively complex shapes are possible. The reader is advised to obtain specialist advice prior to designing any component. This is not a technique involved in treating metals.

Chapmanizing
A salt-bath form of *surface hardening*.

Application range STEEL

It is a form of *nitriding*, using a liquid salt which is heated by electrodes passing through the salt. This same electric current is used to decompose compounds within the salt, producing nitrogen in the atomic state and releasing hydrogen which burns at the surface of the salt.

The treatment temperature is about 800°C, and a case depth in the order of 0.002 in (0.05 mm) can be produced in approximately 2 h. Deeper case depths up to 0.1 in (2.5 mm) are claimed with this process.

The components, on removal from the salt bath, are hard and no subsequent quenching or other *heat treatment* is necessary. This is, therefore, similar in many respects to *nitriding* and has some similarity to the 'Tufftriding' process. It has, however, been largely replaced by the gaseous *nitriding* process which, while expensive, is cheaper than Chapmanizing. For most components, the process of *carbo-nitriding* will give comparable results at a much lower cost. An alternative name for Chapmanizing is *wet nitriding*.

Charpy Test (Charpie Test)
A form of *impact testing*. The specimen is machined to a square section, usually 10 mm^2 and 55 mm long, with a notch machined at the centre. The standard and shape of this notch are critical, and unless tightly controlled the results of the test will be meaningless. The specimen is located horizontally at both ends and the pendulum load strikes the specimen at the centre. The energy absorbed will be a function of the impact resistance of the material. Described fully under *Impact Test* in Part Two **Mechanical Testing**.

Chasing
A machining operation, not strictly a metal process or treatment where surface forms already produced (commonly threads) are re-formed (chased) to ensure an accurate condition.

Application range ALL METALS

Little or no difference in dimension is implied, but the cutting operation, usually carried out on a turning machine, ensures that the existing form is true. It is commonly carried out on threads and similar shapes. It is sometimes applied to the production of shapes in sheet metal using hand tools as a form of *embossing*.

'Chemag' Process

A proprietary process using hot, strong alkaline agents containing chromate as the oxidizing agent.

Application range STEEL

It is a form of '*Black Oxide*' treatment.

Chemical Deburring

See *Chemical Polishing*.

Chemical Machining

The removal of metal by the use of chemical attack.

Application range ALL METALS

The components are usually masked to prevent attack on the areas not requiring metal removal. The material is then immersed in the chosen chemical where the unmasked metal is dissolved. The rate of chemical attack must be rapid, particularly where extensive metal removal is required, as it will be appreciated that as metal is removed from the surface, undercutting of the masked area will occur.

Considerable skill and control is essential in the choice of chemicals and temperature used, and specialist advice is essential.

The process is useful where intricate shapes are required, particularly with metals with machinability problems. With some metals, massive areas of material are dissolved, for example aluminium components in aircraft construction. The reader is advised to obtain further advice on this process.

Alternative treatments would be *electrochemical machining* (ECM), or *spark erosion*. With small components the use of electrodeposition to produce the desired shape should be considered.

Chemical Polishing

This term is used in place of *electrolytic polishing*, but should be reserved for the treatment where a surface finish is improved by chemical action alone.

Application range MILD STEEL, LOW ALLOY STAINLESS STEEL AND ALUMINIUM

Special solutions have been devised which attack the surfaces of these metals in such a manner that the peaks or corners are affected in preference to the valleys or concave surfaces. The theory is similar to that used in *levelling*, where the solution gives off a gas having electrical insulation properties. The gas, generally hydrogen, is attracted and retained in hollows in contact with the surface, thus preventing or slowing down the attack in the valleys and allowing it to proceed at the peaks. This, therefore, results in a general smoothing of the surface.

Chemical polishing is generally reserved for those components which for some reason cannot be *electrolytically polished*, or where *barrelling* is not justified because of the small quantities involved.

Chemical polishing is occasionally used in conjunction with chemical attack to reduce the size of components and produce an improved surface finish.

Some of the proprietary solutions used are complex, and interested readers are advised to contact plating supply houses for details of specific solutions used for the metals mentioned.

This is not a technique which should be considered before obtaining specialized advice.

Chemical Vapour Deposition (CVD)

Chemical vapour deposition is a technique which results in a thin coating on the surfaces of metals. It is the result of a chemical reaction at elevated temperatures in the region of 1000°C.

Application range STEEL

The process is limited to substances which can be vaporized at usable temperatures and then capable of reacting with the surface of steel.

In the main at present this is confined to tungsten carbide, titanium carbide, titanium nitride and alumina deposited on to the surfaces of tool steel.

The process involves high temperature chemical reaction between complex, generally organic compounds of the metal being deposited along with other organic materials. The temperature used is in the region 900 – 1200°C and it will therefore be appreciated that specialist equipment and expertise is required.

This is claimed to result in an improvement in wear, and also in improvement in cutting life for high speed tools etc.

The process is still being developed and is competing with developments in techniques such as *plasma nitriding* and *ion nitriding* or implantation.

Chesterfield Process

A specialist *hardening* and *tempering* process.

Application range STEEL STRIP

The coil is heated to the necessary *hardening* temperature, and the strip is then pulled out of the heating furnace between two water-cooled blocks. This results in a full *hardening* of the steel. Immediately after this quench, the strip is oiled, the quantity of oil being controlled, and then ignited. This results in heating of the strip material and *tempering*.

There are modern techniques which result in the same results. Readers are therefore advised to seek advice on this subject.

Chisel Test

A destructive test, used to evaluate the quality of *resistance welding*.

Application range ALL METALS WHICH ARE WELDED

The test may be carried out on components normally required for use but this results in their being scrapped, and more generally, test pieces are produced which are then destroyed. Where test pieces are used, these must, of course, employ exactly the same material, thickness and welding conditions as those carried out on the actual component. Unless complete standardization of all parameters is used, there is little point in producing the test pieces.

Depending on the importance of the components being produced, the number of test pieces will vary. For high-integrity parts there will generally be a requirement that components themselves are tested and the frequency of this test will be specified. In addition to the component, test pieces will also be required. The frequency of the test pieces will vary from one hour, to a certain ratio of components to the test piece required at the beginning and end of each operator shift; for less important components the test piece will be produced on a daily basis or perhaps at less frequent intervals.

Where *resistance welding* is used, it is strongly recommended that test pieces are produced at least once per week, and certainly when any adjustment is made to any of the welding parameters. The test itself uses a chisel to prise apart the weld. Once the weld has been parted, it is visually inspected and a slug of the material must be able to be torn out of one of the components for the welding to be considered satisfactory.

If the weld is found to part along the join line, this indicates faulty welding. Further examination of the faulty weld should indicate whether the problem has been caused by too much heat, causing fusion instead of forging of the weld, or too little heat, with incomplete forging. With this information, it is possible to make adjustments to the weld settings and further testing must then be carried out until a satisfactory chisel test is obtained.

Chromate Coating

A *corrosion protection* technique to increase corrosion resistance which has many variations.

Application range STEEL, ALUMINIUM, MAGNESIUM, CADMIUM, ZINC

It is well known that chromates are materials which resist further oxidation. Generally, chromating results in the formation of metallic oxides of the metal being considered, and this then absorbs into itself some of the chromate solution. This results in a sealing of the oxide and formation of some metallic chromate.

With steel, chromating has little permanence, but will add to other forms of corrosion prevention and is commonly used following *phosphating* and as an additive to zinc-rich primer paints.

The chromating of aluminium and magnesium considerably enhances their resistance to corrosion and improves to a high degree paint adhesion to the metal surface. With zinc and cadmium, chromating will also considerably increase resistance to corrosion, and this is particularly the case with *electrodeposited* metals. Almost without exception high-quality *electrodeposition* of zinc and cadmium on steel will specify that the deposit is finally *passivated* using one of the many proprietary chromating processes. *Galvanizing* and zinc die casting are also commonly chromated.

This process is often referred to as *passivation*, but it should be noted that other techniques unrelated to chromating also use this term, and it is advised that a clear definition of the process is obtained when *passivation* is being discussed.

Chromating will almost invariably improve *corrosion protection* but does not withstand impact damage. There is now considerable data available on the standard of *corrosion protection* achieved by chromating, and the user is advised to obtain detailed information where necessary.

Alternative proprietary names used in chromating include: *'Alochrom'*, *'Alochrome'*, *'Alocrom'*, *'Alrac'*, *'Bauer Vogel'*, *'Bonderizing'*, *'Cromodizing'*, *'Cronak'*, *'Gramoseal'*, *'Protal'*, *'Protolac'* and *'Pylumin'*.

Chromating

See *Chromate Coating*.

Chromic Acid Anodize

A form of *anodizing*, where the solution used is dilute chromic acid.

Application range ALUMINIUM

The technique was at one time popular as the anodizing process is at least as good as the sulphuric acid anodize, and can with certain alloys produce a thicker anodic film. In addition, it can be used for crack detection. For further information on this technique, see *Anodizing*.

Chromium Aluminium Duplex Coating
See *Aluminium Chromium Duplex Coating*.

Chromium Plating
A plating process for decorative purposes on top of nickel, and as a hard, low-friction metal for engineering purposes. Described fully in Part Two **Electroplating**.

Chromizing
A surface diffusion process, whereby chromium is *alloyed* with iron to give a chromium-rich surface layer.

Application range MILD STEEL AND LOW ALLOY STEEL

The process is analogous in many ways to *carburizing*, but is more similar to certain types of *aluminizing* and other techniques where a metal is diffused into the surface of steel.

The process generally involves the thorough *cleaning* of the components, which are then placed in a heat-resistant box with a proprietary powder of an unstable chromium compound. This box is then sealed and taken to above 1000°C. The compound is decomposed, releasing chromium in an active state which will then react with iron, producing the necessary alloy. This will diffuse away from the surface towards the core.

The time involved and, to a certain extent the temperature, controls the depth of the chromium-rich layer. With treatments in excess of 4 h, the layer at the surface will contain up to 30 per cent chromium with an even merge to the core which contains no chromium. As an alloy of steel with more than 12 per cent chromium has excellent corrosion resistance, it will be seen that the chromizing process results in a corrosion-resistant surface. Steel with chromium above about 25 per cent is austenitic and, thus, has excellent corrosion resistance, comparable to that of the 18-8 austenitic steels, and at the same time can have the mechanical strength of low-alloy medium-carbon steel.

On completion of the treatment, the components are removed from the powder which can be regenerated for further use. Under certain circumstances the components can, if desired, be *welded*, and provided distortion is not a serious problem, the component can be *hardened* and *tempered*. In general, however, chromizing is used on components such as studs or fasteners, where corrosion resistance of a high standard is essential. On sheet metal, it is possible to carry out some *cold working* on the treated strip, provided the dies used are in good condition, and thus do not damage the surface.

The chromizing process can result in the production of components with a corrosion resistance at least equal to that obtained from 12 per cent chromium stainless steel and, provided the loading of the components is efficient, the cost can be less than with stainless steel. There is also the disadvantage with the latter method that 'machinability' of 12 per cent chromium steels presents considerable difficulties where high productivity is essential, while with the present method it is possible to efficiently machine intricate shapes using free machining mild steel, with subsequent chromizing. An alternative method of chromizing, giving a less efficient surface, is *chromium plating* and subsequent *heat treatment* of the plated components, using a controlled atmosphere to prevent oxidation. Chromizing, then, is a method where, under many circumstances, it is possible to produce stainless steel components at an economical cost, provided the surface is not subjected to serious mechanical damage.

Alternatives to chromizing include any of the standard methods used for *corrosion protection*, such as *nickel plating, electroless nickel plating, chromium plating* and, under certain circumstances, *zinc* or *cadmium plating* together with various forms of *painting*. *Metal-spraying* of aluminium and zinc may also result in a technically comparable surface.

An alternative name for chromizing is the *Cinera process*.

Cinera Process
An alternative name for *chromizing*.

Cladding
The application of a thin film of metal to the surface of a base metal.

Application range ALL METALS

Probably the most commonly used cladding applies stainless steel to mild steel and pure aluminium to aluminium alloy. In both cases the purpose is to produce a surface with corrosion-resistant characteristics superior to the base material. The reason in both cases is generally that the base material has attractive mechanical properties which cannot be obtained with the corrosion-resistant material and, thus, the bimetal is a compromise.

Cladding is not carried out as part of a fabricating process, as a general rule, but is part of the process of manufacturing the metal itself.

Steel cladding will only be of stainless steel on top of mild or low-alloy steel. The type of stainless steel used is dependent on the corrosion-resistant characteristics required. It must be noted, too, that the cladding will be relatively thin and, thus, may be damaged either during fabrication, or subsequent use. If the clad material is removed locally, the underlying metal will be prone to corrosion, and as the area available for corrosion is relatively small, such corrosion can be very serious.

It is possible to produce a form of cladding either by *weld deposition* or *metal-spraying*; these processes are described separately. The cladding process will generally produce a skin which is much thicker than that obtained by the normal *electrochemical deposition, galvanizing* or *metal-spraying*.

With stainless-steel cladding, there will be no danger arising from porosity. However, it must be noted that the clad material will be relatively simple in shape, and the process of forming and working can result in the clad being distorted, damaged or in some cases removed. With the correct choice of clad steel, it is possible to carry out a range of metal-making processes such as forming, flanging, and bending, but again some care is required to prevent damage.

Aluminium alloys are generally clad with pure aluminium which has a very high corrosion resistance, particularly to acid conditions.

As alloys are added to aluminium to increase its mechanical properties, the corrosion resistance is decreased.

With some of the heat-treatable, highly alloyed materials with tensile strength commensurate to mild or low-alloy steel, the corrosion resistance is sometimes at a low enough level to require permanent protection. If the components are of simple shape, an economical method of improving the corrosion resistance is for the components to be made from sheet in the correct alloy for mechanical purposes and clad with a thin film of pure aluminium. This will be to the order of 0.005–0.025 in (0.01–0.06 mm) thick. This will result in *corrosion protection*, comparable with the alloy itself being *anodized* or *painted* to a very high-quality finish. It must, however, be stressed that considerable care must be taken during any manufacturing process to ensure that damage does not occur on the aluminium-clad surfaces as pure aluminium is extremely soft and thus easily damaged.

Some aluminium materials or components are clad with high-silicon aluminium alloys. This has a low melting point and acts as an integral braze filler metal in addition to improving the corrosion resistance.

This technique is used to a lesser extent with other forms of brazing alloys to a variety of base metal, and can be a very efficient technique if considered at the correct stage of design.

Other examples of cladding are to be found in some of the more exotic materials, where either because of expense or the presence of certain properties, these materials are clad with a specific material to withstand specific atmospheres. Examples include stainless steels of the austenitic variety which may be clad with cobalt alloys in order to improve the wear characteristics. In this case comparison must be made with *chromium plating* or other similar processes. It is also known that materials such as hafnium can be clad with other materials such as zirconium in the nuclear industry.

Sheffield plate is an example of the historical use of cladding where copper sheet was clad with silver on one or both sides and subsequently formed into hollow vessels. This is now replaced by silver plating.

Rolled gold is a further example where a thin film of gold is clad on copper, or gilding metal, subsequently rolled and made into jewellery. A modern alternative to this is gold plating.

In summary, it can be stated that cladding should be looked upon as a possible means of compromise, regarding surface protection of a metal with the correct mechanical properties, when either the material is too large to be protected by normal means or when the corrosion conditions are highly specialist. It should be added that with aluminium alloys, provided they are simple enough and do not have complicated manufacturing techniques, cladding with pure aluminium can be an economic and technically desirable alternative to conventional *corrosion protection*, by *anodizing* or *painting*.

Classical Anneal

This is a technique used to fully soften steel. Described fully in Part Two **Heat Treatment**.

Cleaning

The removal of soil from the surface.

Application range ALL METALS

Cleaning of metals is the basic process considered here, but there is no obvious reason why the same techniques cannot be applied to non-metallic materials, provided the solutions used do not cause chemical attack. Cleaning is most often required as part of a complete process.

Industrial cleaning costs considerable amounts of money, and the lack of correct cleaning is the probable cause of many serious problems. For example, early failure in engines with moving components which are built without adequate cleaning, resulting in excessive wear. Inadequate cleaning prior to *painting* and *plating* is the prime cause of failure in both these areas. There is also the obvious health hazard under some conditions.

It thus cannot be overstressed that cleaning can contribute, first, to improvements in the economics of many processes and, secondly, to the technical improvement of several processes.

When cleaning requires to be considered, it must be realized that there are four forms of soil or contamination. It is seldom that all four of these are found in any one instance, but equally seldom will a single form of contamination be present. The four different types are: grease or oil; loosely adherent soil; adherent soil; and moisture. Examples of each of these are as follows.

1. *Grease or oil* Includes normal lubricating oils, oil used during machining,

grease used in protecting components during storage and natural oils in fingerprint contamination.

2. *Loose soil* Includes dust and fine turning and grinding debris produced during machining. These are very often found in conjunction with machining oil. This form of contamination can be considered as the chalk present on a blackboard.

3. *Adherent soil* Commonly scale or rust, but might include burnt-on oil or paint which has served its purpose and requires to be removed. To continue the analogy of chalk and blackboard, this soil might be considered to be the paint on the blackboard.

4. *Moisture* Normally present in the atmosphere, and appearing on engineering components as condensation. It will also exist when components are machined using soluble oil. Other obvious examples of moisture contamination include natural rainfall, accidental spillage, leaking equipment and buildings and, of course, in any process where components are subjected to water washing in any form. What is often not appreciated is that compressed air contains appreciable quantities of moisture. In the large areas of the world outside desert and arctic conditions, the air contains moisture. When it is compressed, the moisture is not removed, and as the compressed air expands and strikes cold metal the moisture will be deposited as a fine mist. It is relatively difficult to remove moisture from compressed air and experience shows that the standard methods of moisture removal are either inefficient or incorrectly maintained as a general rule.

The only technical method by which moisture can be efficiently removed involves refrigeration of the air and filtering out the frozen particles of moisture. This is expensive and seldom justified. It must, therefore, be appreciated that either components will be in danger of being damp, or considerable effort must be made to remove moisture using the standard moisture-trapping techniques. Under many circumstances, the above remarks apply to oil which is commonly found in compressed air, but in general oil is more easily removed.

Another common source of moisture contamination is fingerprints, and this will normally be associated with oil or grease naturally occurring on the human skin, but will also contain chemicals which can be aggressive regarding corrosion.

To take each of these types of contamination in turn, the methods for removing any one of them appearing singly are described below. As stated above, however, a single form of contamination will seldom be found in practice, and given below is an evaluation of economical methods of cleaning components which are multi-contaminated.

(a) *Oil and grease* The classical method for removal is the *vapour* or *solvent degreasing* technique. (Further detailed information on this will be found under the appropriate heading in this book.) Briefly, it consists of boiling a solvent cleaner, generally one of the chlorinated hydrocarbons, resulting in a heated vapour phase into which the contaminated components are inserted. The vapour then condenses on the cold components, to produce a clean, hot solvent which dissolves the oil and grease, and as further solvent condenses, these will be removed by the normal condensation action.

Immersion cleaning is also used for grease and oil removal. It has the disadvantage that, as the contaminated components are immersed in the hot or cold solvent, this itself becomes contaminated, and as the components are removed solvent containing contamination is left in contact with the surface layer. This contaminant is left behind after the solvent has evaporated. Under some circumstances, this technique, instead of cleaning the component, can result in a uniform contamination, where previously the component was partially clean with only local contamination. The problem is reduced by use of multistage cleaners, where the components are transferred to successively cleaner solvents. The ideal situation is that two or three immersion dips are carried out followed by a *vapour degrease*.

Wiping, using solvents, has the same disadvantage to a much greater degree than the dipping or immersion process. The cloth, brush or pad becomes rapidly contaminated, and there is a considerable history of technical problems which have been caused solely by the misuse of wiping methods.

Choice of solvent will, of course, depend on the type of contamination present. With modern emulsion or alkali-type cleaners, it is possible to remove oil and grease by what is probably not a true solvent cleaner, but which will result in a high standard of cleaning. But these can result in the components being contaminated with water.

(b) *Loose adherent dirt* This is removed by first wetting with the cleaning solution. This generally requires a soap or alkali cleaner but many of the modern emulsion-type cleaners will also succeed. Once the particles have been correctly wetted, then they can be floated from the surface being treated. This will usually require a high level of agitation, preferably mechanical, where the components are moved in relation to the solution, rather than air agitation, where the solution itself is moved. As an alternative it is possible to use multistage cleaners, where the contaminated but wet material is removed from the one solution into a second solution which will have a lower degree of contamination. This can then be followed by water washing which, if correctly carried out, completely removes all trace of the contamination.

Again it is obviously possible to remove this type of contamination by mechanical means. To revert to the blackboard, this is normally cleaned with a duster which removes a large percentage of the chalk but spreads the chalk evenly across the surface. Only by using successively cleaner dusters can the blackboard be shown to be clean.

(c) *Adherent contamination* Normally, there are two methods of removing this contamination: first, by *blasting* or *wire-brushing*; and second, by chemical means.

Blasting is probably the most economical method, everything else being equal, and under most circumstances will be the most efficient. (The different methods are described fully in Part Two **Blasting**), and the reader must evaluate the problem involved and then choose the correct technique, whether this is the sophisticated *Vapour blasting* or the relatively simple and cheap *Grit blasting*, using some form of slag as the abrasive. It is essential that the correct technique is chosen for the material involved, for the contamination present and, probably most important, for the subsequent treatment. Where any form of *painting* or *metal-spraying* is involved, there will be little doubt that *Abrasive blasting* is technically the best method, and at the same time the most economical choice. Where the cleaning requires to be followed by some form of wet process such as *electroplating*, the economics of *blasting* are more doubtful.

Where chemical means are required, in most cases, the contamination will be scale or oxide of some form: the procedure, then, will be that the component is *pickled*. It must be appreciated that with most metals, and steel in particular, there will be considerable danger of corrosion occurring immediately after the *pickling* operation. It is, thus, essential that tight control is maintained between the removal of the component from the active solution, through the subsequent *swilling* operations, and until any subsequent treatment is applied. Wherever possible, the delay between these treatments should be held to minutes rather than hours and certainly must be hours rather than days.

Chemical removal of adherent soil can also apply to the use of solvents where paint and such like materials become a contaminant. It is not possible in a book of this type to list all the possible solvents for each contaminant. It can, however, be stated that there are now available solvents which are almost universally used for paint removal. These tend to be relatively expensive and it will, therefore, be more economical always to use the specific solvent which has proven the most economical for any one type of contaminant and specialist advice should be sought in this area.

(d) *Removal of moisture* Traditionally, removal of this type of contamination used warm sawdust with the component or components being rolled in ths material. The sawdust absorbed the moisture and, being warm, then dispersed it by evaporation. This technique is now seldom used as experience has shown that the danger of sawdust contamination, particularly in oil holes and on various surfaces, was such that failures were found from this cause.

Components can be successfully dried by immersion in any oil or liquid held above 100°C. While this removes moisture, the components remain contaminated with the material used for its heating properties. With many engineering products, this is no disadvantage, but it will be appreciated that there are many occasions when oil cannot be considered.

It is also possible to 'dry' components by heating in an oven. This is often more difficult than at first realized, since, unless the oven has some form of vent to allow the moisture to be removed, very often the oven will merely become a humidity cabinet with the components being held in a warm, high humid atmosphere. While simple shapes will dry rapidly on removal, assemblies and complicated shapes will readily trap moisture which can result in subsequent corrosion. There is also the possibility of using hot water as a means of *drying*. Provided the water is boiling during the process, and that the components are held in the water long enough to reach the temperature of the boiling water, then on removal satisfactory drying can result, particularly with components having a reasonable mass and simple shape. This form of *drying* will not be satisfactory with components having a large surface area and little mass, or components having complicated shapes where moisture will be trapped and thus the heat in the component will be insufficient to drive off this moisture.

There is also the disadvantage that most natural water contains considerable quantities of solid matter in solution. With hot-water tanks, this solid matter will be concentrated in the water itself owing to evaporation from the surface. This contaminated water will be present on the surface of the components on removal, and the water itself will evaporate leaving behind the solid matter which was in solution. This leaves the well-known drying stains which for some purposes will be unacceptable. Added to this will be any solid matter contamination added to the water during use.

There are now available two proprietary methods of stain-free drying both of which achieve highly sophisticated dried components. The first is the '*Trisec*' *process*, which acts basically as a vapour degreaser. This contains an additive which reduces the boiling point of water so that it will evaporate from the component along with the solvent. This is then removed to the outside of the degreaser and the condensed liquid of solvent and water passes through a separator, where the water is rejected and sent to drain while the solvent is returned to the vapour degreaser sump. Handled correctly, this treatment is technically a highly efficient method of removing water.

There are other additives made to chlorinated solvents which act in a like manner to *Trisec* and efficiently remove all trace of moisture. It is essential that chlorinated solvents have this additive present in correctly designed equipment when used for water removal. Serious corrosion damage to components and the equipment can occur if water is allowed to remain in contact with the chlorinated solvent.

The second method uses liquids, recently developed, which have a revulsion to water. These liquids cannot mix with water and will, in fact, reject water in such a manner that, if they are in contact with water, there will always be a gap between the water and the liquids. These water-repellent liquids are now marketed by several companies, and the reader is advised to contact any of the larger plating supply houses for further information.

The technique with these is that the tank is filled with the liquid and the parts to be dried are immersed for several seconds, then removed and allowed to drain. The water-repellent liquid replaces the water almost immediately and the water sinks to the bottom of the tank; on removal from the tank the liquid

68 / CLEANING

will either drain back into the tank or evaporate. It is generally necessary to modify the tank to have a sight glass to monitor the level of water and to drain this from the bottom of the tank as necessary. Additives may be used such that, when the water-repellent liquid itself evaporates, a thin film of corrosion-resistant material is left. As stated, there is a variety of these liquids on the market, some supplying corrosion resistance of a low standard for short periods, others developed to withstand more corrosive atmospheres.

It is seldom found that only one of the above four contaminants exist at any one time, and it will be seldom economic to achieve separate cleaning systems to remove the different types of soil. For example, a rusted steel component which has been oiled and contains grinding soil and soluble oil as the final contaminant, would not be economically cleaned by *vapour degreasing*, followed by *pickling* in hydrochloric acid, then washing with a detergent-type cleaner followed by *drying* with one of the proprietary liquids described above. It would be more reasonable to clean the component using a soap-detergent cleaner then *pickling*, and finally *drying* with hot oil.

Alkaline descaling agents can be used to clean and descale, and sophisticated techniques such as *sodium hydride* are available for the same purpose. Part of the process will ensure *drying* and as a bonus any loose soil will be eliminated.

For most engineering purposes, the choice is generally between solvent-vapour-type *degreasing*, and aqueous-type cleaning. While there are many instances where the vapour degreaser is preferable, because of its convenience and the fact that the components on removal are dry, this is seldom the most economical method of cleaning in a modern engineering factory.

In recent years there has been considerable advantage in the use of detergent, soap, *electrocleaning* and *emulsion cleaning*. All of these techniques can be employed to remove normal engineering soil with its associated solid contaminant and, when correctly handled, result in a surface having a satisfactory standard of cleanliness which will not rapidly corrode.

It is possible, with the more sophisticated types of cleaner, to use additives in the solution, which remain on the cleaned surface and can inhibit corrosion over a wide range of conditions for relatively lengthy periods. These techniques can ensure a high standard of cleaning with less danger of subsequent corrosion than is experienced using the vapour degreasers, which remove all surface oils and leave steel, in particular, in a condition very prone to corrosion.

Electrocleaning is a technique using the electrolysis of water to give a form of gas-scrubbing on the surface of components and can achieve highly efficient cleaning.

Ultrasonic cleaning uses the energy of the ultrasonic beam to impart a surface pressure, followed immediately by suction or vacuum to the surface of the component. This will rapidly remove loosely adherent soil, and will efficiently remove the less adherent loose soil, but will not normally remove material such as adherent paint or scale.

Cleaning, then, is a very important aspect of modern engineering and industry in general. The reader is strongly advised to examine the soil and the mechanism of contamination and to choose the most efficient method of cleaning, first, to remove the soil and, second, to ensure that the resultant surface is in a satisfactory condition for subsequent processing.

There is available a relatively simple technique for assessing the efficiency of cleaning solutions. This uses the fine wire mesh used in laboratories which can be smeared with the contaminant to be removed. The contaminated mesh is then submitted to the technique which is being evaluated and the efficiency of contaminant removal is quantified by measuring the time taken to free all the holes from contamination.

Close Annealing

A specific form of *heat treatment*, to soften strip material in coil form.

The *annealing* involved is the standard *softening* process for the specific material, but because this is in the form of a tightly wound coil, atmosphere will obviously be excluded from any surface except that on the outside of the coil. This means that the finished material has an oxide-free surface and thus, close annealing can be compared to *box* or *pack annealing* or various types of *bright annealing* using specific atmospheres. The *Bright annealing* and *Annealing* processes are described fully in Part Two **Heat Treatment**.

Cloud-bursting

An inspection process used to identify hard spots on soft material or soft spots on hard material.

Application range STEEL

Basically the process is a *shot blasting* operation, where the material being inspected is blasted with hard, abrasive grit. Inspection of the newly-blasted surface will show distinct difference of texture between soft material and hard material, provided the difference in hardness is appreciable, that is at least 200 points on the Diamond Hardness scale or 20 points on the Rockwell C scale. When soft spots are identified on *case hardened* material, they will show as a darker, duller surface on a relatively bright, reflective background. Where the process is used to show hard spots on soft material, these will be indicated as brighter areas on a duller, matt background.

The advantage of this inspection technique is that it can sometimes be carried out as part of the *cleaning* procedure, provided the shot-blast grit is kept in good condition, and this inspection process is made part of the normal production routine. As an alternative, *etching* techniques can be used to show different metallurgical structures, but these can be relatively expensive and require a good standard regarding surface finish prior to the process being applied.

70 / COBALT PLATING

In general, the process is related only to steel, but there is no obvious reason why it may not be used for other materials. There are very few processes, however, which result in non-ferrous materials having hard or soft spots which are not readily visible by themselves.

Cobalt Plating
This metal, when deposited has similar characteristics to nickel. It can be used to increase die life. Described fully in Part Two **Electroplating**.

COD
Described fully under *Crack Opening Displacement* in Part Two **Mechanical Testing**.

Coining
Essentially a material producing method; it is part of the *forging* process where carefully controlled cold work is applied.

Application range ALL METALS WHICH CAN BE COLD WORKED

On occasion, this can be used to replace machining operations. The coining procedure requires that the previously hot-*forged* component has all scale and oxide removed either mechanically, by machining, or by chemical means. The forging is then restruck cold in machined dies which have the exact form required on the component. Using this process, it is possible to produce very tight tolerances on mass-produced components. It is also possible to impress designs and relatively simple shapes into the surface of the component. The importance of the process will depend on the material used, as the harder or tougher the material of the component, the shorter is the life of the dies. For most purposes, therefore, the coining process will not be economical, but with materials like aluminium, titanium, and many of the low-alloy steels, it is possible to have relatively long runs without deforming the dies and, thus, the process is an economical proposition.

Coining, then, is a suitable alternative to machining or engraving where the material is relatively soft and will accept *cold work*, and where the shape is relatively simple.

Cold Drawing
A general term applied to the manufacture of wire rod or tube where the final drawing operation is carried out in the cold condition.

Application range ALL METALS WHICH CAN BE COLD WORKED

This relies on the fact that many materials can be *cold worked*, resulting in an increase in the tensile properties of the material and an improvement of the surface finish.

The term cold-drawing is confined to the manufacture of wire and tube as the final drawing operation which, in most cases, is designed to give an improved, high-quality surface rather than any dramatic increase in mechanical strength. Further details of the process of *cold working* are given under *Work Hardening*.

Cold Galvanizing (e.g. *zinc plating*)

A term which is sometimes used to describe *electroplating* of zinc.

Application range STEEL

This is meant to distinguish it from the *hot-dipping* of steel into molten zinc which is normally known as *galvanizing*. The term cold galvanizing is also applied to a form of *painting* with specialized paints which result in a film of up to 90 per cent powdered zinc, which remains on the surface bound with some form of air-drying paint. Chlorinated rubber in organic solvents has sometimes been used as the material for binding the zinc to the steel surface. The purpose of the process is to apply zinc to steel, thus giving excellent *corrosion protection* of the underlying steel. Further information will be found in Part Two **Electroplating** and **Painting** (where the pigment would be zinc powder).

Cold Rolling

The application of pressure by rolls on strip material when cold.

Application range ALL METALS

It is the final forming operation and is comparable to *cold-drawing* of wire, and is similarly designed to give an improved surface finish rather than any increase in mechanical strength. Further information on *cold working* is given under *Work Hardening*.

Cold Welding

A term covering a form of *forge welding*, where no heating is applied.

Application range ALUMINIUM, LEAD, TIN, GOLD, SILVER, AND WROUGHT IRON

This applies to various types of relatively soft metals which have little tendency to *work harden*. Metals such as pure lead, gold, aluminium, etc. which do not *cold work* under any circumstances can be cold welded by the application of sufficient pressure. The result is a metallurgical joint which cannot be identified and, thus can be seen to be an 'ideal' join. Other materials, such as mild steel and a large range of soft materials including most of the non-heat-treatable aluminium alloys, tin alloys, zinc, etc., can be cold welded to a limited extent. The extent to which this can be achieved will depend on the amount of impurities or alloying elements present, causing the material in question to *cold work*. In order to produce a satisfactory cold weld, it is essential that the level of *cold work* applied is such that a successful weld is achieved before the degree of *work hardening* produces distorted grains, resulting in an appreciable increase in material hardness. Where the materials being welded have not the ability to absorb *cold work* without hardening, then it will be found that the interfaces, instead of forming a satisfactory weld, produce *cold-work* areas, and thus form pockets where satisfactory welding has not taken place.

Unfortunately, the process is of little use with the majority of engineering materials or components since when *cold working* is applied to them, poor welding results. When, however, a suitable material or component can be used, for example pure aluminium, lead or gold, the cold welding produces very high-quality joints, which, as stated above, cannot be identified and, thus, achieves a 'perfect join'.

Cold Working

A term applied to the working of material to cause distortion of the grain structure. This will almost always result in increased tensile strength and hardness, with a reduction in ductility.

Application range ALL METALS

Certain materials such as pure gold, aluminium and lead do not cold work at room temperature, which means that they can accept plastic deformation to an infinite degree without any increase in hardness.

The vast majority of metals, however, when plastically deformed, increase in hardness, and this may, on the one hand, be used to improve the mechanical properties of components, or on the other, may cause difficulties in regard to cold working. The material will increase in hardness, but ductility may be decreased to an unacceptable level. There are materials, notably cobalt alloys and certain copper and nickel alloys, which cannot be cold worked without fracturing as they do not plastically deform to any appreciable degree at room temperature.

To some extent cold working is an alternative to *hardening* and *tempering*, but it will be appreciated that the size, shape and form of the material exerts severe limitations on the application of this technique. With a suitable material, *surface hardening* can be achieved with controlled cold work.

As a general rule for steels, the greater the 'hardenability' by *heat treatment* techniques, the greater the hardness possible by cold working. (Further information is given under *Work Hardening*.)

Colour Anodize
The dyeing of the anodized film.

Application range ALUMINIUM AND ITS ALLOYS, AND TO A SMALL EXTENT TITANIUM

With aluminium the anodic process results in the formation of a tightly adherent oxide film on the metal. When freshly produced, this film is porous and acts as a mordant in a very similar manner to cloths such as cotton and wool.

The procedure used is that the components are *anodized*, using the *sulphuric acid process* and, on removal from the acid at the end of the cycle, are swilled in clean, cold water and then immediately immersed in the dye. The dye is absorbed into the porous anodic film and, on removal from the dye bath, is again swilled in cold water and immediately immersed in hot, almost boiling clean water. This heat immersion seals the anodic film and the surface is thus permanently coloured. Colour anodizing, then, is not an additive surface film, but a true dye; it can be removed only by removal of the actual metal surface.

By the correct choice of dye, and provided that *anodizing* is correctly carried out, an attractively coloured material, capable of withstanding normal abrasion and abuse (for example ashtrays, etc.) is produced. Using standard *stop-off* techniques, and repeated anodize and dyeing, several colours and intricate designs can be achieved.

It should be noted that the term *black anodize* may refer to normal anodize coloured black, or be the alternative name for hard anodize of aluminium.

Titanium does not have the range of colours which are possible with aluminium, and the metal is seldom or never used for decorative purposes except in jewellery.

Compression Test
A destructive, mechanical test, basically opposite to the *tensile test*. Metals will very seldom fail because of compressive loading, thus the test is not commonly used. Some information on the test is given in Part Two **Mechanical Testing**.

Composite Coatings
These are electro-deposited coatings where a metal such as cobalt or nickel is deposited electrolytically and co-deposited with the metal is some non-metallic or metallic oxide. Described fully in Part Two **Electroplating**.

Contact Tin Plating

Application range COPPER ALLOYS

This is an alternative name for immersion tin plating. The component is immersed in a hot chemical solution which is unstable and thus breaks down to produce metallic tin which then adheres to the surface being plated.

This method is commonly used in the printed-circuit and general electronics industries to increase the 'solderability' of components. It is also used to improve appearance, as an attractive matt-white deposit can be achieved. As the deposit obtained is very thin, the improvement obtained will be relatively slight.

Proprietary solutions are available from the plating supply houses, and these are recommended rather than 'do-it-yourself' solutions. For further details of this technique, see *Electroless Plating* and *Immersion Plating*.

Controlled Anneal

This is a name which can be applied to a number of processes, all of which are variations in the full annealing operation but where there is some variation in the rate of cooling from the annealing temperature. There are a number of proprietary processes which come under the term of controlled annealing. Described fully in Part Two **Heat Treatment**.

Controlled-atmosphere Furnace Brazing

A method of *brazing* which uses a furnace to supply the necessary heat and a controlled atmosphere to prevent oxidation of the component.

Application range STEEL, COPPER AND NICKEL ALLOYS

The technique of *hydrogen brazing* is in fact a form of *furnace brazing*. Reference to controlled atmosphere, however, is generally assumed to mean a reducing type of atmosphere. This will usually be of the hydrocarbon type, the atmosphere being controlled, generally by the *dew-point* method, to be neutral or slightly reducing.

With this technique, it is possible to carry out furnace brazing using fluxes which are relatively inactive, and under certain circumstances without flux. This means that the component must be thoroughly *cleaned* and free of oxide, prior to assembly.

Controlled-atmosphere furnace brazing can be used for any of the standard materials and the reader is referred to the section *Brazing* for further information. The use of inert gas such as argon or, under certain circumstances, helium, is normally referred to as *inert-atmosphere furnace brazing*.

Conversion Tin Plating
See *Contact Tin Plating*.

Copper Plating
Copper was one of the original metals to be deposited by electrolysis and has a history of usefulness to engineers. With the high cost of copper relative to other metals, many of these uses are being replaced by other techniques. Described fully in Part Two **Electroplating**.

'Corrodekote Test'
A form of *accelerated corrosion testing*, using a paste rather than the conventional spray.

Application range ALL METALS PLATED

This has the advantage that no special equipment is required, but the disadvantage that accurate reproduction of results is difficult or impossible, and it is time-consuming in labour. The paste uses kaolin, with copper nitrate, ferric chloride and ammonium chloride. It is applied to coat the articles, which are then allowed to dry before being placed in a humidity cabinet at 38°C and 90–95 per cent humidity. The time of the test will be specified. The coating is then washed off and the surface examined.

It is recommended that a series of tests are carried out and that this technique is not for investigative purposes, but as a test for routine control where results can be compared.

Corronizing
A specialist form of *electroplating* of nickel and tin.

Application range STEEL

The technique involves the *electrodeposition* of a coating of high-ductility nickel, of the order of 0.005 in (0.125 mm). On top of this nickel layer is then plated a layer of tin.

This bimetal deposit is heated in either a liquid, or preferably, an inert atmosphere to a temperature above 175°C, but below 230°C, the melting point of tin. This results in the tin diffusing into the nickel, giving a coating with excellent corrosion-resistant properties.

With the advent of co-deposition of nickel and tin, this process is no longer commonly used. Tin–nickel plating is described fully in Part Two **Electroplating**.

Corrosion Protection
The application of any technique to reduce surface corrosion.

Application range ALL METALS

There has been considerable research and development in this area, and it is now known that corrosion takes four forms. These are:

1. *Room-temperature oxidation* By far the most common corrosion is by oxidation. This applies to all metals, although it is much more obvious with mild and low-alloy steels. It is also a fact that oxidation corrosion will be accelerated to a dramatic degree by the presence of comparatively small amounts of contaminants, with moisture, chloride, sulphate and fluoride being high in the list of accelerating media. Quite a small increase in temperature (10–20°C) will also accelerate corrosion. Oxidation is in fact a form of chemical corrosion.

2. *High-temperature corrosion* This will almost invariably be oxidation corrosion, but again it will be accelerated by the presence of other elements or factors.

3. *Chemical corrosion* Here acids or alkalis are the predominant media causing the attack. In laboratories acids and alkalis are used to dissolve metals and this is a form of corrosion.

4. *Electrolytic corrosion* Here two metals in contact with each other have different electrode potentials. It is now realized that electrolytic corrosion is either the main reason, or a major contributory reason, for most of the corrosion of steel. It is also appreciated that the corrosion found on many metals can be attributed either wholly or in part to this electrolytic action. The dry battery or electrolytic cell where two different metals are connected via an electrolyte, resulting in the production of usable electric power, and the corrosion of one of the metals, is an example of this electrolytic corrosion.

Some measure of the tendency of a metal to sacrifice itself (corrode) to protect another metal in contact with it can be derived from the table of electrode potentials given. The 'noble' metals are most positive, the 'base' metals most negative. The more negative the metal, the greater its tendency to sacrifice itself to protect a more positive metal.

Metals	*Volts*
Gold	+1.5
Platinum	+1.2
Silver	+0.7999
Copper	+0.0337
Lead	−0.126
Tin	−0.136
Nickel	−0.250
Cadmium	−0.403
Iron	−0.440

Metals	Volts
Chromium	−0.71
Zinc	−0.763
Aluminium	−1.66
Magnesium	−2.37
Lithium	−3.02

Some confusion can be caused when the above table is used or consulted when *cathodic* or *galvanic* protection is discussed. In an electrolytic circuit it will always be the anode or positive element which will corrode and the cathode or negative element which is protected.

Confusion arose because the early electroplaters believed that the current passed from the anode to the cathode, as the metal appeared to be transferred in this manner, and it is always the cathode which is plated. It is now known that the current, in fact, flows out at the cathode. To prevent confusion, it is best to remember that the anode corrodes, the cathode is protected and the metals at the top of the table will always be protected by the metals below them sacrificing themselves. In this table, the farther apart the metals, the greater the corrosion potential of the lower metal, and the better the protection of the higher metal.

Included in this section is the fact that stray direct electric current can make one component anodic to another. This technique is commonly used to protect large installations – this is called 'impressed current galvanic protection'. When it occurs by accident serious and rapid corrosion will take place.

The four metals most commonly used in engineering are aluminium and its alloys, copper and its alloys, steel and zinc. In this section each of these metals will be discussed briefly, with information on protection against each of the four methods of corrosion.

Application range ALUMINIUM

To take aluminium first, it is seldom necessary to protect this from room-temperature oxidation as aluminium forms a natural adherent oxide. Long-term corrosion resistance can be achieved either by *chromating* or *anodizing*.

Chemical attack will generally be from alkaline or high-pH conditions. Unlike steel, aluminium can resist neutral or slightly acid attack better than it can resist alkali attack. Thus, any material with a pH value higher than 8 will tend to attack aluminium. The oxide formed by *anodizing* will successfully resist some alkali attack but will not resist severe attack. *Chromating* of aluminium will also assist in resisting this chemical attack but only to a limited extent.

High-temperature oxidation is not a serious problem with aluminium, as the higher the temperature the better the oxide which is produced; and aluminium will not normally be used under conditions where temperatures are considered high as the melting point of aluminium is below 600°C with most of the alloys used. The usable temperature, from the mechanical strength point of view, is very much lower than this level.

Electrolytic corrosion is the most serious problem concerning aluminium. Aluminium is low in the table of electrode potentials and thus, will tend to sacrifice itself to protect most other metals.

The methods of protecting aluminium from electrolytic corrosion are the usual methods, either removing the electrolyte or preventing the second metal from making electrical contact with the aluminium. In this field it is, therefore, fortunate that the oxide of aluminium, produced naturally and by the *anodizing* process, is a non-conductor. This, then, considerably aids the ability of aluminium to resist electrolytic attack, provided the contact pressures are sufficiently low not to break down the oxide film. Where any doubt exists, aluminium should be protected by *painting*.

Where aluminium is used in contact with steel, it is essential that either a good film of aluminium oxide is produced, or complete electrical insulation is achieved either by a paint film or plastic. Where vibration is involved, it will seldom be sufficient for the aluminium to be *anodized*.

Pure aluminium has good corrosion resistance over a wide range of conditions but low mechanical strength. All the aluminium alloys have lower corrosion resistance, with the manganese and magnesium aluminium alloys having the best ratio of corrosion and strength.

Application range COPPER

Second, there is copper and its alloys. There is seldom any serious problem of room temperature oxidation with copper and its alloys as the oxides or compounds formed are stable and relatively adhesive. For centuries now copper has been used in thin sheet form as an attractive roofing material, where the copper gradually turns green as it oxidizes and produces complex compounds on its surface. These seal the surface and prevent further corrosion occurring. Where electrical contacts are involved, the surfaces are commonly *silver plated*, and for *soldering* tin or solder is generally applied, either electrolytically or by hot-dipping. With tin containing alloys, known as bronze, there is some resistance to the production of these green colours and the bronzes retain their attractive brownish colour over a wide range of oxidizing and marine conditions.

Brasses, on the other hand, will generally darken and, under some conditions of oxidation, intergranular attack can occur. It is also common for zinc to be preferentially oxidized resulting in surface deterioration, and the formation of a red colour. This is called de-zincification.

For protection, therefore, it is only necessary to coat the surface of copper and its alloys with a clear lacquer to retain the colour. With brass, this is essential to prevent the possibility of intergranular oxidation.

Copper and its alloys can be used at relatively high temperatures and for many years were the sole materials which could be used under conditions of high temperature, for example, in fireboxes and associated components in steam engines. This applied to certain copper alloys, to pure copper and to low alloys of specific types; bronzes were found more useful than the brasses.

Regarding chemical attack, copper is excellent under corrosive conditions where alkalis are involved, and is satisfactory against sulphate, chloride and fluoride corrosion. It is unable to withstand attack by nitrates.

Electrolytically copper is fortunate in being relatively high in the table of electrode potentials. When in contact with steel, iron and steel will sacrifice themselves to protect copper. Thus, it is necessary to insulate copper and its alloys from iron and steel, not to protect the copper, but to protect the iron and steel. This is usually achieved – as is protection from high-temperature and chemical attack – by either *electroplating* or *painting*.

In general, however, because of the high cost of copper in modern industry, it will be found that either copper and its alloys may be used successfully without protection, or some alternative metal is used, which may then be protected from corrosion. Such an alternative might be 12 per cent chrome or austenitic stainless steel, copper nickel or nickel chrome alloys. Where a lower corrosion resistance is acceptable mild or low alloy steel plated with nickel, zinc or cadmium might be used.

Application range STEEL, IRON

The third metal is steel, which will be considered with cast iron. By far the most effort on corrosion protection is applied to steel in all its forms. It is difficult with steel to differentiate between oxidation and chemical and electrolytic corrosion, since in the vast majority of instances all three will be present to a greater or lesser extent.

Methods of protection are wide-ranging. Probably the simplest technique is to increase the alkalinity in contact with steel, which reduces the corrosion potential under normal atmospheric conditions. This can be said to be a form of protection by chemical action in that it has been found that acid conditions, even as slight as to a pH value 6, will result in some corrosion, whereas a pH of 8 and 10 will inhibit this corrosion. This is evident in the differing rates of steel corrosion found in the UK between soft- and hard-water areas.

The various techniques of corrosion protection appear under their respective headings in this book, and the Index lists the available methods. The different techniques take several forms, the simplest being the production of adherent oxides. The best known is '*Black Oxide*'. This will be successful, provided the oxide remains sealed to prevent attack of the underlying steel. Along the same lines is the *chromating* or *passivation* method, where the surface of the steel is given a film of iron chromate or iron oxide which absorbs some chromate. Slightly better than this is the technique of *phosphating*, which forms a measurable thickness of absorbent material on the surface of the steel. This has some corrosion resistance in its own right which is considerably enhanced, first, by *chromating*, and then either by sealing with oil or using one of the many paint systems.

Other methods of preventing oxidation are based on elimination of the atmosphere. The simplest method, of course, uses oil or grease. Some care is necessary to ensure that the component being oiled is relatively clean and that the oil or grease does not trap liquid or moist air in contact with the steel surface. This can result in severe corrosion locally by the formation of high humidity conditions under slight pressure, where the air and water are trapped by the oil or grease. There are now available some very sophisticated oils and greases which give corrosion resistance over a wide range of conditions, including outside storage over a period of time which can be measured in months. It will be obvious that the more severe the conditions, then the shorter the time for successful corrosion resistance. Unfortunately, one of the problems in this area is that when these methods are used to prevent corrosion, they also prevent the ready examination of the steel surface for slight evidence of corrosion.

A series of materials known as *vapour-phase inhibitors (VPI)* is now available which gives some measure of corrosion protection. All give off a vapour which inhibits corrosion, and share the characteristics of having low vapour pressure. Some raise the pH of the atmosphere to make it alkaline, thus reducing the corrosion potential of the steel, others reduce the corrosion potential of the atmosphere adjacent to the steel. They are most used in storage or transport situations where atmosphere change is limited and thus there is little danger of the vapour contact being dissipated or blown away. They are no use in outdoor conditions.

Where these relatively simple methods of corrosion prevention are shown to be unsuccessful, it will be necessary to make use of either *painting* or *plating* to eliminate atmospheric or chemical oxidation, or to consider a change in material. Where electrolytic corrosion is the major factor, it is possible to make use of electrodes to protect the steel. There are two basic types. First is the use of a metal lower in the table of electrode potentials than iron which will sacrifice itself to protect the steel. The most common method here is the use of zinc or aluminium. This 'sacrificial' corrosion protection can be applied either with *galvanizing, electroplating* or *aluminizing*. There are also available a number of paints which are rich in zinc or aluminium and therefore aid corrosion resistance, particularly if the paint film becomes damaged. Cadmium plating is also successful but generally is confined to steel in contact with aluminium.

Second, is the use of anodes – commonly zinc, aluminium and magnesium which are fitted in specially selected areas. These are designed to corrode while protecting the major component, and must be electrically connected to the protected area and an electrolyte must be present. The anodes are commonly used in situations where steel and copper are in close proximity and provided they are correctly applied, will protect the steel. Considerable care is necessary in choosing the correct metal, placing it in the correct location, with the correct ratio of anode to cathode size, and constantly monitoring the efficiency of the system. It is also possible to use inert anodes, generally titanium, which have an *impressed current* applied to ensure that they are always anodic to the areas being protected. Specialist advice must always be obtained before applying this type of system.

The most common method of electrolytically protecting steel is the use of zinc. This is plated, usually to a thickness of less than 0.001 in (0.025 mm). With most modern equipment, as soon as plating is completed, the component is *passivated* by *chromating*. It has been found that a properly applied chromate film will increase the corrosion resistance by a marked degree.

Cadmium plating is still used but to a much lesser extent, and is only technically required when the steel component will be in contact with aluminium.

Zinc and aluminium in contact results in the production of bulky unsightly corrosion products, which are absent, or present to a very much lesser extent with cadmium.

There is some indication that cadmium has slightly better corrosion resistance in marine atmospheres and has better 'solderability' than zinc, but the higher cost of cadmium compared to zinc means that this metal is seldom called for on modern components. It is now known that cadmium is extremely toxic.

Nickel plating is commonly used when the metal is required to operate in relatively severe conditions. Nickel has excellent corrosion resistance in its own right and it is this chemical resistance which is used. The deposit must be completely non-porous and in general requires to be at least 0.001 in (0.025 mm) in thickness. For purposes of appearance, the nickel very often has a bright deposit of chromium on the surface. (Further details of this and other metals used for corrosion protection will be found in Part Two **Electroplating**.)

Steel is also corrosion protected by hot-dip *galvanizing* and by the zinc coating process of *sherardizing*.

Painting is now used for the corrosion protection of the largest surface areas of steel. (Information on the various methods is given in Part Two **Painting**.) The purpose of the paint itself is to prevent the atmosphere contacting the steel substrate. As it is generally acknowledged that most engineering components at least will be damaged during service, it is essential that some positive *corrosion protection* is given, thus excluding the danger of corrosion creeping under the paint film from damage marks. This is generally achieved by *phosphating* or by the use of zinc-rich primers. (Information on these techniques is given in Part Two **Painting**.)

At higher temperatures, corrosion resistance by *electroplating* of zinc and cadmium is of a limited use, and while nickel gives some protection, it will not be satisfactory where cyclic heating is involved. There are certain paints with high temperature properties. They are based on glass frits.

The process of *aluminizing* is commonly used where temperatures up to 500°C or 600°C are used. Temperatures up to 700°C for short periods, can be used with correctly applied *aluminizing*.

The process of *chromizing* is similar in many ways to *aluminizing* but it is more expensive and, in addition to giving some hot oxidation resistance, successfully coats the surface of steel with chromium, giving corrosion resistance at least equal to that of 12 per cent chromium steel.

Above these temperatures it is generally necessary to consider using an alternative metal to mild or low alloy steel.

Application range ZINC

Finally, there is the protection of zinc alloy components. Zinc has excellent corrosion resistance in its own right but has a tendency to form a rather unsightly bulky white oxide. Since this does not have good adhesion to the zinc substrate it can be washed away, leaving the underlying zinc ready for further oxidation. This formation of the loose bulky deposit of zinc is not nearly as serious as that of the loose oxide on steel, and there is much less tendency to cause pitting.

The oxide formation can be prevented or its speed considerably reduced by *chromating*, and many zinc components are now used with the well-known yellowish green sheen given by the chromate process. This will considerably enhance the corrosion resistance of zinc, provided the chromate film is not damaged.

Zinc components can be *electroplated* but this requires some care. There are, however, a considerable number of zinc die-cast components which have been *nickel* or *chromium plated* to give the appearance of similar components in brass or steel.

Zinc can only be painted successfully when the surface is *etched* to give a key to the paint and special primers are used (see Part Two **Painting**). Provided these precautions are taken, excellent paint adhesion can be obtained on zinc.

There are probably more treatments designed to prevent metal corrosion than any other treatment of metals. Many of these treatments can be very successful, provided they are correctly applied, and almost without exception the cost of repairing and salvaging the effects of faulty corrosion protection will be much higher than the cost of correct application in the first place.

Coslettizing

The name given to a form of *phosphating* of steel. This uses a relatively simple solution, with the purpose of forming iron phosphate on the surface of the component.

Crack Opening Displacement (COD)
A technique designed to evaluate the ability of a component to resist crack propagation. This uses a carefully produced fatigue crack to evaluate the ability of the material to resist propagation.

The procedure can supply considerable useful information regarding the properties of the test piece relating to impact and ductility. It must be appreciated, however, that the results obtained will vary considerably depending on the metallurgical structure of the test specimen at the nose of the fatigue crack. In the opinion of the writer, this test is of considerable value at the design stage in evaluating the crack propagation properties of a specific material in the form of a test piece produced under highly controlled conditions. It is, however, doubtful if the technique is economical when applied to production quality control. The information obtained more economically from routine hardness, impact and tensile tests, in my opinion, is more relevant and meaningful.

The technique of crack opening displacement (COD) or full bend test is a recently developed technique, and readers are advised to evaluate it, and its economical usefulness, before making use of this technique. Described fully in Part Two **Mechanical Testing**. Also called crack tip opening displacement (CTOD).

Crack Testing
An inspection technique used to find material defects, generally, but not always, on the surface. Described fully under *Crack-detection Testing* in Part Two **Non-destructive Testing**.

Creep Test
A destructive tensile type test to assess the hot strength of metal over a period of time.

It is designed to identify the ability of the metal to withstand stretching (creeping) under its own weight over periods of time at different temperatures. Described fully in Part Two **Mechanical Testing**.

Cromalin
A proprietary process used for the *electrodeposition* of metals.

Application range ALUMINIUM AND ALUMINIUM ALLOYS

The general technique employed at present for *electroplating* aluminium makes use of an *electroless* deposit of zinc, and it is the zinc coating which is then electroplated in the normal manner with, initially, the deposition of a copper layer. This is generally referred to as the *zincate process*.

Cromcote
The proprietary name given to *chromating* processes supplied by Walterising Co. (UK) Ltd.

Cromodizing
The name given to *chromating* whereby a film of iron chromate is formed on the surface.

Application range STEEL

This gives the surface some *corrosion protection*, but this is of a relatively low order. *Phosphating, 'Black Oxide'* and even oiling will probably give superior corrosion resistance. Further information on the technique is given under *passivation*.

This name should not be confused with chromizing which gives excellent corrosion resistance to steel, but is an entirely different process.

'Cromylite'
A proprietary chromium plating process using a mixed 'catalyst' to give an even deposit. Further information can be obtained from Oxy Metal Industries (GB) Ltd.

'Cronak'
A proprietary form of chromate treatment.

Application range ZINC AND ZINC ALLOYS

With this technique, the parts are *cleaned* and immersed in a hot solution of acid dichromate in order to *etch* the zinc metal. Further information is given under *Chromate Treatment*; see also *Passivation*.

CTOD
Described fully under *Crack Tip Opening Displacement* in Part Two **Mechanical Testing**.

Cupping Test
A form of ductility test, which is applied only to sheet-metal.

Application range ALL METALS

The test economically examines the ductility of the material with particular reference to its ability to accept *work hardening* and, thus, its capability for *deep-drawing*. An alternative name is the *Erichsen Test* and details of the technique will be found under that heading.

The *tensile test* giving *ultimate, yield* (Proof), *elongation* and *reduction in area* is also an alternative, but very much more expensive.

'Cuprobond'
A technique for applying a thin coating of copper on wire.

Application range STEEL

This is a non-electrolytic process, a form of *electroless plating*. The purpose is to supply a lubricant prior to the final drawing of wire and at the same time produce an attractive copper coloured finish for decorative purposes. Further information on the Cuprobond process can be supplied by Pyrene Chemical Services Ltd.

It should be noted that the copper, while attractive in colour, will not supply any corrosion protection. Any break in the film will allow a cell to be produced when moisture is present. This cell will have the steel as the anode attempting to protect the copper surface. There will thus be a tendency for corrosion to be accentuated on the steel. Further information on this will be found in the section on *Cathodic Protection*.

'Cuprodine'
A proprietary name for an *electroless plating* process to produce a thin film of copper on wire prior to the final wire drawing.

Application range STEEL

Detailed information on the Cuprodine process can be obtained from ICI Ltd.

The same comments apply regarding corrosion resistance as those given under Cuprobond.

CVD
See *Chemical Vapour Deposition*.

Cyanide Hardening
A *surface hardening* method which uses cyanide salts to give a case containing carbon and nitrogen. Described fully in Part Two **Heat Treatment**.

Cyclic Anneal
A specialized *annealing* procedure for higher carbon and alloy steels, and applied only to steel. Described fully in Part Two **Heat Treatment**.

'Dalic Plating'
This is a trade name for an electroplating process using portable equipment. For further information see *Selective Plating*.

Damascening Process

A method of ornamentation which obtained its name in the Middle Ages from the swords produced in Damascus decorated by this process.

Application range IRON AND STEEL

Basically the method is to draw out rods or wires of iron and steel and to intertwine these in such a manner that a product is produced with a typical snake-like design. This requires considerable skill in the manipulation of the iron and steel rods as during this initial stage there is no obvious difference between them. These may then be beaten or rolled to form a solid, apparently homogeneous, article. This is then treated in acid, which attacks the steel component at a faster rate than the relatively pure iron, and at the same time will result in an etching or darkening of the steel. The final result, therefore, will be the relatively bright iron in the unetched or polished condition and the darker roughened surface of the steel. Designs using this process can be inlaid with other materials, such as brass or copper, to give further contrast and more intricate patterns.

The process is also known as *kuft work* and has been much copied by other techniques, such as *painting*, *enamelling* and *acid etching* of bi-metal sheet, or metal-coated plastic. Modern craft-workers in metal still use damascening for the production of high-quality jewellery and ornaments.

Deburring

A procedure which covers many different types of finishing, where the burrs produced by any type of machining operation are removed.

Application range ALL METALS

There are five methods commonly employed in burr removal, all of which can, in theory at least, be used on all materials but, in fact, are generally restricted to the materials commonly used in engineering. These are:

1. **Hand deburring** This uses the normal type of equipment, including files, rotary tools, sandpaper, emery paper and similar techniques. Each component is individually handled and the skill of the operator is the deciding factor in the quality of the end product. With this technique, in addition to removing burrs, it is possible by operator training to ensure that an adequate radius is produced.

2. **Mechanical deburring** This uses the range of equipment in a normal polishing shop, including rotary and flat grinding, wire brushing, and polishing mops of various types. In this case, while the skill of the operator is still of paramount importance, much of the effort is transferred to the machine. Again

this technique can be used to ensure that an adequate radius is produced in addition to the removal of all burrs.

3. Barrel deburring This uses the rumbling or tumbling abrasive effect achieved by rotating or shaking the components in contact with abrasive chips, abrasive dust and water. This method can be used to produce a satisfactory homogeneous finish on all available surfaces. The technique is now highly developed and is discussed under Barrelling.

4. Abrasive blasting Using one of the several types of blasting techniques now available, it is possible to successfully deburr components. This will not normally achieve any form of radius but will usually remove burrs in a technically efficient manner. Blasting can be used to alter the surface finish on the component and a controlled, non-directional type of finish is possible. The techniques are described fully in Part Two **Blasting**.

5. Chemical deburring Using carefully chosen chemicals, and in some cases, electrolytic action, it is possible to selectively attack the burrs on components. This method in general relies on the fact that burrs will, first, be more *work hardened* than other adjacent areas of the same metal, and this generally means that they will be chemically attacked more rapidly than materials with a lower standard of stress. Second, the burrs will usually occur in areas where the chemicals, particularly when electrolytic action is used, will be able to concentrate. Third, chemical attack will often concentrate naturally at corners, and this effect will be accentuated by the electrolytic attack. (See also *Chemical Polishing*.)

Deburring using mechanical aids, while efficient technically, is open to considerable variation. To some extent, the same applies to the use of *blasting* techniques where, unless the burr is standardized, which is seldom economically possible, the time will be variable and, thus, inadequate deburring may occur under certain circumstances.

With the use of *barrelling* techniques, it is possible to lay down control parameters covering batch production, which will cope with a variety of burr types and sizes and result in a reasonably controlled end-product. This technique, with the vast majority of engineering batch-produced components, will give the most economical and technically satisfactory method.

Chemical deburring in general can be applied only to specific components in certain materials but, given the correct combination of circumstances, and particularly if subsequent processing is involved, it can be the best method technically and economically.

Deburring, then, can be successfully used to remove burrs and can sometimes result in the formation of a better radius. Further information on this technique is given under *Electrolytic Polishing* and *Chemical Polishing*.

Decarburizing

The removal of carbon, partially or totally, from the surface.

Application range STEEL

The process is the exact opposite of *carburizing* and results in a surface with a lower carbon content than the remainder of the component. There are very few occasions when this is a desirable condition, but there are occasional requirements where a surface of softer, more ductile condition than the core is required.

The process can also be used as a recovery or salvage operation where, for some reason, the surface of a component has more carbon than is desirable. This could be because of faulty control during *carburizing* or other *heat-treatment* processes, for example the use of a carburizing salt bath in place of a neutral salt bath, resulting in a high-carbon skin. Alternatively the defect could be caused by an undesirably high carbon gaseous atmosphere in normal electric or gas furnaces.

Before the decarburizing process can be correctly carried out under these circumstances, it is necessary to identify the quantity of excess carbon which exists. This is essential in order to control the process to ensure that the resultant surface does not suffer from excess decarburization. Where production parts are decarburized for design reasons, and not as a result of excessive *carburization*, then the investigation will not normally be required, but will be controlled by a written procedure produced after controlled tests.

The decarburizing process itself is carried out in a similar manner to *carburizing*, with the atmosphere in contact with the components being treated as oxidizing. In order to ensure that the carbon removed from the surface of the component is fixed, and thus does not return to the component, the process will be carried out in a gas furnace, or, with pack-decarburizing. The material used must be carefully chosen such that the carbon is combined and thus fixed in the material in contact with the component when it is removed from the surface.

The process is carried out above 900°C in a gas atmosphere in equipment very similar or identical to the *gas carburizing* furnace. Gas carburizers are available which can be used to control the degree of decarburization. They operate by altering the carbon potential of the atmosphere. This can be controlled automatically by various means one of which is to monitor the electrical resistance of a pure iron wire continuously passing through the furnace. The resistance will vary with the *carburizing* and can also be used to alter the carburizing potential of the gases.

Continuously monitoring infra-red analysers can also be used to control the carbon potential, but normally this technique will be found only on equipment designed to *carburize*, not for controlling decarburization.

This process is relatively unusual and should not be attempted without considerable investigation regarding the capabilities of equipment and personnel involved. It can be used as a method of *softening*, but as the surface only will be affected, it cannot legitimately be compared with normal *annealing* operations (see Part Two **Heat Treatment**).

The process of *Malleabilizing* cast iron is a decarburizing process, where the carbon is extracted throughout the section of the component.

The term 'decarburized' is much more commonly applied to the defect of accidental loss of surface carbon than to any process to control on a routine basis the carbon content of steel.

Deep Anodize

An alternative name to *hard* or *refrigerated anodizing*, using sulphuric acid at a low temperature; further details are given under *Anodizing*.

Deep-drawing

A metallurgical production process applied only to those metals such as sheet metal in mild steel, copper, brass, aluminium and other ductile metal resulting in considerable quantities of *cold work*. The result will always be a vessel of appreciable depth.

To some extent this is a relative term, for example, a component in brass or aluminium can be considerably more deep-drawn than any steel component, that is steel in general cannot produce as deep a vessel as materials such as pure aluminium or certain forms of brass.

Degreasing

This is part of the general *cleaning* process and is described in some detail under that heading in this book.

Application range ALL METALS

Technically and simply, it is that part of the *cleaning* process which removes grease, and a common form of degreasing uses hot chlorinated solvents. Very often this is in the form of *vapour degreasing*.

Demagnetization

An operation to remove all directional magnetism from steel components.

Application range STEEL

Demagnetization is carried out by passing the component or assembly through a solenoid-type coil carrying an alternating current.

It is necessary to demagnetize for four reasons:

1. As part of the *magnetic crack test*, where between operations it is essential that all magnetic force is removed prior to remagnetizing in another direction.

Unless this is carried out, it is possible that the remagnetization will result in lesser magnetic forces than are desirable and, thus, some sensitivity will be lost regarding crack detection.

2. To ensure that all magnetic forces are removed from components which will be in service where these magnetic forces may cause undesirable reactions with electronic or other types of equipment. This is the most common reason for demagnetization and its importance is obvious, as electronic equipment is relatively widespread in aircraft, ships, defence equipment, etc.

3. When components are magnetic – however slightly – this will result in the attraction and adhesion of minute metallic particles. These particles can result in excessive abrasion and wear. Where components must be maintained at a high standard of cleanliness, demagnetization is always important.

4. To remove magnetic forces which interfere with welding. When present in sufficient strength magnetic flux will cause the electric arc to wander giving porosity problems.

There are several methods which can result in components being magnetized. The obvious method is by using *magnetic crack test* equipment, where components are held either on purpose or accidentally, in an electric field. This will result in some directional magnetism being induced and the degree of induction will depend on a large number of variables.

In addition to this method, it should be realized that vibration of steel which is held or placed within the earth's magnetic field, that is, the north–south direction, will also result in magnetization. This means that any static engine which has a vibration will result in certain components gradually becoming magnetic. This also applies to components being machined, for example where there is long-cycle machining, causing vibrations in the components being manufactured. It is, therefore, essential that with components where magnetism must be absent, these should be demagnetized after all machining. There are also components which require demagnetization at intervals during their life cycle.

'Deoxidine'

A proprietary name for a series of solutions which are de-rusters and also inhibit the metal surface generally by producing a phosphate coating.

Application range STEEL AND ALUMINIUM

The same solutions are used as etchants for aluminium to produce a satisfactory surface prior to *painting* or the '*Alochrom*' process.

The solutions can be used neat or diluted for dip, immersion, brushing or spraying. It is preferable that they be washed and dried by oven, hot air or hot water, but in general the importance of removing these solutions is not as critical as with normal de-rusting solutions since they are based on phosphoric acid, which is not a serious corrosive agent.

Further information will be found under *Cleaning*. Detailed information can be obtained from ICI Ltd.

Depletion Gilding
A method of gold finishing whereby the component has all the metals apart from gold removed from the surface by controlled acid etching.

Application range GOLD ALLOYS

This technique was practised by some ancient civilizations and is no longer used.

Descaling
The term covers any method by which scale or surface oxide is removed.

Application range ALL METALS

The scale will generally be produced during the manufacturing process or in service and may either be of the type which is visually extremely obvious in the form of rust or millscale, or visually unobtrusive.

The various methods of descaling are *blasting*, *pickling*, either *acid* or *alkaline*, *sodium hydride treatment*, *polishing* and *barrelling*. Each of these processes is described under the relevant headings in this book.

Dew-point Control
A method applied to certain processes where the composition of a hot gas requires control. It is also a means of controlling humidity, and thus where corrosion conditions are strictly monitored, the dew point is recorded.

Application range ALL METALS

The dew point is the temperature at which water vapour in the atmosphere precipitates, and, thus, dew begins to form.

As the gas involved – whether air or a controlled atmosphere containing moisture – is cooled, the dew point is noted with the visible presence of droplets of moisture, and is recorded in degrees centigrade.

The method is of particular use in controlled atmospheres during any *heat treatment* which is designed to prevent oxidation. These atmospheres will invariably contain water vapour and it is the ratio of water vapour to other constituents in the gas which very often decides whether or not the atmosphere is oxidizing.

This form of control, therefore, has use in certain metallurgical processes. It requires special equipment and will often be included as part of the control equipment of the furnaces.

'Diabor'
A proprietary name for the process of *boriding*.

Application range STEEL

Described under *Boriding*; details of 'Diabor' can be obtained from ICI Plc.

Diamond Pyramid Hardness Test
A *hardness test*, using a diamond-shaped indenter (usually made from a diamond).

The load is commonly applied via levers, and the resultant impression measured – the larger the mark, the softer is the material. The test is also known simply as the *'Pyramid hardness test'*. Described fully in Part Two **Hardness Testing** under *Pyramid Diamond Test*.

Die-casting
A method of casting which, briefly, covers the process where molten metal is poured, sometimes under pressure (injection or pressure die-casting) into a mould or die.

Application range ALL METALS WHICH CAN BE CAST

The die will be made of metal and immediately on solidification will open to eject the casting. It is only of use on long production runs but can produce castings of a high standard of integrity and surface finish. It is a production process and readers are advised to obtain specialized technical information if necessary.

Die Welding
A form of forge *welding* generally confined to sheet metal.

Application range STEEL

The process achieves overlapping of the metal sheets, using heated dies which hold the sheets together during *welding*. This is, therefore, a form of *forge welding*, where the heat to raise the metal temperature is by conduction from the dies. The necessary pressure to cause the weld is applied by conventional means to the heated dies. Described fully under *Forge Welding* in Part Two **Welding**.

Differential Heating
A specialized form of *heat treatment* in which one portion of the component is heated either at a different rate, or to a different temperature, and where the rate of cooling following the heating will generally control the type of treatment

being applied. This specialist procedure will usually require an individual specification being written for each component being treated. It can cover items such as *hardening*, *normalising* and *induction hardening* which will be found in Part Two **Heat Treatment**.

Diffusion Bonding

A method of joining similar, or specifically different, metals, generally using pressure.

Application range ALL METALS

The materials to be bonded are thoroughly *cleaned*, then have the surface oxide removed either by chemical or mechanical means. To ensure a good bond, the surfaces to be joined must have a reasonably good surface finish, the standard of which will depend on the metal being bonded. The surfaces are brought into contact with each other and pressure applied. This will either be static or rolling, depending on the components and materials involved.

With materials such as pure lead, gold, tin, etc., diffusion bonding takes place at room temperature, at atmospheric pressure, provided the surfaces are clean and oxide-free, and sufficient time is allowed. By increasing the temperature, the bond will be achieved in a shorter time. The temperature involved will be relatively low with most metals as the formation of an oxide layer inhibits the bonding. The process is only applicable to metals which can be cold worked without becoming too brittle. It will be appreciated that metals which resist oxidation – such as gold – will bond more readily and can use higher temperatures than readily oxidized metals.

Diffusion bonding results in a form of *welding*, covering *cold welding* and *pressure welding*, and can also be applied to any *forge welding* operation. It is, however, generally reserved for operations where either static or rolling pressure is involved.

The resultant bond, when applied to identical metals, cannot be identified by any metallurgical method – in other words, it is an 'ideal joint'. When dissimilar metals are joined, the result is a homogeneous, diffused merge of the two metals involved. It will, therefore, be seen that this process produces a high-standard weld, but the technique can only be applied to components of relatively simple shape. The process of *cladding* sheet by rolling is a variation of diffusion bonding.

Dinanderie

A form of metal craftsmanship which takes its name from an area in Belgium.

Application range COPPER, BRASS AND BRONZE

It consists of decorating the metal by working with specially shaped hammers and anvils, resulting in a raising of the metal surface, which is then decorated. The name applies more to the type of design rather than to any specific type of metal working.

Dip Brazing

A form of *brazing* where the component or assembly is immersed in a molten bath. The liquid may be molten salts but, under certain circumstances, can be molten metal, or some other liquid such as a high temperature heat transfer liquid may also be used. The process is described in more detail under *Brazing*.

Dip Moulding

A technique where the component is dipped into liquid plastic. On removal either by the action of air, with air-curing plastics, but more generally, by heating, the plastic solidifies to give a plastic coating to the component. There are a large number of variations of this basic technique and many purposes are served by dip moulding.

Application range ALL METALS

Many high-quality components, particularly tooling, are protected during transit and storage by a thin, tough plastic coating which is applied by dipping the tooling and then allowing the solvent to evaporate. In this case, the plastic material generally has an additive oil which gives protection from corrosion and at the same time prevents the plastic material from adhering to the metal. When required, the material is cut with a knife and removed.

Other forms of dip moulding use more conventional plastics and may require that good adhesion is necessary between the metal and the plastic. Depending on the quality of this adhesion and the plastic chosen, it may be necessary either to blast or etch the surface and to use a special primer.

Dip moulding has many similarities to the *dry-powder painting* technique, especially where the heated components are dipped into a fluidized bath of plastic powder. This same technique of heating the components prior to dipping in the molten plastic, can achieve excellent results regarding control of thickness and adhesion.

Dip moulding can also be used with processes where a pattern or series of patterns is used, these being dipped to give the required thickness of plastic, and the finished article after curing is stripped from the mould. These will be hollow plastic components. The thickness of the coating can be controlled either with the viscosity of the plastic, the heat of the component dipped in the plastic, or by application of several coatings.

Dip Soldering
A method of joining; it is similar to *dip brazing*. (See *Soldering*.)

Dispersion Hardening
An alternative name for *age hardening* or *precipitation hardening*. Described fully under *Precipitation Hardening* in Part Two **Heat Treatment**.

Dot Welding
A *fusion welding* process generally for salvage purposes, which can also be used for joining.

Application range STEEL

It uses a pistol type of equipment which controls the feed of the electrode, the pulse of the relatively low current required for the weld 'dot' or 'spot' and also the subsequent air supply which rapidly cools the molten weld. Advances in the use of standard *electric welding* equipment has made this specialized technique unusual with modern equipment. The standard achieved is very dependent on the skill of the operator.

It is, in effect, a *spot* or *tack weld* process. Described fully under these headings in Part Two **Welding**.

Double Refining
A term occasionally applied to the two *hardening* operations which follow *carburizing* in high-quality *case hardening*.

The first operation *refines* (hardens) the core of the component, and this is followed by a second operation to harden the case. This second operation results in the core being *tempered*. In modern industry it is more common to use the terms 'refine' and 'harden' rather than 'double refine'. Further details of these techniques are given in Part Two **Heat Treatment**.

Double Temper
A *heat-treatment* operation which can act to give 'secondary hardening' (described fully in Part Two **Heat Treatment**). There are theories that a two stage temper will give better dimensional stability than a single temper. The second temper can be below, the same, or slightly above the temperature of the first temper.

Drawing
This process has two distinct meanings in the metallurgical context. The term is still correctly used in the modern sense, where wire is drawn to a smaller

diameter. Wire-drawing takes large-diameter wire and passes it through a die to produce a wire of smaller diameter. Additionally the term is used, although it is relatively old-fashioned now, as an alternative to *Tempering* (described fully in Part Two **Heat Treatment**). Here the term 'draw back' means to reduce the hardness.

Drawing-back Process

An old-fashioned name for the *heat treatment* which reduces hardness and increasing ductility. This is now referred to as *Tempering*. Described fully under this heading in Part Two **Heat Treatment**.

Drifting Test

An inspection process to measure the ductility of the material.

Application range MILD STEEL, COPPER, ALUMINIUM AND THEIR ALLOYS

The test consists of punching sheet metal with a punch or drift of predetermined size. The amount or size of the 'cup' produced is a measure of the ductility of the sheet metal involved.

The test is comparable to the *cupping test* and is similarly subject to individual interpretation regarding details of the test and results obtained.

The test is also sometimes applied to a similar ductility test for tube or pipe. With this, a punch or 'drift' is used to expand the mouth of the pipe. The amount of expansion which must be achieved without cracking or bursting will be stated as part of the specification, as will the angle of the 'bell mouth'.

This is an economical, simple, destructive test to give an indication of ductility. Tensile testing would give similar plus additional information at much greater cost.

'Dri-loc'

A proprietary process and a variation on the '*Loctite*' process, where liquid resins harden in the absence of air and are, thus, used as a method of retaining nuts on threaded studs, or in other forms of glue jointing.

Application range ALL METALS

With the 'Dri-Loc' process, the resin is encapsulated in microdrop form. This means that minute drops of the '*Loctite*' liquid are surrounded by a second resin which prevents the droplet being affected.

The result is a liquid which can be applied to threaded components when the liquid on the outer surface of the microdroplet will dry to give a form of lacquer which is dry to the touch. When the screw is assembled, the action of the thread being twisted inside the nut results in this touch-dry, encapsulating material fracturing, thus releasing the resin which then acts in the same manner as a glue holding the nut in position.

This has the advantage over the normal '*Loctite*' process and other forms of gluing that the components are already pretreated and there are, thus, no problems regarding correct application, which under many circumstances can be in difficult conditions. The lacquer can always be applied under the desired conditions. Further information can be obtained from the Loctite Co.

Drop Forging

A manufacturing *forging* process and, therefore, outside the remit of this book. Briefly, the bottom half of the die is static and the component is forged when it is forced into the die by the hammer dropping. Alternatively, force can be applied by the moving section itself being a die and, thus, imparting form to the component. The term drop forging is also used where anvil and hammer are involved and the desired shape is produced by the skill of the operator.

Drop Stamping

A manufacturing process similar in many respects to *drop forging*; the user is advised to obtain further information from a specialist.

Drop Test

A quality-control operation, usually applied to finished components or assemblies.

Application range ALL METALS

With one technique, a known weight is dropped from a standard height on to the test piece or component, and the result is noted. Generally no cracking or fracture will be accepted and the maximum degree of deformation is specified.

More generally, with modern equipment, the component or assembly itself is dropped, and the test is required to show any weakness identified by subsequent inspection.

With wheel-type components, the drop is carried out twice, with the wheel turned 90 degrees between the two tests, and the subsequent examination carried out for distortion as well as cracking.

With moderate electronic equipment, the drop test is carried out to find any mechanical weakness and to identify any component which has poor electrical conductivity or broken connection. The height of the drop, the number of tests and the angle or angles of impact will all require to be specified, together with the subsequent electrical and mechanical details.

As stated, the drop test should always be a quality-control operation. It is not an inspection method, but should be carried out in random samples at a rate agreed.

There are now available a number of detailed specifications covering the drop and bump testing of electronic equipment, and users are advised to contact the British Standards Institution or the American Society for Testing and Materials for further information.

Dry Blasting

The general name given to any form of *blasting* where the abrasive grit is dry; the term should be qualified with the type of grit or shot used. Described fully in Part Two **Blasting**.

Dry Cyaniding

A process of locally *hardening* the surface.

Application range STEEL

The component is heated either in a furnace, or by flame, to above red heat, then immersed in a dry, cyanide-rich powder. The component should be rubbed in the powder, when it will be found that some of the powder fuses and adheres to the surface. The component is then removed and the process repeated.

The process is essentially identical to that described under *cyanide hardening* but can be carried out on site and does not require the relatively sophisticated equipment of the latter process.

It must be emphasized that the salts or powders used are rich in cyanide and are, thus, extremely poisonous. The gases given off contain hydrocyanic acid, and are also poisonous. The water used to wash the components on completion of the process will be high in cyanide and will be unacceptable for disposal to local authority drains. It will, therefore, be seen that this local hardening method is not to be attempted except under highly controlled conditions.

This process is generally confined either to treatment of individual components or very small-batch production, or to local areas which require to be hardened. It is difficult, if not impossible, to achieve accurate control of case depth, and generally temperature control is non-existent or of a low standard.

The term 'dry cyanide' can also be used to refer to the gas process of *carbonitriding* as a means of distinguishing this from the normal cyanide process which uses molten salt.

Dry-drawing

A finishing technique where wire is drawn through a solid or semi solid lubricant. This results in a high-quality finish, which is generally bright. The name is used to distinguish it from the *wet-drawing* process where the wire is drawn through a liquid.

It is a form of cold drawing.

Dry-film Lubrication

A relatively modern technique – a form of *painting* – which became available with the advent of high-strength paint films with excellent adhesion.

Application range STEEL

The term can also be applied to certain *surface* treatments such as *phosphate, borax, 'Sulphinuz'* and *'Tufftride'*. These coatings are generally applied only to steel.

The purpose of dry-film lubrication is to coat the surface of components subjected to wear or light abrasion where standard lubricating techniques for some reason are not possible. There are two basic types of paint both using the same vehicle, generally some form of phenolic resin. These are colloidal graphite and molybdenum disulphide. These materials are produced as extremely fine powders which are then used in the same manner as a pigment with the resin being used as the vehicle.

The method of application is that the components are first thoroughly *cleaned* to a high standard and are then subjected to carefully controlled *blasting* using a fine grit in the range of 120–220 holes per inch mesh size. Immediately afterwards, the paint is applied in the form of an extremely thin, even coating. This must be by the standard method of *spray-painting* with considerable skill on the part of the operator where high-quality dry-film lubrication is required.

When the technique is successfully applied, the film will have excellent adhesion and will withstand considerable wear, provided this is within the hardness range of the paint film. If abrasive particles or pressure using a harder material than the paint film are applied, the film will be cut. Because the pigment, either graphite or molybdenum disulphide, is a self-lubricating material, disastrous breakdown may be delayed but further deterioration will not be prevented. Provided the loading applied does not result in cutting, extremely long life can be obtained using this technique.

The components to which dry-film lubrication is applied include items such as ball or roller bearings used in corrosive atmospheres or temperatures above that at which normal oil or grease operate. Provided the loads applied are within the limits stated above, then long service can be obtained. Other instances are moving components which cannot be lubricated for any reason such as the danger of contamination in food handling equipment.

For room temperature purposes, the graphite impregnation is probably superior, with the molybdenum disulphide being more useful at the higher temperature range. There is considerable overlap and the user is advised to obtain specialist advice in this area. As with other processes, it has been shown conclusively that an incorrect treatment here will result in early failure. Dry-film lubrication to some extent is competing with the '*Sulphinuz*' and '*Tuff-tride*' processes, but is probably used at a lower temperature than either of these treatments. *Phosphating*, particularly when lubricated, is also a competing process.

There are a number of proprietary processes available.

Dry-film Painting

A method of coating articles with powder. It is an alternative name for *Dry Powder Painting*. Described fully under this heading in Part Two **Painting**.

Drying

Drying means the removal of moisture. Various standards of drying are used either regarding the extent of water removal, or the standard of cleanliness which remains following the operation.

Application range ALL METALS

Water boils at 100°C and its vapour pressure at lower temperatures is relatively low. Thus, components raised above 100°C will dry, because any moisture in contact with the surface will evaporate. It must be noted, however, that closed containers do not dry as on cooling any water vapour trapped within a chamber will condense and reappear as moisture. This can also occur on hollow, cup-shaped articles which are incorrectly loaded during the drying operation.

It should also be realized that with components placed in a closed oven at temperatures slightly above 100°C, any water will be converted into steam or water vapour and unless this is forced out of the oven, some condensation will result on removal and cooling. Thus, where drying is important, the oven should have ventilation and preferably a fan.

Other methods of drying are carried out by warming in absorbent material such as sawdust. This can be extremely efficient but care is required that sawdust does not block oil passages or become trapped in the mechanism of assemblies.

Drying by immersing in hot liquid is extremely efficient regarding the removal of moisture but will, of course, result in the component being contaminated with the liquid. It is also essential that the liquid and moisture are not soluble as this results in the surface being contaminated with the liquid which will contain some water.

Drying is also possible by the use of boiling water. Provided the water is actually boiling, the components being dried will rapidly attain the temperature of the water, and on removal, the surface water immediately evaporates. It is obvious that this method is possible only where the components have, first, a reasonable mass compared to the surface and, second, where they are of a relatively simple shape with no pockets which will trap water.

One of the problems of drying using either water or hot air is that water evaporates from the components and leaves behind any solids which are in solution in the water. For most engineering purposes, this very slight surface contamination is of no importance.

With many parts however, a water-free, chemically clean surface is essential. Use should be made of modified chlorinated solvent degreasing equipment, such as '*Trisec*', which is a proprietary method (discussed in this book under its own heading), or water-repellent liquids which have recently been developed. These liquids are completely incompatible with water and, when any component containing water is immersed, the water is immediately removed and falls to the bottom of the vessel. On removal from the vessel, the water-repellent liquid evaporates, leaving a surface free of contamination. Drying is discussed further under *Cleaning*.

Ductility Measurements

Ductility is the mechanical property showing the ability to accept plastic deformation without cracking or fracture.

Application range ALL METALS

A large percentage of the research and development carried out on metals is searching for an increase in ductility, or retention of the available ductility while improving other mechanical properties, such as the tensile strength. Almost invariably as tensile strength increases, so ductility decreases. The importance of ductility in engineering components is their need to withstand slight amounts of stretching or deformation without cracking, and this means that when any material is taken outside its yield point or plastic limit, it will not immediately crack or fracture, but that some stretching will take place.

Many manufacturing processes also rely on the fact that metals are ductile, and the choice of a metal may in fact be based on the fact that it can be shaped by working. There are many materials which would find considerably more use in engineering if they possessed measurable ductility.

The practical method of assessing ductility is to make use of the tensile test piece by measuring the 'elongation' or the reduction in area. These tests show the amount of stretching, or plastic deformation between the elastic limit and fracture. This test is expensive and for rapid assessment or ductility a variety of other tests based on bending have been used. When any metal is bent to cause permanent deflection, then the outer surface of the bend has been plastically deformed. Depending on the type of test, the shape of the test piece, the radius applied to the edges of the test piece and, in some instances, the rate of deformation, information can be made available regarding the ductility of the material. There is no doubt that while *bend tests* are extremely useful, the results obtained should be assessed with some caution. There are no circumstances in engineering where *bend testing* will exactly reproduce the stresses applied. Other forms of ductility measurements are the *wrapping test* and *cupping test*. These ductility tests are usually of service for sheet metal which will be used for *cold-forming* operations. (The methods of assessing ductility and the *bend tests* are described fully in Part Two *Mechanical Testing*.)

'Durboride'
This is a proprietary process, several varieties of which exist, where boron diffuses into the surface producing very hard borides with excellent wear resistance.

Further information from Degussa Ltd.

'Durferrit'
This is a trade name covering a number of salt bath processes which can be used for case hardening steels.

There are various types available identified by specific number references. Some of these contain cyanide but there are cyanide-free salt baths which carburize. This applies to all salt baths.

Further information from Degussa Ltd.

Durionizing
A *chromium plating* process, giving a hard deposited coat.

It is generally applied only to steel but, under certain circumstances, can be applied to copper alloys and, under very limited circumstances, to aluminium. Described fully under *Chromium Plating* in Part Two **Electroplating**.

Durometer Test
This is a hardness testing technique used for rubbers and plastics and not on metals.

It measures the resistance to penetration with a known load and thus is a quite different technique from any metal hardness technique.

The various techniques used for testing hardnesses on metals are given in Part Two **Hardness Testing**.

Dye-penetrant Crack Testing
A *non-destructive* form of *crack test*.

It uses a penetrating dye to highlight surface cracking. Described fully in Part Two **Non-destructive Testing**.

Eddy Current Sorting

A technique used for metal sorting where use is made of two balanced coils through which is passed an electric current.

Application range ALL METALS

The output from the coil is fed to an oscilloscope with each coil feeding a different plane. This results in a balanced picture on the oscilloscope. When any metal object is placed in one of the coils, this will upset the balance. If, however, an identical object is placed in the second coil, then the output will again be balanced. Any variation between the objects in each coil will show on the oscilloscope.

Further details of this and other methods of determining differences in metals are given under *Metal-Sorting*; this eddy current technique can also be used as a means of *flaw detection*, and details of this will be found in Part Two **Non-destructive Testing**.

Eddy Current Testing

A method of *flaw detection*. Described fully in Part Two **Non-destructive Testing**.

Edgewick Hardness Test

A Vickers-type *hardness test* with an applied load rapidly variable throughout the complete range. Described fully in Part Two **Hardness Testing**.

'Efco-Udylite' Process

A proprietary *bright nickel plating process*. Described fully under *Nickel Plating* in Part Two **Electroplating**. Further information can be obtained from Oxy Metal Industries (GB) Ltd.

Elasticity Test

A test seldom carried out in isolation, and a stage in the *tensile test*. Described fully in Part Two **Mechanical Testing**.

Elcoat CRC

This is a proprietary process whereby, using the plasma spray technique, chromium carbide is applied to the surface.

This can be used for any metal capable of withstanding the temperature of application and gives excellent wear resistance and considerable corrosion resistance.

Further information can be obtained from Elbar B.V. or PMM Surface Coatings.

Elcoat MZ

This is a proprietary code given to a coating of magnesium zirconate which is applied by plasma spraying on to high temperature components to protect them from high temperature oxidation.

Further information can be obtained from Elbar B.V. or PMM Surface Coatings.

Elcoat TC

This is a trade name for a technique whereby tungsten carbide is applied by the plasma spray technique. This gives excellent wear and abrasion resistance up to relatively high temperatures.

Further information can be obtained from Elbar B.V. or PMM Surface Coatings.

Elcoat 37

This is a proprietary process for the two pack technique of chromizing plus aluminizing.

Further information can be obtained from Elbar B.V. or PMM Surface Coatings.

Elcoat 101

This is the proprietary process covering the sophisticated diffusion aluminizing process.

Further information can be obtained from Elbar B.V. or PMM Surface Coatings.

Elcoat 240

This is the proprietary name for a chromizing process used on nickel alloys.

Further information can be obtained from Elbar B.V. or PMM Surface Coatings.

Elcoat 360

This is a proprietary siliconizing process which first using a pack process diffuses titanium into the surface of the material being protected, generally nickel or cobalt based alloys. This titanium enriched surface is then subjected to a siliconizing process which is a pack process, that is the components are packed in a mixture of silicon powder with proprietary activators at elevated temperature.

This produces a complex layer which it is claimed gives excellent protection against sulphides and also hot maritime conditions.

Further information can be obtained from Elbar B.V. or PMM Surface Coatings.

Electric Arc Welding

A general name given to any of the *welding* processes where the source of heat is obtained by an arc of electric energy.

Application range ALL METALS WELDED

The electric arc, most commonly, will be struck between the electrode and the component being welded, and the arc will result in the ionization of air. The term electric arc welding, however, can also be used to cover *carbon dioxide welding* and any of the other techniques such as *argon arc welding* or even *submerged arc welding* all of which make use of the electric arc.

The terms electric arc welding, *arc welding* and electric welding are generally taken to cover the same process and, as stated above, in general, will be taken to mean the relatively simple air arc. Described fully in Part Two **Welding**.

Electric Cleaning
See *Electrocleaning*.

Electrochemical Machining
The controlled removal of metal by electrolysis.

Application range ALL METALS

This makes use of all the principles of *electroplating*, with the components in this instance being made the anode in the electric circuit and, thus, having some metal removed. The cathode will be the 'tool' which may or may not be destroyed or altered during the electrochemical procedure, but in general will be non-active. The common electrolyte used with modern electrochemical machining is brine.

The technique was developed to produce relatively intricate shapes on materials difficult to machine, and it appears that it will remain restricted to these fields because of the expense of the equipment, and the high degree of control necessary.

Briefly the procedure is that, using a development programme, a shape is produced and an electrode formed exactly to this shape. This electrode is then made the cathode in an electric circuit, the raw material, sometimes rough-formed by conventional machining, being the anode. The electrolyte is brine and is circulated past the gap between the anode and the cathode at a relatively high velocity. A direct-current electrical circuit is then completed with the distance between the anode and the cathode being controlled electronically, usually by the electrical resistance in the circuit. This results in metal being removed from the anode and, provided the correct material is chosen for the cathode, it will be removed in the fast-flowing electrolyte and not deposited on the cathode.

Under many circumstances, however, there will be some dissolution of the cathode, hence the necessity to commence with the anode of a shape different from that of the finished article. At the end of the process, the anode will have the desired shape – it will be appreciated that this, in general, will be on one surface only, and thus components of a complex nature require two or more operations to produce the finished article.

This procedure, being a true electrochemical action, leaves the surface with a generally smooth finish and, except in exceptional circumstances, the surface finish will be suitable for use on components under fatigue conditions.

Comparable processes are conventional machining and, as stated above, electrochemical machining will seldom be economically justified if a conventional machining technique can readily produce the required shape. To some extent *spark erosion* is competitive, but while this is considerably cheaper, it will not produce intricate shapes, and leaves a surface which often requires further machining to remove the poor fatigue surface.

Electrocleaning

The *cleaning* process in which an electric current is passed between the components being plated and an inert electrode.

Application range ALL METALS

Solutions used vary widely depending on the type of soil involved and, in particular, the metal being cleaned. The principle is that the components have electrical conductors attached either by wiring or jigging, and are then immersed in the cleaning solution and the contact connected to the power source. This often will be capable of being reversed, so that components can be altered to become the anode or the cathode.

When the components are anodic, oxygen will be produced at the components' surface and released. This is the result of the electrolysis of water. This release of oxygen gas acts in a manner not unlike mechanical scrubbing and, as the gas will be released at the interface of the metal and the contaminating material, there will be a tendency to blow off the contaminant from the surface. The solution used will generally be alkaline when steel components are involved and, thus, oil and grease will tend to be saponified and taken into solution at the same time.

When the component is made the cathode, then hydrogen will be given off at the interface. This will generally be a more active scrubbing action than the oxygen given off when the component is anodic. There will also be attracted to the surface many of the contaminants contained in the cleaning solution. These will not normally adhere to the surface, but will result in an unsightly, badly contaminated surface, with the contamination very loosely adherent. For this reason it is essential that, when electrocleaning is used, the cathodic stage is always followed by a final anodic clean.

Correctly used, this technique will result in higher standards than simple *immersion cleaning*, and under many circumstances will be comparable with *ultrasonic cleaning*; the process of *Cleaning* is discussed under its own heading.

Electrocolour Process

A technique used to deposit oxides, to give an attractive finish.

Application range STEEL

Low-voltage, direct current is used and the colour of the resultant film will vary, depending on the surface finish of the component being treated, and the thickness and type of the film produced. These parameters will vary with the solution chosen, generally an organic copper solution, the strength of the solution and the current density used.

The deposit produced can be pink, blue and green of various shades, or a dark grey almost black. It will not protect steel from corrosion, and unless lacquered will have a very short life.

Modern techniques of coloured lacquers, *electrodeposit* of metals with various shades, and the use of *colour anodizing* of aluminium, are more easily controlled, giving a wide range of colours and achieving much better corrosion resistance. These have supplanted electrocolouring as a production process.

Electrodeposition

The general term used to describe the process where direct electrical current is used to electrolyse a solution with the resultant plating or depositing of a metal as an adherent film on the article being treated. See Part Two **Electroplating**.

Electroforming

The production of specific shapes by electroplating of metal.

Application range ALL METALS WHICH CAN BE PLATED

Electroforming is normally confined to use with special-purpose materials where intricate shapes are required and relatively thin metal deposits involved.

With the process, a mould of the article required is produced. This may be in the form of a plastic or in some cases a wax impression is produced or it can be metal. This mould is then made conductive by any of several methods. In the case of metal moulds, special coatings can be applied which will ensure that *electroplating* will occur but the adhesion of the plating to the metal substrate will be relatively poor. Metals such as aluminium which are conductive, but have very poor adhesion to the deposited metal can also be used.

In the case of plastic or wax or other non-metallic moulds, these can be made conductive by spraying with carbon or a metallic powder or by one of the *electroless plating* processes. Once the mould has been prepared and is conductive, then the metal to be electroformed will be deposited under the identical conditions described under the metal plating process.

The most common metals used for this method are nickel and copper. Nickel is used for many purposes the best known being that of mass-produced gramophone records. With this, the original sound is encapsulated on a form of wax. This wax is then *nickel plated* to produce an exact replica of the form cut into the wax during the recording session.

This is then used to produce the die from which the mass produced record is made.

Other intricate shapes can be produced in a similar manner, in many cases using materials such as thermosetting or chemically setting plastics or plaster of Paris. There is no reason why materials such as wood cannot be used as the mould for electroforming, and it is now possible using this technique to reproduce the exact form of leaves, flowers, babies' bootees, etc. The metal plated will depend on the purpose to which the product is to be put. Thus, nickel will be used where excellent corrosion resistance with limited mechanical strength is required. Precious metals such as gold or platinum can be used, and there are very attractive intricate jewellery pieces which are produced by this method.

On completion of the plating cycle, the components are removed from the standard *electroplating* solutions, washed and dried in the normal manner. Depending on the material of the mould, will also depend the subsequent operations. It is generally necessary to remove the mould completely, and in the case of the master gramophone record, this is readily achieved by a low-temperature oven treatment which results in the wax melting and being removed. The nickel master mould suitably stiffened is then used to mass-produce the plastic records.

Other materials will have a variety of methods of eliminating the mould, depending of course on the mould material. Thus, there is a jewellery process which uses low-melting point, lead-based alloys as the mould. These are plated with gold or platinum and the mould is then removed by heating to above the melting point of the lead-based alloy, approximately 200°C–300°C. The efficiency of the electroforming will to a large extent depend on the choice of the moulding material and the readiness by which this can be removed after *electroplating*.

The technique can be compared to the competing process of *electroplating* directly on top of a cheaper base metal which has been previously formed by *casting* or pressing operations. It is also competitive, under many circumstances, with modern types of *casting* such as the *lost wax* process or *shell moulding*. For many purposes, however, electroforming will have no real economic competitor. This is where thin metal of low strength is required in an intricate shape difficult to machine, and here electroforming will be a uniquely suitable process.

Electrogalvanizing

An alternative term for *zinc plating*.

'Electrogranodizing'

A specific and proprietary form of *phosphating* where electrolysis is used.

Application range STEEL

With modern solutions, a high standard of phosphate coating can be achieved using conventional *phosphating*; thus, this process is now seldom employed. Further details can be obtained from ICI Ltd.

Electrography

An inspection technique to show microporosity in metallic or non-metallic coatings.

Application range ALL METALS

The component being tested is made the anode in a direct-current electric circuit, with a moistened, sensitized paper supplying the electrolyte, and a metallic electrode clamping the paper to the component as the cathode.

This technique can be used as a relatively simple device to show porosity, when the paper will generally be sensitized with dilute sulphuric acid. The existence of any porosity will be shown by the presence of dark spots or marks on the paper where the base metal has been allowed to react with the sulphuric acid.

The current density used will vary depending on the sensitivity required and the type of coating being tested. Higher current density can be used with non-metallic coatings, such as paint than with *electroplated* deposits. With these metal coatings, there will be some reaction with any acid electrolyte, and while this will generally be less than with steel, clear indications of fine porosity will be difficult to identify.

This inspection method is only possible with components of a simple shape; it requires some skill and is relatively slow. It is, thus, seldom applied as a routine means of inspection, but is commonly applied as a quality-control technique to articles of high value, or where technical integrity is essential.

Alternative inspection methods are visual inspection, which requires to be carried out at high magnification to be meaningful; thickness checking – below a specified thickness, most coatings will exhibit porosity; and *accelerated corrosion testing* such as the *humidity* or *salt-mist tests* which will destroy the sample being tested but clearly show the presence of pores. Often all of these techniques are used in modern quality-control sample checking.

The *Holiday testing* of paint is based on the same principle, using water as an electrolyte. This technique generally feeds the current through an amplifier to ring a buzzer. Thus, any low resistance in the paint film from pores will increase the current and sound the buzzer.

Electroless Nickel Plating
The deposition of nickel by chemical means only.

Application range ALL METALS WHICH CAN BE PLATED AND MANY PLASTICS

The deposit plated is an alloy of nickel and phosphorus.

The procedure is that the component to be plated is thoroughly cleaned to a very high standard and is then *etched* or otherwise prepared, if it is not a normal metal. This preparation may take the form of plating on plastic. Immediately after the *etching* or *sensitizing*, the component is immersed in the electroless nickel plating solution. This is a relatively complex solution, but there are satisfactory proprietary solutions on the market. It is possible for manufacturers to produce from raw chemicals their own plating solution. If this procedure is followed, it is necessary to ensure that competent laboratory personnel and equipment are available at all times. It should be appreciated that under these circumstances the manufacturer will not have recourse to the research and development which is being carried out in this field and which has resulted in considerable technical advances over recent years.

It must also be realized that the plating process is carried out at a temperature in excess of 80°C. If the component being plated is kept cool during the previous processes of *etching* or *sensitizing*, with subsequent swilling, then there will be considerable chilling at the plating interface which will result in faulty deposits and poor adhesion. It is, therefore, advisable that all components with any appreciable mass, particularly where the surface area is relatively large, should be preheated during the *etching* or *sensitizing* and subsequent swilling. This preheating must not interfere with the rapid transfer of the component from the etch to the plating solution; as with *electroplating*, this is essential for good-quality adhesion. Thus, it is not possible, for example, to use hot air as a means of preheating.

The plating process itself will immediately commence when the component reaches the temperature of the solution. This is a true electroless process. The solution is so formulated that not only will the nickel be deposited on steel, etc., but that the procedure continues once the base metal has been fully covered by the nickel deposit. Most solutions will plate at a speed of approximately 0.0001–0.001 in (0.0025–0.025 mm) per hour. The rate of deposit, to some extent, will depend on the solution analysis, but in the main is controlled by the temperature of the plating.

The problem encountered with electroless nickel plating is that, by definition, the solution is unstable and, therefore, tends to deposit nickel on all materials with which it comes into contact, such as filters, pumps, holding tanks, etc. Modern solutions are more stable, and provided that they are correctly handled and kept filtered and free of all suspended material, and in addition, when not in use, are kept below the critical temperature, then it will be found that solution stability is generally satisfactory.

In some cases, it will be suitable to have what can be termed 'one-shot' solutions. For components such as intricate moulding dies, it is possible to use the die itself as a plating tank, to heat the die to the desired temperature, to *clean* and *etch* the surfaces and then to pour in the required quantity of electroless nickel solution. This will have the necessary strength and volume to deposit the required thickness of metal. On completion of the process – which will be when the solution has been depleted below a certain level – this solution is rejected.

While this procedure has certain obvious advantages, it must be used with caution as another of the problems with the electroless nickel solution is the difficulty of chemically stripping the deposit, which is much less chemically active than the steel substrate. Thus, if the calculations regarding the surface area are wrong, or if the temperature of the solution drops, a faulty deposit may be obtained. Also as the solution is depleted in nickel when plating takes place, the conditions for satisfactory plating become more critical and so there is a greater danger of a faulty deposit.

The deposit as plated has a hardness and adhesion similar to electrolytic *nickel plating*. Hardness, however, can be considerably increased by an *ageing* or *precipitation hardening* treatment. This requires that the component, after plating, is heated to a temperature of 400°C for 1 h. This results in an increase in hardness to approximately 700–860 DPN, Rockwell C 59–64, from a hardness of approximately 200 DPN. If a lesser hardness is required for any reason, this can be achieved either by a lower temperature for the same time, or continuing heating at 400°C. This *precipitation* treatment is identical in all respects to that described under this heading in this book. While a dramatic increase in hardness is achieved, from approximately 200 DPN to above

850 DPN, there is an accompanying reduction in ductility, as would be expected. This treatment also acts as a very successful adhesion check and any faulty adhesion will be shown by bursting or peeling during or immediately after the treatment. It must be realized that only the deposit has this hardness and testing must use very light loads to achieve any satisfactory result.

While electroless nickel plating is a comparatively recent innovation, it is finding considerable use in two distinct fields. First, it is being employed where good corrosion resistance is required on intricate shapes or in crevices, blind holes and deep cavities where normal electrolytic plating is either not possible, or uneconomical. While the corrosion resistance of electroless nickel plating is less than that of the electrolytic process, it is still vastly superior to normal steel and is comparable to austenitic stainless steel. In addition, there is the considerable advantage that the electroless nickel deposits in a homogeneous, uniform manner on all surfaces, and here it differs from electrolytic plating, where there will always be a discrepancy in the thickness deposited between areas of high-current density such as edges and points and areas of low-current density such as recesses. For this purpose, it is finding use and replacing other forms of *corrosion protection* such as electrolytic nickel and, in some cases, *nickel*, *chromium* or *zinc plating*. Under certain circumstances, mild-steel pressings can be electroless nickel plated as an economical substitute for austenitic stainless-steel components.

Second, electroless nickel plating is being put to use today in situations where, subsequent to plating, *heat-treatment* is employed to give a high surface hardness. Tests have indicated that this deposit is comparable to *hard chromium plating*, consequently there is now a choice where wear resistance is required, particularly with corrosion resistance, between the two forms. Electroless nickel in the hardened condition can be compared under many circumstances to *carburizing*, although there is some evidence that the friction characteristics for the same surface finish are that the latter will be superior to the present process. However, where any corrosive atmospheres or liquids are involved then normal *carburizing* will break down from corrosion, thus resulting in poor frictional characteristics which in a very short time can become much worse than with electroless nickel, which will not be affected by normal engineering corrosive atmospheres.

'*Kanigen*' *plating* is a proprietary form of electroless nickel. Briefly, it can be said that this deposit should be considered at the design stage, where corrosion resistance of a high standard is required on intricate shapes difficult to plate by conventional methods. It should also be considered where surface hardness is required. Where superior corrosion resistance and hardness are necessary, electroless nickel has very little competition. This will be 12 per cent chromium steels in the hardened conditions and *metal-spraying* or *weld deposit* of high cobalt-nickel type complex alloys.

The process is now commonly used as a salvage technique where unacceptable wear or undersize components with a hardened surface are involved. As the maximum build up is less than 0.010 in (0.25 mm) there are obviously severe limitations when it is compared with deposits of hard metal by welding or metal spraying. It is generally competitive when compared to salvage hard chrome plating.

Electroless Plating
The chemical deposition of metal without electrolysis.

Application range ALL METALS AND PLASTIC

There are two basic mechanisms under the general heading of electroless plating. The simplest of these is the displacement of one metal by another. On the table of electrical activities of metal, gold and precious metals appear at the top and the base metals such as aluminium, magnesium, lithium, etc. at the bottom. If one metal is immersed in a solution of a metal appearing higher in the table, then electroless plating takes place; the higher metal will try to displace the lower metal. This explains the well-known phenomenon where a piece of steel immersed in a copper sulphate solution will be copper plated. For this form of electroless plating to be successful, certain conditions are necessary regarding the material in solution. The range of conditions is relatively wide and it is, thus, comparatively easy to produce deposits of a higher electrochemical metal on top of a lower one. However, adhesion under these circumstances will generally be of a low order.

Processes which make use of this electroless deposit involve gold, silver, platinum, etc., at the top of the range, which are deposited on to lower metals, the most common of which is copper on steel. The deposits invariably will be very thin and have porosity. Tin is also a common deposit for cosmetic and technical reasons and is used on top of solder to a considerable extent in the electronics industry. In engineering, copper sulphate solution is employed to deposit a very thin coating of copper on steel. At one time this was a popular method of colouring steel, since it could be readily marked by scratching for 'marking-off' purposes during machining. One problem with this is a tendency for corrosion to occur. This is partly caused by the acid nature of the solution, added to which is the fact that copper and steel in contact with each other will increase the corrosion of steel. As stated, electroless coatings will normally be very thin and easily damaged. The exception is electroless nickel plating, where solutions have been developed to give coatings up to 0.010 in (0.25 mm) and thicker.

The other form of electroless plating to some extent overlaps, but in this instance much more complex chemistry is involved in addition. With this form, the base material is generally a non-conductive plastic or similar material. The initial process requires that the plastic is first chemically *etched* so that a key can be produced. After thorough washing, an unstable solution deposits very thin palladium. This deposit is then used to absorb tin ions in the form of stannous chloride, and it is this tin-compound deposit which is subsequently used in a similar manner to the electroless deposit described above.

By far the most common metal deposited is copper, this is used to a considerable extent in the electronics industry in the production of printed circuits, and under similar circumstances in photoprinting; in addition, many useful and decorative plastic components are now made to appear metallic, generally being chrome plated, using this technique. After copper the most common metal deposited is nickel, usually an alloy of nickel and phosphorus which can be hardened by a *precipitation ageing* treatment, to give relatively good mechanical properties. This electroless nickel has unique characteristics and applications in the electroless plating field. Further details of the process and its uses are given under *Electroless Nickel Plating*.

Electroplating on to an aluminium substrate generally commences by depositing electroless zinc on the aluminium, and it is the zinc film which is then electroplated.

Electrolytic Etch

The use of electrolysis to etch metal surfaces.

Application range ALL METALS WHICH CAN BE ELECTROPLATED

In general, it is confined to steel components and is differentiated from *electrolytic polishing* in that no attempt is made to smooth the surface, but rather the etching is designed to attack the surface to produce a clean, oxide-free material and at the same time to accentuate any surface defects.

Electrolytic etching is sometimes used prior to *electroplating* and is a popular etch prior to *chromium plating*. It is also used as an inspection technique, where minute surface cracks are enlarged by the actions of the etch. This is because the etch will concentrate at the corners and, thus, in effect open up a minute crack. The method is also occasionally used to identify abuse of hardened steel surfaces. This abuse will commonly be the result of faulty grinding, and the electrolytic etch, in a manner identical to that of metallurgical etches, will show up the different metallurgical structures which have been produced. A correctly hardened and ground surface will have a homogeneous martensitic structure which will show as a matt even grey. Any abuse at grinding, or for any other reason, will result in softening of this martensite, or in extreme cases result in rehardening of the surface martensite. This local *tempering* of the surface will show as a different colouration with the electrolytic etch. A similar technique is to use dilute nitric acid. This and other techniques are described under *Etching*. The two forms of electrolytic etching, anodic and cathodic, are also separately discussed.

Electrolytic Polishing
The use of electrolysis to improve the surface finish.

Application range ALL METALS WHICH CAN BE ELECTROPLATED

The purpose is to produce a bright surface with a high reflective finish. In most instances this is for decorative purposes and often is associated with some other form of metal finishing such as *anodizing* or subsequent *plating* or *lacquering*.

The principle of electropolishing is, that when a metal is subjected to a direct current, in an electrolyte, and is made the anode, the *electro-plating* process will result in metal being removed from the anode, which in this instance is the component.

It is also a fact that the current applied will concentrate at corners and peaks. This is a well-known phenomenon and is one of the problems associated with metal plating. This concentration of the current at peaks is made use of in electrolytic polishing, since this means that more metal will be removed from peaks than from valleys.

By the judicious choice of electrolytes, this smoothing action can be accentuated and the resultant components on completion of electrolytic polishing will have a smooth finish with a bright, reflective finish. The solutions used vary considerably as do the current and voltage requirements. Generally, however, the solutions are acid, commonly being phosphoric acid, sulphuric acid and sometimes chromic acid. There are a large variety of proprietary solutions, most of which are based on one of these acids or acid mixtures but with additives which improve the efficiency of the process. Users are advised to contact one of the plating supply houses for further information.

Alternatives to electropolishing include *barrelling* or mechanical polishing, and consideration should be given to processes such as *vapour blasting*, which can achieve high-quality surface finishes of an attractive visual nature.

Electron Beam Welding
A specialist weld method, using vacuum chamber and high-energy electron beam. Described fully in Part Two **Welding**.

Electro-osmosis
Similar to the *electrophoresis* process, but applied where solutions of liquids are involved, whereas *electrophoresis* is applied to colloidal solutions, which contain solids in suspension. Electro-osmosis is used to separate similar liquids from each other, or to separate a solvent, generally water, from solution, leaving a more concentrated solution.

This makes the process of considerable use in the medical field, which is no part of this book. Electro-osmosis, however, is now being commonly used in the preparation of process water and, under certain circumstances, as part of the effluent-treatment process.

The theory of electro-osmosis is that ions in solution are attracted to an electrical charge which is located on the other side of a membrane, generally a plastic. These selected ions will be persuaded to pass through the membrane, thus separating them from the other ions. Under specialized conditions, ions of solids in solution can pass through this plastic membrane. Thus, it is possible to take water which is saline, or salty, and by correct choice of membrane, to achieve water on the other side of the membrane which is of much lower salinity.

With electro-osmosis, it is not yet possible economically to achieve water of high purity from high-salinity liquid, but brackish water, which is not suitable for drinking, can be used to produce a liquid which can be safely used by humans and other animals.

In industry, the use of electro-osmosis at present is mostly in the water preparation field, where water of extremely high purity is required. The electronics industry and certain branches of the food processing and pharmaceutical industries also require high-purity water. Under these circumstances, it is usual for electro-osmosis to be used in conjunction with the more conventional ion exchange.

This process, then, is in competition with ion exchange and flash distillation in the production of large quantities of potable water. By itself, at present, it cannot be used for the production of high-purity water, but in conjunction with other processes such as ion exchange, high-purity water can be produced.

Electro-osmosis is the principle involved in the painting system known as electropaint or electrophoresis.

Electropercussion Welding

A form of electric *Resistance Welding* used on high-conductivity metals. Described fully under this heading in Part Two **Welding**.

Electrophoresis

The use of *electro-osmosis* in paint application.

The industrial application of this technique is confined to *painting*. Described fully under *Electrophoresis* in Part Two **Painting**.

Electroplating

This covers all aspects of the deposition of metals. The techniques used and the metals deposited are listed and described fully in Part Two **Electroplating**.

Electroslag Welding
A high productivity type of welding.

Applied only to relatively simple welds such as straight vertical runs; it is similar in some respects to *submerged arc welding*.

Described fully in Part Two **Welding**.

Elongation
One of the mechanical properties evaluated during the destructive *tensile test*. Described fully under *Elongation* in Part Two **Mechanical Testing**.

'Eloxal' Process
An *anodizing* process for producing an oxidized film.

Application range ALUMINIUM

Details of this are given under *Anodizing*. It has been impossible to discover any information on the source of the name 'Eloxal'.

'Elphal' Process
The coating of steel with aluminium.

Application range STEEL

A proprietary process developed by the British Iron and Steel Research Association, where steel strip which has been degreased is passed through a suspension of fine aluminium dust, in a methylated spirits–water mixture.

Using a direct electric current, the aluminium particles are attracted and made to adhere loosely to the steel surface. The liquid is then evaporated by heating and the loosely adherent powder is then rolled in contact with the steel. The finish achieved at this stage will also depend on the type of rolls used. The coating is then *heat treated* to cause *sintering*, but no *alloying* of the aluminium and the steel. A thickness of up to 0.0015 in (0.040 mm) is possible.

This process results in a coating of aluminium, almost free of oxide on the surface of the steel strip. The steel strip can be bent and drawn without breaking the aluminium film. While similar in many ways to *aluminizing*, this method differs in that it is a coating of aluminium rather than aluminium oxide, and no diffusion occurs. It results in excellent corrosion resistance, but is relatively expensive, and is applicable only to sheet metal.

Alternatives are zinc or tin plated sheet or painted or plastic coated steel strip.

Embossing

A method of producing surface designs on sheet-metal components.

Application range ALL SHEET METALS

The technique of embossing means that the sheet metal is *cold worked* by indentation, resulting in a raised design appearing on the reverse side. The material is stretched, thus it is necessary that the metal used must have the necessary ductility to cope with this stretching without cracking or being too brittle.

In general the term embossing covers relatively slight work, often applied by passing the sheet through rolls which have the design raised or cut on the rolls. Frequently, these rolls are 'embossed' themselves and this can be by *electro-deposition*, *metal-spraying* or *etching*. The term is not normally used to cover these techniques or the methods of machining or engraving to produce raised designs.

Emulsion Cleaning

A *cleaning* technique, using special liquids, which are themselves emulsions, and very often act because they emulsify the contaminant.

Application range ALL METALS

Emulsion cleaners have two separate liquids which clean selectively. These are highly technical materials and for correct use the manufacturers' instructions must be carefully followed. These instructions will vary depending on the type of cleaner involved, the metal being cleaned and the soil to be removed. It is not possible, therefore, to generalize on the techniques required. It can be stated that correctly used emulsion cleaners can have an extremely long life and result in cleaning of a high technical standard. The user is referred to the heading *Cleaning* for further information.

Emulsification results in oil or grease being broken up into fine droplets, which are then suspended in the aqueous solution, and thus readily removed by an efficient water swill.

If a scrubbing action can be achieved this will increase the cleaning action. Ordinary soap acts in this manner.

There is a relationship between emulsions and colloids. A true colloid is a 'suspension' of a solid in a liquid. An emulsion is a 'suspension' of a liquid in a liquid. A mist or fog is a suspension of a liquid or a solid in a gas.

Enamel Plating

A term which was at one time applied to *colour anodizing*. It is no longer used.

Application range ALUMINIUM

With this, the aluminium is first *anodized* then the fresh surface is immediately dyed. It is obviously possible to use this term to cover other forms of surface treatment, generally where a colour change is involved.

Enamelling
A painting technique, requiring stoving.

True enamel is a glass frit which is fused to give a vitreous finish. The term 'enamel' has now been accepted to have a much wider meaning and, thus, it is included with other forms of *painting*. Described fully in Part Two **Painting**.

Under most circumstances some form of heating will be involved. Some paint manufacturers now use the term 'enamel' to indicate a degree of abrasion resistance of the coating applied, but almost invariably the 'paint' will have been heated or 'stoved' after application.

'Endurance' Process
A proprietary name, covering a *metal-spray* or *weld-deposit* technique. Further details can be obtained from Dewrance & Co. Ltd, and the process is described under *Metal-spraying*.

Endurance Testing
This term can be applied to any form of testing designed to find the life of a component, or more generally, an assembly or complete unit or engine.

Application range ALL METALS

The condition of testing, length of test and loads applied will be given in the test schedule. Endurance testing applied to metals is an alternative name for *fatigue testing*. Described fully under *Fatigue Test* in Part Two **Mechanical Testing**.

'Endurion'
A form of phosphating.

Application range STEEL

A conversion coating not applicable to stainless steels, which will considerably improve the corrosion and wear resistance of a phosphate coating.

The component is cleaned and *phosphated* in the normal manner and, after cold and warm water rinsing, immersed in the 'endurion' solution. The immersion time is between two and five minutes. The solution is used boiling, hence the necessity to pre-heat the components prior to immersion. No delay should occur between *phosphating* and 'endurion' treatment and the phosphate process should not include any chromate seal.

The 'endurion' process results in the conversion of the phosphate coating to a harder, more homogeneous surface. It can be used as the normal pretreatment for paint, but will only be necessary when a very high corrosion resistance is required in the event of paint damage.

'Endurion' surfaces can be made to absorb oil; it is claimed that the resultant surface will have low friction characteristics and the ability to withstand galling and pick-up. This characteristic is used on fasteners which require torque loading, as there is some indication that the 'endurion' process, correctly carried out, will give superior results to other forms of *phosphating*.

The 'endurion' process will also accept thin films of dye or lacquer which can be used to colour the surface. Because the phosphate is relatively dark the tendency will always be to use a reasonably dark shade, and some use has been found for this process where a matt non-reflective material which has also high corrosion resistance is required. Further details on this process can be obtained from the Pyrene Chemical Services Ltd.

Equi-tip
This is a portable hardness testing technique recently developed. Described fully in Part Two **Hardness Testing**.

Erichsen Test
A destructive test to evaluate the ductility of sheet metal, it is also known as the *cupping test* or the Erichsen cupping test.

Application range ALL DUCTILE SHEET METALS

The procedure is that a ball of known diameter is impressed under a standard load on to a sample of the sheet metal. This results in a cup being formed, of the approximate diameter and depth of the diameter of the ball. On completion of the test, the cup produced is carefully examined for bursts or cracks or any wrinkling of the material locally which would indicate failure. Since this is a standard test, a system of comparing the material under test with previous batches is available to the inspector.

The test is a cheaper method than *tensile testing*, giving slightly more specialist information under certain circumstances. It can be compared with the procedure now carried out by some users of high-quality sheet-metal in mass-producing difficult pressings. There are occasions when such users reach an agreement with the material suppliers that the material in question must make a specific pressing. Each delivery is then tested by producing a pressing, and in the event of any defect appearing, the material will be rejected. Another method of assessing the suitability of sheet metal for *deep-drawing* purposes is the ratio of 'yield' or 'proof stress' to the 'ultimate tensile strength': the greater the difference, the better the material will be for deep drawing. Other ductility tests such as *reverse bending* and *drift testing* are comparable under specified conditions and are generally cheaper but are not as readily evaluated between different batches as the cupping test.

Etching
A term covering a variety of different techniques which achieve completely different purposes, but all involved with metal removal or chemical action on the surface involved.

Application range ALL METALS

Briefly, the term covers the following seven techniques:

1. *As a surface preparation for electroplating, etc.* In this context the term denotes that metal is chemically removed, and generally means that the surface is left in an active condition. This is alternatively referred to, under certain conditions, as *activation* (see Part Two **Electroplating**).

2. *For removal of metal.* This is used, in the printed-circuit industry in particular, to denote that stage where material not required on the finished printed circuit is removed by chemical solution. Again, then, etching here is a chemical means of removing metal.

3. *To highlight surface defects.* This again is a chemical process which, under some circumstances, may be electrochemical. The technique here makes use of the fact that corners or edges will normally be more active than flat surfaces. This effect can be accentuated by the electrolytic action and is used as a means of opening cracks or surface defects. This means that minute surface defects such as cracks or slag stringers will be highlighted because the edges are attacked, and the width of the crack increased, making it easier to identify by visual examination.

4. *To show the metallurgical structure.* This can be on a 'macro' scale, where the component is subjected to one of the several etches to highlight the surface condition. After use the surface of the component is examined, and this *non-destructive* testing technique is to highlight several types of defect, most of which are the result of grinding or similar processes where the surface layer of the material, generally hardened steel, has been affected very locally (see Part Two **Non-destructive Testing**).

Alternatively, this is also used as a *destructive* test, where the components are cut and the surface prepared and then subjected to etching. This technique indicates the method of *forging* or *welding* and any defects present. It will also show any gross lack of homogeneity in a metal structure (see *Macroetch Test*).

5. *Microetching.* This is a laboratory technique, used on the polished microspecimen to show the different structures which indicate the metallurgical state of the material. Details of these etches are readily available and do not come within the scope of this book.

6. *For identification or marking.* This technique uses needles, electric-spark marking, acid etching or, under some circumstances, stamping. The purpose is to mark the surface for identification purposes. Acids are seldom used in modern industry because of the danger of corrosion but have been used and would be etches. Some marking materials are still available but it must be ensured either that the material is self-destroying, or that it is obvious that any liquid capable of etching or marking the surface must also be capable of causing corrosion, and appropriate protection then applied.

Where marking is achieved by simple *painting* or stencilling, this is not normally referred to as etching, but inks are available which have a corrosive content and, thus, etch or stain the surface giving permanent marking.

Acid etching for marking was replaced, first, by the electric-discharge method in which a hardened steel or tungsten-carbide needle was used to produce sparks using an electric current. This method was shown to be dangerous where hardened steel components were involved as the high temperatures required for the spark discharge also softened and, in some cases, rehardened and cracked, the surface of the steel. The modern technique of marking uses a tungsten-carbide steel which is connected to a vibrator, and the surface of the steel is marked by the vibration against the tungsten-carbide steel. It will be appreciated that this method is not possible on hardened steel surfaces, and this is considered an advantage as marking on these surfaces will always present the danger of forming stress-raisers and the probable propagation of cracking. There is a fact of life that no material can scratch or mark a material harder than itself.

7. *For artistic purposes.* This is the selective removal of metal by chemical action. Historically the surface of the metal was waxed and the artist prepared the design on the wax coating. The metal, generally copper or copper alloy, was then subjected to a solution which attacked the wax and the metal.

Where no wax remained, the metal attack would commence immediately, whereas when the solution had first to attack the wax the metal would be affected less. There would therefore be selective chemical removal of metal.

The modern version uses silk screen printing and photo-mechanical techniques to achieve the same result.

A variety of metals including stainless steel and aluminium are now used.

'Eutectrol' Process

A continuous *gas carburizing* process.

Application range STEEL

With this, there is a furnace using some form of moving belt or track which continuously moves the components being treated. These enter at the cold end of the furnace, are gradually heated in a *carburizing* atmosphere, which may or may not vary with the temperature requirement, and with the track to give complete control for the carbon potential. During the whole of the cycle, the components are moving and are in the carbon-rich gas. Described fully under *Carburizing* in Part Two **Heat Treatment**.

Evaporation Process
An alternative term for *vacuum deposition*, where the metal to be deposited and the article to be treated are taken to high vacuum. The metal is then heated, boils and on cooling is homogeneously deposited on all surfaces.

Explosive Riveting
A method of *riveting* where an explosive charge is used to form a rivet head. It is obviously of highly specialized significance and is generally used only when it is impossible to have access to the portion of the rivet which forms the head. The explosive charge is shaped to the desired head-form and inserted against the shank, normally on the inaccessible inside of the two skins to be *riveted*. The method of initiating the explosion, the amount of explosive charge, etc., will be a function of the materials and geometry of the parts involved.

Explosive Welding
A form of *forge-welding* where, using explosive forces, the process is achieved by high-impact force. It can only be applied to a limited number of materials. Described fully in Part Two **Welding**.

Extrusion
A manufacturing process, not a metal treatment. It is a specific type of *forging* where the material is forced through a die of requisite shape. Extrusion can be carried out in various stages but, in general, the process is limited to materials which are readily forged and the economics of the process largely rely on large usage, relatively simple, easily forged materials and the production of intricate shapes.

It will be appreciated that the more intricate the shape, the higher the cost of the dies, etc., but the greater the saving in subsequent machining. The materials most commonly extruded at present are austenitic stainless steels, aluminium alloys, copper alloys, with a limited number of more exotic materials such as nickel–chrome alloys. Use of the relatively simple steels is unusual, since these are quite readily hot-worked using conventional rolling techniques. Extrusion can be carried out cold or hot. Users are advised to obtain specialist advice prior to designing components which are to be extruded.

Fadgenizing
A mechanical *surface treatment* applied prior to *electroplating*.

Application range ZINC DIE CASTING

With this, each casting is individually mounted and rotated in a slurry of silica flour (fine sand) with other additives. The purpose is to polish the surface without breaking the surface skin, prior to plating. Experience has shown that normal *buffing* and *grinding* causes damage to the surface, resulting in faulty plating.

The modern *barrelling* process, correctly controlled can be used instead of Fadgenizing. There are also chemical and electrochemical treatments for zinc-alloy casting which give satisfactory results in plating.

'Fafcote'
An alternative name for the '*Endurion*' process.

Falling Weight Test
An inspection or quality-control technique which, in its simplest form, consists of dropping a weight on to the component and examining the result. The shape and weight of the material being dropped, the height and acceptable damage, require to be specified. A variation is that the component being tested is held at different angles and the height and weight also varied. The number of drops can also be varied, and the impact surface altered.

This test can also be applied to assemblies and is now a common part of the testing carried out on electrical and electronics assemblies. These are then subjected to electrical testing. These are generally specialized tests, carried out after all normal *non-destructive tests* have been completed. It is comparable in many ways to the *drop test*, where the component itself is dropped.

'Fasbond'
A proprietary name for *phosphating*. Further information can be obtained from Walterising Co. (UK) Ltd.

Fatigue Test
The application of cyclic tensile loading.

A long-term mechanical test which is destructive. It is not a normal part of the material specification, but the fatigue strength of components is of vital importance to engineers. Described fully in Part Two **Mechanical Testing**.

'Feroxil' Testing

A form of *non-destructive testing* which ensures freedom from surface ferrite.

Application range AUSTENITIC STAINLESS STEEL

During manufacturing processes which include any heating, such as *welding* or *annealing*, the surface of austenitic stainless steel components is converted to oxides which contain iron in the ferritic state. This means that any corrosive liquid such as water or foodstuffs which come in contact with the surface of the stainless steel can become contaminated, or the surface can give the appearance of rusting. In order to prevent this, stainless-steel containers must be treated by *'passivation'*. This can either take the form of nitric acid treatment, or some lesser acid such as citric acid, the purpose of which is to dissolve the free ferrite from the surface. Abrasion by mechanical polishing or blasting using iron-free grit can also be used.

There are now available pastes which contain acids designed to remove ferrite. These can be applied locally to welds. Considerable care is necessary in ensuring complete removal of all traces of the paste. The passivation treatments are discussed in more detail under that heading.

The Feroxil test is designed to show that *passivation* treatment has been successful and that the surface is free from ferrite. The Feroxil test is not designed to identify the ferrite content of austenitic steel, but only surface contamination. The *Ferrite Test* is described under this heading below.

The test solution consists of potassium ferricyanide in a solution of water which has been slightly acidified with hydrochloric or acetic acid. The technique is that the surface being tested is cleaned using a solvent such as trichloroethylene and, using a glass dropper, spots of the test solution are dropped on to the test spot. If any free ferrite is present, the test solution will immediately turn to a bright dark blue – Prussian blue. If no such colouration is shown after 10–30 seconds contact, the surface is free from contamination. The use of filter or blotting paper will often be necessary to ensure surface contact.

The test solution must be made up immediately before each test. Equal quantities of potassium ferricyanide, hydrochloric (or acetic) acid, and sodium chloride are mixed and applied by the dropper or filter paper.

Usually, it will not be necessary to test complete surfaces, but areas such as welds must always be tested and edges or drilled or punched holes should also be carefully examined.

If austenitic stainless steel has been *annealed*, it will require *pickling* to remove the free surface ferrite, and in this case, checks at random areas on both sides of the surface, and in particular edges, should be carried out.

Electronic instruments are available which can determine the amount of ferrite in austenitic stainless steel. These ferrite meters are very useful for determining whether or not a weld or a mass of stainless steel is in the fully ferritic condition. However, the instruments are not sensitive enough to identify the extremely thin film of free ferrite which may exist and can cause serious contamination. The use of meters is discussed under the heading *Ferrite Testing*.

Ferrite Testing
A test to identify the presence of iron in ferrite form.

Application range AUSTENITIC STAINLESS STEEL

Stainless steel is a single phase material, where by adjusting the chemical analysis with the addition of chromium and nickel in particular, it is possible to depress the change point of steel to below room temperature. This means that the austenitic phase will exist at room temperature, whereas in the case of plain carbon and low-alloy steel, it is necessary to heat to above at least 650°C to produce this austenitic phase.

The austenitic phase is non-magnetic, as the change point from magnetic iron to non-magnetic iron occurs at a lower temperature than the change from ferrite to austenite. This change in the magnetic properties allows the use of an instrument to assess the percentage of ferrite which exists in any austenitic stainless steel. The necessity for this is that the corrosion resistance of austenitic steel relies largely on the fact that it is free from ferrite and is, in fact, a single-phase material. Therefore, where corrosion resistance of a high standard is required, it is commonly specified that the ferrite percentage must be at a low level.

There may also be occasions when materials must have no magnetic properties, for example when navigation instruments are involved, or with some nuclear applications.

The ferrite test makes use of equipment which measures the amount of magnetism in the metal. The same equipment which is used to measure the thickness of paint or electroplating deposits on steel can be calibrated and used to estimate the ferrite content.

It is usual that an agreement is reached between supplier and client regarding the method of testing, equipment used and level of magnetism (ferrite content) which is acceptable.

It is possible using chemical analysis to accurately estimate the composition of the alloy and, thus, to predict the amount of ferrite theoretically present. This makes use of the Schaeffler diagram which shows the ferrite content related to the nickel, chromium and molybdenum analysis.

'Ferrostan' Process
A method of continuous electrolytic tin plating.

Application range STEEL STRIP

The finally cold, reduced strip is continuously fed through cleaning, etching, plating and swilling processes required in electroplating. The solution is generally an acid sulphate which gives a matt finish. This can be brightened by flowing the coating by subsequent heating to above the melting point of tin, but more commonly, use is now made of brightening additives included in the plating solution, giving a bright deposit as plated. This is a standard tin electroplating system, carried out continuously with the strip moving through the necessary solution at up to 30 mph.

Ferrous Blasting

A term sometimes applied to the use of steel or cast iron shot in blasting. It is used to differentiate from iron-free grit which is specified for some purposes. It would not be expected to cover the use of slag or iron contaminated grits.

'Fescolizing'/Fescol' Process

A proprietary name for hard chromium plating, which may or may not include an underlay of nickel.

Application range ALL METALS WHICH CAN BE ELECTROPLATED, BUT GENERALLY STEEL

The process is used on new production parts to produce hard-wearing, low-friction surfaces. It also finds considerable use as a salvage process where hardened parts have become worn in service. Details of this process can be obtained from Fescol Ltd. and further information on chrome and nickel plating are given in Part Two **Electroplating**.

File Test

A form of hardness testing using the scratch principle, where the test material can only be cut by a material harder than itself. Described fully under *Mohs Test* in Part Two **Hardness Testing**.

Fillet Welding

A weld join where two pieces of metal at right angles to each other are joined by a weld run.

Application range ALL METALS

This can be achieved by a single weld run on one side, or a double fillet weld on each side of the vertical member. Unless care is taken to weld at 45° to each member, then one side will have good fusion, while the other side or leg of the weld will have little or no fusion. Many of the welding techniques described in Part Two can be used to produce the fillet weld.

Fingerprint Testing

A destructive test which is applied to plated components, most generally zinc plating, to assess the ability of the deposit to withstand corrosion caused by handling.

Application range ZINC AND CADMIUM PLATING

A drop of the solution is placed on the test surface and the time taken for gassing to commence measured. Plating deposits which can withstand longer than 30 seconds without gassing are considered satisfactory.

The solution used is as follows:

Lactic acid (80% solution) 25 ml.
Sodium chloride 5 g.
Distilled or demineralized water 75 ml.

Care must be taken to avoid contamination with copper and nickel. Solutions are only stable for one week.

Alternatively two solutions may be used which are mixed prior to use.
Solution A sodium chloride 10 g dissolved in water (distilled or demineralized) to 100 ml.
Solution B lactic acid (80% solution) 50% by volume with water (distilled or demineralized).
Mix equal volumes of solution A and B immediately before use. Solutions A and B are stable for up to 6 months.

This is a quality assurance test which is only of real significance when it is applied as a routine check on production. It is normally used in conjunction with salt spray, thickness and adhesion tests where high quality electroplating is required.

'Fink' Process

A form of *aluminizing*.

Application range STEEL

The steel components are, first, treated in a hydrogen atmosphere to remove any oxide, and then immediately immersed in molten aluminium. This results in an alloy forming between the steel surface and the aluminium.

With this process, *corrosion protection* is given by the alloy, rather than production of aluminium oxide as in most forms of *aluminizing*. The use of hydrogen increases the cost and difficulty of control, thus other methods of aluminizing are generally considered more satisfactory for most purposes.

'Fire Cracker' Welding

This is a welding technique used only for fillet type welds where the fluxed electric arc welding rod is laid along the metal to be joined. One end of the rod is connected to the welding generator and the far end is then electrically connected to cause an arc to be struck.

Application range ALL METALS

The technique uses a copper shield to reflect the heat of welding into the fillet being welded. The copper shield is not electrically connected to the welding rod but acts merely as a means of ensuring that the maximum heat is concentrated in the fillet welds.

The weld proceeds along the rod at a rate determined by the energy involved.

This technique should only be used where access is impossible with standard welding techniques.

The difficulty of ensuring a reasonable standard of fillet welds is high and thus it must never be used where any important welds are involved.

It must always be possible to design high integrity welds where access is available, and thus this technique should not be necessary where quality welding is essential.

Alternatives would be the use of *Thermit welding*, or *tack welding* using short electrodes or *brazing*. All of these techniques are described fully in Part Two **Welding**. Mechanical methods, such as bolting, riveting or gluing should also be considered.

Fire Gilt Process

A method of producing a coating of gold on a metallic substrate.

Application range COPPER ALLOYS AND STEEL

The process uses the fact that gold can be dissolved in mercury (gold amalgam) to give a paste, which can be wiped or smeared on the surfaces to be treated. When the article is heated, the mercury is driven off leaving a gold film. The process is confined to the jewellery trade, and with the advances in technique of electroplating, is becoming less common.

The amount of gold deposited will be a function of the percentage of gold in the amalgam and the thickness of the amalgam applied. The adhesion of the deposit will always be suspect, and depends on the cleanliness of the substrate and to a large extent the metal used.

A very considerable health hazard arises from the use of mercury, and this process is not advised except under carefully controlled conditions.

Firth Hardometer Hardness Test

A *Brinell*-type test. Described fully in Part Two **Hardness Testing**.

Flame Annealing

A term used to describe the method of softening using a gas flame for heating.

Application range ALL METALS

All the comments which apply to annealing, apply here. Very often this process is used to locally soften or stress-relieve, since this heating method can be extremely flexible.

Care must be taken as the flame can heat to well above the critical temperatures required for the hardening of steel, and it is essential to ensure that if the steel is heated to above red heat, it is slowly cooled otherwise the part will fail to anneal. This can be extremely dangerous if local areas are heated to above the critical hardening temperature, leaving the adjacent metal cold. When the flame is removed the cold adjacent metal will quench the hot metal causing hardening and possible cracking. Described fully in Part Two **Heat Treatment** under *Flame Hardening*.

Flame Cleaning

A method of removing paint and oxide from steel using a gas–air flame (See *Flame Descaling*).

Flame Cutting

A method of cutting metal, using a high-temperature flame.

Application range STEEL

The term may be applied to the use of the oxyacetylene hand-held torch method, but is generally applied to the production of shapes from metal plate, using equipment which automatically guides the nozzle.

Sophisticated pieces of equipment are now available, using laser beams to accurately follow lines on drawings, and flames which utilize modern technology with a variety of gases from hydrogen lancing, propane, oxyacetylene to plasma cutting. Multihead cutters can produce simultaneously a number of identical shapes. Plate thickness in excess of 4 in (100 mm) can be cut to produce intricate shapes with surface finish and tolerance comparable to that produced by machining techniques. With the plasma torch metal thickness in excess of 40 in (1 m) can be cut.

However, an oxide will always be produced on the surface, and this may be adherent and abrasive, thus it will aid corrosion resistance, but may interfere with subsequent *welding* or *painting*.

With some steels, flame cutting can result in local surface hardening which may cause cracking. Where high-integrity components are involved, it is always advisable to consider pre- and post-heating. This can usually be achieved with a second flame, either before or after the cutting nozzle. The importance of this will depend on the steel used, the thickness involved, the speed of cutting and the uses to which components are put.

It is not advised that flame cutting of steel which is at a temperature below 20°C is carried out.

Flame Descaling/Flame Cleaning

This uses oxyacetylene or other gas burners which are specially designed to produce a flame with a large area, and used with a high-velocity flame in the highly oxidizing condition. This is used to burn off all existing paint and, if any oxide scale, or rust is present, it also will be removed. While the process can sometimes be automated, it is generally used for salvage or cleaning purposes and, thus, the ability to automate is severely limited. Usually the torches are hand-held, and may be used in conjunction with wire-brushing.

The procedure is that the equipment is brought on site and the flame lit and adjusted to a predetermined limit to a hard oxidizing, very hot flame. Specially designed burners can be used to give flames of considerable area. Oxyacetylene or butane gas give equal results, provided the correct burners are used.

The process can conveniently *descale* or clean structures such as mild steel bridges or other large areas which have been allowed to deteriorate. The resulting surface is of a high standard, free of all paint and corrosion products and, provided the flame is in the correct condition, the surface will have a reasonably adherent oxide film which can be sealed by treatment with the modern paint systems (see Part Two **Painting**). These must be applied immediately after the flame cleaning process.

This method of *descaling* is usually cheaper than *blasting* but does not achieve a satisfactory blast pattern, thus the subsequent paint system will not have as good adhesion. It has the advantage over blasting that inaccessible areas can be reached. Other methods such as acid *pickling* will be cheaper and more efficient when they can be applied.

Flame Hardening

A form of *surface hardening*, very similar to *induction hardening*, which can be an alternative to other forms of *Case Hardening*. Described fully under this heading in Part Two **Heat Treatment**.

'Flame Plating'

A form of *metal-spraying* which is carried out by the Union Carbide Co. It is not possible, at the time of writing, to purchase any equipment for the Flame Plating process, but it is possible to have the process carried out by the Union

132 / FLAME SCALING

Carbide Co., who have set up equipment at strategic points, for the engineering industry.

Application range ALL METALS AND SOME NON-METALS SUCH AS CERAMICS

The process can be used for the *metal-spraying* of any of the common metals, alloys, oxides or ceramics on to any common metal. The process, in many respects, can be likened to the firing of a gun. The spray material in powder form is loaded into the breech, together with the propellant which can be in the form of an explosive gas.

The process is controlled so that exactly the same amount of material is deposited with each explosion, and the explosive gas can be monitored to be such that the combustion products are inert; the explosion is caused by electrical discharge. The process is capable of being repeated several times a second.

The distance of the material being sprayed from the nozzle of the gun is predetermined, thus the impact force of the powder on to the metal is tightly controlled. Since the gas is of an inert nature, it will be seen that the sprayed metal impacting on to the component is free of oxidation, which is the usual cause of porosity with conventional spraying. The amount of metal deposited with each explosion can be controlled with narrow limits, and in general will be 0.0075 in (0.2 mm) thick by approximately 1 in (25 mm) diameter. With this method, it is possible to achieve high spraying temperatures, and at the same time high impact pressures with excellent control of the propellent gases. Whereas with more conventional forms, as impact pressures rise, control of the propellent gases is usually lessened and the danger of oxide trapping is increased.

With Flame Plating, it is possible to deposit all the normal metals. However, it is unlikely that the process will be economical where low-quality metal-spraying for purposes of corrosion protection is required, but with ceramic-type materials or material such as tungsten carbide and the cobalt alloys (Stellite–Haynes etc.), where the resultant product is required to be of very high integrity, the method will have considerable economical and technical advantages. The only comparable technique is the plasma method, which is described under *Metal-spraying*.

With Flame Plating then, it is possible to deposit a controlled thickness of metal from under 0.010 in (0.25 mm) to over 0.025 in (0.6 mm) with outstanding adhesion and an absence of porosity or oxide trapping. The process is expensive and, thus, only justifiable when high quality is essential.

Flame Scaling

An alternative name to *flame cleaning*, where the fierce, oxidizing flame is used to remove scale from any component. The process can also be used during the steel-manufacturing process to remove scale formed during previous processes and, thus, eliminate the danger of scale being trapped at subsequent *rolling* or *forging* operations.

The term is also used to describe a process for *zinc coating* on wire. The wire is *annealed*, *pickled*, *fluxed* and *galvanized* in the normal manner by passing through a bath of molten zinc. While molten, the zinc galvanizing is wiped to give an even coating and is then passed through the oxyacetylene flame which re-melts the zinc to give better adhesion.

Flash Butt Welding (Flash Welding)
A form of *forge welding* which uses electric resistance heating with some arcing to supply the necessary heat. It can be used for a large variety of metals and is one of the welding techniques used to join dissimilar metals. Described fully in Part Two **Welding**.

Flattening Test
A form of *bend test* which is applied to tube.

Application range STEEL

The test uses a length of tube, and sometimes the length will be related to the diameter, or the wall thickness of the tube, with some specifications. The test piece is then squeezed to cause flattening. For tube manufactured from longitudinal welded materials, the weld must be at the point where maximum bend will occur, and this will be at 90° from the part being flattened.

The specification will detail the degree of flattening required, and this can vary from as little as one-third of the diameter to complete closure with thin walled tubing. The specification may also require that a record is kept of the difference between the size with the load applied and the size when the load is removed. This difference will give some information on the elasticity of the material.

On removal from the device causing flattening, the tube is visually examined for evidence of bursting or cracks at the area of maximum bend. Some specifications will require that all edges are radiused, while others will allow slight bursts at the edges.

It will thus be seen that this test, like the other *bend tests*, requires considerable specification and it should not be carried out unless all this information is available, otherwise the results obtained may well be meaningless, and can result in argument regarding interpretation.

Flaw Detection
An alternative name for *non-destructive testing*. Described fully in Part Two **Non-destructive Testing**.

Flex Testing
A method of testing sheet metal to ascertain its capability to accept *bending* and *drawing* stresses.

Application range ALL SHEET METAL

In its simplest form this test consists of turning back a corner of the sheet metal to be used, the operator assessing its stiffness. A 'flex tester' can be used to quantify this resistance to bending and will measure the load required to produce a specific amount of movement.

Metal which can be readily bent round a small radius will be more capable of *deep-drawing* than materials which are stiff or crack. The radius must be such that stretching of the metal surface occurs.

This is a specialist, locally destructive, but highly economic test, and is a cheaper alternative to the *Erichsen cupping test*, and certainly much cheaper than a full *tensile test*. It does, however, rely heavily on the skill of the operator and this has obviously disadvantages. Used as a quality assurance test on a routine basis, it can supply considerable information on sheet metal.

Flow Soldering
A method of joining by soldering.

Application range COPPER AND ITS ALLOYS

A high-production technique, where assemblies to be *soldered* pass across the molten solder. The solder will be in the form of a wave, and the components are generally conveyorized and preheated and fluxed prior to passing over the solder wave.

It is confined to high productivity assemblies such as printed circuit boards where shapes are quite simple, but many components are involved.

Fluidized Bed Heating
A method of heating components using hot gas and fine powders.

Application range ALL METALS

The technique is that the hot gas is blown through the fine powder by carefully designed and located nozzles.

When the correct pressure and gas distribution is achieved the powder acts in an identical manner to a liquid.

The technique is now also a method of feeding solid fuel such as powdered coal to a furnace, and also to coat articles with plastic powder.

When used for heating metals the powders used are inert and are heated by the products of combustion of fuel. The gases can be inert, for example hot argon could be used if desired, but are more commonly hot carbon dioxide with water vapour, carbon monoxide, etc. which can if desired be controlled to give a reducing atmosphere and thus prevent scale formation, and are in fact products of combustion.

Considerable control is then required to ensure the atmosphere is consistent at all times.

The components are immersed in the hot powder when the necessary heat transfer takes place.

The technique has the advantage over salt baths that on removal from the fluidized bed the components are clean and dry.

Fluxing

This is seldom a process in its own right but is commonly part of other processes. These processes will almost invariably involve heating of metals and the purpose of fluxing is, first, to remove or prevent oxidation; second, to alter or control the oxidation products so that they do not interfere with the end-product; and finally to reduce the melting point of the material to give improved fluidity.

Application range ALL METALS

There are four main processes using fluxes:

1. *Casting* Fluxes are used to reduce or eliminate oxidation, and at the same time to confine the products of oxidation and reduce their melting point. They may also be used to improve the fluidity of the molten metal and thus allow the undesirable solids to float to the surface. Fluxing may also be used to de-oxidize, where additives may be included in the flux.

2. *Welding* Fluxes are very commonly used during *welding* and *brazing*. The purpose here is first to remove surface oxidation and to prevent further oxidation occurring. This means the flux must be able to attack, and dissolve surface oxides, and to prevent further attack must exist as a liquid, non-porous blanket covering the affected area at the weld temperatures.

Second, fluxes are designed under some circumstances to increase the fluidity of the cast weld metal. This allows the solid oxide particles and other impurities to rise and float to the surface, and thus not become trapped in the finished weld. The increased fluidity may also result in a reduction of gas porosity as the trapped gases can also readily rise to the surface.

Third, under some circumstances, fluxes are used as an insulating blanket, to reduce the rate of cooling. This is often desirable in *welding*.

Finally, the flux can be used to add any necessary alloy element to the weld metal. It is sometimes easier to weld with a simple filler rod than an alloy material but, for purposes of mechanical strength, the weld itself requires to be alloyed.

3. *Soldering* Fluxes are used in *soldering*, first to scavenge any remaining

oxide from the surface and, second, to prevent oxidation occurring. This means that the choice of flux will depend on the state of the surface. With normal engineering-type *soldering*, and to some extent *brazing*, the flux will be an active agent containing free acid which attacks oxide and dirt on the surface. For high-integrity *soldering* in the electronics and electrical industries, these fluxes are not permitted, since a serious problem with corrosion will exist and washing of electronic equipment is not always possible. The electronic fluxes are, therefore, much less active and this means that the component must be presented for the *solder* or *braze* thoroughly clean and free of oxidation.

4. *Galvanizing* Immediately after *pickling* and prior to hot-dip *galvanizing*, components must be fluxed to give the necessary alloying of the zinc and iron. The most common flux used is ammonium chloride, which may be applied as a liquid solution or as a hot-dip or, in some cases, as a blanket on the molten zinc.

From the above, it will be seen that fluxes are used for cleaning, for removal of oxides and for reducing the melting point to act as a 'blanket' to prevent oxidation.

Footner Process

This is a method for removing relatively severe corrosion, or light scale, and producing a surface suitable for *paint* adhesion.

Application range STEEL

The technique is that the components are first *pickled* in hot 5 per cent sulphuric acid for up to 30 min. They are then removed, washed and rinsed in hot water, then immersed in phosphoric acid usually containing some iron. This results in the production of iron phosphate on the metal surface.

It will, therefore, be seen that this process is a combination of standard sulphuric acid pickling and phosphating. There is no reason why pickling in other acids should not give the same success, if followed by any of the standard phosphating methods.

Foppl Test

A hardness testing technique now seldom used. Described fully in Part Two **Hardness Testing**.

Forge Weld

The original method of welding carried out by the skilled blacksmith. With this, the metal being joined does not melt or fuse, but is joined by genuine forging pressures on the metal. The resultant join very often cannot be detected by any means and is thus the 'ideal joint'. Described fully in Part Two **Welding**.

Forging
A manufacturing production technique which can be applied to most, but not all, engineering components; it is thus not a metal treatment.

Application range ALL METALS

Briefly, forging covers any method where work is applied to the material. Ingots of cast material can be forged but forging is also commonly applied to bar material which is already in the forged condition. The techniques of rolling, drawing, extruding, are all forms of forging.

Forging should eliminate all evidence of cast structure, and is carried out over a range of temperatures, from room temperature to just below the liquefaction temperatures.

Certain materials, notably cobalt alloys, and some pure metals such as chromium, cannot accept any form of working without shattering and, therefore, cannot be considered for this technique by design engineers. The user is advised to obtain specialist advice, if the choice between the present method and casting is important.

Forward Welding
Also called 'forehand welding'. This is the name given to that welding which proceeds in the direction most suitable to the welder, most commonly from left to right. The term then is applied only to hand-welding and in general is restricted to the electric arc process; but there is no reason why it should not be applied to gas welding.

'Foslube'
A proprietary, organic-type lubricant which is impregnated into the phosphated surfaces.

Application range STEEL

In order to be successful, this impregnation must be carried out immediately following that operation. The technique is described in more detail under the heading *Phosphating*.

Fracture Test
A simple, destructive mechanical test, designed to furnish information on *ductility*, and can be used as an assessment of welding. Information can also be

obtained on serious material defects by examination of the fracture face. Described fully in Part Two **Mechanical Testing**.

Fremont Hardness Test
A hardness test. Described fully in Part Two **Hardness Testing**.

Fremont Impact Test
A former type of *impact test* where a known weight was allowed to fall from a known height on to a test piece to cause fracture. The test piece was 3 cm long, 0.8 cm by 1 cm in section, with an underside notch 0.1 cm deep and 0.2 cm wide. The test piece was supported in a horizontal manner over a gap 2 cm wide, with the notch on the underside.

The test relied on a copper cylinder being struck immediately after the fracture of the test piece by the same weight used to cause fracture. The more brittle the material, the greater the compression of the copper.

The test has been replaced by the *Charpy* or *Izod* method. Described fully under *Impact Test* in Part Two **Mechanical Testing**.

Friction Welding
A form of *forge welding* where the heat required is obtained from friction when two surfaces are rubbed together under pressure. At the appropriate time, upset pressure is applied. Described fully in Part Two **Welding**.

Frosting
A type of metal finishing where a fine matt finish is produced.

Application range ALL METALS

The term is general and includes the use of single techniques such as *etching*, *blasting* or, under certain conditions, *scratch-brushing* or even *barrelling*.

The finish produced may be made more permanent either by the anodizing method on aluminium or, in some cases, by a thin *electrodeposit* very often chromium but sometimes nickel or zinc on steel. *Painting* can also be used to produce this finish on any metal.

Fuess Testing
A method using the spark discharge technique of identifying different metals or alloys.

Further information will be found on this technique under *Metal-sorting*.

Full Anneal

The term now generally used when steel components are heated to above the upper critical temperature and slowly cooled. Described fully under this heading in Part Two **Heat Treatment**.

Furnace Brazing

A form of brazing which uses a standard furnace set at the required temperature to cause melting of the braze material.

Application range ALL METALS JOINED BY BRAZING

Since no attempt is made to alter the atmosphere of the furnace, conditions will be oxidizing and assemblies must be fluxed to allow brazing to proceed. Consequently the method is very similar to hand brazing using either a gas torch, or the induction technique, and the same problems exist (see the section on *brazing*).

The advantage here over the above methods is that a batch of components can be brazed simultaneously. This means that the components must either be jigged or, more commonly, located in the assembled condition by using a tight fit. Copper brazing is commonly carried out with this technique and very often the copper braze material is applied by electro-plating. The components can then be assembled using the copper to give an interference fit and thus hold components in the correct position. It is important to ensure that the correct clearance, less than 0.001 in (0.025 mm) exists prior to plating or to the application of filler metal.

On removal from the furnace, the components must be treated in some manner to remove all evidence of flux, which being corrosive will result in corrosion of the assembly during service. While blasting under certain circumstances can successfully clean components from flux, it is possible that the blast will not reach all recesses where flux may be trapped, and washing with a dissolving liquid is recommended.

Furnace brazing may also be carried out with a controlled neutral or reducing atmosphere. Provided the components are clean, no flux is required. This method will be similar to hydrogen or vacuum brazing.

Fluxes are now available which are active at low temperatures. They therefore successfully prevent oxidation and remove slight traces of surface contamination, which above a certain temperature become non-active, thus leaving no corrosive deposit. Provided that these are used, no washing or cleaning for flux removal is necessary after furnace brazing. This removal of corrosion products from the heated braze area is an advantage over hand or induction brazing, but flux which has not been taken to above the critical temperature will, of course, remain corrosive. With furnace brazing, all the components are taken to the brazing temperature.

Aluminium and many of its alloys can be furnace brazed using 10–12 per cent silicon aluminium alloy which has a low melting point. It is possible to obtain some sheet alloys clad with this material, which is designed to simplify aluminium brazing, including furnace braze.

Fusion Welding

The original welding method was carried out by the blacksmith without any melting or fusion. With the advent of oxyacetylene and later electric arc welding, metals were fused together by melting the filler material and the adjacent metal. Described fully in Part Two **Welding**.

Galvanic Protection

A term used in *corrosion protection*, where the surface coating applied will sacrifice itself to protect the underlying metal.

Application range ALL METALS

The surface is protected by a metal which has a lower electrode potential than itself. By far the most common metal used is zinc to protect steel. The process of coating steel with hot molten zinc is known as *galvanizing* and the *electroplating* of steel with zinc is often referred to as *electrogalvanizing*.

Technically, any metal used to protect a metal higher than itself in the electrode potential series would supply galvanic protection but, in practice, only zinc is referred to in this context. Further details of this method will be found under *Zinc Coating* and *Corrosion Protection*.

Occasionally the term may be used when anodes of zinc, aluminium or magnesium are used. See *Cathodic Protection*.

Galvanizing

This is the application of zinc to prevent corrosion.

Application range STEEL

The temperature of molten zinc is above 500°C, and this precludes the use of galvanizing on any steel which is tempered below that temperature, since the process will then result in softening.

The procedure requires that the components are in a moderate state of cleanliness, that is all excess oil and grease are removed, and any oxide or scale which is not eliminated with light acid *pickling* must be removed by *blasting* or some other method.

The components which are to the required standard of cleanliness are subjected to an acid pickle either with hydrochloric acid, generally with a strength of 10–20 per cent, used cold, or hot sulphuric acid, generally with a strength of 25–50 per cent. The choice of acid for the pickle will depend on various factors, such as economy, capital equipment, etc. (These are discussed in more detail under *Pickling*.)

On removal of the components from the pickling-tank after the required time, which will depend on the degree of corrosion, scale or oxide which is present, the parts are washed in cold, running water. This wash is not as critical as with other processes, but it is important that excess acid is removed as quickly as possible and the parts passed to the next operation of *fluxing*. Under certain circumstances, for example, where work is of a relatively high standard regarding surface contamination, then no *fluxing* will be required. The components will be immediately immersed in the molten zinc. In general, however, it is necessary that all parts are fluxed, the purpose being to remove the last traces of oxide or corrosion, these then appearing as oxides or slag on the surface of the molten zinc.

The method of *fluxing* will vary under many circumstances and can be applied as a powder to dried, warm components or to cold, damp components. It can also be applied as a solution in water or, under some circumstances, the molten flux is floated on the surface of the molten zinc and the components are fluxed as they pass through this blanket (see *Fluxing*).

It is vitally important that the components are thoroughly dried to remove the last traces of moisture prior to entering the molten zinc. This is generally achieved by the components being removed from the flux solution and oven-dried. This can be varied by the components being preheated in the flue of the molten zinc equipment, or with small components of a simple shape, by slowly immersing in the flux blanket floating on the molten zinc.

As the galvanizing process is carried out at temperatures above 450°C, any moisture which enters the molten zinc will immediately be converted to steam with explosive results, which are highly dangerous as the molten zinc is ejected from the bath. The dried, *fluxed* components are gently lowered into the zinc, with the operator positioned behind a protective barrier.

The components are left in the molten zinc for a specified time. This will generally be long enough for them to reach the temperature of the zinc. If the components are removed before they reach this temperature, the zinc coating will be relatively thick and can be patchy with poor adhesion. If the components are left too long in the molten zinc, there will be a tendency for the iron–zinc alloy, which forms at the interface and is relatively brittle, to be excessively thick with resultant poor-quality galvanizing.

On removal from the zinc, considerable skill is required to reduce the tears which form on the bottom corners. Apart from being unsightly, these tears can cause considerable problems at subsequent machining or fitting operations, and with high-quality galvanizing, must be reduced to the minimum. The components, under some circumstances, can be quenched.

The temperature at which galvanizing is carried out is relatively critical and should be controlled. Various additions can be made to the liquid zinc to improve adhesion and appearance. One of the common additions is that of small quantities of aluminium which, when correctly controlled, give a bright, even zinc deposit. Unfortunately, excessive aluminium can result in brittleness.

High silicon steels – above 0.5 per cent silicon – give problems to the galvanizer as this encourages the formation of the brittle zinc–iron alloy. When it is suspected that high silicon steel exists the galvanizer should be advised to allow time for trials on test pieces.

Galvanic coatings will have a corrosion-resistance life depending on the thickness and the atmosphere under which the component is used. It is generally agreed that galvanizing gives a high standard of corrosion resistance, particularly when used under rural conditions. When used in a marine or industrial environment, corrosion resistance is reduced but this applies to other comparative processes in a proportional manner. Galvanized coatings of 0.005 in (0.125 mm) will give a life of over 10 years in industrial and marine atmospheres under most conditions, and over 20 years of rural life is normal.

The use of a chromate seal on the zinc surface will enhance the corrosion resistance.

When it is considered necessary to paint galvanized components, special etch treatments or etch primers are necessary otherwise paint adhesion will be found to be extremely poor.

Galvanizing results in corrosion resistance superior to *electroplating* of zinc, sometimes called 'electrogalvanizing' as the coating is appreciably thicker. The protection is comparable to that achieved by a good *painting* system using, for example, zinc-rich primers applied to a *blasted* surface and finally sealed with a high-quality top coat. The appearance is not particularly attractive and as stated above the evenness of the coating can be variable, thus precluding the use of galvanizing for components which require to have good visual finish and fine limits. Galvanized components should not be *welded* or *brazed* as this will destroy the zinc film. There is also a health hazard to operators as zinc fumes are toxic. Zinc will also result in the weld being brittle if the weld pool is contaminated with zinc.

Galvannealing

A modification of the *galvanizing* process.

Application range STEEL

On removal from the molten zinc, in place of the cooling, the components are retained at a temperature of approximately 450°C. This results in the zinc forming an alloy with the surface of the steel. The time for this to take place will be short and is a function of the zinc thickness, and to a lesser extent the *galvanizing* conditions, the steel or iron being treated and its surface condition. Excess time will be more damaging than the retention of free zinc on the surface from too short a time.

The end product has a more homogeneous deposit than with *galvanizing*, and being matt grey is more visually attractive than the normal finish. The deposit is brittle, however, and will crack if any bending moment is applied. Where prior *cleaning* and *fluxing* have been correctly carried out, the adhesion of the zinc–iron alloy will be good and flaking should not occur. The cracking will reduce the corrosion resistance of the coating, which is slightly inferior to the conventional galvanized zinc deposit.

The process is more expensive than *galvanizing* and results in a lower standard of corrosion resistance, but the homogeneous deposit is more attractive. The deposit can be coated with standard paints, the surface cracking and oxides ensuring good adhesion.

Gas Carburizing

A specific type of *carburizing*.

The process uses furnaces into which is fed the necessary gas atmosphere. The process results in a case very similar to that achieved by the *pack carburizing* process but with certain advantages, particularly economical, but also technical. Described fully under *Carburizing* and *Gas Carburizing* in Part Two **Heat Treatment**.

Gas Welding

A method of welding using the combustion of gas as the heat source. This will usually be fusion welding.

Application range ALL METALS WHICH CAN BE WELDED

The gases used most generally are oxygen and acetylene giving the well-known *oxyacetylene process*. There is, however, no reason why any other gas or gases which can burn to give the necessary energy in the form of heat should not be used. The heat of town gas and air for welding materials such as lead or tin is quite common, and the use of hot air for *welding* plastics is possible. The use of propane, butane etc. is now becoming quite common.

See also Part Two **Welding**.

Gator-card

A form of metal and ceramic spraying using a modification to the plasma system.

Application range ALL METALS

The technique uses an inert gas which is ionized and passed through a nozzle. Particles of the metal or ceramic to be deposited are fed into the gas stream and are given a high velocity.

The high impact force thus obtained gives excellent adhesion to normal steel or other relatively soft metals. Surface roughening prior to spraying is not essential.

The process is generally applied only to sophisticated components where high integrity regarding adhesion, lack of porosity and controlled thickness is essential.

Gilding

A process in which gold is coated on the surface of another metal.

Application range MOST METALS

There are several techniques used:

1. Gold leaf can be applied to metals and other materials. Pure gold is normally used but certain alloys of gold are possible alternatives. The gold is rolled or beaten into very thin sheet, and this can be of such thinness that it is transparent. It is one of the properties of gold that even at this extreme thinness the material retains its colour. There are two methods of gold leafing. Most commonly the leaf is glued to the surface of the article. The second method makes use of the ductile properties of gold and the leaf is beaten on to the article being treated. Obviously the second method generally requires a thicker leaf and greater manipulative skill. As stated any metallic or non-metallic material can be gilded but considerable skill is required to give a high quality treatment.

2. Another traditional method uses very fine gold powder which is then mixed with a liquid glue which is flammable. This mixture is then applied in a similar manner to paint and, on completion, the glue is allowed to evaporate or in some circumstances can be ignited.

3. Electrodeposition of gold or gold alloys. Described fully in Part Two **Electroplating**.

4. The application of gold using the amalgam technique. See *Amalgamating*. The component is heated to vaporize the mercury leaving the gold.

5. The chemical deposition of gold. This makes use of an unstable gold solution which breaks down in contact with certain metals and deposits a very thin layer of pure gold. This is *electroless gold plating* or *immersion plating*, and is described under *Electroless Plating*.

The gilding process can be compared to rolled gold where a very thin layer of gold is rolled or forged to the surface, generally copper alloy, and the resultant material is used to manufacture articles such as jewellery. The gold may be rolled on to the copper and the bimetal is then reduced in size, thus giving a very thin gold layer. It will generally be an alloy of copper which is used as the substrate.

Gilding can be a high-quality process and the quality will be a function of the method and the thickness applied. However, particularly with electrodeposition of gold, there is no obvious advantage to deposits above a certain critical thickness: below this thickness the deposited gold should be free of pores to achieve a satisfactory corrosion-free surface. Obviously, where articles are designed to be handled or used continuously, the thicker the deposit, the longer the life of the coating. Gold is a very soft metal which will not stand abrasion. It does not oxidize, and thus requires no polishing.

The quality of the process will depend to a large extent on the skill of the operator. It can, however, be stated that gold leaf correctly applied will generally give a higher-quality product than any other method of applying gold metal directly. Electrodeposition results in gilding of a high standard, while the electroless deposit will invariably result in a relatively lower standard since the deposit is extremely thin and the process is difficult to control. The technique using amalgams or gold paints will give inferior gilding.

Glass-bead Blasting
A specialized technique using glass balls. This is a gentle type of blasting. Described fully in Part Two **Blasting**.

Glueing
This term covers the method of joining where adhesives are used.

Application range ALL METALS AND OTHER MATERIALS

The technique is to clean and prepare the surface and then to apply the glue in one of several forms to either a single or both surfaces. The surfaces are then brought together and either left alone, or pressure applied or heat and pressure. There are some very sophisticated techniques now available where high vacuum is applied along with heat.

There are a large variety of substances used as glue or adhesives. These vary from the simple animal products, very often based on gelatine, to very sophisticated organic materials which have very high strength and very high adhesive properties. Many modern glues are two-pack requiring mixing immediately before application.

Without exception the preparation of the surfaces is of considerable importance. Unless the surfaces are correctly prepared, the standard of adhesion can be expected to be very low. In many cases the glue will contain a solvent or additives which attack the surface increasing the adhesive properties. The reader in general is advised to follow the instructions very closely under all circumstances.

Modern glues can be applied to join metals giving a joint strength which can be compared in many circumstances to that of soldering or brazing. Certainly where heat cannot be applied for any reason, a glue can be expected to give

146 / GOLD PLATING/GILT PLATING

comparable or better results than other methods of joining. The *'Loctite'* process is a form of glueing.

Gold Plating/Gilt Plating

Gold is plated for decorative purposes, but also for its excellent corrosion resistance and constant electrical-contact conductivity. Described fully in Part Two **Electroplating**.

Goldschmidt Process

A method of fusion welding using heat from chemical reaction.

Application range STEEL

This process uses oxidation of aluminium powder to aluminium oxide in the presence of iron oxide or, in some cases, magnesium and chromium oxide to achieve a high quantity output of concentrated heat. The process is used by mixing the finely powdered aluminium with the finely powdered oxide and igniting the mixed powder. Depending on the process used, the mixed materials are held in a refractory mould or by some similar method. The process is commenced by igniting the mixture either by gunpowder, or by using electrical resistance or spark discharge.

Once ignition has commenced the oxidation of the aluminium powder results in the release of considerable energy as heat. As the powders are intimately mixed the oxidation of the aluminium makes use of the oxidized iron, chromium or magnesium, and these metals are therefore reduced from the oxidized state to that of a relatively pure metal. The aluminium oxide formed will rise to the surface of the molten metal, but it is impossible to obtain the metal in a relatively high state of purity with little or no contamination by its own oxide or the oxidized aluminium. Very little use is made of the process with chromium or magnesium or other metal oxides. This is a method of joining steel under certain circumstances and has the advantage of being extremely portable.

This is a rather exotic technique of achieving a very concentrated heat using a chemical reaction. It therefore has the advantage of being portable but there are now many more attractive alternatives for portable heat. The process using iron oxide is described fully under *Thermit Welding* in Part Two **Welding**.

Graduated Hardening

An alternative name for *interrupted quenching*, *austempering* or *martempering*. With this the steel, which will generally be a tool steel of high or medium-high carbon with alloying contents, is quenched into a hot liquid, usually salt, but very often a molten metal such as lead. The steel is held in the quenching medium for a specific time and is then cooled to room temperature either in air or oil. Described fully under *Austemper* and *Martemper* in Part Two **Heat Treatment**.

Grain Refining

A heat treatment operation designed to produce fine grains.

Application range STEEL

With other metals, the grain refinement will generally require some form of working in order to break down the grain structure. With steels which are capable of being recrystallized by *heat treatment* alone, the procedure is that the material is heated to above the critical temperature when crystallization will take place. Whenever steel is heated to above this critical temperature grain refinement will occur.

It is only, however, the process carried out after the carburizing operation that is referred to as 'refine'. An alternative name would be 'core harden'.

When steel is *carburized*, it is held at a temperature of approximately 900°C for several hours. During this, grain growth is a continuing process, and at the end of the *carburizing* cycle, it will be found that the steel has an excessively large grain and will, therefore, be brittle. This brittleness will apply to both the case and the core, but if only the case is hardened, then the core is not taken above the critical temperature which causes recrystallization to occur. It is, therefore, essential that a separate operation is carried out to improve the ductility of the core where high-integrity components are involved. Described fully under *Refine* in Part Two **Heat Treatment**.

The *normalizing* process, too, is used for grain refining, but this operation will in addition remove any evidence of *work hardening*, and the prime reason for using the method is to achieve a homogeneous grain size rather than a small grain. Grain refinement apart from the above is not normally carried out in its own right, but will almost invariably be achieved when heat treatments such as *hardening* or *solution treatment*, are carried out.

By holding the steel too long at the correct temperature, or using too high a temperature, grain growth will take place. This can only be recovered by re-treating at the correct temperature for the correct time.

'Granodizing'

A proprietary name given to a *phosphating* process and the complete associated processes, covering *cleaning*, *phosphating* and *chromate* sealing. Information can be obtained from ICI Ltd. Further details on the *Phosphating* process are given under that heading.

'Granosealing'

This is the proprietary name given to a *chromate* seal which can be carried out following the *phosphating* process. This generally follows the *'Granodizing'* process.

Application range STEEL

The purpose of the *chromate* seal is to impregnate the phosphate film as this has been found to considerably enhance corrosion-resistance properties of the phosphate without in any way affecting the quality of paint adhesion. Information on this process can be obtained from ICI Ltd. See also *Chromating*.

Graphitizing

This is designed to convert iron carbide to graphite.

Application range CAST IRON AND SOME SPECIALIST STEELS

Described fully in Part Two **Heat Treatment**.

Green Gold

An *electrodeposit* of gold; it is an alloy containing small amounts of cadmium or silver.

Application range ANY METAL WHICH CAN BE PLATED

Addition of silver to gold tends to convert the reddish-yellow into green, and gold containing 25 per cent silver is a definite green. Twelve per cent cadmium will give the same effect and an alloy of gold, cadmium and silver is also green. It is also possible to achieve a green colour by alloying with nickel.

The solutions used are generally proprietary, and the user is advised to obtain specialist advice regarding the conditions of plating.

Information on *Gold Plating* will be found under that heading in Part Two **Electroplating**.

Grit Blasting

A technique of abrasive cleaning or surface preparation using sharp particles.

This is a relatively vague term. Described fully in Part Two **Blasting**.

Haig Prism Test

A test for hardness, seldom used today. Some details are given in Part Two **Hardness Testing**.

Hammer Welding
This is an alternative name for *forge welding*.

Application range ALL METALS WHICH CAN BE FORGED

Described fully under *Forge Welding* in Part Two **Welding**.

Hand Brazing
A form of *brazing*, usually carried out with hand-held torch to supply the necessary heat. The term, however, may refer to the *induction brazing* technique, where the induction coil is used as a means of heating individual components, these being fed through the coil by hand. This term, then, does not attempt to specify the brazing material used or the method of heating. The reader is referred to the heading *Brazing* for further information.

Hanson–Van Winkle Munning Process
A method of descaling.

Application range STEEL

The component, or material, is first made the cathode in a warm sulphuric acid solution. The temperature and amperage are not critical, but the strength of the solution must be maintained within the required range. If the acid strength falls, the attack on the steel will increase and may cause pitting. Any increase in the strength will reduce the attack, and may not remove the scale.

When the scale has been removed, which takes several minutes, the parts are removed from the solution, drained or swilled, then immediately made the anode in a 40–50 per cent sulphuric acid solution at room temperature. This results in a matt-silver finish, which will rapidly corrode unless thoroughly washed immediately, neutralized, dried and protected.

As a method of producing an attractive surface finish, this process has been superseded by various *grit blasting* techniques, notably *dry blasting*, with fine-mesh grit, and *vapour blasting*. *Barrelling* techniques will produce a comparable finish.

However, this technique of double *electrolytic etching* in sulphuric acid is still practised as a method of preparing steel surfaces prior to *electroplating* where adhesion of a high standard is essential.

Hard Chrome Plating
The term used to denote the deposit by electrolytic action of chromium which is used for engineering purposes. This distinguishes the deposit from the decorative bright chrome plating. It is very often deposited directly on to the component, but can have an undercoating of nickel.

Hard chromium plating will generally be deposited to give a thickness of at least 0.001 in (0.025 mm) and can be to a thickness of 0.030 in (0.75 mm).

Flash plating with 0.0001 in (0.0025 mm) chrome plating can improve where light shading loads are involved.

Described fully under *Chromium Plating* in Part Two **Electroplating**.

Hard-drawing

A manufacturing process, and not a metal treatment which applies cold work causing hardening.

Application range ALL WROUGHT METAL

The process is generally applied to *wire-drawing* but can be loosely applied to any wrought form.

Briefly, the material is 'drawn' through a die or in some cases may be 'extruded' through a die, or rollers can be used. As the material is reduced in section, it is *cold worked* and, thus, increases in hardness. Depending on the material used, this increase in tensile strength will be a function of the *cold work* applied. This will always be accompanied by a decrease in ductility, and with some material this can be unacceptable.

The process can result in comparable tensile properties to that obtained by *hardening* and *tempering*. With steel products, there is a relationship between the 'hardenability' by *heat treatment* and the ability to accept cold work by the 'drawing' process. Additional information is given under *Work Hardening*.

Hard Facing

A process usually applied to either *metal-spraying* or, more commonly at present, to *weld deposition*. The term is occasionally used to describe the *chromium plating* and *electroless nickel plating* process. In addition, the deposition of a corrosion-resistant film is sometimes referred to by this term, since most hard facing deposits are also corrosion-resistant.

Application range ALL METALS

Hard facing using *Metal-spraying* techniques is described in some detail under that heading.

Hard facing *weld deposition* is most commonly applied to materials such as mild steel or stainless steel which have either relatively poor corrosion resistance or, more generally, have low surface hardness, and the purpose of hard facing is to give a wear-resistant surface. The process is, in essence, a *welding* process, usually applied by hand and using the skill of the welder, although it is possible to use automatic techniques for higher productivity, provided the shape of the component lends itself to this type of operation, and there are sufficient quantities involved to warrant the capital costs involved.

The technique will usually use *oxyacetylene* or *electric arc welding* methods with hand-held rods of the desired composition. The most commonly applied hard facing materials are the cobalt metals under the general name of 'Stellite', or 'Haynes alloys', but the use of 12 per cent chrome steels is now possible.

With *oxyacetylene welding*, the base metal is brought to temperature and *fluxed*, and the hard facing metal is then melted and deposited on the surface of the component. Under many circumstances, it is undesirable to have excessive or even limited weld penetration, since it is found that excessive weld penetration results in dilution of the hard facing material by the base metal and, thus, the technique requires that the hard deposit is laid on to the metal with the minimum amount of puddling or working. As this is generally recognized as being poor technique, it is often found that the use of skilled welders for hard facing operations is unsatisfactory. These remarks apply to the relatively thin hard facing deposits commonly found in engineering, where the purpose is to act as a bearing diameter or a sealing diameter on relatively small components.

There are now large chemical and civil-engineering components in the form of vessels and heavy components of various types where hard facing is required, and the deposition will be by the higher-productivity *electric arc welding*. Using this technique, the rod will be flux-coated in the normal manner and deeper penetration welds than with *gas welding* are normally accomplished.

To obviate the problem of diffusion of the base material into the facing material, it is common to apply two layers of the hard facing material. In some circumstances the first layer will be of a different chemical makeup from the top layer, or layers. This initial deposit is sometimes referred to as *buttering*, and on occasions the complete process is said to be 'buttered'.

The hard facing technique is identical to that of depositing a corrosion-resistant surface, and in many instances the same materials are used, the purpose of the deposit being to improve resistance to corrosion and wear. However, it is possible to deposit an austenitic-type stainless steel with this technique and, since this does nothing to enhance wear resistance, it should not be referred to as a hard facing.

Distortion problems exist with hard facing using the *welding* technique, and this distortion will be greater than that found with the *metal-spraying* technique. The control of surface dimensions on the as-deposited component will also be better with spraying, but it is more economical to deposit a thick layer with *welding*, and the adhesion will generally be better than with metal spraying.

Hard facing, then, is an economical method of achieving a wear-resistant surface, either locally or over a complete surface. It is generally cheaper than other methods of producing a hard surface, such as *carburizing*, *nitriding*, *chrome plating*, etc., but is generally chosen either because the size of the component precludes these methods, or because some corrosion resistance is also required. It can be applied on site and, thus, has the advantage of portability. Since it is a *welding* process and, thus, prone to all the disadvantages of that method regarding stressing and cracking, it should be helpful to refer to the remarks on these problems under Part Two **Welding** before it is contemplated to subject an unknown material to hard facing by *weld deposition*. Serious cracking can occur if an existing hard steel has a weld deposit applied, unless stringent control is applied.

Hard Plating

A term which generally refers to the *hard chromium plating* deposit, but can cover any form of plating where surface hardness is increased.

Application range ALL METALS

It includes *facing* by one of the *metal-spraying* techniques. This, then, can be applied to any metal which can accommodate the plating process, and the term is seldom used except to cover *electro-* or *electroless deposition*. Additionally it may be found in specialized use or related to a local area of a company.

The most common methods of hard plating, in order of popularity, are as follows: *hard chrome plating*, *electroless nickel plating*, '*flame plating*' using any of the hard metals deposited by this technique and deposition by *metal-spraying* and *welding*. Further details of each of these techniques will be found under the relevant headings.

Hard Soldering

A term used for a higher melting point solders than the common variety.

Application range COPPER, ALLOYS AND STEEL

Solders melting at about 250–270°C are termed 'soft solders', whereas those melting in the region of 300°C or upward are termed 'hard solders'. The actual increase in hardness is not to any great extent. These solders are the tin–lead variety with additions of indium and other elements.

They are used in the same way as normal tin solder and have two prime uses. First, when two or more solder joints are required with the joints relatively close together, the first joint will be made with the hard, higher-melting-point metal, and thus when the second is made using the soft or lower-melting-point solder there will be less danger of the first joint failing during the *soldering* operation. The second use is where joints of slightly greater strength than the tin–lead are required. The hard solders are generally found in use in the electronics industry, or to a lesser extent in engineering assemblies.

'Hardas' Process

A proprietary hard *anodizing* process, also known as *deep anodizing*, *refrigerated* or *black anodizing*.

Application range ALUMINIUM

The process uses refrigerated sulphuric acid as electrolyte and increases the surface hardness of aluminium and its alloy, with a resultant dark grey or black surface finish. The process is described in more detail under *Anodizing*.

Hardening

The most important of the *heat treatment* processes, which increases the tensile properties of metals. Described fully in Part Two **Heat Treatment**.

Hardness Testing

This is a test aimed at identifying by non-destructive, or very locally destructive, techniques the strength of a material and its ability to resist abrasion. There are a large number of techniques and tests available. Described fully in Part Two **Hardness Testing**.

'Harperizing'

A proprietary form of *barrel deburring*.

Application range ALL METALS

This uses the technique of having two relatively small barrels of the conventional type which are held between revolving parallel large-diameter plates. The barrels are each made to revolve epicyclically in the opposite direction, thus achieving conditions of high centrifugal force. The barrels contain the component to be deburred together with conventional ceramic chips, abrasive material and water.

The process is the fastest method of *barrel deburring* normally carried out, and further information on the 'Harperizer' can be obtained from W. Canning and Co. Ltd. (See also *Barrelling*.)

Hausner Process

A *hard chrome plating* process.

Application range STEEL AND COPPER

This uses the conventional solution, but has low-voltage alternating current superimposed on the direct-current plating. This has the effect of reducing the polarizing effect of the hydrogen deposited at the plating surface, and thus speeds up the rate of chromium deposition.

This codeposited hydrogen produces an electrical insulation effect on the surface which slows down the rate of chromium deposition and increases the thickness of deposit ratio between the high current density areas (corners, etc.) and the low current density areas (centre of flat surfaces). The alternating current helps to reduce the quantity of hydrogen at the plating surface; this improves the rate of deposit and at the same time reduces the difference between the deposit at corners and centres of flat areas. The technique requires special equipment and control. There are now several 'self regulating' proprietary solutions which achieve the same effect by chemical means.

The reader is advised to obtain specialist advice in this area.

Haworth Test

A quality-control technique to estimate the resistance of a surface to abrasion.

Application range ALL METALS

The component being tested is brought into contact with a revolving disc of metal or rubber and the chosen abrasive is fed on to the contact faces. The contact pressure, abrasive type and quantity are rigidly controlled. Use can be made of either an abrasive slurry or dry abrasive powder. The assessment will be carried out on a time basis for any one set of variables and uses visible damage as the method of assessment.

This test can be used to control any of the techniques for improving surface wear. It is, however, more expensive and time-consuming than either *hardness testing* or thickness measurement, and thus is not normally used when these methods are possible, as with *carburizing, chrome plating*, etc. With techniques such as *anodizing*, '*Sulphinuz*', '*Tufftriding*', etc., or to assess the effect of altering surface finish, this type of abrasive test can give excellent control. It is essential that the parameters of pressure, abrasive quantity, grit size and time are controlled for meaningful results. The *Taber abrasive test* is similar but less flexible.

Heat Tinting

A technique for producing a coloured surface.

Application range STEEL AND COPPER ALLOYS

Other materials such as silver and nickel are very occasionally coloured using this process.

The process uses coloured oxides, independent of the colour of the materials themselves, resulting from temperature application and to some extent the texture of the surface of the components. In addition, the heat tinting process can on occasion be used to improve corrosion resistance.

The surface is cleaned to a relatively high standard and is then heated either in air or a relatively simple controlled atmosphere such as steam. (Heat colouration achieved by other means is generally referred to as a specified treatment. The *'black oxide'* treatment of steel is a form of heat tinting even though it is carried out by immersing the component in a liquid medium.)

The colour achieved by the process varies for the following reasons:

1. Temperature used.
2. Atmosphere achieved: if the material is heated in an air-circulating furnace, the colour achieved varies to a limited extent from that achieved under conditions of static air at the same temperature. It will be appreciated that the use of gas heating can give considerable variations to the atmosphere in contact with the component's surface, thus affecting the colour achieved.
3. Time at temperature: the time and temperature are generally related such that the higher the temperature the lower the time, and vice versa, but the range of temperature will in fact be relatively slight. However, visual differences will occur in components with a rapid change in section, where a thin section achieves the desired temperature quicker, and results in slightly more intense coloration than the remainder of the component.
4. Surface texture: a very rough machined surface will not result in the same visual appearance as a polished surface finish, and this again may considerably differ from a finely ground finish.
5. Surface cleanliness: this can have considerable effect on the colour achieved for two reasons. First, under some circumstances the surface contamination may inhibit the formation of the oxide tinting, and second, the contamination may contain some constituent which contributes to the surface oxidation colour. This will commonly be the case when oil is a contaminant on the surface.
6. Time-lag: the time between preparing the surface and carrying out the treatment may have some effect, as the surface itself may become slightly oxidized, resulting in a passive or semi-passive surface. Reaction will not be as rapid as with a freshly ground or polished surface.

With steel, the process is that the clean component is heated to temperature under controlled conditions. The colour range is from straw or very light yellow to dark blue, almost black. The latter colour is most easily controlled and when *'steam blueing'* is used a permanent surface colour with some degree of corrosion resistance is obtained.

With copper alloys, the most common materials for colouration are the bronzes where, by the judicious use of heat, attractive shading can be achieved, and this again will result in some increase in corrosion resistance.

Since the heat tinting process is essentially used for decorative purposes, the finish is not of a high quality. The improvement in corrosion resistance is relatively slight and applies only to indoor conditions with relatively well-controlled humidity. Where *'black oxide'* and to a lesser extent *'steam blueing'* treatment is used, corrosion resistance is superior to heat tinting but, again, this cannot be compared with any form of *plating* or *painting*.

Under ideal, controlled conditions, i.e. with a high-quality, clean surface in a free oxidizing atmosphere, the temper colour range for steel is as follows:

°C	
220–230	pale yellow
240	dark yellow
255	yellow/brown
265	brown/red
275	purple
285	violet
295	light blue
315+	dark blue

From the above it can be seen that critical control is essential, if a homogeneous colour is required. Heat tinting is more commonly used as a method of assessing the temperature that the surface of steel has reached.

Heat Treatment

This term is used to cover all the various techniques used for thermal treating metals. These have all been collected into one section (Part Two **Heat Treatment**) where they are listed and discussed in some detail.

HEEF

This is a trade name for high-efficient hard chromium plating.

It is claimed that with this process deposits of high-quality chromium can be homogeneously produced at much higher rates than with conventional chromium or with present high-speed deposits.

Further information from M and T Chemicals.

'Heliarc' Welding

The term given when helium is used as the inert gas.

Application range ALL METALS WELDED

This is, thus, identical to *argon arc welding*, and the term is sometimes applied to that process. Described fully under *Inert-gas Shielded Metal Arc* in Part Two **Welding**.

Herbert Hardness Test

A rebound type of hardness test seldom used. Described fully in Part Two **Hardness Testing**.

HOT WORKING / 157

High-frequency Induction Welding
A type of heating used for some forms of *forge welding*. Described fully in Part Two **Welding**.

Holiday Test
A term applied to the *porosity testing* of paint or plastic films. Described fully under *Spark Test* in Part Two **Non-destructive Testing (NDT)**.

'Homocarb Process'
The name given to a proprietary process which is a form of *gas carburizing*, where the gas is produced by dripping a fluid at a controlled rate into the furnace. This furnace will be similar in many respects to the normal retort-type furnace but instead of a gas feed to supply the active gas which is then burned off at the exit, the gas here is produced within the furnace by the liquid entering as drops at the top of the furnace, and the carburizing potential is controlled by the feed rate of the liquid. The excess gas will be burned off in the normal manner. For further details of *Gas Carburizing* see Part Two **Heat Treatment**.

Homogenizing
Generally taken to be an alternative term for *normalizing*. Described fully in Part Two **Heat Treatment**.

It should be noted that this term can have a specialized meaning within certain industries or even large companies.

Hot Quenching
The name formerly given to the treatment of steel whereby steel was quenched and held at a high temperature prior to cooling to room temperature. This is now more generally known as *austempering* or *martempering*. Described fully under *Austempering* or *Martempering* in Part Two **Heat Treatment**.

Hot Working
A manufacturing process involving forging at high temperature.

Application range ALL METALS WHICH CAN BE FORGED

Briefly, this is carried out above the temperature at which recrystallization occurs immediately and, thus, no *work hardening* or *cold working* is achieved. Certain metals therefore, such as pure aluminium, gold, lead and tin are 'hot worked' at room temperature. This means that they can be extruded in the cold or beaten into very thin sheets as very little oxidation occurs at room temperature. Other materials such as steel, copper alloys, nickel alloys, aluminium alloys and most of the materials commonly found in engineering are *hardened* by *cold working* and become brittle. Above a certain critical temperature, which varies with each metal or alloy, this cold work is eliminated by recrystallization occurring and this is the range in which hot working takes place.

The upper temperature is defined by the temperature at which grain boundary liquefaction will occur. Again this is a well-defined temperature and results in a cast structure at the grain boundary. This is known as 'burning' and cannot normally be eliminated by any method, except considerable hot working within the correct temperature range, or by re-melting the complete component.

There are some materials where the range of hot working is so small that they are not considered to be *forging*-type alloys. These alloys invariably will accept very little cold work without fracture, the most notable being the cobalt alloys.

The reader is advised to obtain specialist advice. This is a basic manufacturing process and thus not within the remit of this book on metal treatment.

Hot-dip Coating

A term generally applied to *galvanizing* but strictly speaking can be applied to any of the processes where heat is used to coat the material.

Application range STEEL

The *aluminizing* process, using molten metal could, therefore, come within this description. In general, however, hot-dip coating is the process in which steel components are *cleaned* and *pickled* and, after *fluxing*, are immersed in molten zinc. This results in an adherent film of zinc which is an extremely efficient means of preventing corrosion. Further details are given under *Galvanizing*. The term generally is used to differentiate between this process and either electroplating or metal spraying of zinc.

Huey Test

A means of rapidly assessing corrosion resistance.

Application range STAINLESS STEEL

The test consists of heating a specimen in boiling, 60 per cent nitric acid for 48 h and finding the loss in weight. The test is repeated five times, the loss in weight being measured after each immersion.

This, then, shows the rate of corrosion which can be expected and gives some indication of its type. It will usually be found that the first and second immersions result in a greater weight loss than the subsequent tests, and generally an average of the last weight losses is taken to indicate a consistent corrosion rate. The initial immersions will result in the removal of loose or semi-adherent material and the normal surface-oxidation layer.

This is a reasonably long-term test. The *weld-decay test* (72 h) will give a more rapid indication of any tendency to intergranular corrosion, but is not comparable regarding general corrosion affecting the material. It may be necessary to carry out corrosion testing for specific substances, and the Huey test is a general indication of rate of corrosion under aggressive oxidizing conditions.

The test can be varied by using different corrosive media and conditions.

Hull Cell Test

A technique for the quality control of *electroplating* solutions.

Application range ALL PLATING SOLUTIONS

The technique is that a known, measured quantity of the actual plating solution is placed in the cell. The cell is manufactured in plastic and is thus unaffected by any plating solution used. The Hull cell is of specific shape having approximately half of one of the long sides in the form of a wedge. The test piece is placed on the angular wall while the anode is located at the short end of the rectangle. This means that the cathode, or test piece, has one end relatively close to the anode, the other end is remote from the anode. Using controlled voltage and current, it is therefore possible to produce a test piece which has a variation of plating conditions across its length. After the specified plating time, which can be as short as 2 minutes or up to 30 minutes depending on requirements, the test panel is removed and examined.

The Hull cell test enables comparison with previously plated test pieces, and the purpose is not to produce an individual test piece with unique characteristics. Used on a daily, or even twice-daily basis where necessary, it will show up any variation in plating before the effects are seen on the plated components. Conditions must, of course, be carefully standardized regarding temperature, current, voltage and the conditions of the anode and cathode.

Many plating shops use this technique to maintain a very high standard of control. Minor changes in the solution characteristics are identified and remedied before production components can be affected.

Standard test pieces plated under carefully controlled conditions in the actual plating tank can be used, but it is generally found to be difficult or impossible to standardize all conditions, whereas this is reasonably simple with the Hull cell technique.

Humidity Testing

A form of testing which is commonly used to assess corrosion resistance but can also be used to measure the dimensional stability of materials under humid conditions.

Application range ALL METALS

Many electrical and electronic devices are subjected to humidity testing, which is then followed by electrical tests to ascertain the effect, if any, that moisture has had on the assemblies. It is also possible to carry out humidity testing and electrical tests at the same time, using specialized equipment.

Humidity testing is carried out inside a cabinet where the atmosphere is controlled at specified temperature and humidity. Many tests require that the temperature is varied over a period of time in a cyclic form. The test is readily achieved at over 90 per cent humidity by ensuring a large surface area of water within the test cabinet. This will result in the atmosphere being saturated with water vapour for the temperature used.

It is this high-humidity form which generally requires use of a temperature cycle; this will vary over a wide range depending on the type of component being tested and whether the test is designed to simulate actual conditions or is an accelerated test. A common cycle would be 8 h at 50°C and 16 h at 20°C. Variations exist using higher and lower temperatures. However, when variations in the percentage humidity are required, this is much more difficult and use is made of saturated salts. These chemicals ensure that the laid down humidity at the test temperature will be maintained. Typical cycles as above are used, although less variation is demanded with the controlled humidity test, as there is less flexibility.

Examples of the salts used at varying humidities are as follows:

Saturated salt solution	Temperature (°C)	Relative humidity (%)
Sodium hydroxide	2	5.5
Lithium chloride		16
Magnesium chloride		35
Sodium chloride		75
Sodium hydroxide	10	5.5
Zinc chloride		10
Calcium chloride		38
Sodium nitrate		78
Sodium hydroxide	20	5.5
Zinc chloride		10
Calcium chloride		32
Potassium nitrate		45
Sodium nitrate		66
Sodium chloride		76
Potassium chloride		86
Zinc chloride	50	10
Sodium nitrate		67
Sodium chloride		75.5
Potassium chloride		80.5

The humidity test when applied to *accelerated corrosion testing* is a relatively mild method of testing. Where *salt-mist testing* is specified, this will produce a very high humidity at the controlled temperature. Where *accelerated corrosion testing* is the requirement, there is little point in running *salt-mist tests* in parallel with the present testing. Where, however, electronic assemblies require testing, or where the dimensional stability or material is being tested, then there is no alternative to humidity testing.

There are a number of specifications covering the various 'humidity tests' for the different materials and the reader is advised to consult the British Standards Institution or the American Society for Testing and Materials for specific information.

Hydraulic Pressure Test

A method of pressure testing which can be used on components and assemblies to prove that no leaks are present; it can also be a method of general integrity testing to show that the parts are strong enough for service. Described fully under *Hydraulic Pressure Test* and *Pressure Test* in Part Two **Non-destructive Testing**.

Hydrocarb Process

A specific form of *gas carburizing*.

Application range LOW CARBON STEEL

With this, the carburizing gases are preheated prior to entering the gas carburizing chamber. It was claimed that this increased the rate of carburizing, but it is now seldom used. Described fully under *Gas Carburizing* in Part Two **Heat Treatment**.

Hydrogen Annealing

A method of *softening*, using hydrogen as an atmosphere to prevent oxidation. Described fully in Part Two **Heat Treatment**.

Hydrogen Brazing

With this process, the *brazing* is accomplished in an atmosphere of hydrogen, generally using electric heating.

Application range COPPER, STEEL, NICKEL ALLOYS

This has the advantage that it will chemically reduce any slight oxide on the surface and will positively prevent oxidation occurring, thus, provided the components are fed into the furnace in a clean, dry condition, no oxidation can occur and *brazing* will take place without the use of any flux.

Hydrogen brazing was developed for the high-temperature braze alloys, for use with stainless steel and nickel alloys. With these, it was found that high-quality joints were not obtained economically, using conventional techniques. Hydrogen brazing is now commonly used with conventional brazing materials where high-production, high-integrity joints are required.

Most commonly high-production hydrogen brazing uses the humpbacked furnace, where the assembled components are placed on a conveyor which enters the furnace and rises up a slope. Along this slope hydrogen is fed in, and as it is lighter than air, it rises up and leaves at the peak of the hump where it is ignited on leaving the furnace. The components increase in temperature as they ascend the slope, and the maximum temperature is achieved at the peak where brazing occurs with the melting of the braze alloy. As the components are carried down the slope the temperaure drops and the braze metal solidifies, all under a hydrogen atmosphere, preventing any oxidation.

This equipment is commonly used on mass-production components with copper as the *brazing* material. This is often applied as an electroplated deposit and is used to form an interference fit, to hold the assembled components together until *brazing* occurs; it is probably one of the most economical methods of high-production joining. The brazing temperature will be approximately 1080°C–1100°C when copper is used as the braze metal.

Other forms of hydrogen brazing use batch-type furnaces, where the hydrogen is fed under pressure and is ignited on leaving, and the *brazing* cycle occurs as the furnace is heated and cooled. This can be compared to furnace brazing where fluxes are used, or to *vacuum brazing* where the only disadvantage is the need for a much higher standard of cleanliness and elimination of surface oxidation. *Argon brazing* is similar to the latter process and again does not result in positive reduction of surface oxidation, thus very high standards of cleanliness and freedom from oxidation are essential.

Hydrogen Embrittlement

A defect which occurs during the *electroplating* process, and in *electric arc welding*.

Application range STEEL

The modern theory is that, during the plating process, atomic hydrogen will be produced at the cathode of the component being plated. This atomic hydrogen is extremely active and has the capability of entering the interstices of the metal. Being unstable in the atomic state, the hydrogen will combine as rapidly as possible with other atoms to form molecular hydrogen, and this molecular hydrogen having a higher unit volume than atomic hydrogen results in an internal pressure in the metal being plated.

HYDROGEN EMBRITTLEMENT / 163

The same defect occurs during *arc welding*, where any dampness either in the atmosphere or, more commonly, in the flux coating will be broken down into atomic hydrogen by the electric arc, and again this atomic hydrogen will diffuse into the weld.

With high-ductility materials of relatively low strength, the stresses imposed by the hydrogen will not cause fracture, but will relieve themselves by stretching the material. With high tensile material or lower ductile material then, the result of the stresses may be the cracking known as hydrogen embrittlement cracking.

Hydrogen embrittlement which has not resulted in cracking can always be eliminated by a stress-relieving process, which will drive out the hydrogen before it can cause fracture or failure. With plated components, the ruling is that above 600 N/mm^2 (40 tonf/in^2), it is mandatory to stress-relieve at approx. 200°C before and after *electroplating*. Under some circumstances, e.g., with high-integrity aircraft-type components, the use of electroplating is forbidden above a tensile strength of approx. 1250 N/mm^2 (80 tonf/in^2).

Hydrogen embrittlement is generally eliminated from the *welding* process by the use of low-hydrogen rods. These are flux-coated rods where there is no water of crystallization or other moisture included in the flux coating. It is essential that these rods are kept at above 100°C as soon as the seal is broken, and thoroughly *dried* for at least 2 h above 200°C if they become damp or are left in the atmosphere for any time longer than about 1 h. The manufacturers' instructions must be followed in detail.

There is a long history of serious failure from hydrogen embrittlement. This takes the form of a brittle fracture at stresses considerably below those specified by the design and they can, thus, be of a disastrous nature.

It is not readily possible to prove the presence of hydrogen, since it is a mobile gas which is difficult to trap for analytical purposes. The fact that it can be eliminated by a simple *stress-relieving* process is fortunate. It is essential that all plating of high-tensile steel at or above 600 N/mm^2 (40 tonf/in^2) is always stress-relieved. Considerable care must be taken during welding to ensure that, with high-tensile materials, there is no danger of moisture contaminating the weld, either from the atmosphere, from contamination of the surface, or from the flux itself.

Hydrogen embrittlement is thus not a metal treatment which should ever be carried out on purpose. It is felt that engineers should have some understanding of the theory behind the problem, and the importance cannot be overstressed either of preventing hydrogen embrittlement occurring, or of *stress-relieving* high-tensile components which have been subjected to *electroplating*. Hydrogen embrittlement cracking will probably occur immediately after *arc welding*, thus *stress-relieving* is too late.

Gas welding and brazing do not present any problem from hydrogen embrittlement as the heat energy is insufficient to produce atomic hydrogen. With some metals, notably copper, there can be a reaction between the moisture resulting from the gas flame and the metal. This is sometimes referred to as 'hydrogen embrittlement'.

Hydrogen Welding

A form of *shielded arc welding* which uses the energy of ionized hydrogen. It is not commonly used. Described in Part Two **Welding**.

'Hynac'

The general name of a range of proprietary coloured coatings which can be applied directly to metals.

These coatings are water based, have an etchant included and can produce extremely thin attractively coloured coatings on aluminium in particular, and under certain circumstances on steel. Further information can be obtained from Pennwalt Ltd.

Ihrigizing

A surface impregnation process to improve corrosion resistance.

Application range STEEL, CAST IRON

The surface is treated in such a manner that it becomes high in silicon, to form a surface layer of silicon carbide. This is resistant to many acids and general corrosion. The components are heated in a closed container in contact with silicon carbide or, in some circumstances, ferrosilicon in an atmosphere of chlorine at a temperature between 950°C and 1025°C.

The depth of the silicon impregnation will be dependent to some extent on the surface finish of the material involved, and the analysis of the material being treated, but to a much greater extent on the length of time at which the material is held at the correct temperature. The result is a surface layer containing approximately 14 per cent silicon, which will reduce towards the centre where the silicon content will be the nominal amount of the basic material being treated. This technique results in a surface which is extremely hard and brittle, but which can withstand a corrosive atmosphere and will, in fact, be capable of withstanding chemicals such as boiling sulphuric acid over a wide range of concentrations. It is also capable of withstanding oxidation up to relatively high temperatures in excess of 750°C.

The process is extremely expensive and requires specialist equipment. With the advent of stainless steel and titanium, and to a lesser extent *nickel plating*, there is now very little application for the Ihrigizing process.

Immersion Coating

A term which can be applied to any form of coating carried out by immersing the component. It can be used to distinguish between processes requiring electrolytic treatment and those which occur by chemical treatment alone. It can also be applied to a process such as *galvanizing*, where the component is immersed in molten zinc. The term should never be used without indicating the specific type of coating involved and how the process is to be carried out.

Immersion Plating

The plating of a metal which occurs when one metal is immersed in a solution of another metal.

Application range ALL METALS WHICH CAN BE PLATED

The solution must contain a metal which is more electropositive than the metal being plated. The latter dissolves in the solution of the more electropositive metal, and this will then precipitate on the surface of the less electropositive metal. This, then, is a form of *electroless plating*. It is now common usage that the term immersion plating is used where a deposit of the more electropositive metal is obtained and the plating process then stops. The term *electroless plating* is used for the process where the deposition of the metal being plated continues, and there is a build-up on the material being plated. Electroless plating can be applied, under some circumstances, to non-metallic substrates, while immersion plating refers only to the deposition of an electropositive on to an electronegative metal. The electrode potentials of metals are given in the section *Corrosion Protection*. This gives a guide to the metals which may be deposited, but there are other factors involved which may prevent the reactions required from taking place.

Impact Test

This is a mechanical test, generally on a carefully machined test piece which is designed to indicate the capabilities of the material to absorb impact energy, at the nose of a notch.

The tests do not by themselves give an accurate measurement of ductility. The purpose of the impact test is to identify any tendency for an apparently ductile material to be brittle when a notch exists. One value of the test is to identify the temperature at which steel, in particular, ceases to have the capability of absorbing energy and becomes brittle. It can, as a general rule, be stated that any material which is shown to be brittle using ductility testing techniques will show low impact values. It is not, however, of necessity true that materials which are shown to be ductile will also have high impact values.

It will, therefore, be seen that in any series of tests, *bend testing* as a measure of ductility should be carried out prior to impact testing, as this is much more economical than the impact test. The ductility, identified as elongation or reduction in area, at the *tensile test*, can also be used to identify brittle material.

Impact tests of an empirical nature can be carried out on components by striking a sharp blow. Low-impact-strength material will fracture, high-impact material will bend.

Described fully in Part Two **Mechanical Testing**.

Impregnation

A process to prevent leakage of porous castings by filling or impregnating any porosity identified.

Application range ALL METALS

There are various techniques, which can include simple processes such as *painting* the surface with non-porous paint or immersing the component in hot solutions either of plastic or, in some cases, sodium silicate. The modern technique makes use of a range of materials, some of which are quite complex and are designed to react with the actual metals, first, giving a good bond and, second, forming a solid but flexible material.

Most modern techniques use the vacuum pressure cycle. With this, the clean component is placed in an autoclave which is then sealed and the pressure reduced. Under some circumstances, a very low vacuum is required. When the desired vacuum has been achieved, the liquid used for impregnation is pumped in to release the vacuum and the pressure cycle is then commenced.

The theory is that the vacuum will ensure that all gases are removed from the porosity, which will then subsequently be completely or as much as possible filled by the liquid under pressure. The subsequent process, then, will depend on the impregnation fluid used. Many modern materials, on removal into the air, and washing, react and the impregnation material will become solid, adhering to the material being treated.

In general it is not advisable to carry out the impregnation process as a routine, since this would seem to indicate that the *casting* process has been accepted to produce porous casting as a standard. It is generally better to control the casting and to eliminate general porosity rather than to impregnate as part of a production process.

The castings should be *air-pressure tested*, and those showing no leakage at the specified pressure passed as satisfactory. Any casting showing leakage should be examined and only those with microporosity submitted to impregnation. There is a considerable history to show that macroporosity can be sealed by impregnation processes, but that the sealant works loose during service with consequent failure. It should, therefore, be a quality decision regarding the degree of porosity which can be salvaged by impregnation. This decision can only be made after *pressure testing*, before impregnation, not by visual examination after 100 per cent impregnation of all castings.

Where high-integrity castings are required this process should not be used.

Impressed-current Corrosion Protection
A development of *cathodic* or *galvanic protection*.

Application range STEEL

The technique is generally limited to large structures, commonly in the marine environment and on underground pipes.

The theory is that, using anodes which are non-active, an electric current is made to flow between the anode and the material. This will ensure that the steel being protected is cathodic and thus should have no corrosion or corrosion at a very slow rate.

Considerable care is necessary to ensure that the protected component remains at all times cathodic to the artificial anode. When conditions alter, there is the possibility that the protected component can itself become anodic to some other material or even that certain areas of the protected assembly can become anodic. Any steel which becomes anodic will have the corrosion rate increased by a considerable amount and, thus, the effort to protect can result in severe corrosion. For this method to be effective, it is necessary to find the resistance of the electrolyte and to carefully space out the anodes being used and monitor the anode and the cathode to ensure that correct conditions of current flow are always present.

With modern *corrosion protection*, impressed current is sometimes used in conjunction with paint systems, but will never be used in conjunction with *galvanic protection* using anodes which are active and sacrifice themselves to produce the necessary protection. The user is strongly advised to obtain specialist advice before attempting to protect expensive assemblies using impressed-current corrosion protection.

'Imprest' Process
A proprietary method whereby aluminium and aluminium-alloy sheets are cold-rolled in contact with materials such as fabrics or wire in order to give a delicate embossed pattern. The pattern will not be an exact replica of the master and, thus, some development will probably be necessary to find the pattern necessary to give any exact impression. Further information can be obtained from the British Aluminium Company.

Inchrome Process
An alternative name for *chromizing*.

Application range STEEL – GENERALLY LOW CARBON

With this, the surface is impregnated with chromium by a hot diffusion process to give 30 per cent chromium at the surface.

Indium Plating

This is deposited on top of electroplated lead for high-duty bearings. No other reason for the use of this relatively rare metal is recorded. Described fully in Part Two **Electroplating**.

Induction Brazing

A method of joining using a lower melting point alloy than the metals being joined.

Application range STEEL, CAST IRON, COPPER AND ITS ALLOYS, ALUMINIUM UNDER CONTROLLED CONDITIONS, NICKEL AND TITANIUM ALLOYS

This obeys all the rules laid out in the section *brazing* which are required for high-quality joins. The induction technique is the method used to bring the components being joined to the *brazing* temperature. Under many circumstances, the induction heating coil will be used manually, that is, the component will be held within the coil and, either using a hand or foot switch, the high-frequency current will be applied, the joint will heat up and the operator can watch the metal melting and then remove the component from the coil.

This technique lends itself readily to automation or semi-automation, where parts are continuously fed to the coil or can be held by some fixture which automatically switches on the electrical current, which is automatically switched off after a pre-set time cycle.

Induction brazing is in competition with normal *gas brazing* and will seldom be as economical as *furnace*, *salt-bath*, *hydrogen* or *vacuum brazing* as all of these techniques can be designed for use with large quantities simultaneously, whereas induction brazing produces only a single join at each operation. Technically, when properly controlled, an equal standard of *brazing* can be achieved from any of these procedures, thus the only parameter to be examined is that of economics.

Brazing will often be an alternative to either *soldering* or *welding* as a method of joining. With modern glues there is also the possibility that joints of a satisfactory standard can be produced by gluing.

Induction Hardening

A *hardening* technique which uses *induction heating* as the means of heating the components. It is comparable in many ways to *flame hardening* which is a more flexible, but less controllable, method. Described fully in Part Two **Heat Treatment**.

Induction Heating

A method of heating, using high-frequency electrical current. The heat obtained can be used for *casting*, *welding*, *brazing* and basic *heat treatment*. The *casting* techniques do not come within the scope of this book, but *Induction Welding*, *Brazing* and *Hardening* appear under these headings in this book.

Application range ALL METALS

Induction heating uses coils which act as conductors for the high-frequency electricity and when any metal is placed inside the coil, there will be an induced effect. The molecules in the metal are forced to change direction with each reversal of the high frequency, and this heats the metal. With very high frequency, the effect is confined to the metal surface and the 'core' is only heated by conduction. As the frequency is reduced, so this 'skin' effect becomes less and it is more difficult to achieve a surface heat only. It will thus be appreciated that for melting or general purposes, high energy and low frequency is necessary. For the skin heating used for *hardening* and *brazing*, high frequency is necessary. The high-frequency method in general uses valve-type oscillators, the lower frequency makes use of rotating generators.

Inert-atmosphere Furnace Brazing

A method where the *furnace brazing* technique is used, thus allowing a batch-type process, but here the furnace atmosphere consists of an inert gas. In the UK this gas will be argon, but any of the inert gases may be used and, where economics justify, helium will be used.

Application range STEEL, COPPER, NICKEL ALLOYS

Using the inert-gas atmosphere means that no positive *cleaning* or removal of oxide is achieved and thus the component to be brazed must be thoroughly cleaned and free from oxide.

It is most unlikely that *fluxing* could be used as most fluxes would result in contamination of the inert atmosphere. However, there are certain types of fluxes which are relatively inert which could be used with this method. All other details of the process will be identical to *furnace brazing*, and the user is referred to this heading, and the heading *Brazing* in this book.

The use of hydrogen brazing and, to a lesser extent, *controlled atmosphere brazing*, results in an atmosphere which can be used to remove the last traces of oxygen and oxides and, thus, under many circumstances superior results will be obtained. No such oxide removal will be achieved with the present process and, in addition, there can be considerable difficulty in purging the furnace or chamber prior to *brazing* in the inert gas. Any air which is drawn into the atmosphere or is trapped in the assembly can result in the formation of surface oxide which will reduce the standard of *brazing* achieved.

Minute quantities of oxide or surface contamination can result in less than perfect wetting of the surface and thus poor brazing.

Vacuum brazing is very similar in all respects to this process and may be more economical.

Inert-gas Shielded Metal Arc Welding

A technique which shrouds the weld with argon or helium which is then ionized to give the *welding* heat. This heat will always be less than that achieved with air, but overcomes the problems of weld oxidation. Described fully in Part Two **Welding**.

It is now often used to describe welding processes which use gases such as CO_2, mixed air and argon, and CO_2 and argon. These gases will give higher speed welding but will not eliminate oxide and slag formation in the weld and thus some care is necessary in the use of this description.

Inertia Welding

This is very similar to *friction welding* which is described fully in Part Two **Welding**.

Integral Welding

Any form of *welding* where no filler metal is used, but the components themselves are designed to have some portion which fits with the mating component.

Application range ALL METALS WELDED

The *welding* action melts this metal which thus becomes the filler in the process. It will thus be seen that integral welding must only refer to *fusion welding* as no form of *forge welding* requires filler material.

Integral welding was introduced by the aircraft industry and is a technique whereby using, in particular, the *inert-gas shielded metal arc* weld TIG joins can be produced with a minimum of weight and of a very high integrity.

Interrupted Ageing

A technique of *ageing* or *precipitation hardening* where the process is carried out in stages, with the components being cooled to room temperature between each stage. It is used when the exact time or temperature to achieve maximum hardness is not exactly known.

Application range ALL METALS WHICH CAN BE AGED

The parts are heated to a predetermined temperature for a stated time. This time and temperature are calculated not to give the full hardness, but from the hardness achieved it should be possible to predict more exactly the time and temperature to give full hardness. With modern, more detailed knowledge of the alloy content, this method should not be necessary. Described fully under *Precipitation Hardening* in Part Two **Heat Treatment**.

Interrupted Quench
A specialist type of *hardening*.

Application range STEEL

With this the component, instead of being left in the quench until it achieves the temperature of the quench medium, is removed after a specified time and is allowed to cool in air. The purpose is to temper the material, using the heat remaining in the component at a certain stage of quenching.

If components are all of identical size, and the *heat treatment* batches are all of exactly the same size, with the components located in the same manner so that the heat removal is always at the same rate, then interrupted quenching will be identical to *austempering* or *martempering*. More generally, interrupted quenching is carried out by an individual heat-treater, who pits his judgement against the variations which can occur during the *heat treatment* of steel. If the operator has the necessary skill, then successful results can be achieved, but it is much more likely that variations from component to component within the same batch as well as between different batches will occur and this method of *hardening* and *tempering* is not to be recommended where treatment of any importance is required.

Inverse Annealing
A term which has, on occasion, been used in place of *precipitation hardening* or *ageing*, and is liable to cause confusion. The derivation would appear to be from the fact that with steel the temperature in this process results in *softening*, whereas in alloys capable of being hardened by *ageing*, the same temperature results in *hardening*. See Part Two **Heat Treatment** for details of the various methods of *softening* and *hardening* metals.

Ion Nitriding
This is an alternative name for *Plasma Nitride*. Described fully under this heading in Part Two **Heat Treatment**.

Iron Plating
Iron is the cheapest of all metals and, therefore, finds very little application as an electrodeposited film. There are, however, some characteristics of iron which can be made use of in conjunction with more expensive alloys such as copper. One use of an iron deposit is on copper soldering bits. This prevents the solder from alloying with the copper and considerably increases the life of the bit. Iron at one time was used in the printing industry but with the advent of modern techniques this is no longer used.

Described in Part Two **Electroplating**.

Isothermal Annealing

Applied only to steel; it is a form of *softening*. Described fully under *Isothermal Annealing* in Part Two **Heat Treatment**.

Izod Test

A form of *impact testing* which can be applied to all materials. The test piece is round with a notch machined at right angles to the length and at the mid-section. Round test pieces without notches are also used. Some test pieces have a notch machined in relation to each end. That is, a single test piece can have two notches and thus used to give two tests on the same material.

To carry out the test, the specimen is held in a vertical position with the notch at the top face of the vice facing the hammer used for the impact test. A pendular hammer is swung to strike the vertical half of the test piece and the arrest of this hammer is measured; the greater the arrest, the higher the impact strength of the material being tested. It is recommended that at least two specimens are prepared and tested and that at least two of the specimens give comparable results.

The problem with the Izod test is that the notch is difficult to reproduce accurately and, thus, there can be some scatter. Many specifying authorities are replacing this test by the *Charpy impact test*, but it is still commonly used with castings, and non-ferrous metals.

Described fully under *Impact Test* in Part Two **Mechanical Testing**.

Jacquet's Method

An *electropolishing* technique used in the production of metallographic specimens.

Application range ALL METALS

The piece of metal to be examined is mounted in a special conducting material and then mechanically polished in the normal manner to produce a flat surface with a good finish. This will have a layer of work-hardened material which distorts the grain pattern.

Electropolishing is then carried out by making the specimen the anode in a solution whose composition will depend on the metal being polished. The polishing removes the work-hardened layer and etches the surface.

There are now available specialized pieces of equipment which can be portable, which allow the production of a high metallurgical type of polish on local areas. This gives much greater flexibility to the metallurgist regarding the possibility of carrying out a detailed examination of metal structures on site.

Using this technique of *etching*, the skilled metallurgist can identify many different structures and, under many circumstances, can supply information on whether or not the material complies with the specification without destroying the component being examined. Described fully under *Etching* in Part Two **Non-destructive Testing**.

This is an economical technique of finally polishing and etching micro specimens if a sufficient number of specimens in the same material are required. It is technically comparable with the standard micro preparation and etching technique.

Jagger Test

A form of *hardness test*. Described fully in Part Two **Hardness Testing**.

Japanning

A term used to describe a form of *stove enamelling*.

Application range ALL METALS

It is derived from an ancient Japanese process in which the component is coated, then stoved at about 90°C with a form of pitch or tar and thinned turpentine containing other additives. The result is a shiny, black surface. Japanning can be applied to many materials and 'patent leather' is a form of this process. The term is now applied to a wide range of paint finishes, covering many colours, which in many cases do not require stoving. In general, however, there is the connotation that the finish achieved should either be black, or at least have a high shine.

This term which is applied to many surface finishing treatments at one time had a well-defined, specific meaning, but which with modern finishing, giving apparently the same result, now has a very blurred meaning.

Jet Test

A *thickness testing* technique described under *British Non-ferrous Jet Test (BNF)* in this section.

Jetal

A name given to the '*black oxide*' treatment of steel, using hot, oxidizing alkaline solutions. This results in the formation of an adherent, dark oxide with good corrosion resistance under most conditions. See *Black Oxide* for more details.

Jominy Test
A quality control test to give a measure of 'hardenability'.

Application range STEEL

A standard test piece is machined from the steel being tested, and normalized to remove any previous *heat treatment* and all cold work. It is important that the *normalize* is carried out as the previous *heat treatment* history of the steel can have some effect on its 'hardenability', and thus could give variable results. The normalize should be carried out on the rough machined test piece and final machining carried out after normalizing.

The test piece will be 4 in (10 cm) in overall length, 1 in (2.5 cm) in diameter with a head 1.125 in (2.8 cm) at one end 0.125 in (0.32 cm) in length. This test piece is carefully heated to the *hardening* temperature for the steel in a neutral or controlled atmosphere. The time at temperature will vary for the steel being tested but will be in the 15–30 minute range. The test piece is then rapidly removed from the furnace and inserted in the quench. This is a collar through which the parallel diameter of the test piece is placed with the collar holding the 1.125 in (2.8 cm) diameter head. This head is then sprayed for 10 minutes with a controlled jet of cold water, the collar preventing the remainder of the test piece from any contact with the spray, thus ensuring that only the head is water-quenched. The test piece is then removed and a flat carefully ground to remove 0.020 in (0.5 mm) along the complete length and a hardness survey carried out along this flat.

With steels of high 'hardenability', there will be a longer length of high-hardness steel from the test piece 'head' than with steels of low 'hardenability'. Graphs can be readily constructed showing the depth of hardness obtained on the test piece. Provided care is taken, excellent reproducibility of the hardness pattern will be achieved.

This end-quench 'hardenability' test is seldom carried out as a routine quality control, but charts and tables have been published of the results obtained on Jominy test pieces for most common steel specifications. These are used for design purposes and some steel specifications now use this parameter as the basis for control. It is not advised that this test is used for random control, as the number of variables which can exist would tend to make the results meaningless. For high-production, high-integrity *heat treatment*, where economy is essential, and thus low-alloy steel is used, routine Jominy end-quench control is advisable and can be used in place of chemical analysis for the control of raw material. For investigation purposes, normal quenching, micro examination and hardness surveys will probably give more useful information than a 'one-off' Jominy test.

The use of a wedge-shaped test piece which is quenched in air, oil or water, ground along one length which is then subjected to a hardness survey is a useful alternative, provided the conditions of test, and the test piece are standardized.

K4
This is a code name sometimes applied to Kolene K4.

Further information from Kolene Corporation, USA, or ICI Mond Division, UK.

'Kanigen Plating'
A proprietary name which was given to the first patented process of *electroless nickel plating*. This is a deposit of nickel and phosphorus which requires no electric current and which can be hardened by subsequent treatment. As this was a leader in the field, the name 'Kanigen plating' is very often at present used as the generic name for *electroless nickel plating*. The 'Kanigen' process is now carried out by Fescol Ltd.

Karat Clad
Selrex decorative gold.

A proprietary decorative gold plating process. Further details from Sel Rex Ltd.

Kayem Process
A procedure which gives economical press tools for small production runs.

Application range MILD STEEL – DEEP DRAWING AND OTHER HIGH DUCTILITY MATERIALS SUCH AS COPPER, BRASS, ALUMINIUM ETC.

The normal press tool for high production runs is produced by sophisticated machinery from sophisticated steels with complicated *heat treatment*. It is seldom possible to predict exactly the shape of a pressing which will be produced from a press tool. Where long-run production is necessary, the press tools will be extremely expensive to produce and difficult to modify when they are completed.

The Kayem process allows a small quantity of pressings to be produced, using soft zinc dies, and these are used to give the necessary information to produce the final press tools.

One procedure for this is that the initial pattern, either in wood or plastic or any other material, is used to produce a shape in plaster of Paris. This plaster of Paris pattern is then used to produce zinc press tools. The temperature at which zinc is cast is low enough for the use of plaster of Paris as the mould, and the two halves of the press tool can be made in separate operations. These two portions of the press tool can then be employed to produce the necessary preliminary pressings and any difference in the shape finally desired can be made by either filing the zinc to remove any excess metal, or by using special solders to build up any depressions which have been found. In this way the zinc press tools can be used economically to produce a limited number of components for trial purposes, and when a satisfactory press tool has been achieved, the zinc press tool can be employed as the master for sinking the final hardened steel die. Kayem press tools have been used to produce short-run production pressings as well as initial trial pressings.

Keeps Hardness Test

A form of *hardness* test which uses the depth of hole produced by a rotating drill of standard hardness as the means of measuring hardness. Described fully in Part Two **Hardness Testing**.

Keller's Spark Test

A method which can be used to differentiate between steels using the spark produced by grinding as the means of assessing the difference.

Application range STEEL

This makes use of the fact that the carbon content of the steel will affect the type of spark produced, and that certain alloying elements will result in the spark having a characteristic colour.

Any metal when it is heated above a certain critical temperature gives off lightwaves. The wavelength of the light in question will be specific to the metallic element involved and the intensity of the light will be proportional to the amount of that element present. The same principle is used for some of the more sophisticated chemical analysis instruments which can quantify and accurately assess the amount of metallic elements present in an alloy. The carbon content is estimated by the intensity and shape of the spark produced, and requires considerable experience to give meaningful results. The fact that some information on the carbon content can be obtained makes the Keller test more useful than other portable spark analysis techniques. It is seldom that the Keller test can be correctly applied under field conditions.

With the Keller test, use is made of a grinding wheel to supply the necessary energy to produce the temperature required. While it is possible to examine the sparks to a limited degree with a normal grinding wheel, for more sophisticated results it is advisable that the direction of the grinding wheel is reversed so that the sparks rise up in front of the operator. If the lighting in the area is reduced and the background painted a matt-black, then with practice considerable accuracy can be achieved in determining the alloy content of a steel. The alloy content best estimated is chromium and molybdenum, with tungsten and nickel at less accurate levels. This technique, and others for the same purpose, are listed and described under *Metal-sorting*.

Kenmore Process
A process confined to the production of a plated deposit on a wire core, for further wire drawing, and decorative or corrosion resistance.

Application range STEEL

The most common metals involved are copper or nickel which are electro-deposited on the surface of a steel wire, which is then drawn to increase the tensile strength and to reduce the thickness of the material in question. The drawing will act as an inspection process to ensure that only a high-quality plating deposit will pass.

This process is used as a means of high-production, economical coating of steel wire and as such is to be compared with hot-dip coating. Where the coating material can be applied by means of a molten dip, for example, tin, lead or zinc, this will be more economical than the present process, which can be used for metals such as copper or nickel which cannot be readily applied by hot-dipping.

'Kephos'
The proprietary name of a form of phosphating using a non-aqueous solution normally applied to spraying which can also be applied by dipping or brushing.

One advantage of this technique is that it does not require washing, but can be left to air dry prior to further treatment such as painting. Further general information will be found under *Phosphating*; detailed information can be obtained from ICI Ltd.

Kern's Process/Kern's Test
A method of testing the thickness and, to some extent, the mechanical properties of a paint film.

Application range STEEL

The method is that the surface is *blasted* with an abrasive powder, usually carborundum, until the coat of paint has worn through to the base metal.

The principle of the test is that any paint film will be abraded and removed by the powder, whether carborundum or aluminium oxide. The test makes use of standard conditions of nozzle diameter, air velocity, particle size, hardness and impingement of the grit, and the assessment of the paint film is in the amount of abrasive material used or time taken to achieve removal of the paint film.

Since this test, in addition to assessing the thickness of the paint, can be used to assess the abrasion resistance of the paint film chosen, it thus has a certain advantage over either *hardness* or *thickness testing*. The test can be compared with the more conventional method by which a loaded scribe is abraded against the paint film and the degree of wear is measured against time. This *Taber test* is probably more reproducible but will give less information than the Kern's test properly carried out.

Kirsch Test
A portable *hardness test*, using a hardened steel punch. Described fully in Part Two **Hardness Testing**.

Knoop Hardness Test
A form of *hardness test* used for thin sheet or very light loads. Described fully in Part Two **Hardness Testing**.

Koldweld
A form of *forge welding*.

Application range ALUMINIUM, COPPER, GOLD, TIN, LEAD, SILVER

This is a form of *pressure welding* where no heating is involved and the metals are joined by the application of pressure alone. This necessitates that the metal must be capable of accepting a considerable amount of cold work, and that the interfaces of the parts being joined are not oxidized, otherwise good *forge welding* will not be achieved.

For a correct Koldweld, the shape and size of the dies used are very critical. Described fully under *Forge Welding* in Part Two *Welding*.

'Kolene'
This is a proprietary technique using electrolytic molten salt baths to remove sand and non-metallic contaminations.

Application range STEEL, CAST IRON, NICKEL ALLOYS

The process makes use of proprietary salt at temperatures between 400 and 500°C. The components to be treated are connected to a direct electric current and immersed in the molten salt.

The polarity of the component can be altered when the work is charged negatively, which is the most common technique; in approximately 30 minutes sand, oxide and scale even adherent to the surface will generally be removed.

By reversing the polarity and making the component positive, non-metallics such as graphite will be removed from the surface. This will result in the surface being clean and ready for further processing.

The process is comparable in many ways to 'blasting' but has the considerable advantage that internal passageways and suchlike areas can be chemically cleaned.

The process is similar also to sodium hydride but can only be used on metals which are not attacked by caustic or hydrogen but is very much more economical than sodium hydride.

Further information from Kolene Corporation or ICI Mond Division, UK.

Korel Method

This is a method of applying paint with the paint gun being used for pre-heating and curing the paint.

Application range ALL METALS

Basically it is a standard spray gun but in addition there is a propane compressed air burner in the form of an annular ring round the front of the spray gun.

This can be used to pre-heat the component to be painted. When pre-heating is complete the paint is sprayed through the heated air of the annular ring with the temperature controlled to between 60 and 70°C.

It is claimed that this technique has the advantage of firstly pre-heating the surface, thus eliminating any danger of moisture, and because of the high volume of moving gases involved it also acts to blow off any loose particles. Secondly, the paint is applied hot and thus the solvent content will be reduced drastically or eliminated and finally the paint will cure much more rapidly than at ambient temperatures.

The process is claimed to be intrinsically safe and to be ideal for outdoor applications.

The equipment can use Korel approved paints in the epoxy and modified epoxy ranges and can apply relatively thick coatings in a short period of time as the intercoat adhesion is claimed to be excellent. Certain powders can also be sprayed using this technique.

Further information can be supplied from the manufacturers, Korel Korrosionsschatz – Electronik.

Kuftwork

An alternative name for *Damascening*.

Kynar

This is the trade name of a stable plastic material which is available as a dispersion capable of supplying a coating to metals which, when correctly applied, has excellent corrosion resistance, electrical resistance and reasonably high temperature properties it is claimed above 100°C.

Kynar is a thermal plastic in the fluoropolymer range.

Further information can be supplied by Pennwalt Ltd.

Lacquering

An alternative name, under many circumstances, for *painting*, to mean specifically coating with a clear varnish. Described fully in Part Two **Painting**.

Lap Welding

A term referring to a specific design of join, not any specific method of applying the weld. A lap joint is one where the two joined faces overlap each other and are held together by gluing, riveting or bolting.

Application range ALL METALS

Lap welding is generally two fillet welds, one at either end of the pieces being joined, but may be a single fillet weld, and in some cases the complete edge of the join can be achieved by *fusion welding* on all four surfaces. The method also includes *forge welding*, where the lap joint is heated and hammered to give a forge join. Described fully under these headings in Part Two **Welding**.

Laser Welding

A method of *fusion welding*, using the energy obtained from a laser to supply the necessary heat.

Application range ALL METALS WHICH CAN BE WELDED

The laser energy is obtained from a light source which is concentrated inside a crystal, and then released as a high-energy beam of light. This will not affect the atmosphere through which it passes, but any solid object in the path of the light-wave will be heated as the energy is released. The advantage of laser welding is that the energy beam can be extremely narrow and, thus, the heat-affected zone of the weld is kept very small.

Welding can be carried out in normal atmospheres, but will require either flux, or that the weld metal is protected by an inert gas. The technique is comparable in many ways to *electron-beam welding*, but does not require the use of high vacuum, thus is very much more flexible. It can also be used in a similar fashion to *inert-gas shielded metal arc welding*, with the laser energy used in place of the metal arc.

This is a modern technique and the capital cost of equipment is high with high operating expenses. Where high-integrity joints by *fusion welding* are required, particularly with thin sheet material, laser welding can have an advantage in flexibility and cost over the *electron-beam process* and, in the production of technically superior joins, over the conventional techniques. It is advised that up-to-date information is obtained from specialists such as the Welding Insititute.

The laser beam can be used for many purposes in addition to *welding*, and laser cutting of a variety of materials, metallic and non-metallic, is now quite common. The laser beam is also used for control purposes and many inspection techniques make use of this energy source. The beam has all the advantages of a light beam, with the addition that it has a measurable energy, capable of doing work a long distance from the source of the beam.

Laxal Process
A method of inhibiting corrosion.

Application range STEEL

With this, the components are immersed in a hot oxalic-acid solution. The resultant film has properties not unlike that achieved by *phosphating*, which has now largely replaced this process.

Lead Annealing
A form of *softening*, where molten lead is used as the heat-transfer liquid and also prevents surface oxidation.

Application range STEEL – COPPER ALLOY, ALUMINIUM

Lead melts at 327°C and it will, therefore, be appreciated that it is not possible to use lead annealing at any temperature below 330°C. As the temperature increases, the rate of oxidation or drossing will increase, and above approx. 600°C or 700°C, this rate will generally become unacceptable. Thus, the lead annealing process can only be of use when the *annealing* or *softening* temperature lies between these temperatures. It is, therefore, of limited use for steel components where the *softening* temperature is usually accepted as being in the range 600°C and upwards. Where materials are lead annealed, care must be taken that no alloy is produced between the metal being treated and the lead.

Lead has been used in the *heat treatment* of steel where *tempering* must be carefully carried out, and there is the process of *martempering* where the steel is quenched into a molten bath, traditionally a lead bath. However, with the advances in salt-bath technique and the continuing rise in the price of lead, there are now relatively few applications where molten lead is used. Health hazards associated with metallic lead have also contributed against any increase in its use.

Lead annealing was commonly used to soften steel wire after *cold-drawing*. The hard wire passed through the molten lead at the chosen temperature. This resulted in an increase in ductility, a reduction in hardness with no danger of any surface oxidation. There was the added advantage that a thin film of lead remained on the surface. This could act as a die lubricant for further drawing or as a means of improving the corrosion resistance of the final wire.

Lead annealing will be seldom used today, as salt baths and conventional furnaces have improved and are more attractive technically and economically. *Martempering* and the various methods of *annealing* and *softening* are listed and discussed in Part Two **Heat Treatment**.

Lead Patenting
A term used in *heat treatment*.

Application range STEEL

With this the parts are quenched at a controlled temperature. It is thus an alternative name for *martempering*, with the quench medium being molten lead. *Martempering* at present generally uses a salt bath, and is largely overtaking the use of lead on technical, economic and health grounds. Described fully under *Martempering* in Part Two **Heat Treatment**.

Lead Plating
A seldom-used electrodeposit. Described fully in Part Two **Electroplating**.

Lead–Tin Plating
A seldom-plated alloy which should not be confused with *solder plating*, where the tin–lead alloy with 60 per cent tin is commonly plated. Described fully in Part Two **Electroplating**.

Levelling
A term used in *electroplating* to denote a deposit which has resulted in an improvement to the surface finish by smoothing or levelling the substrate surface. Described fully in Part Two **Electroplating**.

Lime Coating
This covers two processes, firstly as a coating to act as a lubricant for wire-drawing, and secondly a coating to neutralize an acid treatment.

Application range STEEL

The most common is where a coating of lime is applied, very often in conjunction with some other process, to ensure reasonable adhesion of the lime to wire during a wire-drawing. To a lesser extent the term is applied to the use of lime or other alkali materials on steel which has been *pickled*, and the purpose here is to produce a surface less prone to subsequent corrosion.

It is a known fact that the *pickling* process itself will result in a surface more prone to corrosion than a normal surface. If in addition to this the *swilling* operations have not been ideal, subsequent corrosion can become serious. It is, therefore, sometimes the practice that following *pickling* the steel is immersed in some solution which will neutralize any remaining acid, and then ensure that some of the alkaline neutralizer is retained on the surface.

Lime coating is not common within modern industry. Where this was traditionally used during wire-drawing, use is now made of *phosphating, borax*, or glass treatment, or other lubricants such as graphite, molybdenum disulphide or metal soaps, all of which tend to be more efficient and convenient than the coating of lime.

Where corrosion inhibition is required, there are available many alkali solutions, some of which are proprietary and are added to the swill water, or used as a separate dip after the first swill. Lime is still used under some circumstances, but more commonly a caustic-soda solution is employed, with additives to improve its wetting ability and stability.

Linde Plating

This is a trade name covering the process given in more detail under the heading of *Flame Plating*.

Liquid Honing

An alternative name for the *vapour blasting* process, where abrasive grit in slurry form is used as a blast medium. The term can also be used to cover the *barrel deburring* process. Both are described under their own headings.

Liquor Finishing

A term which can be applied to a variety of metal treatments where the process uses a liquid. It is, however, specifically applied to a wire-drawing process where after *cleaning* and *pickling* the wire is treated with a mixed acid copper and tin sulphate.

Application range STEEL

The process results in a deposit by chemical reaction without any electrolysis of a thin film of copper and tin, giving a brownish-red colour. This is an *electroless* or *immersion plating* process, further details being given under these headings.

After washing, and while still wet, the wire is drawn to a smaller size. The deposit acts as a die-lubricant and the drawing action stabilizes the deposit. The final drawn wire has an attractive colour, but the corrosion resistance achieved is of a low standard. This process is seldom used by modern industry.

The use of *phosphating, borax* treatment or metallic soaps are all modern aids to wire-drawing, which leave a deposit giving corrosion protection, and lubrication superior to the process described.

'Lithoform'

A proprietary form of *phosphating*.

Application range ZINC – GALVANIZED, ELECTROPLATED or ZINC RICH PAINT FINISHES

The coating is applied as a brush system and the purpose is to etch the zinc to allow it to accept paint. Without some form of treatment a zinc coating will normally have poor adhesion for paint films.

Detailed information on the 'Lithoform' process can be obtained from ICI Ltd. General information will be found in Part Two **Painting**.

'Loctite'

A proprietary method used to prevent nuts working loose in service, or for other similar applications.

Application range ALL METALS

The materials used are specially formulated resins which in contact with air remain liquid, but which solidify when air is excluded. The technique, therefore, is to paint the liquid either on the stud or the nut, and when the nut is tightened, the threads where the studs are in contact will have no air present. Under these conditions the resin will solidify, thus holding the nut.

This technique is now being successfully used in place of lock washers or, in some cases, locking wires or pins. It has the considerable advantage that, when torque-loading is specified, the exact load, allowing for the 'Loctite' liquid, can be specified and, thus, the exact load on the stud will be achieved. With lock washers and locking pins, it is sometimes necessary to overtighten or undertighten to locate the holes. Further information can be obtained from the Loctite Co.

Lost Wax Process

A *casting* technique, and as such not a metal treatment.

Application range ALL METALS WHICH CAN BE CAST

Briefly, the process is that a pattern is produced in wax. Any wax can, in fact, be used, but with modern investment-type castings where high integrity of detail is necessary and where dimensions must be held to very tight limits, special waxes with specific characteristics, particularly regarding expansion, are used. These can given an excellent surface finish to the finished article.

This wax pattern is used to make a mould. There are various techniques of producing the mould on the wax, including forms of spraying and dipping. It will be seen that the mould using these techniques will faithfully follow any details on the wax pattern. The mould will then be hardened, the wax then being removed by melting, leaving the mould ready for *casting* with the desired metal.

This process was used by the Egyptians and although ancient has been refined by modern technology to produce components which are intricate in shape and which can be made to hold extremely tight tolerances. It is also known as investment casting, or the *cire perdue* process.

Ludwig Test

A form of hardness test, using a conical indenter. It is now seldom used, and is listed in Part Two *Hardness Testing*.

Luminous Painting

A process which can be applied to any material in which light can apparently exist under black-body conditions.

Application range ALL MATERIALS

Two basic methods are used:

1. The use of radioactive substances, such as radium and thorium salts among others. These materials retain luminosity continuously until the end of their radioactivity. Since radioactive waves are given off, these paints are no longer used except for very specific purposes.

2. The use of materials such as calcium, barium and strontium sulphides, which have the ability to absorb light and which appear luminous in the dark. The degree to which this effect will be apparent obviously depends on the intensities to which the luminous paint is initially subjected. The effect is reduced and eventually ceases over time. The surface can then be reactivated on presentation to light. The active surface can be readily contaminated to prevent reactivation.

It will be appreciated that the above paints have limited applications, for example, the hands and figures on watches, the lubber line on compasses and other instruments used for navigation, etc. Their application will make use of standard techniques such as hand painting with fine brushes, *silk-screen printing* and transfer system. They are all expensive paints and, thus, will only be used when necessary and where wastage can be controlled. In the case of radioactive materials, it will be necessary to comply with the regulations and laws covering the handling of these materials. The vehicle or medium in which the active ingredient is suspended for application must be very carefully chosen and specialist advice is necessary.

Macroetch Test

An inspection technique used to indicate gross defects and to show the direction of flow lines in *forging* and other similar characteristics, for example, whether or not the material has been welded or is a bimetal type of constructions.

Application range ALL METALS

The technique is that the area being examined must have a reasonably smooth surface, and this must be produced by a free cutting action and not by smearing or *work hardening*. In many cases, for instance, the examination of welds or bimetal materials, the area being examined must be a cross-section. In general macroetching will be destructive but some information can often be obtained from surface examination.

The prepared surface is chemically etched, the etch chosen depending on the material involved. Each metal or alloy will have a specific type of etch, a few of which are as follows:

1. *Aluminium and its alloys* 10 per cent sodium hydroxide or, under certain circumstances, sodium fluoride in caustic soda.

2. *Copper and its alloys* Dilute ferric chloride will give successful results. Ammonium persulphate with ammonia will also give good results but is a more gentle etch and thus requires a higher standard of finish, a longer time, and perhaps warming.

3. *Mild steel and low-alloy steels* 4 per cent nitric acid in water or, under some circumstances, alcohol. Again this etch gives a more gentle effect and thus requires a higher standard of finish. Dilute ferric chloride can also be used and is faster.

4. *Stainless steels* These require the use of ferric chloride in hydrochloric acid. In some instances hydrofluoric acid is used. Very often the etch or specimen will require to be heated for the etch to be effective.

5. *Magnesium and its alloys* Acetic acid with tartaric or nitric acid is the most common etch.

6. *Nickel alloys* These make use of ferric chloride in hydrochloric acid and will probably require this or the component to be heated to give any meaningful result. Alternatively copper sulphate with hydrochloric acid is used.

7. *Titanium and its alloys* These require dilute hydrofluoric acid to which can be added nitric acid. The component may require to be heated to achieve an etch. Titanium and its alloys are extremely difficult to etch, and swab etching with hot solutions is often necessary.

8. *Zinc and its alloys* These are etched with chromic acid to which should be added some sodium sulphate or nitric acid.

The above is not a comprehensive list of etches, but is supplied to give basic information. The technique of etching varies depending on the equipment available and the skill of the operator. There are many more etches, some used electrolytically for specific purposes.

In general the prepared surface should be cleaned and thoroughly washed to present a surface with a film of water. Use can be made of hot water to warm the component and, thus, increase the speed of *etching*. The test piece is then immersed in the etch solution and tilted at frequent intervals to present fresh solution to the surface being attacked. Another technique is to immerse the test piece and then with cotton wool on stainless steel tongs swab the surface.

The technique often used in laboratories is to warm the test piece under running hot water and then using a swab of cotton wool soaked with the etch solution to wipe the surface of the test piece. This technique allows the operator to see the changes taking place and to control accurately the degree of etch achieved, but requires some skill.

It is sometimes necessary to etch the test piece and then to repolish with fine emery paper, repeating the etch and polish alternately. This has the effect of giving acceleration to the effects normally seen with a single polish and etch.

The interpretation of the results of the macro test piece requires considerable skill and experience. The evidence of a weld will be readily seen, but whether or not the operator can identify the presence of undesirable characteristics in the weld will depend upon the training given. Likewise the direction of *forging* flowlines can be readily identified, and obvious defects such as re-entrant flowlines, but the presence of less obvious defects requires that the operator has been correctly educated.

The macroetch test should be used as a means of controlling quality either on the premises of the producer of the material or as part of the incoming inspection techniques. This test is now a common part of the procedure used to approve a *welding* process and its operator, and as a routine test on welds. The macroetch is, in addition, a routine part of metallurgical investigations and is generally carried out as the first part of those investigations following visual examination.

In general this is a destructive test and no real alternative exists if the maximum information is to be gained. It is, however, possible using *radiographic (X-ray)* or *ultrasonic* techniques to identify some of the gross defects which would be found during the macro examination, such as *casting* porosity and slag or oxide porosity in *welding*. It is a debatable point whether macro examination is more economical than these non-destructive techniques, but in many instances it will be more economical and will supply considerably more information than the previous two tests but this will only be on one plane and thus the choice of the area to be examined is important. It also results in the component generally being scrap.

As components reach their finished stage, the machining content makes them expensive and the fact that the macroetch is generally destructive precludes the use of this technique at this point of production. The information which can be gained from this technique is very often a function of the skill in choosing the location, and in the experience of the inspector in *etching* the specimen and interpreting the visual results obtained. It is a very common technique used by laboratories investigating material problems.

Madsenell Process

A method of removing scale or surface oxide.

Application range STEEL

This is a *sulphuric acid etching* technique, where the part is made the anode, and its principal use is the final *descaling* of any remnants of oxide prior to plating or, in some specific instances, for other purposes. The main idea is that the anodic etch, with the production of oxygen at the acid–component interface, will remove or encourage the diffusion of occluded hydrogen from any previous operation and, thus, prevent this being trapped at subsequent *electroplating* processes.

The process uses sulphuric acid at approximately 75–80 per cent strength at room temperature and a voltage potential of 10–20 V.

For the removal of hydrogen, this procedure has now largely been replaced by other forms of preplating treatment, notably the stress-relieving at approx. 200°C of any component which has had any treatment liable to result in hydrogen occlusion, or any component above approx. 750–950 N/mm^2 (50–60 tonf/in^2) tensile strength, particularly those which have operations resulting in stress being applied.

A 50 per cent *sulphuric acid etch*, where the component is the anode, is commonly used where adhesion is of prime importance, as the final etch prior to *electroplating*.

Magna Flux

An inspection process for identifying cracks.

Application range STEEL

Briefly, the process is that the component is made magnetic by any suitable means. Any interruption in the magnetic flux pattern caused by a crack or any other discontinuity can then be shown, because fine iron particles will adhere to this area.

The term 'magna flux' has now been replaced by '*magnetic crack testing*' (MCT) and '*magnetic particle inspection*' (MPI). Described fully under this heading in Part Two **Non-destructive Testing**.

Magnet Test

A form of inspection, requiring considerable skill in its finer form and, in essence, uses a magnet to separate different types of metal.

Application range STEEL – NICKEL AND NICKEL ALLOYS

For ordinary mild steel and low-alloy steels including the 12 per cent chromium stainless steels, there is no normal method of sorting using magnets. However, with stainless steels of the austenitic variety which are nominally non-magnetic, it is possible using skill and experience to separate the high-molybdenum type from conventional stainless steels. Certain nickel alloys can also be sorted using the magnetic sorting test but, again, considerable skill is required. It should be appreciated that the fully non-magnetic austenitic stainless steel can be made magnetic by severe *cold work*, and this effect may confuse the magnetic sorter.

To some extent this is similar to *eddy current sorting*, which uses sophisticated electronic equipment. Even so, in experienced hands it is quite possible to carry out reasonably accurate sorting magnetically. This and other techniques are further discussed under *Metal-sorting*.

It has largely been replaced in routine sorting by the portable equipment described under *Metal-sorting*.

Magnetic Anneal
A specialist type of *heat treatment* to produce certain characteristics in magnetic materials.
Further information given in Part Two **Heat Treatment**.

Magnetic Particle Examination (MPE)/Magnetic Crack Test (MCT)/ Magnetic Particle Inspection (MPI)
A form of *non-destructive testing* for surface cracking. Described fully in Part Two **Non-destructive Testing**.

Malcomizing
A method of surface *hardening*.

Application range STAINLESS STEEL

This is a form of *nitriding*. Processes such as '*Tufftriding*' and '*Sulphinuz*' should be evaluated as they can show considerable advantages in reducing friction wear, and seem to have certain advantages with stainless steels. The use of *dry-film lubrication* should be examined.

Malleabilizing
This term means the production of a ductile material from a brittle material.

Application range CAST IRON, NICKEL – SOME STEELS

In the case of malleable iron, this is produced from normal cast iron which is relatively brittle and the process involves the removal of the embrittling material, that is carbon in the form of graphite.

The components are packed in airtight boxes which contain some oxidizing material. It is possible using modern furnace equipment to seal the components in a furnace and to pass an oxidizing atmosphere over them to achieve the same effect. The components are then taken to above 900°C when the reverse of the *carburizing* process occurs. Here the graphite is oxidized in preference to the iron. As the surface is denuded of graphite, the graphite remote from the surface will migrate towards the surface in an attempt to regain equilibrium. As this carbon reaches the surface it is oxidized; the process results in a component free of graphite without the oxidation of the iron.

For economic reasons, the process is applied only to components of relatively thin section and relatively large surface area. For technical reasons, this is necessary as it is unlikely that the furnace conditions can be controlled for periods long enough to ensure that no oxidation of the steel occurs. Brittleness will return to the surface, by the formation of iron oxide either as a complete surface layer, or more dangerously as a grain boundary effect. With high-quality malleable iron, there is often some evidence of graphite, but this must be well distributed and in relatively small amounts.

Malleable iron is normally an expensive method of producing ductile material when compared with, for example, mild steel in the wrought condition. When it is considered, however, that highly intricate shapes can be produced very cheaply in cast iron, then provided these shapes have a thin section and relatively large surface area the malleabilizing process can be economical.

It should be noted that SG iron, that is spheroidal graphite cast iron, is often termed malleable iron, and has properties as good as, and often superior to, iron which has been malleabilized as described above.

Malleabilizing of nickel requires that during the *casting* operation additions are made to the molten nickel to remove any embrittling materials such as hydrogen, nitrogen and the oxides of carbon.

When this is applied to steel it generally means that an annealing operation has been applied. It should, however, only be used when an improvement in ductility is involved.

Maraging

A form of *heat treatment* applied to a specialist ferrous alloy which results in a very high tensile strength material. Described fully in Part Two **Heat Treatment**.

Marquenching

A form of *heat treatment* applied only to steel, where the components are quenched from the hardening temperature into a hot, molten liquid and held there until the desired transformation has occurred. Described fully in Part Two **Heat Treatment**.

Martempering
A specialist form of *heat treatment* which uses a liquid salt bath or metal as the quench medium. Described fully in Part Two **Heat Treatment**.

MBV Process
The *modified Bauer Vogel* process, which is a form of *anodizing*.

McQuaid Ehn Test
A quality control technique involving the metallurgical properties of the material.

Application range STEEL

This is extremely useful in assessing the *heat treatment* characteristics. The test gives information on the grain size of the steel, which is useful in determining the inherent brittleness and also supplies data on the hardenability of the steel involved.

The steel to be tested is *carburized* for sufficient time to ensure that free cementite is produced at the surface. The test piece is then cooled very slowly and a micro section is produced. This will show well-defined grain boundaries at the surface, where the free cementite exists. The size of these grains can be measured and used as a means of grading the steel. In addition, the type of structure achieved gives information on the hardness characteristics or brittleness and ductility of the material after *hardening*. Where conventional *heat treatment* is to be carried out and the temperature control of the furnaces is of a high standard, the test will be of limited significance. Where high-temperature treatment is involved, such as *carburizing*, with no subsequent *refining* operation, then it is essential to ascertain that undesirable grain growth will not occur. Thus, the McQuaid Ehn test is a desirable means of quality control where steels which are not grain-size controlled are used for high-integrity production parts which have been high-temperature *heat treated*. The *Shepherd test* is comparable in that it also uses a standard test piece, hardened under controlled conditions and examined metallurgically.

Briefly, this test is a measure of the grain size of steel, and gives some indication of any tendency for grain growth. It is not commonly used as a routine quality control.

MCT
See *Magnetic Particle Examination/Magnetic Crack Test/Magnetic Particle Inspection*.

Mechanical Alloying
A recently developed technique whereby metal particles are ground to a very fine powder and impacted to produce a true metal alloy.

Application range METALS SUCH AS IRON, NICKEL, COPPER, COBALT

Where conventional melting and alloying is possible, mechanical alloying will never be economical. There is also a range of metallic elements which do not lend themselves to this form of alloy manufacture.

The technique uses a ball mill with very high velocities and special techniques which ensure that the balls are reciprocating and thus impacting each other. In order to prevent oxidation of the material being treated, use is made of inert gases such as nitrogen, or under certain circumstances argon or helium.

The materials to be alloyed, in a relatively fine powdered form and in the correct proportion, are added to the ball mill which is rotated using the desired gas. As the balls impact each other they first produce a fine powder of the materials used, with the particle surfaces being free of oxide. As the ball mill speed increases impact load eventually forces together various particles producing a true alloy. The end product will therefore again be a powder but whereas the powder used as a raw material had particles of the separate metallic elements, the resultant powder has particles of the alloy involved.

The technique is of considerable use where elements with a wide difference in melting point produce a valuable alloy. Conventional alloying is difficult under these conditions without the danger of serious segregation.

The end product can be sintered or melted in a conventional manner.

Mechanical Refining
An alternative term for *hot working* in which the grain structure is reformed and thus *refined* by the mechanical action of the work applied.

Application range STEEL

When steel is heated above a certain critical temperature, recrystallization occurs and the existing grain structure is destroyed and replaced by a completely new crystallized structure, starting as a nucleus and growing from this pinpoint. During *hot working*, then, the material must be above the recrystallizing point if the process is to conform to its own definition. The work applied will result in a breaking down or distortion of the crystal structure which exists, and immediately this new structure is formed it will again start to grow. If controlled cooling is carried out, grain growth will be inhibited. In general the term mechanical refining is not used, '*hot working*' is more commonly applied.

Mechanical Testing
This term covers all the techniques which are used to specify and test for the properties of metals.

The tests are listed and described in Part Two **Mechanical Testing**.

Mechanical Working
A term which is applied when material is taken above its yield point, when it will deform in a plastic manner and a new shape is produced.

Application range ALL MATERIALS WHICH CAN BE HOT WORKED

The term is usually applied to differentiate the shaping of material where cutting is not involved. It does not, however, differentiate between components which have been strengthened by work hardening and those which have been hot worked, where the material recrystallizes with no alteration in mechanical properties.

Mellozing
A method of zinc coating.

Application range MILD STEEL

The process is relatively uncommon and uses molten metal which is atomized by spraying through a nozzle under pressure. This has largely been superseded by methods whereby the material being sprayed, in the form of powder, rod or wire, is melted or atomized in a gas stream, where the particles are given the energy of the gas stream.

Melonite
This is the trade name used in America for the processes known in Europe under the general name of Tufftride.

Further information from Degussa Ltd.

Mercast Process
A variation of the *lost wax* casting process.

Application range ALL METALS WHICH CAN BE CAST

In this case mercury is used to fill the master mould. The mould with mercury is then taken to below −60°C at which temperature the mercury is solid and can be removed from the mould and used as a pattern. This is then treated with a slurry in the identical manner to the wax in the *lost wax* process. The coated mercury is then allowed to return to room temperature when the mercury will, of course, be liquid and can run out of the mould, which is then used for casting in the normal manner.

This has very limited application because of the health hazard involved in handling mercury, in addition to the high cost of mercury itself and the expense of cooling to the low temperature. It has the obvious advantage that mercury is a highly fluid liquid which will thus faithfully reproduce intricate designs.

The same degree of precision can be achieved using low-melting-point types of alloy, such as the bismuth alloys. These are cheaper than mercury, are solid at room temperature but liquid at relatively low temperatures in the 40°C upwards range, and have not the same serious health risk of mercury.

The process has been replaced by the *lost wax* technique.

'Merilizing'

A proprietary phosphate process used on steel; see *Phosphating* for further details.

Mesnager Test

A form of *impact test* which uses a shallow notch with a rounded bottom.

Application range ALL METALS

The dimensions of the test piece are 60 mm long, 10 mm square, with the notch being 1 mm wide and 2 mm deep with a rounded bottom. The fracture is carried out on a *Charpy* type of machine. There is a tendency to standardize on the *Charpy* specimen for *impact testing* and some use still is made of the *Izod test* but it is very seldom that the Mesnager test is called for. Details of *Impact Testing* are described fully in Part Two **Mechanical Testing**.

Metallic Arc Welding

A general term which can be applied to any form of *electric arc welding* where a filler rod is used. The term is not generally applied to the sophisticated forms such as *inert-gas shielded metal arc welding* but there is no technical reason why this term should not in addition cover these processes.

Metallic Cementation

A term, not now commonly applied, covering the process where any metal is diffused into steel.

The *cementation* process itself is an alternative name for *carburizing*, and hence has been applied where objects are treated in a manner similar to *pack carburizing*, that is, the method of surrounding components with a compound or gas which will affect the surface. Processes which can be said to come under the present heading would be *sherardizing, aluminizing, chromizing* and other similar methods, in addition to carburizing.

Metallic Coating
A general term covering all forms of coating with metals but which, under many circumstances, has quite specific meaning in that a process such as *chromium plating* may be known to the operators and their immediate colleagues as the 'metallic coating' process. Therefore, it will be seen that it is not possible to precisely define the process here, and readers are warned that the term can mean different processes in different areas or industries.

Metallic Painting
A general term covering any process where a component is painted with a metal-bearing paint. It need not be a metallic component as this method can be carried out on plastic or wooden parts, and there is no difference in technique from *painting* as a process in its own right. The most common metals applied are zinc and aluminium for *corrosion protection*, and copper alloys and gold for decorative purposes.

The term can also be applied to any metallic coating where the application is by a 'painting' technique.

'Metallization'
A proprietary *metal-spraying* process.

With this, the metal being sprayed, either in powder or rod form, is fed into a high-intensity, high-velocity gas flame. It is atomized there, and picks up some of the energy from the flame, the metal particles then being impinged on to the surface being sprayed.

Metal-Lock
See '*Metalock*' *Process*.

'Metalock' Process
A proprietary procedure for the salvaging of cracked or porous castings or other fabrications.

Application range CAST, RUN – GENERAL CASTINGS

The procedure can be used on any metal and consists of drilling small overlapping holes around or along the area to be repaired. These holes are then filled by hammering in a soft plug slightly oversize to the drilled hole.

196 / METAL-SORTING

The procedure relies on considerable skill in drilling the correct sized holes at the correct point and in choosing the correct material to plug the holes. Very often the material used is pure nickel, but this can vary depending on the material being plugged.

There is no alternative procedure of an identical nature, but cracking or porosity on occasion can be remedied by the *impregnation method*. Serious cracking will require repair by brazing, soldering or welding.

With the 'Metalock' technique cracks in relatively thin-walled assemblies can be made pressure-tight to high pressures and a high standard. Further information can be obtained from Metalock Ltd.

Metal-sorting

A term covering many procedures which can be adopted to segregate materials which are known or suspected to have been mixed or to identify unknown metals.

Application range ALL METALS

The following are the procedures applied.

1. *Magnetic sorting* This can be applied only if there is a difference in the magnetic characteristics of the mixed material. With skill, however, different groups of materials can be sorted using a hand-held magnet. These will be the group of austenitic stainless steels which contain some molybdenum. Other materials such as nickel and some of the cobalt alloys have different amounts of magnetism and, provided information is available and the sorter has the necessary skill and experience, correct segregation can be achieved. Magnetic sorting will of course readily separate ferrous materials from non-ferrous.

2. *Eddy current sorting* There are instruments available which can assess the effect of eddy currents on any material. The procedure here is that two identical coils have the output from the coil fed to an oscilloscope, with one coil feeding the X plates and the other the Y plates. A direct current is passed through the coils and, provided this is identical in each coil, the resulting trace on the oscilloscope will be a circle.

When any metal object is placed in one of the coils, this will affect the magnetism or eddy current in the coil and the trace will no longer be circular. If, however, an identical object is placed in the second coil, then the coils will again be balanced and the trace will return to a circle. Modern equipment makes use of small probes, to give the same effect.

This technique can be used for differentiating between a wide range of characteristics. These include *heat treatments* in addition to different materials themselves. The disadvantage of this method is that the mass and shape of the component or material in the coil will affect the eddy current, and it is commonly found that unless the dimensional tolerances are very tight or the materials being sorted have gross differences, no clear difference in pattern will be seen.

There is now equipment available, where it is claimed that specific parameters such as hardness, metallurgical structure, surface conditions etc. can be identified. The reader is advised to obtain further information on this subject. It is known as Electro-Magnetic Detection (EMD).

3. *Spark testing* It is possible with skill and experience to segregate a number of materials purely by examination of the spark produced. This is based on the fact that any metal when heated above a certain level will give off a lightwave or series of lightwaves which are exact to themselves. This property can be seen in the different coloured sparks obtained by different materials, and with experience it is possible to spark a large variety of materials and identify them with a degree of accuracy. In addition to separating materials such as tungsten or nickel from steel, it is possible to differentiate between relatively similar types of alloys with this technique.

To be carried out properly, the grinding wheel should be reversed from the normal to throw the sparks upwards, and they should be examined against a black or dark background under conditions of subdued lighting.

4. *Instrumental spark testing* There are a number of instruments on the market notably the '*Fuess*', the '*Steelascope*' and the '*Metascope*' which are reasonably portable instruments. With these, an electrical discharge spark is made between an electrode and the metal being sorted. The lightwaves from this spark are then divided by a series of prisms and it is possible to examine them in an eyepiece. These machines, with calibration, can indicate clearly the lightwaves present, and thus positively identify the presence of an element in an alloy. With skill and experience, it is further possible to judge from the intensity of the light the amount of element present. The more sophisticated machines photograph the output and, thus, permanently identify the material.

While these instruments can be used to give approximate analysis of metals, they are more useful in metal-sorting. Here, with a minimum of skill, examination of a large number of bars or components can be made and, using a standard of known analysis, rapid sorting of mixed materials can be achieved.

This equipment in sophisticated form is available for laboratory use where accurate comprehensive analysis of metallic elements is achieved.

These spark emission type instruments only identify metal elements, thus cannot be used for carbon, sulphur, silicon etc. This reduces their usefulness, and this serious limitation in their application limits their practical application.

5. *Pyrometry effect* This instrument makes use of the fact that when two dissimilar metals are joined, and the join heated, an electrical current is produced. This electrical current is dependent on the metals used and is a function of the temperature to which the metals are heated. With two known metals at any one temperature, an exact electric current is produced for any pair of metal elements. It is this which allows the use of this principle for temperature measurement. When this procedure is used for metal-sorting, a

probe of a standard metal is heated to an accurately controlled temperature. This is brought into contact with the metal being examined. A reading will be obtained on the output meter and this must be noted. The probe is then brought into contact with a piece of the known metal. If this has the same metallic elements present, then the pyrometry effect will mean that the materials being compared are the same if the same reading is obtained.

Provided the probe contact surface is kept clean and the point of the contact is in the same relative condition, this technique does not require that the material is in the same metallurgical condition or that the mass of the metal is similar.

The machine has the defect, however, that carbon is not a metal and, thus, this instrument cannot differentiate between different steels having the same basic alloy content but dissimilar carbon content. This machine is known as the 'Metal Monitor'.

6. *Chemical analysis* This is the 'referee method', which consists of chemical analysing by any of the techniques available in the laboratory for the requisite elements. It will always be more expensive, but more accurate, than any of the above methods, although these in skilled hands can give results which are completely satisfactory.

Chemical analysis should always be planned carefully, since often it is only one or two elements which are the critical materials in the composition of an alloy. There is, therefore, little point in carrying out sophisticated and expensive analysis for elements which cannot supply the necessary information.

Many of the chemical reactions which are used in the laboratory to give accurate results have been modified to give spot checks which will identify specific elements in an alloy. These can, in some cases, be used with relatively unskilled personnel, but unless conditions and solutions are carefully standardized the results will be suspect.

It is now known that different chemical analysis of a metal can achieve the same mechanical results. It is often possible to identify the metallurgical condition required to produce the mechanical property, more economically than by chemical analysis. The information available from micro (metallurgical) examination is often much more valuable economically than from chemical analysis alone. It, however, requires to be evaluated by an experienced skilled metallurgist.

Metal-spraying

The general term given to the procedure whereby one of several metals is deposited on a substrate.

Application range ALL METALS

Metal-spraying is the subject of several publications and it will, therefore, be seen that it cannot be fully covered in a book of this kind. Briefly, metal-spraying is carried out for three reasons; these are:

1. *Corrosion protection* The metal being sprayed will require to satisfy the necessary technical objectives to give the corrosion resistance required. The most common metals applied are zinc and aluminium. Metal-spraying is now the usual method of ensuring a high-standard long-life corrosion protection on structural-steel assemblies. The sprayed metal is very often sealed by *painting* or the application of some other top coat. In addition to structural steel, this method is used when corrosion protection is required on certain engineering components, and is a suitable method of protecting high-tensile parts which cannot be *electroplated* because of the danger of hydrogen embrittlement, and cannot be painted because of the operational environment.

2. *Hard facing* This uses a wide range of metals. Many high-tensile metals are sprayed to give hard facing. The materials used fall into several well-defined types:

   (a) Tungsten-bearing or tungsten-carbide materials.
   (b) Cobalt and nickel hard-facing materials, generally containing some chromium.
   (c) High-manganese chrome facing materials.
   (d) Miscellaneous materials used for specific purposes, including non-metals.

The term 'hard' is relative to the requirements and the substrate, and can vary from materials with a hardness of very little more than 250 DPN to materials in excess of 900 DPN.

3. *Salvage purposes* When engineering components are found to wear in service, or for economical, technical and time reasons it is necessary to salvage these components, then metal-spraying is a technique which has been successfully used.

In this case, the choice of the deposited material should be carefully made, and knowledge of the type of surface involved, and the reason for excessive wear, is obviously necessary before the choice of a suitable metal-spray material is possible. The choice, then, should take into consideration the substrate material, the fact that the original component may have been hardened by some process and whether or not the wear can be attributed to any specific cause.

The materials sprayed range from mild steel to the hardest materials known to engineering. It is not good practice to assume automatically that the application of a harder material than was previously known to be present will result in a better component, as this increase in hardness will invariably result in an increase in brittleness and this will often be accompanied by technical difficulties which can arise when spraying harder materials.

A good compromise will very often be to use the patent process known as

'Sprabond', which is a thin film of molybdenum, and to this surface can then be applied a material in the same hardness range as that on the original component, but making use of materials which are known to be eminently suitable for metal-spraying.

While, then, the techniques of metal-spraying are quite varied, the above are the usual reasons for metal-spraying. The prior preparation is generally standardized. This requires that the material should be reasonably clean, although not to the high standard required in the case of *painting* or *electroplating*. Following this the material will invariably be required to be roughened. This is necessitated by the fact that no chemical or metallurgical bond is produced during the metal-spraying process, and consequently it is necessary to provide a key for the satisfactory adhesion of the metal-to-metal bond. This key can be most successfully achieved by *grit blasting*. A high standard blast is carried out, the type and sharpness of the grit being carefully controlled. No attempt, however, should be made to metal-spray on top of a surface which has been blasted using blunt shot, where *peening* or *cold work* will be carried out, thus detracting from the adhesion. It must also be understood that a satisfactory blast process will result in a surface which has a high activity and can rapidly corrode. Any corrosion present will detract from the ideal metal-to-metal bond. Use may also be made of carefully controlled machining, where the surface contours can achieve the necessary key.

As soon as possible after *blasting* or machining, and this within minutes rather than hours for high-integrity components, the metal being sprayed must be applied.

The various techniques of metal-spraying are as follows:

1. One of the original methods was to make use of a standard oxyacetylene welding torch with a hopper in which the powdered metal to be sprayed was attached to one of the gas streams and controlled with a valve. The technique was that the operator used his welding torch to preheat the prepared surface, and when this had been carried out to his satisfaction, the valve was opened to allow the powder to enter the gas stream. This would then deposit on the heated metal surface.

This technique gives little or no adhesion of the metal being deposited but can be used as a two-stage process where the deposited metal has two distinct materials, one of which is the hard material, and the other a lower-melting-point material, generally of a high nickel or cobalt type. The technique used here is that the deposited metal is applied with a relatively soft or low-temperature flame and flux, and when this has been deposited to the area to the necessary thickness, the powder control valve is closed and the temperature of the flame increased. This results in melting or fusion of the lower-melting-point component of the powder which results in a form of *brazing* to the metal surface, this brazed material holding within itself the hard component. This technique is still commonly applied and is known, among other things, as the *endurance process.* It is, in fact, very little different from the technique in which a welder deposits the hard facing material in the form of a rod or stick.

2. *Flame impingement* This is the most commonly used method of metal-spraying, using a specially designed flame gun with either a continuous rod or a hopper feeding powder into the flame.

The technique is that the flame has a high velocity, much higher than that achieved with the standard *welding* technique. This is used to atomize the powder or rod and to surround the small particles of metal which will then generally, but not always, be in the molten state. These tiny droplets of molten plastic or hot solid metal are then projected with the energy of the flame on to the prepared surface.

The problem with this method is in the necessary tight control of the flame conditions. If too high a temperature is used, or too oxidizing a flame, then the droplets of molten or plastic material will tend to become oxidized and there will be a deposit of metal oxide rather than the metal itself giving a porous unsatisfactory deposit. Of all the types of metal-spraying, this is the one where the skill of the operator is of greatest significance, for it is with this that most of the technical problems are known to exist.

It is therefore inadvisable that this technique is specified when high-integrity components require to be metal-sprayed without at the same time ensuring that the conditions are predetermined and controlled. This technique is perfectly suitable for structural-steel components, where corrosion resistance is the prime object, but again some reasonably tight control of the flame conditions must be assured.

A modification to this technique is that an electric arc is struck between two inert electrodes, and this is used as the heat source in place of the gas flame. An inert gas can then be used, or even a reducing gas to supply the velocity to the metal being deposited.

3. *Plasma coating* This technique is essentially the same as the above with the exception that the energy is supplied to the metal particles using the *plasma heating* technique. This produces a high-intensity source of heat which can achieve temperatures to the order of 30 000°C. This temperature can be produced in any gas and this is essentially a technique of disturbing the electron configuration of the gas. With this, the advantage over the more conventional forms of metal-spraying is that it is not necessary to use an oxidizing gas, and gases such as nitrogen or even argon, which is completely inert, can be used.

The technique is that the plasma flame is produced and then diluted to a manageable temperature. At the same time the metal is fed into this very high-velocity flame, and can be either in the form of powder or a rod.

The *plasma* method is expensive and is seldom or never used when the conventional techniques would give satisfactory results. It is reserved for high-integrity components where excellent adhesion or relatively sophisticated materials are required. These are generally of the tungsten-carbide type but can also be metals of the high-cobalt hard-facing variety.

The procedure is noisy and while this need not be serious, it must be taken into consideration when planning to use this technique. When correctly applied, the plasma-spray will give high-density, low-porosity (low-oxide) deposits with excellent adhesion. While in many cases the use of *preblasting* is not necessary, the same criteria will exist with this technique as with the more conventional metal-spraying, in that the use of a good-quality *grit blasted* surface will enhance the adhesion of the deposit.

4. *'Flame plating'* A proprietary process which makes use of an explosive discharge. The procedure is basically that a rifle is loaded with the powdered spray material and aimed at the component, the deposit being carefully metered with each loading. At the same time the propellant, invariably a combustible gas, is loaded again in metered proportions. The breech is then closed and the charge is fired by an electric spark, resulting in impingement, under very high velocity and carefully controlled conditions, of the deposit metal on to the component. Each discharge results in an area of approx. 3.75 cm (1.5 in) in diameter and 0.02 mm (0.00075 in) in thickness. Immediate reloading of the gun after each discharge is possible and four or six discharges a second will be achieved. The gas supplying the velocity to the metallic particles is of a known analysis and is controlled to ensure that a neutral or reducing composition is achieved, thus preventing oxidation of the spray metal particles.

Of the available metal-spraying techniques, this will probably achieve the most consistent standard of adhesion and control of oxidation or porosity. Further details can be obtained from the Union Carbide Co.

Metascope Testing

A method by which different metal alloys can be approximately identified using the spark discharge technique.

More information on this technique will be found under *Metal-sorting*.

'Metcolizing'

A form of *aluminizing* which is generally applied to cast iron. The technique is that aluminium metal is sprayed on to the prepared cast-iron surface and oxidized in a furnace atmosphere. This results in an adherent layer of aluminium oxide with some occluded aluminium. Further information will be found under the headings *Aluminizing* and *Metal-spraying*.

Method X

A term sometimes applied to the *spark erosion* technique, and further details will be found under this heading.

Microcharacter

A specific *microhardness testing* technique, where instead of measuring the actual size of the indentation the area being examined is scratched with a diamond indentor with a known small load, generally 3 g. This technique is used when different grains within a metallurgical structure require to be assessed for hardness.

The method is that the instrument is set up and the grains to be assessed are scratched. The width of the scratch is measured, and it will be appreciated that the wider the width the softer the grain. In general no attempt is made to quantify different hardnesses obtained but rather to assess the harder or softer grains within one metallurgical structure. With improvements in the standard micro hardness tester, it is now possible to carry out actual testing on all but the smallest grains.

'Micro-Chem'

A proprietary *electrocleaning* process used for brightening and *'passivating'* stainless steel, particularly that in pipe form as the process will successfully treat both inner and outer surfaces.

The process is a form of *electropolishing* and can give considerable improvement to surfaces, resulting in a smooth, shiny finish. Its advantage over normal methods is that inner surfaces of tubes of relatively thin diameter can be successfully treated. The process requires that the components undergo thorough *cleaning* and *descaling* prior to entering the 'Micro-Chem' bath. Further information is obtainable from the Rath Manufacturing Co., Janesville, USA.

Microhardness Test

A *Vickers*–type test, using very light loads on a standard microscope. Described fully in Part Two **Hardness Testing**.

MIG Welding

A welding technique, using the *inert-gas shielded metal arc*. Described fully in Part Two **Welding**.

The term is an abbreviation for 'Metal Inert Gas Welding'.

Modified Bauer Vogel Process

The Bauer Vogel process is the chromating of aluminium. The modified process (MBV) is that the parts after being *anodized* are immersed in a chromating solution. Further information is given under *Anodizing* and *Chromating*.

Mohs Hardness Test

A test using standards of varying hardness to scratch the component. It is used by geologists. Described fully in Part Two **Hardness Testing**. It can be a useful technique for checking surface hardness of engineering components.

Mollerizing

An alternative name for *aluminizing* and in general is applied only to mild steel components. The process is used to improve the oxidation resistance of steel.

'Molybond'

This is a proprietary process for use with threaded components.

The process relies firstly on a high-quality manganese phosphate coating being applied followed by the 'Molybond' process.

The 'Molybond' coating is applied by spray and brush within 4–8 h. This is cured after application and has a final thickness in the region of 0.015 mm (0.0005 in).

If the coating is given a final oil dip it is claimed that it has excellent low friction properties along with excellent corrosion-resistant properties.

Further information on low friction coatings of this type will be found under *Dry Film Lubrication* and *Dry Film Painting*.

Further information from Molybond Laboratories, Division of Placer Exploration Ltd.

'Molykote'

A proprietary form of *dry-film lubrication* which uses molybdenum disulphide as the active ingredient in the varnish (see Part Two **Painting**). Further details can be obtained from Fescol Ltd.

Monotron Test

A seldom-used method of *hardness testing*, based on the *Brinell* principle, where the pressure required to produce a certain size of impression is noted. Described fully in Part Two **Hardness Testing**.

'MPI'

See *Magnetic Particle Examination/Magnetic Crack Test/Magnetic Particle Inspection*.

Muschinbrock Hardness Test

An obsolete form of *hardness test*.

Natural Ageing

An alternative name for *room ageing*, where precipitation hardening types of alloy will show an increase in hardness when the component is left for some time at room temperature. This can be after *casting*, or after the *solution treatment* process. Described fully under *Precipitation Hardening* and *Solution Treatment* in Part Two **Heat Treatment**.

NDT

See Part Two **Non-destructive Testing**.

Needle Descaling

A form of *descaling* using specialized tools which vibrate or revolve to impact the steel surface. These tools have a bunch of hardened needles set in holders which allow the points of the needles to remain in contact with the contours of the component. A variety of tools of different shapes and types are available.

This method, then, can reproduce to some extent the process of *shot blasting* to remove scale, and has the advantage that it is portable and can be used under circumstances where a high-quality blast is difficult. It has the disadvantages that it involves arduous labour and is extremely noisy and dusty under most circumstances; it also relies on the tools being kept in good condition. It is the sharpness at the point of the needles which supplies the cutting action to give the surface profile required for good paint adhesion. Blunt needles will remove brittle, loose scale and paint, but will not remove adherent scale or paint. The contact area is quite small, thus in addition to the work involved being hard and unpleasant, it also requires considerable care on the part of the operator to cover all surfaces adequately.

Negative Hardening

A form of correction or salvage, where a material has been embrittled by incorrect *heat treatment*. It is a loose definition, covering several possible treatments, the most common one being a *normalize*, where brittleness is removed by ensuring a homogeneous grain structure. However, it can, and is, probably more correctly applied, when a component is rehardened after being overheated, resulting in undesirable grain growth. Thus, the term negative hardening can be equated with *refining* or *normalizing*, both of which are described under the appropriate headings in Part Two **Heat Treatment**.

The term should be queried, to ensure that all concerned understand the implication.

Negative Quenching
A term which is occasionally used to describe any method of cooling which is accelerated, but results in no metallurgical changes occurring. An example of this would be the coooling of plain carbon steels from their *tempering* temperatures. The term is seldom applied in modern industry, and should always be queried.

Nertalic Process
A form of *argon arc welding*, where a consumable electrode is used. It is now commonly called the *metallic inert-gas* (MIG) process, where the *argon arc* process uses a continuous consumable electrode as the filler material. Described fully under *Inert-gas Shielded Metal Arc Welding* in Part Two **Welding**.

'Nervstar'
A proprietary process producing a black deposit of chromium.

This process is used for decorative purposes only. Where a non-reflective black surface is required for technical reasons such as non-reflectivity, consideration should be given to *black nickel* or some form of *painting*. Information on Nervstar can be obtained from Oxy-Metal Industries (GB) Ltd.

Ni-Carbing
The original name given to the *carbonitriding* process. This makes use of a gas atmosphere, rich in carburizing but also containing ammonia, for the *case hardening* of steel. This results in a case containing carbon and nitrogen. Described fully under *Carbonitriding* in Part Two **Heat Treatment**.

'Nichem'
A proprietary form of *electroless nickel plating* produced by M. L. Alkan Ltd.

Nick Break Test
A quality control test to assess the standard of *Welding*. Described fully under this heading in Part Two **Mechanical Testing**.

Nicked Fracture Test
A destructive test to assess rapidly and economically, the notch sensitivity of bar and sheet material. Described fully in Part Two **Mechanical Testing**.

Nickel Ball Test

A method of assessing the efficiency of an oil used for quenching at heat treatment. As the '*hardening*' of steel is achieved by quenching, and the degree of hardening is related to the speed of quenching it is sometimes necessary to measure the rate of quenching of different oils. The nickel ball test makes use of the fact that nickel, being a magnetic material, will lose its magnetism at a specific temperature on heating, and regain its magnetism at the same temperature on cooling.

A 1 in (25 mm) diameter nickel ball is heated to 850°C in a tube-type furnace. When the ball has attained this temperature the furnace is tilted to allow the ball to roll out and fall into the test quench oil. As it leaves the furnace the ball triggers a timer. In the oil the ball is held on a tray below which is a magnet. When the ball reaches the temperature of 350°C it regains its magnetism, this being the Curie point of nickel. The ball is attracted to the magnet, tilts the tray and stops the timer. The test is repeated and the results averaged for each oil test. The more efficient the quench oil the more rapidly the nickel ball will cool from 850°C to 350°C. A nickel ball is used as it resists oxidation at 850°C, and has a lower Curie point than iron. It thus can be used repeatedly and the temperature range gives meaningful results. This is a relatively simple test requiring easy to produce equipment and gives rapid results. The *silver ball test* requires more sophisticated equipment but gives additional information.

An alternative method of assessing oil efficiency would be to harden test pieces of steel under controlled conditions and accurately measure the hardness. This can be a lengthy procedure and unless carefully controlled can produce variable results.

'Nickelex'

A proprietary process for the plating of a tin–bronze alloy deposit. Further information can be obtained from Silvercrown Ltd. Discussed fully under *Bronze Plating* in Part Two **Electroplating**.

Nickel Plating

Seldom used in its own right at present, although it has excellent corrosion resistance. All *chromium plated* parts, where the plating is for decorative purposes, will have an underlay of nickel. Discussed fully in Part Two **Electroplating**.

'Niflor'

This is a composite coating of electroless nickel co-deposited with PTFE particles. It results in a thin film which can be tightly controlled and thus can be used on tight-limit engineering components. It has low friction, self-lubricating properties which are also corrosion resistant and operate over a reasonably high temperature range.

Further information on dry-lubrication processes will be found under the heading *Dry-film Lubrication*. Further information on Niflor can be obtained from Fothergill Engineered Surfaces Ltd.

'Nigrax'

This is a proprietary nickel plating solution which is modified to give a black dull deposit. It is therefore of use in camera components and where heat absorption is required. It is basically a standard nickel plating solution with additives to give the black effect. The corrosion protection given is not as good as that provided by a standard nickel plating solution. Further information can be obtained from W. Canning and Co. Ltd.

'Nirin'

An alloy electroplate of nickel and iron, producing a bright deposit. The process is used for decorative and corrosion resisting purposes. Further general information is given under *Alloy Plating*; detailed information can be obtained from Oxy-Metal Industries (GB) Limited.

'Nisol'

This is the proprietary name for a nickel plating solution used for barrel plating. The solution is the basic sulphate/chloride type and can have brighteners added. Further information can be obtained from W. Canning and Co. Ltd.

Nitralizing

The process is designed to produce an active surface ensuring an excellent bond between steel and glass coating.

Application range SHEET STEEL

An intermediate process used in the production of *vitreous enamel* components. After the normal standard of *cleaning* and *pickling* to remove all soil and surface oxide, the sheet metal is then immersed in molten sodium nitrate at about 500°C. This produces the active surface which reduces the adhesion problems which can exist in *vitreous enamelling*. There is no obvious alternative to this particular treatment.

Nitrarding

An alternative name to *nitriding*. Described fully under this heading in Part Two **Heat Treatment**.

Nitration

An alternative term to *nitriding*. Described fully under this heading in Part Two **Heat Treatment**.

Nitriding

A type of *surface hardening* where atomic nitrogen combines with the surface iron to form the inherently hard iron nitride. Described fully in Part Two **Heat Treatment**.

Nitrogen Hardening

An alternative term to *nitriding*. Described fully under this heading in Part Two **Heat Treatment**.

'Nitrox'

This is basically the sursulf process, a nitro-carburizing process involving a subsequent oxidizing treatment.

Application range STEEL

The 'nitrox' process basically supplies the surface hardness of the sursulf process, which is essentially carbonitriding, and then oxidizes it. The oxide is adherent and thus considerably increases the corrosion resistance. The process, therefore, has the considerable advantage of supplying a hard, wear-resistant compressive, therefore fatigue resistant, surface, which is also corrosion resistant. It is comparable to techniques such as hard chromium plating or techniques such as metal deposition where hard, wear-resistant and corrosion-resistant materials are added to the surface.
Further information from Degussa Ltd.

'Nitrox P'

This is basically the same as 'nitrox' with a superior surface finish.

Application range STEEL

'Nitrox P' involves the process described in 'nitrox', followed by polishing of the surface to give the desired surface finish. This is followed by a pre-heat and then a re-oxidation of the surface. It therefore gives the carbonitrided layer, that is the sursulf layer, which is then oxidized and the oxide layer partly polished off to give the superior surface finish and then re-oxidized. It has all the advantages claimed for 'nitrox' plus the addition of a good surface finish.

'Nivo'
This is the proprietary name for a nickel-plating solution which is dull and low stress. It is recommended for use for heavy depositions, for build up, for salvage purposes etc.

Non-destructive Testing (NDT)
This covers the various tests and techniques which do not affect the components or material being tested, but are designed to identify whether specific defects are present. Described fully in Part Two **Non-destructive Testing**.

Normalizing
A *heat treatment* process which removes the variations and stresses resulting from *forging* or *casting*.

Application range STEEL

Further information on this and other forms of *homogenizing* are described fully under this heading in Part Two **Heat Treatment**.

'Noskuff'
A method of improving the friction properties.

Application range STEEL

The process is a salt-bath procedure which impregnates the surface. It is a relatively modern technique comparable in some ways to '*Sulphinuz*' and '*Tufftride*' and will give superior results to *phosphating* or other forms of *dry-film lubricant*. Further information can be obtained from ICI Ltd.

Notch Bar Test
A general term applied to *impact testing*, where a machined specimen is used with a notch to ensure failure at the specified point. Described fully, and also under the headings of *Fracture Testing* and *Impact Testing*, in Part Two **Mechanical Testing**.

'Nubrite'
A proprietary bright *nickel plating* deposit. Described fully under this heading in Part Two **Electroplating**.

Oil Hardening
This is part of the process of *hardening*.

Application range ALLOY STEEL

The steel is taken to above its upper critical temperature, held for a specified time and then quenched in oil. This technique results in the formation of the metallurgical structure known as martensite, provided the material is of the correct composition.

Oil quenching is a gentler method of cooling than water quenching, and is more severe than air quenching. Materials which would shatter under water quenching can use this process to obtain full 'hardenability'. Components high in carbon and chromium, which might crack when oil-quenched, require air quenching. Described fully under *Hardening* in Part Two **Heat Treatment**.

Olsen Test
A test to measure in a reasonably accurate manner the ability of sheet metal to accept *deep-drawing* operations.

Application range ALL METALS

The material is clamped securely and a ball is then loaded and pressed to form a 'cup' or depression. The depth of depression achieved before bursting occurs gives an indication of the ability of the steel to be deep-drawn. In addition the method of bursting will indicate to the experienced personnel the material's suitability for certain types of *deep-drawing* operation. The test, also known as the *cupping* or *Erichsen test*, is a relatively simple method of assessing the ductility of sheet material. Discussed under *Elongation* in Part Two **Mechanical Testing**.

'Onera Process'
A form of *chromizing* which is a proprietary French process for steel only where the surface is impregnated with chromium to give a layer of high chromium content. Further information is given under *Chromizing*.

'Osronaut'
A proprietary name for a barrelling technique. There are 'Osronaut' processes covering all the conventional barrelling systems. Further information will be found in this book under the heading of *Barrelling*; detailed information can be obtained from Osro Ltd.

Oven Soldering
A method of solder joining discussed under *Soldering*.

Overageing

Ageing is the traditional term applied to the *precipitation hardening* where, following *solution treatment*, the material is hardened by the precipitation of intermetallic compounds, which causes strain in the lattice with a consequent increase in tensile properties.

Application range ALL ALLOYS WHICH AGE HARDEN

An overaged component will either have been held at too high a temperature or at the correct temperature for too long. Both cases will result in a reduction of the maximum tensile strength achieved at the peak of the ageing cycle. On most occasions, overageing is undesirable in that the ideal mechanical properties have not been obtained. There are, however, instances when overageing is specified. One of these is with beryllium copper alloys, where at the peak of the ageing curve maximum tensile strength is obtained but electrical conductivity is relatively low. Overageing can obtain an increase in the electrical conductivity of the alloy without reducing the tensile strength by an unacceptable amount. There may be other occasions where overageing can achieve a similar compromise. *Precipitation Hardening is* described fully in Part Two **Heat Treatment**.

Over-drawing

A term applied to the *wire-drawing* process, and refers to a defect where the wire receives excessive *cold working* and thus is less ductile than specified. Over-drawing is, therefore, a defect which should be avoided.

Oxalic Acid Anodize

A seldom used method of *anodizing* aluminium. With this, oxalic acid is used as the electrolyte.

The acid is highly toxic. Sulphuric acid is now by far the most common electrolyte.

Oxyacetylene Welding

A form of *welding*, using the heat from the combustion of acetylene and oxygen to supply the necessary energy.

Because of the difficulty of controlling the combustion, the oxyacetylene heating is not normally used for materials where the pick up of oxygen to form oxides would be a disadvantage. There is, however, no doubt that given the necessary skill, it is possible to make joins in materials such as aluminium, stainless steel and certain of the copper alloys. Considerable care must be taken to ensure that the joins achieved are of a reasonable standard and are not contaminated with oxides. Described fully in Part Two **Welding**.

Pack Anneal
A form of *annealing* where the material is packed in an inert compound, or formed into packs. Described fully in Part Two **Heat Treatment**.

Pack Carburize
A method of *case hardening*, using solid chips of treated charcoal. This has now been generally replaced by *Gas Carburizing* and *Carbo-nitride*. Described fully under these headings in Part Two **Heat Treatment**.

Painting
The name given to a number of quite different techniques which result in a non-metallic coating being applied to the surface being treated.
   Historically a paint is a liquid which converts to a solid after application. The subject is discussed in some detail in Part Two **Painting**.

Palladium Nickel Plating
This is an alloy deposit which has recently been developed.

Palladium Plating
A platinum group metal used in the manufacture of jewellery. Described fully in Part Two **Electroplating**.

'Pallnic' Plating
This is the proprietary name for a plating solution which produces a deposit of palladium–nickel alloy by electrolysis.
   It is stated to have excellent corrosion resistance, comparable in many ways to gold regarding corrosion and contact resistance and to have excellent ductility.
   Further details can be obtained from Engelhard Industries Ltd, and under *Palladium Nickel Plating*.

Paraffin Test
A form of *non-destructive testing* which has been replaced by the more modern technique listed and described in that section. The paraffin test was the early method of carrying out *dye-penetrant crack testing*.
   With this, the component was immersed or soaked with paraffin which was then wiped from the surface, the surface being examined for paraffin seeping

from cracks. Modern *dye-penetrant* methods are quicker and much more sensitive.

'Parkerizing'
A proprietary name for phosphating steel. It is a method of *phosphating*, and is discussed generally in that section in this book. There are various types of 'Parkerizing' and further information can be obtained from the Pyrene Chemical Services Ltd.

Passivation
A term used to denote one of several techniques which increase the corrosion resistance of the surface, or remove some active constituent which might corrode.

Application range STAINLESS STEEL, NICKEL, CADMIUM, ZINC, to a lesser extent CARBON STEELS AND ALUMINIUM

The term '*passivation*' is used for two distinct purposes.

(a) Probably the most common now is for the technique which will produce a surface film, generally on electroplated deposits, which will increase by a marked degree the corrosion resistance.

A very similar process is used for aluminium. These make use of the oxidizing properties of chromates.

Steel can be chromate treated with slight improvement in corrosion. Phosphated steel is commonly chromate treated to give considerable corrosion improvement.

In a similar field are the use of other oxidizing liquids to produce an adherent oxide on the surface of metals. Commonly used is nitric acid for stainless steels and nickel and its alloys.

(b) The second process is for the removal of activated material. This is normally applied to austenitic stainless steels. For various reasons the surface of austenitic stainless steel may be contaminated with ferritic material. This can be removed either by mechanical means or more generally by chemical attack. The term '*passivation*' is commonly applied to this technique although the more acurate description would be '*deactification*'.

The various processes can be described as follows:

1. Chromate passivation of electrodeposited metals. This is generally applied only to cadmium and zinc deposits. Immediately following the electrodeposition and the initial swilling operation the components are immersed in a chromate solution. This converts the freshly plated surface to cadmium or zinc oxide and some chromate is absorbed into this surface.

The components are then removed after a very short immersion, measured in seconds, and swilled in cold water. Careful drying is necessary to give the best corrosion protection and it must be appreciated that the freshly applied film is extremely delicate and will be damaged if handled. After 24 h the film is much more able to withstand normal handling.

There are two basic types of chromate passivation on zinc and cadmium. Firstly, the colour passivate resulting in a khaki or yellowish coloration. This is the colour of the chromate solution. Secondly, there is the clear passivation where the formulation is such that the colour has a very slight bluish tinge but should not have any evidence of yellow or khaki. While the yellow or khaki chromate colour is less attractive visually it has superior corrosion resistance properties.

It should be noted that in addition to the slightly unattractive appearance it is impossible to achieve a homogeneous colour passivation from batch to batch or even within the one batch being processed.

When tested by the *salt–mist accelerated corrosion test* it will be found that correctly applied, colour passivation will have improved the corrosion resistance of the cadmium or zinc deposit by at least 100 per cent.

Further information will be found under the heading *Chromate Coating*.

2. Passivation of metals using the chromate process will also increase the corrosion resistance. Cadmium will never be used in the massive state but occasionally zinc components are used in service without painting or electroplating. The corrosion resistance can be markedly improved by *chromate coating*. Again the process results in the formation of an active oxide into which is absorbed some chromate. The same remarks apply to hot dipping of zinc – that is galvanizing.

Aluminium is also commonly treated in this manner. Both these metals can have the chromate applied by spray as well as immersion. The chromate passivation can also be used for both zinc and cadmium as a method of pretreating prior to painting.

Normal low alloy and plain carbon steel will have a very slight improvement in corrosion resistance. This is not normally sufficient to justify the use of the process.

3. Oxide passivation of metals. Nickel, its alloys and to some extent stainless steel, both martensitic and austenitic, can have a passive film produced by immersing in concentrated nitric acid. They must be thoroughly washed and dried under controlled conditions, following the nitric acid immersion. The components must be clean prior to the treatment, and concentrated nitric acid must be used. Any dilution will result in no passivation or can cause very active attack.

The film will be extremely thin and susceptible to mechanical damage at a low level.

4. Deactification (passivation) of austenitic stainless steel. When any heating process is applied to austenitic stainless steel some free ferrite will be produced on the surface. This will apply to castings and during any welding operation. The ferrite will be present even when no visual evidence of blueing or other discoloration can be seen.

The second reason for ferrite contamination on the surface will be the use of either iron-containing abrasive dust or the use of wire brushing or tooling with free iron present.

All of these will result in stainless steel with a very thin film of free iron which can be attacked by substances with low corrosion potential. Foodstuffs, particularly fruit-bearing and whisky or beer, will be contaminated if in contact with this free ferrite.

It is therefore common that components subjected to any of the above processes are passivated, that is deactivated, prior to entering service. This can be with dilute nitric acid, citric acid, or there are now available pastes containing acidic compounds which can be applied locally to welds.

In all instances it is essential that all traces of the chemical are removed as the result of this contamination would be more serious than the basic cause, that is the free ferrite.

The *Feroxil test* should be applied following passivation to ensure that the process has been correctly carried out. This is a simple test, described in this book, to identify the presence of free ferrite.

It is also advisable to test for traces of the substance used to remove the contamination, such as nitric acid etc.

Patenting

A form of delayed quench heat treatment.

Application range COLD-DRAWN STEEL

It consists of heating the material to above the upper critical temperature, probably 900°C, followed by quenching in liquid metal or salt at about 500°C. It will thus be seen that patenting is a method of delayed quenching similar to *austempering* or *martempering*. The result is a material which is more ductile than hard-drawn wire. It may be used in this condition when toughness is a major requirement, but the wire or rod is often given a further drawing operation to increase the tensile properties.

This is a common meaning of the word 'patent' when applied to metals. It is sometimes used as an adjective with other materials where a specific process has been patented, for example 'Patent Leather'.

It is advised, therefore, that some criticism is applied to the term if no specific information is available.

Peel Test

A destructive test which is commonly applied to paint and, on occasion, electroplated deposits. Further details of this and other similar tests are given under the heading *Adhesion Testing*.

The name is on occasion given to the *chisel test* used to assess spot welding.

Peen Plating

A technique which is a combination of *barrelling*, and *peening* resulting in the deposition of the chosen metal.

Application range STEEL, COPPER ALLOYS

No electric current is required and the method is to load a rotating barrel with the components to be treated, the metal to be plated in the form of a very fine dust suspended in liquid, generally water or paraffin, and the material to cause impacting. This will vary depending on the articles to be peen plated but is generally in the form of balls or of irregular shaped, solid material. The barrel is rotated, resulting in the metallic dust being sandwiched between the components and the impacting material and the components coated with the metal. This is generally zinc, but can be cadmium, brass or tin.

The adhesion of the plating does not approach the standard of conventional *electroplating*. It will also be appreciated that the metal will only be adherent where the impacting material can make contact with the components and, thus, internal surfaces will either have no plating or an even lower standard. The method does have the advantage that, since no hydrogen is evolved, there is no danger of hydrogen embrittlement. This process, therefore, finds some application for high-tensile components such as springs which require *corrosion protection*. Such components will require to be varnished or lacquered following plating to achieve appreciable life under normal service conditions. It is unlikely then that the process can compete economically or technically with modern zinc-rich etch primers. Nevertheless, large batches can be treated for minimum cost of labour and materials. Thus, low-cost ornaments and similar articles can be peen plated, for example, with brass, to achieve the desired appearance.

Peening

A form of surface cold work to improve fatigue strength.

Application range STEEL, ALUMINIUM AND ALLOYS

The process is designed to produce a compressive surface and can be carried out using a peening hammer. In modern industry the process is generally carried out using a *shot blasting* process. As stated, the purpose is to compress the surface layer and the method can be used to improve the fatigue strength of components. When any bending stress is subsequently applied, a certain tensile stress level must be produced before the surface will be at a state of nil stress.

The process takes its name from the peening which was commonly carried out to close any surface defects on castings or cracks on wrought material. This operation is still in existence but to a much lesser extent as the danger of fatigue propagation has become better appreciated. The modern practice is to 'stop' the crack by drilling small holes at either end, or more commonly to gouge out the crack and repair by *welding* or *brazing*. The technique of *Shot Peening* is described fully in part Two **Blasting**.

Pellin's Test
Described fully in Part Two **Hardness Testing**.

'Penetrol Black' Process
A proprietary form of 'black oxide' where an artificial oxide is produced on steel surfaces. This oxide is formed in a hot alkaline solution containing oxidizing agents, generally chromates. This process is described in more detail under the heading '*Black Oxide*'.

'Penybron Plating'
A proprietary *bronze plating* solution, giving a copper–tin deposit, normally for decorative purposes. Information is obtainable from the suppliers, W. Canning and Co. Ltd.

Percussion Welding
A form of *resistance welding* used on components which have good electrical conductivity, such as copper and aluminium alloys. Described fully in Part Two **Welding**.

Pfanhauser's Plating
A specific acid solution used for plating platinum. The material to be plated must be *cleaned* and prepared in the normal manner for plating. The solution is a chloroplatanic acid, with complex phosphates used hot, with platinum anodes.

The as-plated deposit is bright and finds uses for optical purposes in addition to jewellery and for *corrosion protection*. *Platinum Plating* is described fully in Part Two **Electroplating**.

Phosphating
A surface treatment for pre-painting and corrosion resistance.

Application range STEEL, ALUMINIUM, ZINC

Phosphating is the most common pre-treatment prior to *painting* of steel apart from shot blasting.

The process is designed to convert the surface of steel components into iron phosphate, with or without other metal phosphates being present. The process requires that a chemical cleanness is achieved similar to that for *electroplating*. If necessary, parts must be treated to remove any oxide or rust. An acid pickle may be sufficient but it is preferable that any heavy rust or scale is removed mechanically by *wire-brushing* or *blasting*.

Several proprietary phosphate solutions contain de-rusters, and these solutions can be satisfactory when components have an active but thin rust or oxide coating. It is not advisable to phosphate components which are heavily rusted or scaled: apart from the fact that these oxides will make a heavier demand on the chemicals, there is some danger that the iron oxide may not be fully converted to phosphate, thus leaving rust or scale which can propagate below the paint film. The proprietary solutions used for phosphating can be applied by spraying, sometimes at room temperature, and may not require subsequent *swilling*. Most phosphating has an aqueous base and is used at above 60°C, with some solutions requiring temperatures in excess of 90°C.

All the dip solutions require *swilling* in clean, running water.

The process is relatively short, measurable in minutes. Some solutions require only 3–5 min immersion, others 20–30 min. On removal from the phosphate solution, the components should be washed in cold water followed either by immersion in boiling water or very commonly in a chromate-sealing solution. Where the phosphate is used as a pre-paint process, the first coat of paint must be carried out within as short a period as possible (certainly within 24 h and preferably within 4 h).

The ideal is that the paint and phosphate processes are used continuously. A delay of up to 4 h would not be serious and up to 12 h is generally acceptable, but a delay in excess of this should not normally be considered. The reason for this is that the phosphate coating is designed to absorb the liquid paint. Excessive delay after phosphating will result in moisture from the atmosphere being absorbed into the phosphate film. In addition, this can result in the serious defect of a water film being trapped adjacent to the steel. This will obviously contribute to the corrosion of the substrate if at any time the paint film is damaged or broken.

Phosphating is now quite commonly used as the corrosion protection on steel components where corrosion potential is low. Phosphating can be successfully used in tooling, where the components are stored under low humidity, heated conditions. The phosphate film is then sufficient to prevent corrosion by normal atmospheric and fingerprint handling. This phosphating should be sealed with chromate and immediately afterwards immersed in oil. The oil should preferably be above 100°C in order to drive off any water absorbed in the phosphate film. This oil film will be found to drain off the hot components, leaving a very thin absorbed film of oil. As an alternative to oil, various lacquers are used which are also absorbed into the phosphate surface.

Phosphating is not suitable for outdoor corrosion protection.

Phosphating cannot be used as the key to paints which will be heated to above 150°C. The phosphate film contains water of crystallization which is driven off about 150°C, causing failure of the phosphate film.

There is no real alternative to phosphating as a key to normal paint for steel. With some modern paints the adhesion obtained on clean blasted steel is excellent, but corrosion can still be a problem when the paint film is damaged, allowing the atmosphere to cause corrosion which will creep under a satisfactory paint coating. For medium-term corrosion protection, *'black oxide'* will give equal or better protection, and *chromating* approximately comparable protection. Proprietary solutions for phosphating include *'Bonderize'*, *'Granodize'*, *'Merilize'*, *'Parkerize'* and *'Walterize'*.

Physical Testing

A term covering the variety of tests carried out on all metals to assess physical properties, and including *mechanical testing*, and *hardness testing*. The term is also applied to assessing other properties such as coefficient of linear expansion and conductivity, also electrical or heating tests. The latter physical tests are specialist laboratory forms where normal routine procedures do not apply, and consequently are not described in this book.

Physical Vapour Deposition (PVD)

This is the general name given to a number of techniques involving the deposition of a coating from the vapour phase using vacuum conditions. This allows reaction to take place at temperatures usually below 300°C giving thin layer alteration to the base metal.

Information on processes which fit the description of physical vapour deposition will be found under *Sputtering, Vacuum Coating* and *Vacuum-coating Deposition or Plating*.

Pickling

This is the term commonly applied to any chemical treatment used to remove oxide or scale from the surface.

Application range ALL METALS

The term is most commonly applied to the use of sulphuric or hydrochloric acid for the removal of scale formed on steel during hot-forming operations, less commonly to the acid *etching* process which is part of conventional *electroplating*.

Where any metal has its surface oxide removed chemically the process can be called pickling. The treatments applied to the common metals are given below.

The material must first have excess oil or grease removed, but a high standard of cleanliness is not normally required. The parts are then lowered into the pickling solution. The chemical used will depend on the metal being treated and the amount of scale involved. For mild and low alloy steel, this will generally be hydrochloric or hot sulphuric acid, but it is possible to use other chemicals for specific purposes, though invariably this will be more expensive. Phosphoric acid and sodium hydroxide (caustic soda) with additives are examples of the more specialized pickling solutions for steel. Where pipes or hollow components are involved, they must be moved at frequent intervals to prevent airlocks. It is desirable that mechanical agitation is used for other parts. With stainless steel or high nickel alloys, hydrofluoric acid, with or without additives, is used. This is a particularly hazardous acid and must be handled with great care. Ferric chloride in solution with hydrochloric acid can be used for pickling any of the above materials, but it is expensive to use and can lead to effluent problems. Copper and its alloys are pickled in a variety of solutions ranging from concentrated nitric acid to dilute nitric acid, with additions of hydrochloric and sulphuric acid.

Depending on the alloy involved, any of the above acids can be used with a variety of additions.

Aluminium and its alloys seldom require pickling as the oxides involved are adherent and do not form a massive scale, thus there is not normally a need for any removal. Where pickling is involved, caustic soda is the chemical most frequently employed; sodium fluoride with hydrofluoric acid is an alternative. Sulphuric acid is also possible but this requires to be heated.

Magnesium is de-scaled or pickled when necessary with sulphuric acid in an electrolytic process, which is essentially a form of *anodizing*. It is, however, more generally mechanically de-scaled.

Titanium and its alloys are pickled with hot hydrofluoric acid with various additives. It is stated again that this is a hazardous acid and must be handled with extreme care.

It is probably more important to pickle or de-scale titanium than any other metal. Above approximately 300°C, titanium combines with all atmospheres, except inert gases. Thus nitrogen, hydrogen and oxygen, which are commonly used, result in the formation of surface films which must be removed to prevent a serious reduction in the fatigue strength of the component. These surface films are adherent and not readily identified; they can be inter-granular in nature and thus present a considerable fatigue hazard.

Zinc will not normally be pickled but, when necessary, will use hydrochloric or sulphuric acid with additives.

Additives employed with the above chemicals fall into two categories of use, the least important of which is to improve the activity of the chemical involved. With some stainless steels, nickel alloys, and in particular titanium, this is common. More importantly, additives are required to smooth the action, commonly to direct the attack on the oxide film and inhibit any activity on the metal itself. There is now some very clever chemistry involved with these proprietary inhibitors, and complete processes, where the components can be left for an infinite time in the acid without the metal itself being seriously attacked. This, however, requires that the proess is controlled regarding temperature, additive and strength. The reader is advised to obtain specialist advice in this area.

The *sodium hydride process* is a universal method of pickling, which can remove all oxide scale in a technically efficient manner, but requires expensive capital equipment and has high running costs. In addition, there are certain inherent dangers arising from the use of hydrogen and metallic sodium at high temperatures.

Pickling is normally an alternative to mechanical methods of de-scaling such as *blasting, wire-brushing, needle de-scaling* or *grinding*. Where immersion in liquid is possible and high quantities are involved, it will generally be the most economical and technically superior method. It has the advantage that all surfaces are treated, whereas alternative methods concentrate only on outer surfaces. It has the disadvantage that noxious chemicals are used, and the component will generally corrode after treatment, unless considerable care is taken.

With large components or awkward shapes, *blasting* or other mechanical methods will be cheaper and, in many instances, the only possible method. Users are advised to contact a plating shop supply house for further information on pickling solutions.

'Pillet Plating'

A proprietary *palladium plating* solution, containing palladous chloride, disodium phosphate, diammonium phosphate and benzoic acid.

Palladium is commonly plated from a proprietary, ready-to-use solution supplied by one of the supply houses. Some information on *Palladium Plating* will be found under this heading in Part Two **Electroplating**.

Planishing

A *work hardening* technique which can be applied during the manufacturing cycle when the material, generally in sheet form, will be subjected to a limited amount of cold work.

Application range ALL METALS

During manufacturing, planishing will be a hand operation where, by use of hammers and considerable skill, surface blemishes are removed. This technique may require the use of soft bolsters on the underside of the material, and hammers are used which have round heads. Planishing can also be employed where a limited shaping operation is required and where the surface is to be altered or improved.

It is also possible to use planishing rolls in sheet metal operations, where again a limited amount of *work hardening* will be applied; this will often shape the component and always result in some alteration to the surface finish.

Plasma Cutting
A form of metal cutting where *plasma heating* is used.

Application range ALL METALS

The plasma heating technique uses a high-energy, high-frequency electric current to 'interfere' with the electron configuration of the chosen gas. This takes place over a period of time and when the 'interference' ceases the energy absorbed over the period of time is immediately released as heat, giving temperatures up to 30 000°C.

The gas or gases involved will generally be relatively simple, for example, air, oxygen, argon or nitrogen and depending on the use, the flame may be diluted with cold gases to give a more controllable cutting action. It is possible to cut sections of considerable thickness. Using inert gases little or no scale is formed. Further information will be found under *Plasma Heating* and *Plasma Welding*.

Plasma Heating
This uses high-frequency, high-electrical energy to affect a stream of gas.

In essence, this energy is used to disturb the electron configuration of a portion of the gas stream. The electrical energy input is then stopped whereupon the electron configuration immediately reverts to the stable form, instantaneously releasing the high energy which has been absorbed by the disturbed portion of the gas stream. Plasma heating, then, is a method of concentrating and releasing electrical energy in the form of heat which is absorbed by the gas stream. Any gas, including the inert gases, can be used as the means of utilizing this energy.

The electrical high-frequency, high-voltage source of heat results in heat energy of a very high order, above 30 000°C being possible.

Plasma heating is a relatively modern technique which, because of the very high temperature possible, has presented considerable problems in control. When it is considered that the temperature from the oxyacetylene flame is in the region 3000°C–3500°C, and the electric arc struck in air is to the order of 5000°C, it will be appreciated that the heat available with the plasma flame is considerable.

In its raw state, plasma heating can be successfully used for cutting, and this technique is commonly applied in steelworks and heavy engineering for the efficient cutting of thick steel sections. Plasma heating can also be used as a *welding* method. *Metal-spraying* using the plasma flame is now quite common.

It is now possible to control the plasma flame to give oxide-free surfaces when cutting all metals, and to give a high-quality surface finish on quite thin sheet, in addition to massive metal sections.

Plasma Nitride
A form of *nitriding* carried out under vacuum using the *plasma* heating technique. It is a specialist form of case hardening. Described fully in Part Two **Heat Treatment**.

Plasma Plating
A term occasionally used to indicate the *plasma heating* technique used for metal spraying. The correct term is *plasma spraying* and further information is given on this under *Metal-spraying*.

Plasma Welding
The welding technique which uses the plasma heating of metal. Described fully in Part Two **Welding**.

Plating
This term has two distinct meanings. In heavy engineering, 'plating' and 'plater' are applied respectively to the manipulation of steel sheet or plate, and the skilled personnel who are involved. These do not come within the scope of this book. Secondly, the term is used to indicate the 'electrodeposition' of metals. This covers the whole area of *electroplating*, and information on this and the metals plated will be found in that section in Part Two.

Platinizing
A relatively general term which is used to indicate any article which has been coated with platinum. In general, however, it is confined to the technique of *immersion plating* of platinum, usually on copper or bronze components, to give an attractive silvery sheen. This is a typical *electroless* or *immersion* type of coating which, because of the high expense of the solutions required, is seldom used.

Platinum Plating
Platinum is a very expensive metal. The electrodeposition is possible using very acid solutions. The deposit is used for jewellery and is finding a limited application in industry because of its excellent corrosion and oxidation resistance. Described fully in Part Two **Electroplating**.

Plug Weld

A form of *welding* where a hole is drilled in one component and this is then 'plugged' by the weld which is also fused to the second component. Generally a number of plug welds are involved in any one assembly. The technique is also known as *rivet welding*, and employs the *electric arc* or *oxyacetylene* methods.

It is basically a fillet weld round the perimeter of a drilled hole, which is then filled to complete the plug. To be successful there must be fusion on the vertical walls of the top component and the horizontal surface of the bottom piece. It has all the problems of *fillet welding* plus some access difficulty if the plug diameter is small.

Plumstone Blasting

A form of *blasting*, using ground plumstones as the abrasive. It is used to remove contamination from soft materials such as aluminium. Described fully in Part Two **Blasting**.

Poldi Hardness

A form of portable *hardness test* based on the *Brinell* method, using a preloaded spring to impart the energy to the punch. Described fully in Part Two **Hardness Testing**.

Polishing

A technique to improve surface appearance.

Application range ALL METALS

There are two distinct methods of achieving the desired effect, which is a smooth surface very often but not always having a high reflectivity. The two methods are, first, to use cutting materials generally in the form of grit to cut the surface being polished. Second, there is the method of smearing. Here the surface is not cut but burnished or smeared to produce the desired effect. Most types of polishing are, in fact, combinations of these two methods.

The abrasive type of polishing is generally required for a higher standard finish and will almost invariably be the technique used where there is a rough surface, since it is not normally possible to use the smearing effect to give a satisfactory result. The roughness of the original surface is judged and the size of the grit used will approximate to, or be slightly finer than, the surface finish which exists. In other words, if the scores on the original surface have a depth, either average or maximum of, for example, 0.025 in (0.5 mm), then the grit size which will be used to polish this must be approx. 0.025 in (0.5 mm) or slightly less. Any attempt to polish this finish with a finer grit size will either result in failure, or take an extremely long time. The polishing technique chosen, then, should be used to produce a series of parallel scores or scratches of a homogeneous nature. These will replace the scoring which already exists and which is unlikely to be homogeneous.

Under many circumstances this will not produce a satisfactory surface, and the next stage will be to use a coarser grit size. Polishing is then repeated to ensure the complete removal of all evidence of the original scores, and this can best be achieved by an action at right angles until all the scores are in this same direction. Where metallurgical polishing is being carried out, or where a high standard is required, for example, in an optical application, it is essential not only to remove scores from the previous operation but also the surface material immediately below the scoring to approximately the same depth as the original scores. The reason for this is that each scratch will have affected the surface below itself by *work hardening* and for metallurgical polishing, in addition to removing the scores, it is also necessary to remove the effects of this *work hardening*. The polishing should be continued using successively smaller grit particles at right angles to the previous operation, noting the above comment on the work-hardened layer, until a satisfactory standard has been achieved.

The grit sizes which are available for polishng and the approximate actual sizes are as follow:

Polishing Grit Sizes

Nominal aperture size		
Inches	Microns	Mesh no.
0.0015	38	400
0.0015	45	350
0.0021	53	300
0.0025	63	240
0.003	75	200
0.0036	90	170
0.0042	106	150
0.005	125	120
0.0059	150	100
0.0072	180	85
0.008	212	72
0.01	250	60
0.012	300	52
0.014	355	44
0.0167	425	36
0.0197	500	30
0.024	600	25
0.028	710	22
0.0335	850	18

Aperture size range	Grade no.	Mesh range	
0.0035	90	00	170
0.0042 – 0.0021	53 – 106	0	150 – 300
0.0025 – 0.005	63 – 125	1	120 – 240
0.0035 – 0.0071	90 – 180	$1\frac{1}{2}$	85 – 170
0.0042 – 0.008	106 – 210	F2	70 – 150
0.0059 – 0.0167	150 – 425	M2	35 – 100
0.008 – 0.0197	210 – 500	S2	30 – 70
0.014 – 0.0278	355 – 707	$2\frac{1}{2}$	22 – 44
0.0167 – 0.0335	425 – 850	3	18 – 36
0.0275 – 0.066	699 – 1676		10 – 22
0.055 – 0.11	1405 – 2812		6 – 12
0.066 – 0.13	1676 – 3353		5 – 10

Methods of using the grit particles for polishing are varied. A rotating grinding wheel, or wheels, with successively smaller grit sizes, may be employed. In addition, there is the lapping technique, where a flat band of emery paper rotates against the component. This technique can be modified to polish relatively simple shapes. The application of hand-held paper achieves successful results, and this can utilize strips of abrasive-impregnated cloth. The most common method of abrasive polishing uses mops where the abrasive is impregnated on the mop surface or held with a greasy substance. Alternatively, the abrasive is held in a lubricating material and is continuously applied to the mops. The component is then held against the rotating mop, often lubricated with grease. The technique, then, can be extremely simple, relying on the skill of the operator. Where mass-production techniques are possible, automated machines will be used when the component will again pass through the various stages of abrasion described above.

The second, smearing type of polishing is designed to fill the valleys of a surface by bending the peaks over in such a manner that the resultant surface is smooth and generally highly reflective. It will be appreciated that this method is only possible where the original finish is reasonably homogeneous. Where a surface has several relatively deep scores, it will generally be necessary for the deep scratches to be removed, or at least considerably reduced, using the abrasive method. With polishing, it is quite common that the initial stage is to produce a homogeneous, scored surface by use of an abrasive mop or belt, as described above. This homogeneous finish is then smeared, using the *burnishing* technique described below.

It will be seen that materials most readily polished by this technique are those which are relatively soft and ductile. It will seldom be possible to produce a smooth, reflective surface by the simple one-stage method of filling valleys by bending the peaks. Normally, there will require to be a further smearing effect to ensure that each of the tiny joins is invisible. This method, as distinct from the abrasive, makes use of the *work hardening* properties of the material:

where this can be accepted to a considerable degree, then polishing by smearing can be a relatively simple operation. As the resistance to *work hardening* increases, it will be more difficult to achieve a satisfactory visual standard using this method. The technique uses similar equipment to the abrasive method, the surface having much less free cutting material, or none. This applies particularly to the polishing mop. In addition, the use of *wire-brushing* and mops containing no abrasive can be used. An optically flat surface will not be achieved and, while it will be generally easier to produce a shiny surface, it will be almost impossible to produce one capable of reflection. This technique, however, is much cheaper than use of the abrasive material. *Burnishing*, using a smooth tool, is possible with this technique.

Polishing, then, is a mechanical method of improving the surface finish of materials. With the abrasive method, it is possible to produce optically flat components, giving true reflection. For most decorative purposes, this is neither desirable nor economically justified, and the combination of cutting and smearing to produce a shiny surface is by far the most common of the methods. It will be appreciated that it is not always possible to achieve the ideal technique where the direction of polishing is changed between each grit size. Polishing can make use of other techniques such as *blasting*. *Vapour blasting* achieves a homogeneous smooth surface but the finish will seldom be reflective. *Barrelling* can also be used, and provided it is correctly controlled, the polishing can be a purely abrasive type. *Electrolytic polishing*, and to a lesser extent *electrolytic etching* or *chemical polishing*, are alternatives which are designed to attack the 'peaks' and retain the 'valleys' which exist on most metal surfaces.

Porcelain Enamelling

The application of glass frit to a surface.

Application range CAST IRON OR STEEL

The process is a specific form of *vitreous enamelling*.

With this, the material to be treated is first *cleaned*, generally by a *blasting* process but, under some circumstances, by *chemical cleaning* methods. The material is then heated to a temperature above 700°C, and powdered porcelain is impinged on to the heated surface, this can be by hand or by some mechanical means. During the treatment the areas requiring further powder can be readily identified by the different colour of these areas. The finished component is then returned to the furnace where the porcelain is fused at the correct temperature for the powder. Several coatings can be applied, with fusion between each coating.

The material used is a borosilicate glass in the finely powdered form. The process results in an extremely brittle, relatively poor-quality product, where quite minor damage results in chipping. The process is of limited application at present but at one time found considerable use for domestic kitchenware, mugs and advertisement signs commonly found in railways.

There have been considerable improvements in other forms of *vitreous enamelling* and these have resulted in the process becoming more commonly used.

When required for corrosion resistance, the technique can often be replaced by conventional *stove enamelling*. Many forms of *electroplating*, particularly *zinc plating* and hot-dip *galvanizing* give equal and in some cases better *corrosion protection* without achieving the same high standard of abrasion resistance obtained from porcelain enamelling. Further information is given under *Vitreous Enamel*.

Porosity Testing
An inspection technique to determine whether any pores exist in plated or painted coatings. Described fully under *Porosity Test* in Part Two **Non-destructive Testing**.

Postheating
A term applied to any general *heat treatment* which follows a heating operation. It should not be used in the case of a well-defined operation such as *tempering* or *ageing* but is usually a form of *stress-relieving*. Described fully in Part Two **Heat Treatment**.

Pot Annealing
A process identical to *box annealing*: further information on Annealing is given in Part Two **Heat Treatment**.

Pot Quenching
A technique of quenching components which have been *carburized*, generally *gas carburized*, directly from the 'pot' in which the process was carried out. It is, therefore, a form of direct *hardening*. Described fully in Part Two **Heat Treatment**.

Powder Painting
The application of a paint film using powder.

Application range ALL METALS

This form of *painting* differs from the conventional technique in that no vehicle or liquid is involved. With conventional forms, there will always be a liquid vehicle which is used to carry the pigment, and once this has been used to coat the component, the vehicle by normal oxidation, or by curing at high temperatures, will be converted to a solid and hold the pigment in position.

The powder is applied by one of two methods. First, by immersing the component in the powder, generally in a fluidized bed. Often the component itself will be heated and the powder will thus adhere directly to the component. It is also possible to use electrostatic methods, where the powder is caused to adhere by the electrostatic attraction of the particles to the metal.

Second, a considerable amount of powder paint is now applied using the spraying technique, where the dry powder is blown with a very light application of air as a powder cloud. This cloud of dust particles is attracted to the component by high-voltage electrostatic forces. This method has many of the characteristics of *electrostatic painting* in that the powder will wrap around the components being painted. Some skill, however, is required for the application of powder paint, particularly with intricate assemblies. On completion of the cycle the component, which must be metallic to carry the necessary electrostatic forces, should have an even coating of the powder. If any evidence of unsatisfactory painting or coverage is noted, or if any damage occurs to the film at this stage, the powder can be removed by a simple wiping or blowing action once the component has been removed from the electrostatic connection.

Satisfactory components from either application are passed immediately to an oven which is generally at a temperature of 150°C minimum and can be as high as 250°C. The components do not, of necessity, require to reach this temperature themselves, but must be long enough in the oven for the surface of the component and the paint film to reach this temperature. This results in the powder fusing and flowing to give complete coverage and excellent adhesion to the substrate metal. Where the pre-heating temperature of dipped components has been above 175°C, post-dip curing may not be required. For optimum results, however, curing is recommended.

From the above it will be obvious that powder painting has several distinct differences to conventional *painting*, and the advantage that defects identified during the process are much more easily rectified than with wet paint. In addition, adhesion with powder paint is much superior on bare metal to any of the conventional wet paints. The correctly cured, fused film is harder and tougher than the majority of conventional films and is more abrasion resistant. Considerable evidence has been built up to show that the necessity of *phosphate treatment* prior to powder painting is of less importance than with wet paints.

However, when phosphate is considered necessary, this must be of the light-duty type, such as iron phosphate, as distinct from heavy-duty zinc or manganese phosphates which are normally applied prior to wet-film painting. The reason for this is that, with conventional paints, the phosphate film acts in a manner not unlike blotting paper which is strongly adherent to the steel surface. This, therefore, acts as an excellent key for the wet film. With powder painting, the particles are too large to be absorbed by the phosphate coating and thus the phosphate acts as a barrier between the steel and the paint film. Any distortion during service can result in the phosphate film fracturing, resulting in paint failure.

To summarize, powder painting is a technique more accurately described as the application of a plastic rather than a paint film. The technique uses mainly epoxy powders, but developments are making other plastic compounds available for this technique.

Precipitation Hardening

A process in which certain alloys can have their hardness and mechanical properties improved. This is commonly called *ageing* or *age hardening*. This must always be preceded by *solution treatment*, which may be part of the *casting, forging* or *welding* process. Described fully in Part Two **Heat Treatment**.

Preece Test

An inspection process to determine the thickness of the zinc coating on galvanized components.

Application range STEEL

The procedure is that the component is dipped into a solution of copper sulphate at room temperature. This will attack the zinc and result in the formation of copper on the surface. This copper will be washed off with light rubbing. When the zinc coating has been removed by the chemical action, the copper deposit on the steel will not be removed, and thus the appearance of this adherent copper shows areas bare of zinc.

By standardizing the strength of the copper sulphate, the temperature and the time of dipping, it is possible to use this as a quantitative as well as a qualitative test to indicate the actual thickness of the zinc coating.

The test is now being replaced by the use of electronic or magnetic instruments which instantly show the thickness of the zinc deposit. The test, however, still has some application where porosity might be suspected on local areas. The user is referred to the heading *Thickness Testing* in Part Two **Non-destructive Testing** for further information.

Preheating

A general term applied to any form of *heat treatment* carried out immediately prior to some other form of heating, and covering *welding* and *brazing* as well. Described fully in Part Two **Heat Treatment**.

Press Forging

A manufacturing process, not a metal treatment, and thus not a subject for this book. The term covers any form of *forging* where steady pressure is involved. This can be a hot process or a cold-press process. The required form is produced by a steady pressure. Users are advised to obtain specialist advice when necessary.

Pressure Test

An inspection technique, used basically to identify porosity or leakage in components and assemblies. It can also be used to prove the mechanical strength of components or assemblies.

There are two basic techniques, first, using lower pressures with gas such as air, and second, high pressures using liquids such as water or oil. Both techniques are described under their own headings, and under the general heading *Pressure Test* in Part Two **Non-destructive Testing**.

Pressure Welding

A form of *forge welding*.

Primer Painting

The paint undercoat in contact with the base material, either metallic or non-metallic. In general, it is conventionally accepted that a primer paint has some active ingredient or is specially formulated and is not merely the first coat of a two-coat system.

Application range ALL METALS

Primers can be specially thinned to decrease their viscosity where absorbent materials are being painted and, thus, the primer is absorbed into the base material. Primers may also have active ingredients such as some free acid content, generally phosphoric, to cause *etching* or chromate to convert the activated surface and they commonly also contain metallic particles in suspension, generally zinc but sometimes aluminium, which act in a 'sacrificial' manner when the top coat is damaged. Primers should never be used without the application of surface top coats.

To summarize, primers are the first coat of paint in a multi-coat system, and these must always be compatible with the top coats; they normally have some active ingredient or are formulated in a special manner. It is essential that when a multi-coat paint system is used, all coats are supplied from one source. It should also be noted that many excellent primers are not compatible with excellent top coats even when supplied from the same source. Further information should, therefore, be obtained when necessary.

Part Two **Painting** covers all aspects of the techniques used.

Process Annealing

A specific *heat treatment* applied to steel, usually in the form of sheet or wire. It is a form of *annealing* which removes all *cold work* without actually causing recrystallization, and thus is a form of *sub-critical annealing*.

The purpose of process annealing is to remove *work hardening* prior to further cold work. Where considerable *deep-drawing* is to be carried out, this or *sub-critical annealing* will not be sufficient, and *full annealing* with complete recrystallization will be necessary. *Annealing* is described fully in Part Two **Heat Treatment**.

'Progrega'

A proprietary form of *dry-film lubrication*, making use of the graphite or molybdenum disulphide powder in an epoxy paint.

Further information is given under the heading *Dry-film Lubrication*.

Progressive Ageing

A specific form of *ageing* or *precipitation* treatment which has very limited application with modern alloys.

Application range AGEING ALLOYS

The alloy is heated to progressively increasing temperatures for specified times. With this technique, it is possible to judge the optimum *ageing* temperature where this is in doubt by carrying out a hardness test after each increase in temperature. With modern knowledge and information on the ageing of alloys, the normal procedure is to choose a single temperature for a specified time. Information on *Precipitation Hardening* (an alternative name for *Age Hardening*) is given under this heading in Part Two **Heat Treatment**.

Projection Welding

A form of *resistance welding*. Described fully in Part Two **Welding**.

Proof Stress/Proof Strength

This is the part of the destructive tensile test where the load required to produce a permanent deformation is known as the proof load. With normal testing this is generally 0.1 or 0.2 per cent of the specified gauge length. That is, the proof strength is the stress required to produce a permanent extension of 0.1 or 0.2 per cent of the gauge length.

Application range ALL METALS

The yield stress is very similar and it is generally accepted that this will be 0.5

per cent proof stress. There is a visible indication of yield during the tensile test as the increase in load applied slows or stops as the material stretches or 'yields'. This is readily seen with mild steel and certain other materials but is much more difficult to identify with the majority of materials. Thus, it is preferable that proof strength is identified.

The technique for carrying this out is that the load is recorded at standard increments and the degree of stretch caused for each load is also recorded.

Where a 'yield' is identified it will be readily seen by either the instrument recording the load and amount of stretch applied, or the operator taking the readings, that no, or very slight, increase in load is being applied for a considerable increase in length.

Where this is not readily seen the graph produced following the completion of the tensile test will be prepared. This will show initially a straight line where the extension and load applied are proportional. At a certain point it will be noted that the load applied will give an increase in length to the test piece greater than the previous increase for any unit load. This is the elastic limit and the proof load is then measured either 0.1, 0.2 or 0.5 per cent.

With modern tensile machines the load and stretch are recorded automatically and the proof load identified.

The proof stress is then worked out on the cross-sectional area of the test piece and the load required to give the desired stretch.

Proof Testing

This is the test which identifies the elastic limit of a metal. Described fully in Part Two **Mechanical Testing**.

Protal Process

A surface treatment to improve corrosion resistance and to increase paint adhesion.

Application range ALUMINIUM

With this process, the components are thoroughly cleaned and then sprayed with a solution of chromium and titanium compounds and some alkali fluorides. This converts the surface layer, by chemical action, into chromates and at the same time roughens the surface giving a key to improve paint adhesion. This is comparable to, but more complicated than, the *chromating* process, and is now seldom used.

Protolac Process

A method to produce an artificial oxide for improving natural corrosion resistance and to give a key for paint adhesion.

Application range ALUMINIUM

The process requires that the components are first cleaned and then immersed in an alkaline solution.

The Protolac process can apply to any of the techniques whereby the surface of aluminium is oxidized in alkaline solution, and where the alkali metal is deposited on the aluminium as the stable oxide, thus contributing to the paint adhesion. The use of the *chromating* process now seems to be more usual.

Pull-off Test

A form of *adhesion testing* which is commonly applied to paint films, and can be used for electroplated deposits and, using special techniques, for other surface coatings. The test is employed to evaluate and measure the adhesive strength of the coating to the substrate.

With *painting* and *electrodepositing*, a special test or test area on the component is required. This should be cleaned and prepared in exactly the same manner as for normal production. A small strip, approx. 0.5 in (10 mm) wide and 1 in (20 mm) long, is 'stopped-off', generally using tape, but other methods can be used as long as they do not interfere with the surface immediately adjacent to the *stop-off* area.

The coating is then applied in the normal manner. With electrodeposited coating, the *stop-off* material must be conductive to produce the metal deposit. The purpose of the *stop-off* is to allow a strip of the coating to be readily removed, while still attached to the adherent coating. This strip is then loaded in some manner to apply a tensile pull to the join of the coating to the base material. This load can be applied by a tensile-type machine, or by static weights until failure occurs. The adhesion can also be estimated by using a vice or pliers to hold the strip and then pulling until failure occurs.

This will not supply details of the actual load but, in skilled hands, can give useful information of a comparative type. While the test gives details regarding the adhesive strength of the coating, the method of failure and the state of the underlying material will also supply useful information. The user is referred to *Adhesion Testing* and *Thickness Measurement* for further information on the quality control of coatings.

Punching

A manufacturing process which is a form of cold *forging*, whereby metal is sheared to produce holes or forms in sheet metal.

Application range ALL MATERIALS WHICH CAN BE WORK HARDENED

Considerable skill and experience are required in the design, manufacture and use of punching dies and equipment. With the correct equipment and techniques, it is possible to punch relatively small holes through thick materials and, provided the equipment is kept in good condition, the *work hardening* applied can be held to a minimum.

This can be a much cheaper method of producing holes than conventional drilling. It should be noted that the *work hardening* applied will result in a layer of cold-worked metal adjacent to the cut face. The extent of this will vary with the design of the shape, the efficiency and sharpness of the tooling and, in particular, the material being worked.

Where severe cold work is applied, the surface will be *hardened*, and may have minute cracking. This will seriously affect fatigue strength properties and with high-duty components may require a *tempering* or *annealing* operation prior to service.

Pusey Hardness
A *hardness test* for rubber. Described fully in Part Two **Hardness Testing**.

Pylumin Method
A surface treatment to improve paint adhesion.

Application range ALUMINIUM

It is a form of *chromate treatment* in which a thin film of oxide is formed, and into which some chromate is absorbed. This serves the dual purpose of improving the corrosion resistance achieved by the natural oxide, and at the same time forming a key for subsequent paint.

Pyramid Hardness Testing
A general term for the technique of *hardness testing*, where the pyramid impression is made by a diamond. Described fully in Part Two **Hardness Testing** under *Pyramid Diamond Test*.

'Pyro Black'
A proprietary name for the *black oxide* process. Further information can be obtained from Pyrene Chemical Services Ltd.

QPQ
This is the abbreviated form sometimes used for Tufftride QPQ.
Further information is given under the heading *Tufftride QPQ*.

Quench Ageing

A term very occasionally applied to the *solution treatment* and *precipitation hardening* of the alloys which take part in *ageing*. These are described in Part Two **Heat Treatment**.

Quench Hardening

An alternative name for hardening by heat treatment.

Application range STEEL

This is taken above a certain critical temperature in the region of 700–900°C, depending on the composition, and then held for a specified time before quenching in air, oil, water or brine, for *hardening*. This *hardening* is the result of holding in solution the carbides which would normally exist as a separate phase at room temperature. The term quench hardening is, then, a variation on the normal harden. Described fully under *Hardening* in Part Two **Heat Treatment**.

Quench Tempering

A method of delayed quench. Described fully in Part Two **Heat Treatment**.

Quicking

The term used when a thin film of mercury is deposited on copper. It is achieved by *cleaning* the copper article and then immersing into a solution of mercuric oxide in sodium cyanide. This results in the chemical replacement of the surface layer of copper with mercury. The process stops immediately the surface is covered with mercury, and as there will be some form of alloy with the surface layer of copper, better adhesion is obtained with this deposit than with some chemical coatings.

The process has very limited application in its own right, but is used on occasion when copper articles require to be *silver plated*. Because silver, like mercury, will result in a loose, non-adherent film, it is not possible to electroplate silver directly on to copper. The use of the quicking method is, therefore, one way by which copper articles can be successfully plated with silver.

It is now known that mercury and mercury compounds present a serious health hazard, and neither the metal nor compounds should be handled. Advice must be obtained before using this process.

A more common and technically satisfactory method is to have a nickel flash directly on to the copper, and to plate the silver on the nickel undercoat.

The term quicking is sometimes applied to any mercury amalgam which is formed by rubbing or immersing a metal in mercury. On occasion, it may also be applied to any metallic deposit obtained by chemical replacement, for example, copper on iron. However, in general, the term will be applied to the replacement of copper by mercury.

Radiography

An alternative term for *X-ray*. It is an inspection process using radiant energy in the form of rays for examining solid, opaque components to ensure that they are free of internal defects. Described fully under *Radiographic Test* in Part Two **Non-destructive Testing**.

'Ransburg Process'

A proprietary form of *electrostatic spray painting*.

Application range ALL METALS AND SOME NON-METALLICS

The paint particles are given an electric charge and the component being painted has a negative charge or is at earth potential. The paint particles are, therefore, powerfully attracted to the component. This procedure results in a better wrapround of paint and a considerable saving of paint for articles which have holes or are in the form of a mesh. Further information can be obtained from Henry W. Peabody Ltd. Described fully under *Electrostatic Painting* in Part Two **Painting**.

Recarburization

A salvage process to reintroduce carbon into the surface.

Application range STEEL

This is necessary when for any reason the surface of the component has been decarburized. This may arise during *hot working* such as *forging*, or during a *heat treatment* process which has been carried out in an oxidizing atmosphere. This can result in a reduction of the carbon content at the component surface. If this is unacceptable and it is not possible to machine off the affected area, then this salvage operation can be considered. In essence, this is similar to the *carburizing* process but will seldom be carried out for as long a period or at as high a potential.

If, for instance, a 0.4 per cent carbon steel has been maltreated to reduce the carbon content to, for example, 0.2 per cent carbon at the surface, then it would be unwise to recarburize these components at a potential in excess of 0.5 per cent carbon, otherwise there would be a danger of forming a thin film with carbon higher than 0.4 per cent at the surface. This would almost invariably be a more dangerous defect than the decarburization which it is proposed to rectify.

The recarburizing process is occasionally used in the manufacture of steel where carbon is added to the melt to increase the carbon content of the finished steel. This is a completely different process to the above and will only be applied in steelworks during the manufacturing process, and cannot be used as a rectification process on finished components.

Red Gilding

An alloy of copper and gold which can be electrodeposited. As with other gold alloys, it is advised that the user obtains specialist advice from a plating supply house regarding the solution and parameters requiring control.

'Redux Process'

A proprietary metal bonding or gluing process.

Application range ALL METALS AND MANY NON-METALS

It is particularly useful for the bonding of metallic surfaces or for the bonding of non-metals to metals, examples being the bonding of wood or rubber to metal.

The process is based on phenol-formaldehyde and other resins such as polyphenol in the powdered form. These are mixed and applied to the thoroughly cleaned and preferably slightly roughened surfaces which are then placed under pressure and taken to a temperature of approximately 150°C for a specific required time. The time and pressure involved will depend on the materials and the components being bonded.

The resultant bond has an extremely high strength and was one of the early successes in adhesive science. There are now a number of gluing processes which achieve results as good as with the present process without the use of high temperatures and pressure.

For production use, however, there are considerable advantages in this process, but it is necessary to have the correct equipment capable of controlling temperatures and pressures over a period of time. Users are advised to obtain advice from a company such as Ciba-Geigy Ltd on the specialist subject of bonding by glue.

Refine

A *heat treatment* operation normally carried out after *carburizing* to re-crystallize the grain structure and thus eliminate the coarse structure produced by this lengthy high-temperature treatment. Described fully in Part Two **Heat Treatment**.

240 / REFLECTOGAGE TESTING

The term refine in the metallurgical sense is more commonly applied to the production of metals from ore, but this is not a metal treatment.

Reflectogage Testing

An instrument used in *non-destructive testing*.

Application range ALL METALS

It operates by the use of high-frequency sound waves. These are applied to the component being tested and the resonance is then picked up and fed to an oscilloscope screen. Any defect in the component will be shown as a change in resonance.

This instrument is of a relatively low sensitivity and requires considerable skill to operate; it has been largely replaced by *ultrasonic testing*. There is, however, a new technique based on this principle in which the standard resonance of a component is measured and recorded at the start of its life. Surveys can then be carried out at predetermined intervals and any alteration in the resonance pattern will be indicated and subsequently investigated. This technique is being applied under some circumstances to large stressed structures.

This technique, then, uses the variations in the vibrations induced in components to detect flaws. *Vibration testing* is described fully in Part Two **Non-destructive Testing**.

Reflowing

A technique used in the manufacture of electronic or electrical components, generally printed circuit boards, which have been *soldered*, solder plated or *tinned* using any of the conventional techniques but most commonly by *electroplating*.

Application range COPPER AND ITS ALLOYS

Reflowing is used to ensure that satisfactory adhesion and a surface capable of good quality *soldering* has been achieved. The component is heated, generally by immersion in some form of non-contaminating liquid such as oil, in order to melt the solder deposit and cause it to flow. Provided good control exists induction or radiant heating can also be used, and also with hot air or gas.

With this technique it is possible to produce a bright, attractive-looking material but its main purpose is to ensure good quality. With reflowing any defect on the substrate will not wet, clearly indicating areas where the solder or tin is missing. Pinholes or cissing will also occur, again showing poor adhesion. This technique is, therefore, a quality control system which, at the same time, will improve the visual appearance and can, under many circumstances, improve the resultant solderability.

Refrigerated Anodizing
An alternative term for *hard anodizing*, where the sulphuric acid electrolyte is held at low temperatures. Other names for this process are *black* or *deep anodize*. Further detail are given under *Anodizing*.

Regenerative Quench
An alternative term for *refining* of the core. This process is used to eliminate the large grain size caused during *carburizing*. Described fully under *Refine* in Part Two **Heat Treatment**.

Reheating
A *heat treatment* term which is applied to a variety of processes and, therefore, will mean specific treatments within specific industries or even locally within a certain company. Any process in which the material, having been already heated and cooled, is returned to a furnace, may be termed reheating.

Repeated Bend Test
A destructive test to assess ductility. Described fully under *Bend Test* in Part Two **Mechanical Testing**.

Repousse Process
A decorative technique for sheet metal.

Application range ALL DUCTILE METALS

The process has considerable antiquity.
    The metals mainly used are copper, gold, lead, silver and tin, but any metals which can accept appreciable amounts of *cold work* are also possible.
    The intricate detail of the design is applied by hammering the reverse side of the sheet, the resultant raised portion on the front having a relatively smooth surface. This is not a pressing or moulding process, where the surface of the sheet is forced into a preformed shape; it uses various sizes of hammers and it will be seen that considerable skill is required to produce the necessary detail. The raised portions on the front of the metal sheet may be given further detail by engraving. This process is still used in the jewellery trade but has no significance in industry.

Resistance Soldering
A method of heating for *soldering* which uses the electrical resistance between the soldering bit and the solder, or the component being soldered, to melt the solder.
    Further information on *Soldering* will be found under that heading.

Resistance Welding

A general term, covering several types of *welding* where electrical resistance heating is used. The completed weld is generally of the *forge* type.

Described fully in Part Two **Welding**.

Reverse Bend Test

A destructive test to ascertain the ability of a material to stand severe bending without fracture. It is, in fact, a measure of ductility. Described fully under *Bend Test* in Part Two **Mechanical Testing**.

Rhodanizing

A technique of plating a very thin film of rhodium on to a bright silver or gold deposit to reduce tarnishing or improve wear. Described fully in Part Two **Electroplating**.

Rhodium Plating

A process using this metal of the platinum group which has good corrosion resistance and low contact resistance. Rhodium is cheaper and harder than platinum, but shares its excellent corrosion and oxidation resistance. Described fully in Part Two **Electroplating**.

Rivet Test

A destructive test applied to rivets before they are fitted to ensure suitability.

Application range ALL DUCTILE METALS

There are two variations. In the first test, the rivet is bent through 180°C and closed on itself. Unless this occurs without any evidence of cracking on the outer radius of the bend, the rivet is suspect. This test must be carried out under the conditions in which riveting will occur.

In the second test, again carried out under riveting conditions, a head is formed on the rivet which must be 2.5 times the diameter of the rivet shank. Again, this must result in no cracking.

Both variations are designed to be practical tests which can be carried out by the department responsible for riveting and, thus, no special equipment is required. The tests will show that the material has adequate ductility under the conditions of riveting, not easily provable by normal destructive mechanical tests. It is, therefore, a form of *ductility testing*.

Rivet Weld

A form of *welding*, where the upper component has a hole drilled and this hole is used to produce a weld which will be fused to the lower component. The edge of the hole on the upper component is fused to the lower component, with or without the use of a filler rod. This technique is also known as *plug welding*.

The efficiency of this technique as a method of joining will obviously rely on the number of rivet welds involved, and also on the general standard of *welding*, and the thickness of the material. The weld will always be of the *fusion* type. Described fully under *Fusion Welding* in Part Two **Welding**.

Rockwell Hardness Test

A very common type of *hardness test*, operating on the principle of measuring the depth of impression achieved by a standard load with a standard indenter. Various loads and indenter types are used. Described fully in Part Two **Hardness Testing**.

Rolled Gold

This is a bimetal where a very thin film of gold or more generally a gold alloy is applied to the surface of the component or material.

Application range COPPER AND ITS ALLOYS

This therefore is a material rather than a treatment. Rolled gold is manufactured by taking the base metal, applying a relatively thin film of gold, or gold alloy, and then rolling the material to produce generally one side gold plated or deposited but there may be occasions where both sides of the surface are gold.

This is comparative to gold plating but generally will be used for lower quality products.

Roller Spot Welding

A form of *Resistance Welding* used on sheet metal. Described fully under this heading in Part Two **Welding**.

Roller-tinning

A form of *tin plating* or surface *soldering* applied by roller.

Application range COPPER AND ALLOYS

The technique uses a roller partially immersed in the molten tin or solder.

A second, upper, roller held out of the bath is in contact with this roller with pressure applied to it. The second roller is mechanically driven and the components to be *tinned* or *soldered* are fed between the two rollers. It will be seen that only flat and relatively simple shapes can be treated in this way. The surface to be *tinned* is first cleaned and fluxed and then fed to the roller tin machine. By adjusting the speed of rotation, and the pressure control of the thickness of the deposit is achieved.

In general, this technique is used to produce a visually attractive and bright surface on components, such as printed circuit boards, which have already been *tinned* or *soldered*. The technique is, therefore, an alternative to *reflowing* or, given the correct conditions, could be used as an alternative to *tin* or *solder plating*.

Root Bend Test

A destructive test to prove the quality of *welding*, which in effect is a simple, economical method of assessing the ductility of the weld and any fusion line defects. It should be used in conjunction with face and side bends. Described under *Bend Test* in Part Two **Mechanical Testing**.

Rose Gilding

An *electroplated* alloy of gold, with copper and silver. The colour of the rose will vary with the alloy deposited: the higher the copper content, the darker will be the colour. An increase in silver content makes the rose coloration lighter. As with other methods of gold-alloy plating, it is not possible to give specific information and the user is advised to contact a plating supply house for information on the solutions and their control.

Rumbling

A form of *barrelling*.

Application range ALL METALS

The term is generally taken to include the relatively old-fashioned method of *deburring*, where an open-ended vertical type of barrel is rotated, not unlike a cement mixer. The components, with or without other articles of specified shape, are allowed to contact each other and the material used for *deburring*. In some circumstances, abrasive grit is also added. The barrel is rotated at various speeds from extremely slow to relatively rapid. The result is that edges and burrs are removed from the components.

This process has largely been superseded by the barrel-finishing technique which is much more sophisticated, and where scientific *deburring* and improvement of the surface finish can be achieved under controlled conditions.

Rumbling, in the historical sense, was often carried out dry, resulting in some *peening* of the surface and very often with considerable damage to the components. To overcome this, the process was invariably reduced in speed, and could be lengthy, with *barrelling* periods of up to one week not uncommon to achieve the desired surface finish and absence of burrs. With modern *barrelling* techniques a time in excess of 4 h is considered excessive.

Rustproofing
A general term covering a very wide variety of processes applied to steel. It is used for processes such as *painting, electroplating, galvanizing*, etc., but in general is applied to the processes of *phosphating* and similar low-duty rust-preventative or rust-inhibiting techniques. The term is also applied to oiling when this is used to coat steel for storage purposes. Thus, the term rustproofing should be treated with caution.

Ruthenium Plating
A process using the metal ruthenium, of the platinum group very similar to rhodium. It is only recently that plating of ruthenium has been possible on a commercial scale. Some information is given in Part Two **Electroplating**.

Sacrificial Protection
A *corrosion protection* technique which uses a metal lower than the metal to be protected in the electrode potential series to supply protection.

Application range METALS BELOW IRON ON TABLE IN ENTRY 'CORROSION PROTECTION'

This is achieved because the lower metal acts as the anode, the protected material is the cathode. With an electrolyte present, a cell will be set up where the anode corrodes to protect the cathode. Thus, the anode 'sacrifices' itself to protect the other metal.

In practice, zinc and aluminium are the two metals most commonly used for sacrificial protection. An examination of the table of electrode potentials (see *Corrosion Protection*), shows how a metal lower in the table corrodes to protect a higher metal. This means that gold, silver, platinum, etc. will invariably be protected at the expense of lower metals, that iron will sacrifice itself to protect copper and that zinc and aluminium will be sacrificed to protect iron.

The problem can be complex when more than two metals are involved. Users are referred to *Galvanic Protection, Zinc Coating* and *Corrosion Protection* for further information.

Salt mist/Salt-spray Testing
A destructive inspection technique which is applied generally to surface protection coatings but can be used on components themselves.

Application range ALL METALS

The technique is that the cleaned component is placed in an atmosphere of salt droplets. There are a number of specifications covering the type of salt used, the droplet size and whether or not this is continuous or cyclic. However, it is essential that the salt droplets are given the opportunity to contact the surface being tested. This means that the surface must not become soaked with the solution, thus preventing fresh droplets attacking the surface.

The most generally used technique is based on the American Society for Testing and Materials (ASTM) Specification which calls for a solution of 5 per cent chemically pure sodium chloride at a temperature of 30°C. This is a continuously atomized salt mist where the solution is run to drain continuously and is only used once. Hence, the design of the atomizer must ensure that a constant fog is in existence, but that the surface of the components is constantly draining and the fresh droplets are available to continue any attack. Depending on the material being tested, the specification will quote the number of hours which it must withstand without obvious corrosion.

Other salt-spray tests use the technique of spraying the components at intervals and inspecting the component each time it is sprayed. This can be after 4 h, 12 h or 24 h. Between inspections the components are held in a closed container with controlled humidity. Other test specifications vary the chemical used, the temperature and the cycles. These tests can be more severe than the continuous salt test but can be open to considerably greater variations.

The purpose of the salt-mist test is to compare various types of finishes or materials and their ability to withstand controlled corrosion. It is possible to use this test to monitor and control *electroplating, painting, galvanizing*, etc. for components which during service will never be submitted to this type of atmosphere or condition. The test, therefore, shows a weakness in the specified finish, and while statements can be made regarding the comparisons of hours in the salt mist and years of service life, these will always be open to considerable variations depending on the type of service for which the components are used. There is, however, no doubt that the test is an extremely useful tool in the hands of quality and inspection personnel to ensure that the standard of metal finishing is satisfactory. This can best be achieved by specifying the type of salt mist to which the components will be subjected and the minimum time they must successfully withstand without evidence of visible corrosion.

For '*passivated*' zinc or cadmium plate a time of 144 h continuous salt mist without obvious corrosion is now becoming standard for high-integrity components, although lesser times will often be accepted for components of a less high integrity. A time of less than 48 h will not normally be specified.

A quality paint film will generally specify a minimum of 500 h salt mist resistance.

This test is unique in ascertaining weaknesses in metal finishing but should not be used without simultaneously measuring other parameters, such as thickness and adhesion of the coating, and the ability of the coating to withstand other areas such as light-fastness etc.

It should be appreciated that a high standard coating which has poor adhesion to the base material can successfully withstand continuous salt-mist testing for considerable periods, provided the coating is not broken. However, if the coating has poor adhesion, it is likely that in service it will become damaged, resulting in corrosion. It is not advisable that the salt-mist or salt-spray test should be used as an inspection method to pass or fail a specific batch of components. It is a quality control technique which should be used to continuously monitor the *corrosion protection* system, and should thus show any trend to lose control before this results in components being seen to be faulty. The *cass test* is a specific type of salt-mist test described under its own heading.

Sand Blasting

A term which is still commonly applied to the *blasting* of metals with various forms of grit. In practice, the use under indoor conditions of sand or any silica-bearing material is forbidden in most countries. *Sand blasting* was commonly applied at one time as this is a readily obtainable cheap material which can successfully blast scale and give a satisfactory controlled surface finish. Unfortunately, the health hazard to the operator is not acceptable and thus the use of sand for *blasting* is now illegal. The various forms of blasting are described in the section under that name.

'Sand Ford' Process

A proprietary form of *sulphuric acid anodizing*, which has organic additions to the solution.

Application range ALUMINIUM

This process has all the attributes of *anodizing*. Further information can be obtained from the Sand Ford Process Co. Inc., USA. The *anodizing* technique is described under that heading in this book.

Sandberg Treatment

A proprietary form of *hardening* and *tempering*.

Application range STEEL

The parts to be treated are taken to above the critical temperature necessary for all the carbides to be taken into solution. They are then removed from the furnace and cooled either by spraying with air, oil or water, or in some circumstances steam for sufficient time and under controlled conditions, to result in the transformation to martensite. The quenching is then stopped and the parts allowed to cool to room temperature, giving a form of *tempering*.

It will thus be seen that this technique is similar to *austempering* or *martempering* and will have the same drawbacks and advantages as these two processes, both of which are described fully in Part Two **Heat Treatment**. The process can only be applied to alloy steels or steel with at least 0.2 per cent carbon. This is basically a form of delayed quench hardening of steel.

Sanding

A term commonly applied in metal finishing where a surface finish is obtained using hand-held emery papers or cloth.

Application range ALL METALS

This produces a directional type of finish which will have a variable surface finish, depending on the operator and the type and standard of grit used. It is in essence a form of *polishing*, but with flat components can be used for *deburring*. Some *blasting* methods, in particular *vapour blasting*, can be used to give a comparable type of finish.

Sankey Test

A mechanical bend test.

Application range STEEL

A test piece is produced 0.75 in (20 mm) in diameter and 4 in (100 mm) long. One end, approx. 2 in (50 mm) in length, is firmly clamped in a vice and using a special attachment the free end is bent through approximately 45 degrees, and this bend is then reversed giving a total bend of approximately 90 degrees. The mechanism is such that the operation is controlled in the degree of bending and counts the number of bends obtained prior to cracking or failure.

This, then, is a form of *bend testing*, where the angle of bending is specified and the specification will require to state the number of bends permitted prior to fracture. It is a destructive test which is aimed at indicating the ductility of the material. Further information on *bend testing* will be found under *Ductility* and in Part Two **Mechanical Testing**.

Satin Finish

The name given to a matt surface finish which can be achieved by several methods.

Application range ALL METALS

This in essence is a dull finish of a non-directional type, but may have evidence of circular finish which is non-uniform, having circles of a relatively small radius. It is generally for decorative purposes.

It can be achieved by *hand-polishing*, when different patterns can be produced, but more often by jigged *mechanical-polishing*. With modern equipment, it is often produced by *vapour* or *dry-blasting* with a very fine grit.

All of these finishes may be the final surface on the component, and this will often be the case when the material used is non-corrosive, such as a silver, stainless steel or nickel silver. Where satin finish is used on normal steel, the components are generally plated.

Chromium plating carried out directly on to the prepared finish gives an attractive appearance quite different from the high sheen, blue finish normally associated with this process. Other deposits such as nickel or zinc can also be applied.

It is now possible to *electrodeposit* material, particularly nickel, which will give this type of finish. This is achieved by codeposition of non-metallic particles together with the nickel metal, and is termed *satin nickel plating*. It is also possible to produce a satin finish with modern paints.

There are now paint finishes which produce a fine matt finish which under some circumstances could be an alternative to the above.

'Satin Kote' Treatment

A proprietary sealing solution which is applied to *satin nickel plated* components when no further plating or *lacquering* is to be applied. The procedure is that the plated components, immediately after washing and drying, are immersed in the solution for a few seconds at room temperature. On removal, they are allowed to drain and will dry at normal room temperature. Further information on this process can be obtained from W. Canning and Co. Ltd.

'Satin Nickel' Plating

A proprietary decorative finish, using the normal dull *nickel plating* solution to which is added material in a slurry form. The plating is then carried out under the same conditions as for the conventional process. This is one method of achieving the *satin finish*, discussed under that heading in this book. Further information on solutions can be obtained from W. Canning and Co. Ltd.

Satylite Nickel Plating

A dull nickel deposit achieved by the addition of specially formulated solids suspended in the nickel plating solution.

The control of these solutions can be difficult as the solids are required to remain in suspension and must have specific properties allowing them to be codeposited with the nickel. Extraneous solids must therefore be kept out of the plating solution. This is one method of producing a *satin finish*. Alternative methods are by *blasting, barrelling, etching* or *mechanical polishing*.

Sawdust Drying

A form of *drying*.

Application range ALL METALS

It was commonly used following *electroplating* or some process where components are cleaned using aqueous cleaners or where some other washing was involved. Unless water is completely removed and the parts completely dried, then with steel components corrosion will almost invariably result.

The technique is that the components are immersed in warm sawdust. Some satisfactory drying is obtained with sawdust at room temperature, provided it is rejected before it becomes too damp. It is, however, advisable that the sawdust is heated. This has the dual advantage that the components themselves are warmed and are easier to dry, the moisture is constantly driven off the sawdust and, thus, there is less danger that this will become a source of corrosion.

Provided the articles being dried are simple in shape and have no holes, then sawdust drying is an efficient method, and there will be stain-free results which is important in areas where hard water is used. Drying stains are caused by the substances in solution in the water being left on the surface of the components when normal air drying is used. The slight scouring obtained with the sawdust method generally removes this small amount of deposit.

Sawdust drying must not be used with engineering components where oil holes are present, or where the component has drilled holes of any type which have reasonably tight tolerances. The sawdust will invariably result in these holes being filled or contaminated and there is a history of failure caused by oil holes being restricted or blocked by the sawdust used during drying.

Alternative methods of drying use hot oil or hot air, both of which have disadvantages. There are now available water-repellent liquids which will efficiently remove all traces of moisture from the surfaces. These liquids, however, are expensive. Another form of drying is the *'Trisec' process*, which uses a standard vapour degreaser with modifications. The different techniques for moisture removal are discussed in the entry *Cleaning*.

Schnadt Test

A form of *impact testing*, using a notched specimen; it is very similar to the *Charpy* test, which is described fully under *Impact Testing* in Part Two **Mechanical Testing**.

'Schoop Process'

A proprietary form of *metal-spraying*.

With this, the metal can be pre-melted and the molten metal is then fed into a high-velocity flame, where it picks up the necessary energy and retains the necessary heat to arrive at the component being sprayed in a molten state at high velocity. The process has now been replaced by more conventional forms of *Metal-spraying*, which are discussed under that heading in this book.

'Schori Process'

A proprietary form of *metal-spraying*.

With this, the metal in the form of a powder is fed into the flame, where it picks up velocity and temperature and is then impacted on to the surface being coated. This type of metal-spraying is sometimes called the 'powder process'.

Scleroscope Hardness Test

This test uses the rebound principle. Described fully in Part Two **Hardness Testing**.

Scouring

A method of *cleaning* by mechanical rubbing.

Application range ALL METALS

Normally it is used in plating shops or, on occasion, as a preparation prior to painting. It can, however, be employed as a method of achieving an attractive surface finish.

For *cleaning*, an abrasive material, generally fine silica sand or pumice powder, is used on a moistened cloth, and the surface is abraded or scoured to remove adherent dirt and often adherent oxide.

In *plating* and *painting* preparation, this technique has the considerable disadvantage that the material used for scouring is itself highly undesirable if left on the surface. There is a considerable history to show that scouring can in itself be a source of contamination and it should not be used where other *cleaning* methods are available. With modern chemicals and techniques, it is generally possible to achieve a more efficient standard of cleanliness at much lower cost than with the present labour-intensive method.

Where it is decided that scouring is essential, there are now available powders which have abrasiveness comparable with the old-fashioned silica-type materials, but which are themselves soluble in water and thus can be more readily removed at the subsequent *swilling* and processing. Users are referred to the heading *Cleaning* for general information on soil removal.

This process, when used to produce an attractive surface finish, has limited applications in the domestic market and to some extent in the jewellery trade. With this, carefully controlled grits are used to produce an attractive pattern on the surface of the material. This can be achieved manually by skilful use of abrasive-loaded cloths or, where sufficient quantities are involved, the abrasives can be fed to rotating pads and the jigged components held against the pads to produce the desired pattern. The parts are then washed and, if necessary, passed to subsequent processing.

Scragging

A procedure used in the manufacture of helical-type compression springs. It is a process of compressing to achieve reduction in overall length.

Application range ALL METALS

The process is that the spring, when initially formed, is made longer than the design requirements. By applying the necessary compression load, the length of the spring is reduced and at the same time compressive stresses are applied to the surfaces of the spring.

Shot peening of the springs will also result in some increase in tensile strength, but scragging, in addition, results in stability. Consequently, many high-duty springs are *shot-peened* to improve fatigue strength, and then scragged to ensure stability.

Scratch Brushing

A form of *polishing*, using a brush with wire bristles.

Application range ALL METALS

Modern scratch brushing is generally carried out using hard-drawn stainless steel wire brushes. The technique uses the revolving wire brush at high speed and applies light pressure to the component. The procedure is commonly applied to remove light scale and surface blemishes prior to *painting* or *plating*. In addition, it can be used to produce an attractive type of semi-matt finish. For this, considerable skill is required.

For scale removal, *blasting* or *pickling* will generally be more economical and convenient. Where an attractive finish is required, other techniques such as *etching, scouring, vapour blasting* or *glass-bead blasting* should be considered. See also *Satin Finish*.

Scratch Test

A form of *hardness testing* not unlike the *Mohs test*, described fully in Part Two **Hardness Testing**.

Sealing
A term commonly applied to a process carried out after certain corrosion-resistant treatments.

Application range STEEL, ALUMINIUM, ZINC AND CADMIUM

The most common techniques which are 'sealed' are anodize, phosphate, cadmium and zinc plating.

In the case of anodizing, it is now known that the oxide film formed is in the shape of very fine hollow tubes vertical to the surface. These are extremely short, and the shape and size of the tubes to some extent gives the character of the anodic film. They are extremely adherent to the base material and, in effect, act like blotting paper. They can absorb liquid into themselves, and when this absorption is complete, they are said to be 'sealed'. In the case of normal *anodizing*, this sealing is achieved by immersion in boiling water. The actual chemistry of the reaction is complex, with the adherent oxide increasing the corrosion resistance.

It is this same effect which allows the freshly *anodized* surface to absorb a dye, which then permanently colours the surface, and again this is sealed with boiling water. This *colour anodizing* of aluminium can only be removed by mechanical action.

Other methods of sealing *anodized* surfaces include *chromating*, which would appear to result in the absorption of some chromate into the *anodized* film to give increased corrosion resistance. Oils, greases and sodium silicate are also used under specific circumstances.

The terms sealing and sealing process are also used when zinc or cadmium has been '*passivated*' using the *chromate process*, and when steel has been *phosphated*. In all these processes, if the freshly treated surface is dried from clean warm water and hot air, the surface will be 'sealed' giving a much superior corrosion resistance.

Seam Welding
A form of *resistance welding*.

Seasoning
The term used when components are stored in outdoor conditions for a considerable time.

Application range CAST IRON, STEEL AND COPPER ALLOY CASTINGS

This is a form of low-temperature *stress releasing* which can achieve some stability if the storage time is over a long period, and particularly if during this period there is considerable variation in temperature. To give efficient *stress releasing*, temperature variations should be as great as possible.

There is no doubt whatever that heating at a temperature as low as 100°C for periods as short as 2 h will give better *stress releasing* than with the seasoning process for long periods of time. If the former method is carried out above 200°C, the results achieved will be very much better than long-term seasoning at ambient temperature, and seasoning which is followed by *normalizing* is thus a complete waste of time.

The term *weathering* is synonymous with this process; see also *Stress-relieving/Stress-releasing* in Part Two **Heat Treatment**.

Secondary Hardening

A *heat treatment* operation which occurs on a limited number of alloy steels. Described fully in Part Two **Heat Treatment**.

Selective Annealing

A term used when local areas of components are softened for any reason. Described fully in Part Two **Heat Treatment**.

Selective Carburizing

A term used when only selected surface areas are hardened using *carburizing*. Described fully, with alternatives, in Part Two **Heat Treatment**.

Selective Hardening

A method achieving local *hardening* of components, leaving adjacent areas soft. Described fully in Part Two **Heat Treatment**.

Selective Plating

An *electroplating* process in which any conventional metal can be deposited locally, using portable equipment. The process has been devised and is marketed by the Metachem Company.

Application range GENERALLY COPPER OR STEEL, BUT ANY METAL WHICH CAN BE PLATED

The method uses normal plating techniques, but the electrolyte is present on an absorbent pad surrounding the electrode. Briefly, the equipment consists of a rectifier transformer, supplying plating current at the normal low direct-current voltages. The output from this equipment is fed to two electrodes, one of which is in the form of a clamp or crocodile-clip, while the other is a probe type of stainless steel or other inactive material. This electrode will vary depending on the purpose and the component, but will be designed to hold the absorbent pad in position round the electrode. Built into the equipment is a changeover switch allowing either electrode to be positive or negative in relation to the other.

The procedure is that the crocodile-clip or clamp electrode is connected to the component to be treated. This connection will be on an area which ensures good electrical contact with the area being plated. The second electrode or probe is held in the operator's hand.

Further information from Metachem Company.

Selenious Acid Treatment
A form of salvage, when the protective paint film is damaged.

Application range MAGNESIUM AND ITS ALLOYS

The technique is that the area is thoroughly cleaned, ensuring that all loose paint is completely removed and that the magnesium metal is chemically clean. This is then treated with a dilute solution of selenious acid. This results in a reddish colour forming, and when this colour has stabilized, that is, the metal shows an even dark red colour visible to the eye, it must be washed with water, thoroughly dried and treated with the paint system as soon as possible to prevent atmospheric attack. Normally this will be a heat-cured epoxy or phenolic type of paint system. This salvage treatment is an alternative to the *chromate* treatment which in general can be applied only to the whole article.

Where magnesium alloy components have their protective paint film damaged, rapid and serious corrosion attack can occur unless this type of salvage treatment is applied. In addition to improving the corrosion resistance the technique considerably improves the paint adhesion.

Self-annealing
A technique of *softening* steel. Described fully in Part Two **Heat Treatment**.

Self-hardening
An alternative term for *air hardening*. Described in Part Two **Heat Treatment**.

Sendzimir Process
A process applied to strip-type components.

Application range MILD STEEL

With this, the cleaned components are first passed through a hot oxidizing atmosphere to produce a light oxide film, and then immediately heated for a short period, approximately 2 minutes at approximately 800°C in cracked ammonia, when the recently produced oxide layer is reduced, that is the iron oxide is converted to iron. A slight amount of *nitriding* will occur but this will be of an extremely low order.

The components are then immediately hot-dip galvanized. Because of this prior treatment, a much heavier zinc coating is obtained than is possible with normal *galvanizing*. As the ability of galvanizing to withstand corrosion is directly proportional to the thickness of the zinc layer, it will be seen that this process can improve the corrosion resistance.

It is not commonly used because of its expense, and components with a comparable or better corrosion resistance can be obtained by a normal carefully controlled process of *galvanizing*, with the resultant surface being subsequently chromate treated and then painted with a compatible paint system.

Sensitizing

A relatively non-specific term used to cover a range of metal processes. In some cases this could be as simple as a *cleaning* operation which removes the contaminant preventing a reaction occurring, thus improving the sensitive nature of the surface to be treated. In other instances the term is used as an alternative for *etching* where the metal surface is cleaned and the oxide removed, producing a surface which will react.

Quite commonly, the term sensitizing is used as part of the *electroless plating* procedure on plastics or non-metallic surfaces. One of the stages in this process involves the *activation* or *etching* of the surface and this is then followed by a chemical reaction with an unstable solution whereby a very thin film of a metal or metallic compound is deposited. This stage of the process is commonly referred to as sensitizing.

Austenitic stainless steel is 'sensitized' at approx. 650°C before being subjected to the 'intercrystalline corrosion test', or *Strauss test*. Because of the non-specific nature of the term the reader is advised to obtain technical advice on the particular case in question before assuming that a specific meaning is implied.

'Sermetel'

A proprietary process in which aluminium and other powders are added to a heat-resistant inorganic binder as filler material.

Application range ALL METALS

The process produces a corrosion and oxidation protection film. This is in essence a paint film which is applied to a blast-roughened surface by any conventional *painting* technique. The thickness is approximately 0.004 in (0.1 mm) per coat and after air drying the coating is cured at approximately 650°C.

It is claimed that the correctly applied and cured coating has corrosion and heat-resistant properties comparable to that of *electroless nickel* and can be applied under less stringent conditions, giving greater flexibility. Further information can be obtained from the Sermetel Division of Teleflex Incorporated.

'Sermetriding'

A diffusion technique to improve wear and abrasion resistance. Some corrosion resistance is also claimed for this process.

Application range STEEL

The technique uses conventional paint spraying equipment to apply the coating which is then heated to approximately 600–650°C. At this temperature the coating diffuses into the steel giving a layer between 0.001 and 0.005 in (0.025–0.1 mm) thick. The coating is an iron chromium aluminium nitride complex.

The technique is thus comparable with *phosphating, Sulphinuz*, and '*Tufftriding*'. Little technical experience is available concerning 'Sermetriding' and the reader is advised to carry out some testing and evaluation. Further information on the technique and characteristics can be obtained from the Sermetel Division of Teleflex Incorporated.

Servarizing Process

A variation of the *aluminizing process*.

Application range MILD STEEL

The components are first cadmium plated and, after washing and drying, are then dipped in molten aluminium. This two-stage process appears to give better adhesion of the aluminium, but is expensive and is little used at present. Furthermore, increased use is unlikely with the rising cost of *cadmium plating*, and the health hazards associated with cadmium and its compounds.

Shallow Hardening

An alternative term for *case hardening*. Like many terms used in processing, it can have a specific meaning in local areas, and care should be taken in the interpretation.

Shape-strength Test

The name which is given under some circumstances to the *fatigue test* carried out on actual components.

Application range ALL METALS

It is now realized that fatigue testing of prepared specimens is of limited application and that only the application of stresses, to simulate operating conditions, to the actual component will result in meaningful figures regarding fatigue strength. The shape-strength test is, therefore, used for fatigue testing either of actual components, or mock-up components which have the relevant shapes manufactured and tested. The *Fatigue Test* is described fully in Part Two **Mechanical Testing**.

Shear Test

Described fully under *Torsion Test* in Part Two **Mechanical Testing**.

Shell Moulding

A form of *casting*, and as such not a metal treatment. Briefly, a pattern, usually in metal, is sprayed with mixture, usually a fine silica sand with a resin binder. This is the 'shell' which when hardened is removed and becomes the mould for castings. This fine shell is generally supported mechanically or in sand. With this technique, castings of very fine tolerance and intricate detail can be obtained. It is comparable to the *lost wax process*, and these techniques for producing close-tolerance intricate shapes should be considered by designers as alternatives to conventional machining and forging when the correct mechanical properties are obtainable.

'Shepherd Process'

A proprietary form of sulphuric *anodizing*.

Application range ALUMINIUM

This uses electrolytic sulphuric acid with glycerine or glycol as an additive.

Shepherd Test

A 'hardenability' test.

Application range MEDIUM AND HIGH CARBON STEEL

It is not generally applicable to alloy hardening steels but modifications of this test can be used.

The technique is that a standard machined test bar in the material to be tested, generally approx. 6 in (150 mm) long and 0.75 in (20 mm) in diameter, is prepared. The actual dimensions of the test piece are not critical, provided they are reproduced with succeeding tests. The test piece is heated to the correct temperature for the material specification and is then quenched. The method of quenching must be tightly controlled and exactly reproducible. For this reason use is normally made of spray quenching and often brine is used as the quenching medium.

On completion of the cooling cycle, the test specimen is sectioned and examined metallurgically, if desired with a hardness survey also being carried out. The extent of the hardening, the type of structure obtained and the hardness figures are all used to indicate the 'hardenability' of the material being tested. Usually, one or more of these parameters will be specified.

This test is in many ways similar to other 'hardenability' tests such as the *McQuaid Ehn test*. A simple technique supplying less detailed information, is to quench a wedge shape of the steel being tested and carry out a hardness survey along the surface of the wedge. Provided all parameters are controlled, this simple test can supply sufficient information to the heat treater to indicate serious differences from other batches of the same material.

An even simpler test for hardenability is to quench the same piece of steel, or identical pieces of the same steel in water, oil and air. Comparison of the hardness achieved and metallurgical structure can supply information on hardenability and if any problems could exist at welding.

Sherardizing

A form of *corrosion protection*.

Application range MILD AND LOW ALLOY STEEL

The process is basically the deposition of zinc and zinc oxide on the surface. It makes use of finely powdered zinc with zinc oxide and the components are packed in this powder and heated to 350°C–450°C for periods which will vary depending on the thickness of the coating required.

On completion of the process, the components are removed and under some circumstances may be lightly wire-brushed. They can, however, be merely riddled to separate the components from the zinc powder.

With this process, a coating of zinc and zinc oxide is obtained much thicker than can be obtained by conventional *electroplating*. The coating can, in fact, be compared in thickness to the hot-dip *galvanizing* process and has the advantage that no tears are formed, thus threads do not tend to become blocked with the solidifying zinc. Sherardizing faithfully follows the contours of the component being treated and, provided the correct conditions are used, the resultant deposit, regarding corrosion resistance, is comparable to both *electrolytic* and *hot-dip galvanizing* for the same thickness of zinc. In general, sherardizing will not be used for thin coatings, as electrolytic zinc will usually be more economical, and will give equal if not better *corrosion protection*. Sherardizing is a valuable process where studs, bolts and nuts and similar types of component, particularly threaded components, are required for outdoor use. When using *hot-dip galvanizing*, it is essential that threads are subsequently reformed and, apart from the considerable expense of this operation, there is some danger that the machining operation will locally remove the galvanized coating, thus allowing corrosion to occur.

Shielded Arc Welding

A rather general term which can cause confusion. When an inert gas such as argon or helium is used, the process is termed *inert-gas shielded metal arc welding*.

This also takes in the terms MIG and TIG. When other materials such as carbon dioxide, nitrogen and even certain fluxes, which produce large volumes of a blanketing gas are used, the process is sometimes also termed shielded arc welding. This general term, then, should obviously be treated with caution and further information obtained. Information on the various processes is given in Part Two **Welding**.

Shimer Process

A form of *case hardening* and a modification of the *cyanide hardening* process.

Application range MILD AND LOW ALLOY STEEL

With this, the components are hardened from the conventional type of salt bath which has been modified by the addition of calcium cyanamide. This increases the speed of penetration of the carbon and nitrogen and, thus, either gives the same depth in less time than conventional *cyanide hardening*, or results in a thicker case depth for the same time. Further details of the advantages and disadvantages of the cyanide process are given in Part Two **Heat Treatment**.

With the Shimer process, there is a greater tendency to produce a sharp demarcation between the case and the core, and thus where bending stresses or high loading is involved, there will always be some danger of exfoliation of the case. High-duty components, therefore, should not make use of this process.

The carbonitride process will give comparable or better results, more economically and with fewer problems of effluent etc.

Shock Test

This term is used to cover a number of highly specific tests, and has been used to cover *impact testing*, which is a form of shock test where a well-defined notch is cut in the test specimen. (This form of testing is described fully in Part Two **Mechanical Testing**.) More generally, however, the term refers to those tests where knocks or shock is involved. There are also a number of shock tests which are applied to assemblies and designed to ensure that any impact shock such as might be expected in service will not result in a failure. Electronic and electrical assemblies are tested to a considerable extent using this technique. This is sometimes known as bump testing. The test can vary from a single shock application by hammer or by dropping, to repeated variable cycles of shocks or bumps applied at different angles from different heights. In general this type of shock testing will be an individual specification, written for a specific component. An attempt should be made, however, to ascertain whether similar testing has been specified for similar assemblies. A series of British Standards applying to electronic assemblies attempt to standardize certain shock tests.

Shore Hardness

A *hardness testing* technique using the 'rebound' principle. Describe fully in Part Two **Hardness Testing**.

Short-cycle Annealing

This term was applied at one time to any form of *annealing* which did not involve the complete cycle of loading into a cold furnace, heating, holding at temperature and cooling inside the furnace.

Application range ALL METALS

The modern technique of loading into warm or hot furnaces and controlling the cooling in sand or some other medium has now replaced this very slow technique, and the term is no longer used. The different forms of *annealing* and information on other treatments are given in Part Two **Heat Treatment**.

'Shorter Process'

A proprietary *induction hardening* treatment. Described fully in Part Two **Heat Treatment**.

Shot Blasting

A general term applied to the technique of *blasting* where shot is used. Described fully in Part Two **Blasting**.

Shot Peening

A controlled type of *blasting* used to improve the fatigue strength of components by *peening*, thus compressing the surface layer. Described fully in Part Two **Blasting**.

Shrink Fitting

An assembly technique where one component is reduced in temperature in order to cause it to contract and this is then fitted into the female component. Shrink fitting is also applied where the female component is expanded by heating and, of course, where one of the components is heated and the other cooled.

Application range ALL METALS

Shrink fitting can make use of a normal refrigerator or deep-freeze unit at 0°C–20°C but more commonly uses carbon dioxide to give −60°C or liquid air or liquid nitrogen at temperatures of −190°C and −196°C. Some care is necessary in that as the component is removed from the refrigerator into air, any moisture in the atmosphere will immediately condense on the cold surface. This moisture will then be trapped in the subsequent assembly. Under many circumstances this will result in no problems and can, in fact, be an advantage in causing corrosion to increase the seizure effect of the components involved. There may, of course, be circumstances when this effect is undesirable. When the female component is heated, the most common method is probably the use of a vapour degreaser at a temperature of 86°C. This, in addition to heating the component, will very efficiently remove all grease and oil and, thus, adds to the effects of any corrosion. The use of hot oil is relatively common but must be approached with some care as it immediately becomes a thick grease or wax when contact is made with the cold male component, if this is taken to a low temperature.

It must be appreciated that, unless the design engineer takes considerable care, it is possible to build in considerable stresses with the shrink fitting technique. The following coefficient of linear expansion figures are listed for the convenience of the user. This means that for every inch of the component's diameter or length, there will be increase or decrease in size by this figure in inches for each degree of temperature over the range of room to approx. 250°C. While this might not appear to be appreciable when the difference between liquid nitrogen and hot oil at 150°C is taken into consideration, the stresses involved on components of 10 in (250 mm) and above can be considerable.

Metal	Coefficient
Aluminium alloys	$20-24 \times 10^{-6}$
Cadmium	30×10^{-6}
Chromium	6.5×10^{-6}
Cobalt alloys	12.5×10^{-6}
Copper alloys	$16-20 \times 10^{-6}$
Gold	14.5×10^{-6}
Iron and alloy steel	$10-12 \times 10^{-6}$
Lead	29×10^{-6}
Magnesium	26×10^{-6}
Mercury	61×10^{-6}

Nickel	13×10^{-6}
Platinum	9×10^{-6}
Rhodium	8.5×10^{-6}
Silver	19.5×10^{-6}
Stainless Steel	18×10^{-6}
Steel-alloy	$10-12 \times 10^{-6}$
Tin	24×10^{-6}
Titanium	9×10^{-6}
Tungsten	4.5×10^{-6}
Zinc	31×10^{-6}

Shrink fitting can be an economical and technical alternative to *soldering* and *brazing*, or mechanical fixing. Where the components are pressed out during servicing or overhaul, one or other must be replaced, or some method of gluing or fixing used on reassembly as the interference fit will have been destroyed. It is also possible by electrodeposition of a soft metal such as zinc, bronze or nickel to recover the interference fit.

Sieve Test

A test to evaluate particle size. Almost all *blasting* requires that the size of the grit particles is controlled. The particles used to manufacture grinding wheels and polishing equipment must also be controlled, together with the grit used for many other purposes, including fillers, cement, etc.

The method is that the sample being tested is placed in a sieve which is shaken: the amount of material retained on this sieve, and the amount passing through, will give an indication of the grit size. It is seldom that a single sieving will give the necessary information. In general, particle sizes will be specified as lying between two limits. To carry out the test it is necessary to shake the grit on two sieves: the material collected on the coarser sieves is rejected, and the material which passes through the finer sieve is also rejected. This means that the material will contain neither smaller, nor coarser, particles than specified. The test, therefore, should result in no grit particles whatever being retained on the coarse sieve and no particles able to pass through the finer sieve. It is sometimes possible that more detailed information is necessary and a series of sieves will then be used and the percentage of grit collected on each measured.

The test itself requires some skill and the equipment must be well maintained. The amount of shaking, the direction of shaking and amplitude are controlled for certain tests, and some effort must be made to ensure that repeated tests are carried out in a similar manner.

The grids are manufactured with wire woven mesh and it is obviously important to ensure that they remain undamaged, otherwise the pattern of holes will be disturbed. If the wires are pushed apart, for instance, larger particles will pass through than the standard mesh indicates.

Unfortunately, a number of methods are used to define mesh size. The following table lists the more common methods of identifying sieve sizes together with particle sizes.

Comparative sieve sizes in order of decreasing opening size

Sieve opening		ASTM Sieve designation standard alternative		Tyler equivalent designation mesh number	British standard mesh number
mm	in	mm	in		
107.6	4.24	107.6	4.24		
101.6	4.00	101.6	4		
90.5	3.5	90.5	$3\frac{1}{2}$		
76.1	3.00	76.1	3		
64.0	2.50	64.0	$2\frac{1}{2}$		
53.8	2.12	53.8	$2\frac{1}{2}$		
50.8	2.00	50.8	2		
54.3	1.75	45.3	$1\frac{3}{4}$		
38.1	1.50	38.1	$1\frac{1}{2}$		
32.0	1.25	32.0	$1\frac{1}{4}$		
26.9	1.06	26.9	1.06	1.050 in	
25.4	1.00	25.4	1		
22.6	0.875	22.6	$\frac{7}{8}$	0.883 in	
19.0	0.750	19.0	$\frac{3}{4}$	0.742 in	
16.0	0.625	16.0	$\frac{5}{8}$	0.624 in	
13.5	0.530	13.5	0.530	0.525 in	
12.7	0.500	12.7	$\frac{1}{2}$		
11.2	0.438	11.2	$\frac{7}{16}$	0.441 in	
9.51	0.375	9.51	$\frac{3}{8}$	0.371 in	
8.00	0.312	8.00	$\frac{5}{16}$	$2\frac{1}{2}$	
6.73	0.265	6.73	0.265	3	
6.35	0.250	6.35	$\frac{1}{4}$		
5.66	0.223	5.66	No $3\frac{1}{2}$	$3\frac{1}{2}$	
5.60	0.221				3
4.76	0.187	4.76	No 4	4	
4.00	0.157	4.00	5	5	4
3.36	0.132	3.36	6	6	
3.35	0.132				5
2.83	0.111	2.83	7	7	
2.80	0.110				6
2.38	0.0937	2.38	8	8	
2.35	0.0925				7
2.00	0.0787	2.00	10	9	8
1.70	0.0669				10
1.68	0.0661	1.68	12	10	
1.41	0.0555	1.41	14	12	
1.40	0.0551				12
1.19	0.0469	1.19	16	14	
1.18	0.0465				14
1.00	0.0394	1.00	18	16	16

Sieve opening		ASTM Sieve designation standard alternative		Tyler equivalent designation mesh number	British standard mesh number
mm	in	mm	in		
microns	inches	microns			
1000	0.0394	1000	No. 18	16	16
850	0.0335				18
841	0.0331	841	No. 20	20	
710	0.0280				22
707	0.0278	707	25	24	
600	0.0236				25
595	0.0234	595	30	28	
500	0.0197	500	35	32	30
415	0.0167				36
420	0.0165	420	40	35	
355	0.0140				44
354	0.0139	354	45	42	
300	0.0118				52
297	0.0117	297	50	48	
250	0.0089	250	60	60	60
212	0.0083				72
210	0.0083	210	70	65	
180	0.0071				85
177	0.0070	177	80	80	
150	0.0059				100
149	0.0059	149	100	100	
125	0.0049	125	120	115	120
106	0.0042				150
105	0.0041	105	140	150	
90	0.0035				170
88	0.0035	88	170	170	
75	0.0030				200
74	0.0029	74	200	200	
63	0.0025	63	230	250	240
53	0.0021	53	270	270	300
45	0.0018				350
44	0.0017	44	325	325	
37	0.0015	37	400	400	400
25	0.001				600
20	0.00075				800

Siliconizing

A surface treatment to improve corrosion resistance.

Application range STEEL AND CAST IRON

This is analogous in many ways to *carburizing* in that the steel to be treated is packed in a silicon carbide compound, taken to temperature above 950°C. In an atmosphere containing chlorine there will be a reaction between the surface iron and the silicon in the compound, resulting in the formation of iron silicates able to diffuse into the core of the steel or iron being teated. The surface will achieve a silicon analysis of approximately 15 per cent, and this will decrease away from the surface where the normal silicon content of the steel, less than 0.5 per cent, is present.

The purpose of siliconizing is to produce a surface which has excellent corrosion resistance, particularly towards acid sulphates. This is similar to certain of the high-silicon cast irons, and the process can indeed be applied to cast iron, but it is generally more economical to produce components in high-silicon cast iron rather than cast irons which are subsequently siliconized. The surface will be brittle and hard, with properties inferior to those achieved by *carburizing* regarding mechanical strength and ductility. The process is not now commonly applied as it is now possible to design components using other materials such as stainless steels, titanium, etc., which have comparable properties of corrosion resistance and much better ductility. The process is also known as *ihrigizing*.

Silk-screen Printing

A form of printing, used commonly to print on paper and some cloth; the same technique can be used for printing on metal.

Application range ALL METALS

Silk-screen printing uses a silk mesh or under some circumstances a finely woven wire mesh which is generally of stainless steel though other metals can be used. The process is still called silk-screen printing when other types of screen are used. In the first stage the necessary design in negative form is applied to fill the pores of the mesh. This can be done by hand, using the skill of an artist, but more commonly stencils which have the outline of the required design are used. The mesh is filled with some non-active material, such as wax, though other materials can also be used. In industrial silk-screen printing probably the most common method is to produce the master silk screen by a photographic process; in this case it is the photographic emulsion which fills the pores of the mesh.

Where necessary the silk or wire mesh is dried at this stage to harden the material in the pores. To print, the silk screen, fitted to a frame, is brought into contact with the article being treated and by using a 'squeegee' paint of the required colour and type is spread across the silk screen. Where the pores of the mesh have not been filled the material, either ink or paint, will be forced through and the required pattern will then appear on the article. Silk-screen printing is most commonly used to produce the desired patterns directly. Whether or not the resultant design is used in its final stage will depend on the application. In industry silk-screen printing is a relatively common method of *stopping-off* a surface for further processing. This can be for *etching* or in order to *electroplate*. Where silk-screen printing is used for purely decorative work, multi-coloured designs can be produced by the use of several silk screens each with the individual design for the colour required.

Silk-screen printing can thus be seen to be a method of producing a number of identical designs. It is a very flexible process. In industry, and in the printing trade, silk-screen printing has a competitor in photographic processes, where master negatives can be made and used in a similar manner to the present process with less emphasis on individual artistic skill and the ability to reproduce the finished article much more rapidly.

Silver Ball Test
A method of assessing the efficiency of quenching oil.

Application range STEEL

Hardening of steel requires efficient cooling in oil and this process gives a means of measuring the cooling rate. The ball is made from pure silver and has a thermocouple at its centre. It is heated to 700°C and held at that temperature for five minutes before dropping into the oil being tested. The thermocouple automatically records the temperature of the silver ball on a chart, and the rate of cooling can thus be measured. Silver metal is used as there are no change points in the cooling range, while the metal has excellent heat conductivity and good oxidation resistance.

This test is more sophisticated and supplies more information than the *nickel ball test* but is essentially similar. An alternative method to either would be to quench standard steel test pieces in the test oils and then to measure the hardness. The silver ball test can supply additional information regarding the shape of the cooling curve which may be useful where intricate components are involved.

Silver Plating
Silver is plated for decorative, electrical conductivity and anti-seizure purposes. Described fully in Part Two **Electroplating**.

Silver Soldering
A *brazing* operation, using copper alloys containing silver which have a lower melting point than the normal copper alloy filler materials. Further details are given in the entry *Brazing*.

Sintering

A production process and as such not a metal treatment. Briefly, with sintering, the metal is produced in a granular form and with certain processes can be made of a mixture of metals. These will be blended and the granules are then compressed to form the shape of the article required. This can be extremely complex and gears of very intricate design are now manufactured by this process.

The components removed from the press, known as the 'green stage', have very little mechanical strength and must be handled carefully to prevent damage. They are then processed through a furnace generally with a controlled reducing atmosphere where some fusion takes place. This fusion will not normally be of the metal itself but with mixed material may result in melting of one component which then holds the other ingredients together. With simpler mixes or single metals, the heating is to below the melting point, but some softening will occur which allows the granules to adhere to one another. The process will not result in components as strong as *forging*, but can be comparable to castings.

It is possible to achieve a designed and controlled porosity which can then be used to absorb oil, thus giving a 'self-lubricating' bearing. The process can also be used to blend metals and non-metals, for example, carbon or PTFE plastic can be mixed and sintered with metals to give special-purpose bearings. The user is advised to obtain further information on this subject.

Skin Annealing

The term applied to the short-term *softening* process on some nickel-chromium alloys. Described fully in Part Two **Heat Treatment**.

Skin Pass

A final rolling operation which is applied to sheet metal or components produced from sheet metal.

Application range ALL METAL

On the production side, the skin pass will be used to produce the final surface finish and simultaneously achieve slightly improved mechanical properties. During manufacturing components are sometimes given a final skin pass through special rolls to remove slight visual defects or produce a specified finish. The material will undergo *work hardening* by the skin pass but this will be to a limited extent.

Skip Weld

The form of welding where adjacent areas between welds are 'skipped' or missed out.

Application range ALL METALS WELDED

While this technique can be applied to any type of weld, in general it is confined to lap and fillet welds. It is seldom used with butt welds.

Because of the high local heating effect, and the large mass of adjacent cold metal, there will be a danger of brittleness or cracking with this technique. It is, therefore, important that steel having high 'hardenability' is not used without controlled preheat. Skip welding will not normally be used on areas where high-integrity joining is required.

Smithing

The name given to any form of working, hot or cold, but generally hot, carried out by the blacksmith. Smithing can, therefore, cover the processes of *forging, straightening* and *welding*.

Snarl Test

A specific ductility test applied to wire.

Application range ALL METALS

The wire is looped and then straightened. This is repeated until failure occurs, and the number of times recorded. The test is designed to ascertain that the wire will withstand kinking in service. It is, then, a specialist form of *bend testing* which does not require sophisticated equipment. *Bend tests* are described fully in Part Two **Mechanical Testing**.

Snead Process

The use of resistance heating for hardening.

Application range MEDIUM CARBON AND ALLOY STEELS

The component itself is made the resistance in an electrical circuit and an electrical current of a predetermined amperage and voltage is passed through with carefully designed connections for a specified time. Because of the component's resistance to the electrical current, heating will occur, and this can be high enough to result in *hardening* when the component is quenched on the removal of the heating source.

It will be obvious that only components of a relatively simple shape are suitable for this type of *hardening*. Complicated shapes will result in local excess heating occurring, while other areas with a higher volume will have a much lower resistance and thus will not heat to the same extent. The technique is of limited application and has generally been replaced by conventional heat treatments or the *induction hardening* process.

Soaking
The term applied to the holding of metals at temperature.

Application range ALL METALS

The length of time for soaking will vary for different materials. Chromium steels, for instance, are known to be sluggish in their metallurgical reactions and thus require longer soaking periods than equivalent plain carbon steels.

Sodium Hydride
A method of scale removal.

Application range ALL METALS

This uses a molten caustic soda in which metallic sodium is dissolved, and through this is bubbled hydrogen gas. This hot liquid is strongly reducing in character and will not attack many metals, but the reducing action can be used to remove surface oxide. It has been shown that the same technique will very successfully remove all evidence of occluded sand from the surface of castings and, being a liquid process, internal passages can therefore be successfully cleaned. This would be difficult or impossible by conventional *de-scaling* or de-sanding techniques.

It will be obvious that the chemicals involved require specialist equipment and considerable care in the control, because of the hazards involved in handling hot caustic soda, molten sodium and hydrogen. The technique is that the components to be treated are cleaned to a reasonable standard but not necessarily chemically clean. Cleaning is required more to prevent contamination of the sodium hydride chemicals rather than for technical reasons. The components must then be thoroughly dried to prevent explosion when they enter the hot sodium hydride liquid. The dried, and generally hot, metal is then immersed in the liquid and the necessary chemicals are added. Hydrogen gas is passed through the solution and there will be a visible chemical reaction in the case of heavily scaled components. When this reaction has ceased the *de-scaling* is completed. Provided the sodium hydride chemicals are kept at the correct balance, no harm will come to metal, even aluminium, although immersed in a strong alkali solution, because of the strongly reducing nature of the chemicals. The components are removed, drained and then quenched in water. This is generally the most spectacular part of the procedure.

While this technique can be used on titanium and its alloys, the time involved with this metal must be carefully controlled as hydrogen attacks titanium and can result in the formation of brittle hydrides.

This is probably the most sophisticated method of *de-scaling* and, under many circumstances, it can be the most economical. Where components have very adherent scale which must be removed for technical or economical reasons, or because of difficulty during machining, the present method should be considered.

Alternative techniques include *pickling*, *blasting* and mechanical *polishing*, none of which is as technically efficient as sodium hydride.

Soft Facing

A term covering the application of a soft metal to a harder base.

Application range STEEL IN GENERAL BUT COULD APPLY TO OTHER METALS

The most common application is that of a bearing metal onto steel, this may be bronze, or a white metal. The method of application will usually be by *metal-spraying* or deposition of a layer of weld metal.

The term soft facing is not specific and can be used to cover a range of treatments, including the *electrodeposition* of pure metals. In general, however, it is confined to *weld deposition* or *metal-spraying* to differentiate these from the more common *hard facing*.

Softening

A general term applied to any *heat treatment* process which is designed to result in the material being in a softer condition on completion of the process.

Application range ALL METALS

The specific, technical term is *annealing*. For further information reference should be made to *Full Anneal, Subcritical Anneal, Spheroidal Anneal, Tempering*, and *Stress-relieving/Stress-releasing*, all of which are described fully in Part Two **Heat Treatment**.

Solder Plating

The plating of the tin–lead alloy known as solder. It is used to produce items which will be subsequently assembled by *soldering*. In printed circuit manufacture solder plating is commonly used as the *stop-off* to prevent *etching* of the copper, and remains on the board to take part in the soldering during final assembly. See also Part Two **Electroplating**.

The use of tin plating as an alternative to solder gives greater flexibility. The shelf life of both solder or tin deposits is related to the deposit thickness, and other factors. It will be measured in weeks under many circumstances, and unless all parameters, including storage, are carefully controlled, will seldom be longer than three months where high-speed soldering with low-activity flux is required.

Solderability Testing

An inspection or quality control test which is commonly applied to components to evaluate a desired degree of 'solderability'.

Application range COPPER AND ALLOYS, STEEL

Two basic tests are involved, both of which require careful control and some equipment. The first test is a method of heating the components or test piece to the desired temperature, usually 260°C for normal solder, and producing a blob of solder, either by dropping or placing or other means. Where components have already been soldered, this test will result in the soldier melting on the component surface.

A visual examination is then made of the solidified blob. Special attention is paid to the angle of the meniscus, and where previous solder is being melted, the examination is made for any evidence of 'de-wetting', that is, breaks in the solder film. Any other defects identified must also be reported. Where components are involved, the soldered joint is also carefully examined for the angle of the meniscus. There should always be a smooth transition, with the angle as small as possible.

This test requires considerable skill on the part of the operator to reproduce the conditions, and also on the part of the inspector to evaluate minor changes in the degree of 'solderability'.

In the second test the component or test piece to be soldered is immersed in molten solder. Again this requires careful temperature control of the solder-pot. The component is dipped into the solder, held there for a specified time and then removed under controlled conditions. In its simplest form the test is carried out by hand and timed by a stop-watch. Sophisticated test units are now available, enabling testing under fully controlled conditions of temperature, immersion time and the method of entering and leaving the solder-pot.

In both tests the use of flux is permitted, this must also be carefully controlled. Most tests normally use a standard low-activity flux, or are carried out without flux in order to give the worst conditions. Flux make-up, the method of application, the time of contact, the shelf life of the test piece or component, must all be controlled to give reproducible results.

Good solderability is achieved by control of the solder deposit to 60 per cent tin, 40 per cent lead, ensuring that no contamination or oxidation occurs during the application of the solder deposit and rigid control of the storage conditions. See also *Spread Test*.

Soldering

A method of joining.

Application range COPPER, STEEL, ALUMINIUM, TIN, LEAD

The technique is that the components to be joined are cleaned in the area of the join, the solder material is then applied and heated. Use of fluxes is commonplace, but there are various techniques where active fluxes are not used.

The join is achieved by the alloying of the solder metal to each of the two faces. By far the most common soldering materials are the tin–lead alloys, 60 per cent tin, 40 per cent lead alloy being the most popular.

The term soldering is inexact but by common consent it is applied to the use of tin–lead alloys and is sometimes loosely applied also to any form of low-melting-point joint using metal as the joining medium. Thus, there are in existence aluminium solders which do not contain appreciable amounts of tin or lead and the term *silver soldering* is used for alloys containing no lead or tin. This latter is a form of brazing.

Metallurgically, the process of soldering is identical to that of brazing. The metals being joined do not themselves melt but the joining metal is of a lower melting point than the components. The process, therefore, can be likened to the use of a molten metal glue. The molten metal must wet the surfaces of each of the components being joined, and this wetting means that a metallurgical join has occurred with some alloying taking place at the metal interfaces.

As with *brazing* the amount of metal between the joints will have considerable influence on the strength of the joint. In general the solder alloy will be mechanically weaker than the components being joined. Thus, when the solder used is thick, giving a wide joint, then the strength of the joint will not be more than the strength of the solder. When, however, the solder joint is extremely thin, then a joint of high strength, almost equal to the weakest of the components, can be achieved.

Therefore, there is considerable responsibility initially on the design engineer to specify the correct dimensions, and on manufacturing personnel to ensure that these tolerances are maintained during production of the components.

The ideal clearance for soldering will be in the region of 0.00075 in (0.018 mm) and 0.002 in (0.05 mm), thus it will be seen that little laxity is possible, for two separate components at least will be involved.

As with *brazing*, the soldering technique should be that the components are cleaned, *fluxed* when necessary, then heated with the solder being fed from the bottom of the joint, so that the molten solder rises by capillary action. It is then possible visually to inspect the top surface of the soldered joint, and if this has a good meniscus, evenly applied over the complete surface of the join, good soldering has taken place. Poor, quality soldering, on the other hand, will show areas where the soldering is missing. It is possible with the correct clearances for solder to rise more than 4 in (100 mm) when bottom-fed with correctly prepared components. The different soldering techniques are as follows:

Bit soldering
With this technique use is made of a shaped mass of metal, generally of copper or copper alloy, which is heated by some external source such as a flame or by

internal electric heating. This is known as a 'solder bit' or a 'solder bolt', and by applying this to the pieces being joined, sufficient heat can be transmitted to melt the solder. Some skill is necessary and the parts must be cleaned. The usual technique is that the solder bolt is fluxed and tinned, that is, the surface is given an alloy film of solder. By repeatedly fluxing and transferring this flux to the component eventually this will be '*tinned*' or given an alloy film of solder. The process is repeated on the other component to be soldered, and if possible the two components are then brought into contact. By transmitting the heat from the bolt, the two thin films of solder will melt and alloy with each other. Light hand pressure is then applied to retain the components in position and the solder bolt is removed so that the molten solder solidifies causing a high-quality joint.

Dip soldering
With this, the assembled components are fluxed then dipped in the molten solder bath.

Flow soldering
With this technique the components are assembled and, where necessary, held by jigging. They must be clean and free from oxide, with the necessary clearances between components being achieved to give the desired quality of joint.

The assembly is then *fluxed*, and may be preheated prior to passing over a wave of molten solder. The contact time between the pieces and the molten solder is very short, seldom being in excess of 2 or 3 s, thus the 'solderability' of the piece must be high. Good-quality, reproducible joints are possible with this technique under mass production or large batch conditions.

Induction soldering
This technique uses the heat from an induction coil. This is not a common technique as the heat source is capable of very high temperatures, but where high production soldering is involved the necessary control can be achieved. More information is given under *Induction Heating*.

Oven soldering
The assembled parts, which have been *cleaned* and *fluxed*, are placed in an oven at the correct temperature. The solder melts and flows and solidifies when the components are removed from the oven. It will be appreciated that a conveyor-type oven is essential to prevent joint failure caused by undue movement of the assembly while the solder is molten.

Resistance soldering
This uses the electrical resistance of the component to give the necessary heat to cause melting of the solder. With certain mass-production or high-batch-quantity parts, this technique can achieve rapid high-quality solder joints on large assemblies which could not be soldered by other techniques.

Ultrasonic soldering
This is usually applied only to aluminium, where the ultrasonic energy is necessary to break up the adherent oxide. The ultrasonic beam is applied through a solder bit or bolt which is also heated. Regarding skill involved, the technique is similar to *bit soldering*.

Wave soldering
An alternative name to *flow soldering*.

A good solder joint will have a light, smooth finish, free from pitting, 'cissing' or any evidence of oxide or excess flux. The solder should show a smooth transition from the joint itself to the base material. Technically, the meniscus produced should have a low angle. If this is not present, it is an indication that the molten solder has not flowed in a satisfactory manner.

Soldering is an alternative jointing process to *brazing* and *welding*, usually requiring much less heat. It is also an alternative to mechanical joining and gives superior electrical joints. Mechanically these will be inferior to good mechanical joints unless considerable care has been taken at the design stage and during manufacture.

To summarize, soldering is a joining technique identical to *brazing* but that less heat is usually required. Fluxes are commonly employed, these give rise to corrosion problems unless controlled or considerable care is taken to ensure their complete removal. There is much evidence to show that unless control of cleanliness and the state of oxidation is applied, 'solderability' problems will arise. The control of storage conditions, and the method of cleaning are essential for consistent 'solderability'. There are several tests, such as the '*spread test*', which can evaluate the 'solderability' of components.

Solid Carburizing
An alternative term, not now commonly used, for the *pack carburizing* process.

Solution Treatment
A term given to the high-temperature part of the two-stage *hardening* treatment applied to many alloys, mostly non-ferrous. This carries out the necessary dissolving of the constituents and leaves the metal in its softest possible condition ready for *hardening* by the subsequent *precipitation treatment*. Described fully in Part Two **Heat Treatment**.

Solvent Cleaning
A cleaning technique where the aim is to dissolve the contaminant.

Application range ALL METALS

This is the term given to the use of organic solvents for *cleaning*. Generally it describes the use of wipe cleaning, where the cloth is moistened in the solvent and rubbed across the surface of the component, the theory being that the solvent will dissolve the soil and this will be removed on the cloth. Care must be taken that the necessary amount of solvent is present and that a fresh side of the cloth, free from contaminant, is presented to the surface being cleaned. Obviously, if this is not the case, then what is achieved is merely the spreading of the contaminant.

Solvent cleaning normally uses liquids of the chlorinated hydrocarbon type, but methylated spirits and methyl alcohol are also commonly used. *Cleaning with materials such as carbon tetrachloride is now prohibited as it has been found that this constitutes a health hazard.* Liquids such as petrol, benzene, ether, etc. may still be used under controlled conditions but constitute a hazard to health in themselves and there is the very obvious danger of flammability. Solvents such as toluene and xylene are employed in modern industry, but also have a health risk when used in confined conditions over long periods and also present a flammable hazard. These specialist solvents are commonly employed in the plastics industry prior to the use of adhesives or surface coatings.

Solvent cleaning also involves the dipping of components into the solvent itself, which is unusual as this results in the liquid becoming contaminated and this is then evenly distributed on the surface of the component on evaporation. Solvent dipping can be efficiently used in conjunction with other forms of cleaning where the dip, usually hot, will remove the bulk of any oily contaminant and the remaining thin film is then removed by a subsequent cleaning operation.

Multi-stage cleaners are available which use this principle. Here a series of tanks is required; the components become gradually cleaner as they progress from tank to tank. Cleaning equipment is marketed having as the final stage a vapour degreaser. It is also possible to use other types of cleaning such as aqueous cleaning, with or without electrolytic action, as the final or subsequent stage after solvent rinsing.

Finally, the term solvent cleaning is occasionally applied to the use of the solvent in the vapour phase. With this, the contaminated components are placed at room temperature in a vapour of solvent. This will result in the solvent condensing and, thus, in intimate contact with the contaminated surface there will be a clean hot solvent which will then run off taking with it the oily type contaminant. This is *vapour degreasing*, described under this heading.

To summarize, solvent cleaning is a general term applied to the use of any organic liquid which uses the action of dissolving to remove oil or grease contaminants. Solvent cleaning must always be used with caution as there is invariably a health hazard and there can also be a fire hazard. Moreover, unless care is taken, the effect is often to spread the contaminant rather than completely remove it. It must be appreciated that solvent cleaning will remove only those materials which are capable of solution in the chosen solvent. This will seldom be an efficient method of totally cleaning an article. The techniques of soil removal are discussed under *Cleaning*.

Solvent Degreasing
An alternative term for *vapour degreasing*. Basically the system is that an organic solvent is boiled to produce a vapour and the articles to be cleaned are placed in the vapour, resulting in condensation of hot, clean solvent to remove contaminating grease. Alternative methods of grease removal are discussed under *Cleaning*.

Sonic Testing
A method of *non-destructive testing* which uses any interference in known sound patterns to identify faults.
  Described fully in Part Two **Non-destructive Testing**.

Spark Erosion
A technique used for the production of intricate shapes, or for the salvage of components with broken taps, drills or other hardened pieces which cannot be removed by conventional machining.

Application range ALL METALS

The technique makes use of an electrical spark discharge. The procedure is that the component being machined or salvaged is made one side of an electrical circuit. This is achieved by clamping the component securely in a tank. The other side of the electrical circuit will be the electrode. The material for this will depend on the material being eroded, with the most common material being brass. This will be shaped according to the hole which is to be formed. Some development work or experience is necessary to decide the size and shape of the electrode as there will always be some erosion of this in addition to the component. Thus the electrode will normally start by having a different size and shape to the finished dimension of the component being eroded.
  Once the assembly has been correctly adjusted, the tank is filled with paraffin. It is important that high-grade paraffin is used, having water at an extremely low level. This is necessary to ensure good electrical resistance and a high dielectric (electrical resistance). The electrode is then adjusted to almost contact the component and a high-energy electric current is applied. Provided the settings are correct regarding the gap between the electrode and the component, then as the electric current reaches a certain value it will discharge across the gap through the paraffin dielectric. The energy from this spark discharge will be sufficient to result in local tensile failure of the material. This erosion is possible because at the point of discharge the electrical energy is sufficient to cause melting of the material, or at least sufficient heat to reduce the tensile strength to a relatively low level. It will be seen that this requires specialist equipment and experience is necessary in the choice of electrode

material and design of the shape. There are available a number of spark erosion machines and the manufacturers supply technical information.

As the method of machining using spark erosion relies on the production of high temperatures at the area between the electrode and the point being eroded, it will be appreciated that the resultant surface will have a metallurgical structure different from the component itself. At the best this will be a thin layer of cast material probably with an extremely thin oxide film. Where materials such as *carburized* or high carbon steel are spark eroded the effect can be more dangerous, because in addition to the thin layer of cast material there will be an area beneath where the heat treatment has been altered. It must, therefore, be appreciated that the fatigue characteristics of the surface associated with spark erosion can be considerably reduced. In many cases this will have no significance, but with high-integrity components it is essential that the danger is realized and steps taken to improve the fatigue strength. The danger can be eliminated by the removal of as little as 0.005 in (0.125 mm) but some investigation may be necessary. This can be by *polishing* or, under many circumstances, a simple *grit blasting* operation. Chemical etching is also possible but care may be required to prevent hydrogen embrittlement.

The technique of spark erosion is extremely useful where shapes are required in hardened or difficult to machine materials. It is possible to form intricate shapes on carburized or nitrided surfaces although, as stated above, care is then necessary to remove the surface film which will have poor fatigue characteristics. Other components, for example, high-productivity, low-integrity parts, can now be produced from intricate dies manufactured by spark erosion. The procedure is also useful in removing broken taps or drills, where the technique will be to erode a hole in the tap or drill and use this hole for its removal. It is possible to drill holes of considerable depth and small diameter, and also to drill holes on curved surfaces which would be extremely difficult or impossible using conventional drills.

To a limited extent spark erosion is competitive with conventional machining but in general it is an expensive technique economically justified only when conventional machining involves extremely high tooling costs or is difficult.

The technique of *electrochemical machining* (ECM) is, under some circumstances, comparable.

For removal of broken tap or dies, it is sometimes possible to use chemical attack on the hardened steel by a careful choice of acid to achieve faster attack on the hardened steel than the material of the component. This is generally, though, of very limited application.

Spark Testing
A technique for identifying pores in paint and plastic films. Described fully in Part Two **Non-destructive Testing**.

There are also some specialist spark tests applied to electrical and electronic components.

Speculum Plating
The *electrodeposition* of a copper–tin alloy, containing approximately 40 per cent tin, to give a white alloy not unlike pewter. It is used only for decorative purposes. Described fully in Part Two **Electroplating**.

Spheroidal Anneal/Spheroidization
A specialized form of *heat treatment*.

Application range HIGH CARBON STEEL

It results in the carbides present being in the form of spheres. *Cyclic annealing* can result in spheroidization. Described fully in Part Two **Heat Treatment**.

Spin Hardening
An alternative term to *induction hardening*, where the component has a round shape and is revolved during the process. The term may be applied to processes where components are revolved during *flame hardening*. With the present process it is possible to quench the component while spinning within the induction coil, or to remove components from the heating source for quenching. Described fully under *Induction Hardening* and *Flame Hardening* in Part Two **Heat Treatment**.

Spinning
A method of cold forming, applied to sheet metal which accepts a high quantity of cold work without becoming too brittle.

Application range ALL DUCTILE METALS

The procedure is that the sheet metal is revolved by, for example, being clamped to the head of a modified turning machine. The metal will then be formed by the use of tooling, which can be as simple as a wooden beam with a rounded end which presses the revolving sheet against a mandrel. This mandrel will revolve together with the sheet; it is held in the chuck of the turning machine and will have more or less the shape of the component being manufactured. By using a mandrel made up of segments which can be released on completion of the spinning operation, it is possible to produce components of intricate shape.

It will be appreciated that where material other than pure aluminium, lead, tin, or gold are involved, none of which can be work hardened at room temperature, there will be an increase in the strength of the component caused by the cold working involved. This *work hardening*, in addition to increasing the tensile strength, will reduce the ductility. Thus, care is necessary, first, in the choice of material to ensure that it is capable of accepting the necessary amount of cold work, and second, that the degree of cold work applied is not excessive in order to avoid local brittleness and cracking.

Spinning is a useful method of producing relatively simple shapes of thin-walled, hollow vessels, but because of the limited choice of materials involved this manufacturing process has comparatively few applications. It is comparable with *deep-drawing* and can produce articles of a more intricate shape with less expenditure in tooling, but probably requires a higher degree of skill in manufacture and a longer operational time.

Spot Welding

A form of *resistance welding*, where individual welds or 'spots' are formed. Described fully in Part Two **Welding**.

'Spra-bond'

A specific form of *metal-spraying*.

Application range ALL METALS

With this, the spray metal is molybdenum and a thin film of this metal is applied to the prepared surface. 'Spra-bond' is seldom or never used as a metal deposit in its own right, but is a commonly applied undercoat where high-quality spraying is required.

The technique used follows that of conventional metal-spray, where the metal is in rod form and fed into a flame of controlled combustion. The metal is 'atomized' and each individual particle is given velocity and energy by the gas stream, resulting in impingement of the particles on the metal being sprayed.

In the case of 'Spra-bond' the metal used is molybdenum, which is expensive but has the ability to resist oxidation at low temperatures. This makes it ideal as a means of preventing oxidation of a surface which has to be metal-sprayed, and the theory of 'Spra-bond' is that the surface of the readily oxidized metal, generally steel, receives a film of molybdenum which will considerably increase the bond strength of any metal subsequently deposited. This term sometimes refers to the application of the complete metal-spray process, where the molybdenum coating is followed by a second metal, for example, stainless steel or one of the cobalt alloys. This is a proprietary process, and further information can be obtained from Metco Ltd.

Spread Test
A means of assessing solderability.

Application range SOLDERED METALS

It was initially devised as a means of assessing the ability of a solder to wet a surface. It is now used sometimes under the heading of 'wettability', as a means of assessing the ability of a surface to be 'wet' by solder.

In order for the *soldering* process to be successful, it is essential that surfaces are *alloyed* with the solder. This means that the surface must be capable of alloying with the prepared surface. Depending on the use of the test, is the method of application. Where it is used to check the ability of a solder to wet a surface, then the surface must be standardized and in general will be clean copper together with a standard flux. Control is necessary for every test. The test piece is then heated to the standard condition and a known weight of solder at standard temperature is dropped on to the cleaned, fluxed surface. The source of heat is removed and after cooling the diameter of the soldered area is measured. The greater this diameter, the better the 'wettability' of the tested solder. In general with this test a minimum diameter of solder will be specified.

Where the test is used to assess other characteristics, for example the surface condition, or the ability of the 'as received' surface to be soldered, then again the flux and temperature will be specified, together with the type of solder used, and the resultant solder blob will give an indication of the 'solderability' of the surface.

Methods of assessing will be variable. In some cases this is the angle of the meniscus which has formed, in others the area which has been correctly soldered. It is also necessary visually to examine this area, which should be free of any bare patches indicating that de-wetting has occurred, or that soldering has not taken place. The visual appearance of the surface will indicate which of these two defects is present and supply information regarding the cause of the defect.

This inspection technique in general is confined to the electronics industry and is an addition to the normal control system and essential visual examination. It is commonly applied to surfaces which have been electroplated and which are to be soldered. It is also used to identify the reason for faulty soldering. Further information will be found under *Solderability Testing*.

Spring-back Test
A simple method of assessing yield strength.

Application range ALL METALS

There are two quite different reasons or meanings of this test.

First, it is used traditionally to ascertain that strip material is in the fully annealed condition. In this test a steel strip is bent and should have little or no recovery. That is, the bend, generally through 90°, should result in the material retaining a bend of 90°. This will indicate that the material is in the 'dead' soft condition. Second, the test is applied to cold-worked material and again generally restricted to steel, but there is no reason why the same test should not be applied to other strip materials. With this test the material is bent through a specified angle, normally 45°, but other angles can be used and will to some extent depend on the service requirements. On release of the pressure, the material is returned by a specified amount. This will give a measure of the *cold work* or *hardening* which has been applied, thus indicating whether or not the yield strength is high enough to comply with specification. This test is, in effect, a highly economical method of assessing with reasonable accuracy the *yield strength* of a material.

For the test to be of maximum use, some laboratory work will usually be necessary to enable the tester to equate the spring-back to the yield point of the material. This is more useful than *hardness testing*, which gives only an approximation of the *ultimate tensile strength*, whereas the present method indicates the approximate yield strength, which is normally obtainable only by relatively expensive, destructive *tensile testing*.

Sputtering

A method of applying a very thin invisible surface coating.

Application range ALL METALS

Sputtering uses a high vacuum, inside which the metal is made the cathode and the component being treated is the anode. The vacuum is then released with an ionizable gas and high voltage passed between the anode and the cathode. Voltages of over 2000 are commonly used, which result in the bombardment of the anode with the metal from the cathode. This is carried out at low pressure and, is in effect, a vacuum treatment. By controlling the voltage and other parameters, it is possible to achieve extremely thin deposits, in the region of one molecule thick on the finished components.

Where a thin optical film is involved, the technique uses an ion discharge of the metal which is then deposited on the surface of the component. The object is either to produce an electrically conductive film on the material, which is itself an insulator, or to produce a film which has specific physical properties, generally of the optical type, which differ from those of the material coated.

One use of the first type is the production of an electrically conductive film for the heating of windscreens, where this thin film is used as a high-resistance conductor. This effectively demists and defrosts the glass by the passage of an electrical current, without interfering with the visibility through the glass. This technique can also be used to produce the electrical conductive surface necessary for *electroplating* of non-conductors. However, with the advent of *electroless plating* of plastic and glass, this sputtering technique is not normally used, since alternative methods are technically efficient and much more economical.

The production of a thin film to alter the physical characteristics of a surface is now the most common application of sputtering. High-grade optical lenses often have specific metals sputtered on to one surface in order to give a particular optical characteristic.

Sputtering is also used in the electronics industry as a means of coating non-conductive surfaces with extremely thin films of conductive metals. This conductive film is then either treated separately or used as a means of connecting leads, etc. to electronic devices.

This process has some comparison with *vacuum deposition*, but this will not give comparable control or the extremely thin films.

Stabilizing

A term used for any form of *stress-relieving* operation which results in a stable component. In general, however, it is accepted that stabilizing is the term applied to components which will be subsequently nitrided. Described fully with other forms of thermal treatment in Part Two **Heat Treatment**.

Stabilizing Anneal

A term applied to any form of *stress-relieving* operation designed to give a stable component.

Application range ALL METALS

It is generally used to denote the treatment applied to austenitic stainless steels, to overcome the corrosion problems caused by carbide precipitation. Described fully in Part Two **Heat Treatment**.

Stabilizing Treatment

See *Stabilizing*.

'Stannostar'

A proprietary process using the acid tin electrodeposit to produce a bright deposit of tin. This has an attractive sheen which gives excellent corrosion resistance and is non-toxic as is the normal hot dip or electrodeposit.

The bright deposit has a lower solderability than a good-quality dull deposit, but when all the parameters are controlled the solderability will be consistent and can retain this quality over a considerable time scale. Further information on 'Stannostar' can be obtained from Oxy Metal Industries (GB) Ltd.

Steam Blueing

A *corrosion protection* treatment.

Application range STEEL

It is, in fact, a variation of the '*black oxide*' treatment and is used to produce a blue surface colour.

For treating, the components must have a good surface finish which is homogeneous, and they must be completely clean. The clean, dry components are inserted into a tempering-type furnace, which is then sealed and the components taken to a temperature in the range 350°C–370°C. Before the components reach temperature, steam is injected into the furnace and the components are allowed to come to temperature in this steam atmosphere. They are then held at temperature for approximately 2 h. The result is that the temper blue colour is intensified and made thicker and more adherent by the steam atmosphere.

While this procedure will give a homogeneous, attractive looking component, corrosion resistance will be of a relatively low standard. This is somewhat less than that achieved by the conventional '*black oxide*' treatment but better than that produced by *heat treatment*. The technique is often used in conjunction with the *tempering* temperature and thus can be a highly economical method of producing articles having a degree of corrosion resistance.

There is a technique in which components such as nuts and bolts are tempered in an air atmosphere then quenched in water containing a percentage of oil as an emulsion. This will give a black finish with comparable corrosion resistance to *temper blueing* without the necessity to have specialized equipment.

Steelascope Testing

A method of identifying methods and alloys in a non-destructive manner. More details of this will be found under the heading *Metal-sorting*. The Steelascope uses the spark discharge system technique which is covered in that section.

Step Anneal

An *annealing* technique which is, technically, a form of *stress-releasing*. With this, the parts are heated in stages or 'steps' to ensure that no stresses are imposed by the heating cycle. Described fully in Part Two **Heat Treatment**.

Step Quenching

An alternative name for *martempering* or *austempering*, where steel components are heated to above their upper critical temperature and, instead of being quenched in the normal manner in water, oil or air to bring them directly to room temperature, are delay-quenched by transferring to a second furnace or molten salt baths. Delay-quenching techniques are described fully in Part Two **Heat Treatment**.

Stitch Weld

A form of *resistance welding* which, for all practical purposes, is the same as *spot welding*. Stitch welding is sometimes applied to the 'breaking spot' (see *Spot Welding*), but in general refers to the process in which a series of spot welds are used to 'stitch' or hold together sheet-metal components during assembly. Described fully in Part Two **Welding**.

Stop-off

The process used during *electroplating, painting*, and some forms of *heat treatment* in which areas not required to be treated are protected.

Application range ALL METALS

With the three processes, the common methods of stopping-off are as follows:

1. With *electroplating*, stopping-off uses either waxes, special lacquers, tapes or a combination of these methods. There is a common requirement in that any method of stopping-off must be capable of withstanding the plating process. As *cleaning* and *etching* prior to plating are specifically designed to remove dirt and adherent compounds, so the means used here must be capable of withstanding these processes, and this can give rise to problems. Waxes were at one time the most common method of stopping-off but, with improved lacquers, these are now less commonly used. They still find certain applications, particularly with heavy chromium plating, where lacquers and tapes would tend to be loosened by the very aggressive liquids used at plating together with the high temperature of 50°C and the scouring action of the gases released. The adhesion of the waxes can be varied and controlled by the temperature of the component being waxed. If this is raised to the same temperature as the hot, liquid wax, then excellent adhesion will be obtained. If the cold component is dipped rapidly into the hot wax, allowing a thin layer to solidify on the cold surface, then adhesion will be very limited.

Skilled operators in plating shops make use of this to obtain the necessary adhesion for the job being plated, by varying the time of immersion to achieve the correct temperature of the part for the desired adhesion. The wax temperatures can vary from melting points in the region of 30°C–40°C to over 100°C, with in general the higher-melting-point waxes being used for the more aggressive conditions. It is now known that some waxes present a health risk.

Lacquers are now available based on a number of formulae which give excellent adhesion, are much easier to apply and remove as special solvents are available for this and, in addition, they can be applied with paint-brushes or spraying.

Plastic tapes are available which can be used for stopping-off. These tapes are generally of the PVC type with special adhesives which will withstand the solutions at plating. Adhesive tapes may be used, which are made conductive by having a film of lead or in some cases aluminium or silver on the outer surface. With these, it will be seen that plating can be carried on top of the tapes and these are now employed for important components where, previously, the edge of the stop-off area tended to build up with the plated deposit caused by high-current concentration. This gave a local area of plating of a lower standard than the remainder. With these conductive tapes, it is now possible to eliminate this local area as the run out of the plating will not be on the component surface and the lower standard of plating deposit is removed along with the tape.

It is also possible to prevent a plating deposit by the use of jigs which physically protect the surface.

Finally, by controlling the location of the anodes and the current distribution to the components, it is possible to control the location and thickness of deposit. Sophisticated forms of this technique are now being developed.

2. Stop-off in *stove enamelling* or *painting* is generally confined to the use of tape or blanks. These must be capable of withstanding the solvents used in the paint, and also the paint itself and the temperature at which any stoving subsequent to painting is carried out. It is essential that stoving does not result in the adhesive being baked on to the surface, making it very difficult or impossible to remove the stop-off medium after painting. Tapes are available capable of stripping clean without deterioration of the adhesives.

Special masks or blanks are also used with high-production painting; these can be cardboard, sheet metal, etc. Under some circumstances they are removed prior to stoving.

Finally, it is sometimes possible to use grease, oil or special compounds to prevent *adhesion* of the paint. These are seldom permitted where a high standard of finish is required as they can readily affect the paint medium.

3. With *heat treatment*, the stop-off process is confined to surface hardening such as *carburizing* and *nitriding*. The process obviously must be capable of withstanding the temperature at which the heat treatment will be carried out. It is seldom necessary to consider removal of the stop-off media as invariably there will be a subsequent *blasting* operation. At one time *copper plating* was the most common method of stopping-off for *carburizing* and *nitriding*. This is still used on a limited scale, but for economic reasons it is now more common that copper, tin or bronze paint is used. These specially formulated paints contain the metal in finely powdered form and are applied by any normal *painting* technique. However, it will be appreciated that considerable care is necessary in application: any missed out areas will result in hard spots which

can cause failure in service. If faulty application results in adhesion failure, this can result in hard areas.

With simple components and hollow shapes, stopping-off can be achieved by metal jigs. These will be tight fitting to prevent the active gases contacting the surfaces being protected.

Stove Enamelling
A general term applied to paints which are cured by heating. Described fully in Part Two **Painting**.

Straightening
This is the process whereby any component which has distorted is recovered by mechanical means. The process is also known as *stretching*.

Application range ALL METALS

Use is made of mechanical presses, either to flatten or to bend, and these can be mounted on the normal type of mechanical press. Operators make use of hand-operated hammers, and of vices to grip and bend. In general the term straightening covers any process in which distortion is removed. Considerable skill is required and for high-integrity components the control applied must be stringent. In order to straighten a component, it is necessary to bend it farther in the reverse direction of the distortion. If over-bending occurs, then the operator must reverse the process and over-bend again but in the direction of the original distortion. It will be appreciated that this repeated bending can result in cracking and if the component is either carburized or through hardened, where ductility is relatively low, the danger of surface cracking will be high. With components where cracking will be a serious defect, it is common to control the amount of acceptable distortion, and by the use of stops or other means to restrict the degree of over-bending. It is essential that important components are *crack tested* after a straightening operation.

It must be realized, in addition, that the straightening operation will of necessity involve the local *work hardening* of the component. Some stresses, then, will be 'built-in' and this can result in distortion during service. It is advisable to stress-relieve straightened components at a temperature above that at which they will be used in service, and then re-inspect for further distortion. A further stress-release may be required if re-straightening is carried out.

It has been shown that the necessity to straighten can be reduced or eliminated by careful control during heat treating. This control requires the co-operation of planning, design, quality control and heat treatment departments and laboratory advice is advisable. With this combined experience, it is often found that heat treatment can be controlled to eliminate distortion in the most complicated components. This may require that *stress-releasing* be carried out prior to heat treating, but generally this will be more economical than a straightening operation, and is certainly less prone to cause cracking.

Strain or Stress Ageing

A method which can be applied to a limited number of alloy materials, resulting in an increase in mechanical properties. The technique is to stretch the material while holding it at a specified temperature.

While this improves the mechanical properties by a marked degree, it is doubtful if they are better than those obtained by more conventional *solution treating* and *ageing*, which will normally be much cheaper. In the case of certain alloys this process can be used to produce specific electrical or magnetic properties.

Mechanical properties can also be improved in certain alloys by applying *work hardening* between the *solution treating* and *ageing* operations. This is not the same as stress ageing.

Strain Gauging

A method of measuring the strain applied to components. There are various techniques. Described fully in Part Two **Non-destructive Testing**.

Strain Hardening

An alternative term to *work hardening*.

Application range ALL METALS

This is achieved by rolling, hammering or bending to apply pressure. Strain is imposed on the grain structure, causing increased tensile and yield stress. At the same time there will be a reduction in ductility. This is caused by the fact that *work hardening* results in deformation of the grain structure; as the amount of work applied is increased so resistance to further deformation increases, resulting in a reduction in ductility.

Strain hardening caused by work is fully removed by recrystallization. The temperature at which this occurs varies between different metals and alloys, and certain materials such as lead, silver and pure aluminium recrystallize at room temperature. These, therefore, cannot be strain hardened. Other materials, too, have a grain structure incapable of deforming by any appreciable amount and, thus, cannot be strain hardened as they become brittle and shatter. It will be seen that ability to strain harden is a function of the material itself, and advice should be obtained, or trials carried out, prior to applying measurable amounts of cold work to any material. It is quite common to temper after work hardening. This is analogous to the temper carried out after thermal hardening.

Strauss Test
A destructive inspection technique which is used to assess the potential to the defect known as *weld decay*.

Application range AUSTENITIC STAINLESS STEEL

Weld decay is the precipitation at the grain boundaries of carbides resulting in the appearance of a second phase, which seriously reduces locally the corrosion resistance of austenitic stainless steel.

The Strauss test takes the suspect steel and initially 'sensitizes' this by heating the test piece at 650°C for 30 min which is the critical temperature and time for precipitation of carbides. The test piece is then boiled for 72 h in a solution containing copper sulphate; the piece is then washed and bent. If the test piece can be successfully bent through 180°, then the material is not susceptible to *weld decay*. However, if fracture or cracking occurs, *weld decay* will occur on heating to within the critical range.

This test can be modified not only to show that a material is susceptible to *weld decay* but also to quantify the degree of susceptibility. Three sensitized test pieces are employed, together with a fourth piece in the 'as received' condition. All four test pieces are immersed in copper sulphate solution and one of the sensitized pieces is removed after 24 h and bent. A second is removed after 48 h and bent, and the remaining two (one of which will be the non-sensitized piece) are taken out after 72 h and bent. If all four test pieces bend satisfactorily, susceptibility to weld decay can be ruled out. If the test piece removed after 24 h cracks on bending, then the material is extremely susceptible to *weld decay*. If the two test pieces at 24 h and 48 h do not crack, but the 72 h test piece cracks and the 'as received' test piece does not crack, then this again indicates low susceptibility to *weld decay*. It will be seen that the information gained in this manner can give considerable help regarding the possible use of the material being tested. The test solution contains copper sulphate 15 g/l in a 5 per cent sulphuric acid solution.

Stress-equalization Anneal
A specific form of *stress-relieving* applied to nickel alloys, which can result in some *ageing* occurring. Described fully in Part Two **Heat Treatment**.

Stress-relieving/Stress-releasing
Terms covering the wide range of treatments applied to reduce or eliminate stress. Described fully in Part Two **Heat Treatment**.

Stress-rupture Test

A long-term destructive test to identify the load to cause distortion or failure.

Application range ALL METALS

The test piece can be either in the form of the component to be tested, or a normal tensile test piece. The test is carried out at a fixed, generally room temperature, but can be at elevated temperatures if these represent the conditions in which the component is to be used. The load is applied and maintained until such time as a specified permanent deformation, of either 0.1 or 0.5 per cent is obtained. The load can be increased or decreased and the test repeated for sufficient points to be plotted on a curve and an estimate of the service life of the component or material obtained.

This test can be used to identify any weakness in the designed shape of a component provided the service loading conditions are accurately known.

Stretcher Straining

A method of straightening simple shaped components or, more commonly, bar and sheet material.

Application range ALL METALS

The load is applied as a straight tensile pull until the yield point is reached, when a permanent small deformation will be obtained. This, in addition to straightening any kinks or minor distortions, will result in a slight increase in the mechanical properties owing to the *work hardening* applied. This technique should be used only as part of a well-controlled process, as it will result in some reduction in the ductility of the material. Commonly this follows an *annealing* operation and so the material will no longer be in the *fully softened* condition. Many extrusions are straightened by this technique.

Stretching

An alternative term for *straightening* when used in general engineering. It is obvious that the word 'stretch' could be applied to describe any technique where a tensile strain is used to increase the length of the material.

Stromeyer Test

A destructive test, where the fatigue strength of the material or component is ascertained. With the Stromeyer test, the specimen is rotated and is stressed by means of a crank which gives the specimen a bending movement. Described fully under *Fatigue Test* in Part Two **Mechanical Testing**.

Stud Welding
A form of *resistance welding*, where the parts being welded are studs of some type which are joined to the major components. Special equipment is used and the studs must also be of special design.

Application range ALL METALS WHICH CAN BE WELDED

The welding equipment holds the stud in such a manner that the weld section protrudes to a certain extent. This protrusion is the area of the stud which is specially designed and will generally have some form of chamfer, such that the contact area will have a lower cross-section, thus higher electrical resistance than the stud.

The stud is brought into contact with the location to which it will be welded and an electric current is passed between the tip of the stud and the major component. The equipment will be designed so that the pressure applied to the stud will be at a constant level. As the current passes through the relatively small contact area caused by the chamfer, *resistance heating* of the stud end to the local area of the component occurs. After a predetermined time, to ensure that the heating has resulted in the chamfer of the studs being raised to the forge temperature, the electric current is stopped and by the use of a spring load upset pressure is applied to the stud. A good *forge weld* will be achieved under satisfactory conditions.

The quality of the weld will, of course, depend on the correct equipment being used, correct settings, and correct design of the stud. Any variations of a significant degree will result either in the stud being heated to an excessive temperature, when some cast structure is present and the resultant weld will be weaker than desired; or if insufficient current is applied, the materials will be at too low a temperature and what is known as a 'stuck weld' will be produced. That is, oxide will be present and an insufficient merge between the stud and the major component.

Specific conditions will vary for different sizes of stud, and may also vary because of the geometry of the major components. It will be appreciated that all surfaces to be welded must be free of oxide and other contamination. Stud welding should not be used where high-integrity welds are essential, because of loading on the studs, unless extremely careful control is possible.

Flash butt welding, where possible, will be an alternative; *friction*, or *inertia welding*, is now being used where high-quality welds are required.

Subcritical Anneal
The *softening* or *annealing* process most commonly applied at present to engineering steels. Described fully in Part Two **Heat Treatment**.

Submerged Arc Welding
A form of *electric arc welding* for relatively simple, long-run welds. With this,

high-productivity welds are possible and, with modern equipment, full penetration welds on thick steel plate can be accomplished to give joins of a high technical standard. Described fully in Part Two **Welding**.

Subzero Treatment

A form of *heat treatment* used on complex alloy steels to ensure that all austenite is transformed to martensite, thus full *hardening* is achieved.

'Sulphinuz'

A proprietary treatment which results in an improvement in the surface friction characteristics.

Application range STEEL

The treatment is carried out in a bath of molten salt, at a temperature in the region of 550°C, the surface of the steel being impregnated with high sulphur compounds. These give a slight increase in overall dimensions and there is also some diffusion of the treatment into the steel itself. The resultant film, however, will never be greater than 0.005 in (0.1 mm).

The colour of the components after treatment will be a rather mottled dark grey or black. It is important to realize that because of the relatively thin surface thickness affected, only very light *polishing* can be carried out. If components are polished to bright metal, then there is a high probability that the effective layer will have been removed. The effectiveness of the treatment is not measured in terms of hardness, but by the high sulphur compounds which exist on the surface. It will also be realized that, since the temperature at which the system is achieved lies in the region of 550°C, it cannot be used on steels which are tempered below that range, since the process will then result in a reduction in hardness. The process is, therefore, not suitable for components which have been surface hardened, and the *hardening* process in itself will invariably give better friction characteristics than this method. The 'Sulphinuz' system has been found to have particular application with stainless steels, and steel where the friction loads are relatively low.

The theory of galling which results in severe friction-wear is that the friction load reciprocates, and at the end of each stroke there will be a tendency for *cold welding* to take place. When the next cycle occurs, this weld is fractured and thus minute holes or crevices appear which gradually increase to produce the typical friction failure. With 'Sulphinuz', a higher load will be required to result in the cold weld, because of the better friction characteristics of the high sulphur surface. It will, however, be understood that it is possible to load the surface to cause failure. The 'Sulphinuz' process comes somewhere better than *phosphating* but will be less effective than '*Tufftriding*'. Further information on the system can be obtained from ICI Ltd, or British Heat Treatments Ltd.

Sulphuric Acid Anodize
A general term covering the *anodizing process* which uses sulphuric acid as the electrolyte. It is by far the most common method employed at present.

Sulphur Printing
An inspection technique, used at the receiving stage or during the manufacturing process of steel components.

Application range STEEL

The reason for the test is to ascertain the standard of cleanliness of the steel. The assumption is that the contamination in steel can be related to sulphur as sulphides in the steel itself. The component or material to be tested is cut to produce a longitudinal cross-section. This is then polished to a reasonably high standard of finish and flatness.

A piece of ordinary bromide photographic paper, which has been well moistened with dilute sulphuric acid, is then placed in contact with the prepared surface. Reaction between the acid and any sulphides produces hydrogen sulphide gas, which locally darkens the silver content of the photographic paper. This identifies the presence of sulphides in the steel, and quantifies the size of individual particles, showing their distribution and the amount present.

This technique is used on a lesser scale with modern steels which tend to have fewer impurities present. It can be used still to show the distribution of sulphides in free machining steel, and as the distribution of these sulphides can affect 'machinability', this is an important factor. It should be realized that only high sulphur-free machining steel can be controlled. Other types, such as lead, will not react. Other techniques used for this test include chemical analysis and micro-examination. Both require laboratory equipment and skilled personnel, whereas the sulphur print may be used in relatively unsophisticated conditions.

'Super Gleamax'
A proprietary name given to a bright *nickel plating* deposit. Further details may be obtained from W. Canning and Co. Ltd. See also *Nickel Plating* in Part Two **Electroplating**.

Superficial Hardness Test
A term applied to any of the testing techniques where a light load is used to measure the hardness of the surface only. These are of most use when inspecting *case hardened* components. The Superficial Rockwell hardness test is probably the most common. This uses standard Rockwell technique with a light load. The latter are described fully in Part Two **Hardness Testing**.

Supersonic Testing

An alternative term for *ultrasonic testing*.

With this an energy wave is fed to the surface of the component and the reflection of this is measured. There will always be a reflection from the rear of the component but any defect between the surface and the rear will show. Described fully in Part Two **Non-destructive Testing** under *Ultrasonic Flaw Detection*.

Surface Hardening

Techniques used to locally harden the surface.

Application range ALL METALS

Like many general terms, surface hardening is used loosely and can mean specific processes under many circumstances. It will, however, cover the processes *nitriding, carburizing, cyanide hardening, carbonitriding, induction hardening* and *flame hardening*. In addition, it can refer to *chromium plating* or the deposition of hard surfacing material, and here the term may be used in connection with metals other than steel. It is advisable, then, that the term is used specifically, together with an indication of depth of hardening or thickness of deposit.

Sursulf

This is a salt bath nitro-carburizing process which gives a hard wear resistant surface.

Application range STEEL

The process is carried out in molten salt. The components are thoroughly cleaned, dried then pre-heated before immersion in the nitro-carburizing salt.

Components can be quenched directly from the salt or annealed by slow cooling followed by controlled hardening and tempering.

The process results in a diffused layer of nitrogen and carbon which, because of its compressive nature, improves the fatigue strength of the component and has a hardness above 800 DPN, and it is claimed as high as 1400 DPN depending on the substrate material.

The process is technically superior to cyanide hardening and is comparable to nitriding. Alternative procedures to obtain similar results would be boriding, Toyota diffusion process, or techniques such as chromium plating, plasma diffusion, plasma coating or certain types of metal spraying.

Further information from Degussa Ltd.

Swageing

This is a *forging* operation, where movement of the metal is produced by the use of tooling.

Application range ALL DUCTILE METALS

The process is used for a variety of reasons, including fixing of assemblies. A shaft can be inserted through a second component with a protrusion, and this is then swaged or cold forged in a form of riveting. Riveting itself is, in fact, a form of swageing. Another application would be where a hollow component is formed by drilling a parallel hole and one end of the hole then being closed by a forge operation. Swageing is employed as an alternative to *welding* as a means of joining, or to the use of mechanical plugs such as threaded end pieces.

Sweating

An alternative name for the process of *soldering*. It is, on occasion, used to describe the heating applied to the *shrink fitting* operation. It gets its name because of the hydrocarbon gas flame, used as the source of heat, which results in the formation of droplets of moisture on the surface of a cold metal component. These moisture drops are the products of combustion which will invariably include water and carbon dioxide. The water in the flame will be in the form of water vapour or steam and condenses when it strikes the cold surface of the heated component. As the component heats, the water vapour no longer condenses and remains as a gas. This is a loose term which does not apply to any specific operation, and as such should be used with caution.

Swilling

A process which is employed only in conjunction with other processes in a series of operations involving the washing of components following processing.

Application range ALL METALS

During *electroplating* and other surface treatments the components are covered by the liquid in which they are treated and, on removal from the solution, a film remains in contact with the surface of the component. Swilling is designed to remove this surface layer, replacing it with water. Considerable research has been carried out into efficient swilling, and it is now accepted that the use of a water dip alone is inefficient as it will seldom result in a film free of contaminating material unless a series of running swill baths is used.

It is much more effective to make use of a fine spray, where the contaminant film drains from the surface to be replaced by fresh sprayed water. However, this technique is only possible with components of a type which permit free draining and can be subjected to spraying on all surfaces. If used in conjunction with automatic on/off taps, which spray only when the components are in position to be washed, this technique can result in extremely large savings of water to the order of 75–90 per cent compared with immersion swilling for the same standard of cleanliness.

There is also the technique of steam swilling, where the components are placed under an inverted hood containing steam. This condenses on the surface and runs off, taking with it the contaminated surface film. Again there is the restriction regarding the shape of the components but a highly efficient swill is possible with little water, since it is possible to boil a comparatively small volume of water to produce the necessary volume of water vapour. Provided the contaminant does not contain excessive volatile matter, an adequate swill is possible. This method has the added advantage that excellent recovery of the contaminating material is possible without excessive dilution.

In the UK water costs are such that it is seldom economically justifiable to use these sophisticated swilling techniques. Nevertheless, with the advent of effluent control, it is often good policy to reduce the flow of swill water and alternative methods can then be considered. Contraflow swilling is a first step in the saving of water using conventional methods. With this, the swill tank is divided into two compartments, the second being fed by the previous compartment. Fresh, clean water is fed into the second compartment in the series and this runs towards the contaminating source. The component is first swilled in the most contaminated water; this is the water which is fed to the drain. On removal from this first swill tank, the bulk of the contaminant will be removed or diluted. The relatively clean component is then immersed in the second swill, and this will be considerably cleaner than the first. In general two contraflow tanks or compartments will be found to be sufficient, but there is no reason why a series of three or more compartments should not be used where components require a high standard of cleanliness, or where a contaminant is particularly difficult to remove, or the geometry of the components makes the operation difficult.

Where high standards are necessary, components are often finally swilled in distilled or demineralized water. This practice is common in the electronics industry and is becoming more common where appearance plays an essential part, since in many areas the use of tap water will result in some staining on the dried article.

It will be appreciated that the standard of swilling varies considerably for a variety of reasons. In a multi-process operation for example, components removed from an acid solution should be thoroughly swilled before subsequently being immersed in an alkaline solution; the reverse situation is also relevant. The reason for this is that there will be a reaction between acid and alkali often resulting in the formation of a solid on the surface of a component. There are also considerable differences in the 'swillability' of different solutions, but in general alkaline solutions are easier to remove than acid solutions.

If correct planning which uses all available knowledge is applied prior to the installations of swills, then considerable saving in water, and often money, can be achieved. With the present water costs, it will seldom or never be possible to save money on an existing installation by spending capital to reduce the amount of water used. However, with tighter effluent controls, and the high probability that sewage may be charged by volume, there may soon be more scope for considering water-saving schemes. By reducing water usage, effluent flow will be reduced.

Taber Abraser
An inspection device which is used to indicate the abrasion resistance of a surface.

Application range ALL METALS

A rotating, circular abrasive disc contacts the surface being tested, and the load applied is controlled so that the time taken to cause a predetermined degree of abrasion on the test surface can be measured. This test is designed to measure abrasive resistance only, and care must therefore be taken in assessing the results.

Additional tests, use reciprocating action, allow the contacting materials to be altered to stimulate different practical alternatives. These can be lubricated and the wear products removed as they are produced, if desired. Further information will be found under *Abrasive Testing*.

Tack Weld
A *welding* technique which can be applied to a variety of weld types. Most commonly it makes use of either *electrical-resistance spot* or *electric arc welding*. A limited amount of *oxyacetylene welding* is also used.

Application range ALL METALS WELDED

Tack welding is the technique of producing a local, relatively weak weld which holds, or 'tacks', the components to be welded in position. The assembled components can then be transported to a welding jig for the final joining, using any of a number of techniques, or it can be inspected in position to ensure that the correct location of the various components has been achieved.

The process known as the 'breaking tack weld' is a weld specifically designed to have poor fusion, so that if in subsequent inspection the components are found not to be correctly located, they can be broken down to their constituent parts, re-located and tack welded a second time without causing damage.

With high-quality components a tack weld will not remain on the final assembly. It is not acceptable that tack welds are 'over-welded' but they must always be ground out before completing the final weld. There is a serious danger that minute cracks will always be present with tack welds.

The term tack welding is sometimes used to describe a series of short discontinuous welds holding components together during their service life. These discontinuous welds will generally be produced by the *electric arc* or *oxyacetylene* processes. Tack welds are also used to locate temporary fitments for a variety of reasons. These fitments are then removed, either mechanically or by burning.

Care must be exercised in tack welding as it will be commonly found that the mass of metal on the component is such that with the very local, intense heat involved, local hardening results. This is caused by the mass of cold metal adjacent to the weld acting as a quench, resulting in a brittle zone around the weld. With high-integrity components, tack welding should be carried out using the same pre- and post-heat controls, and heat input as for the major weld.

Where any shock or heavy stress or bending is applied to a tack weld, the steel should be stress released to ensure removal of this brittle zone. Tack welding must always be carried out in the knowledge that a brittle zone, or cracking, can exist, and thus where lifting eyes, etc. are tack welded to large masses, some pre- or post-heating must be applied. It is now common that weld procedures are produced and often tested for tack welds on important components.

Taylor-White Process

A form of *hardening* applied only to high-speed steels of medium-high tungsten-chromium type. It is a form of *martempering*, where the steel is heated to approx. 1400°C and then quenched in liquid at a temperature of approx. 800°C. It is then subsequently quenched, and this is followed by *tempering*. (See also *Austempering*.)

'Techrotherm Rokos' Process

A *fusion welding* technique, using a carbon electrode in contact with the welded edge, and immersed in the weld pool. The technique requires gas preheating of the area local to the weld, and electric current to complete the fusion. *Submerged arc*, and modern heavy-duty *electric arc welding*, would appear to have replaced this technique.

Temper Blueing

A process which results in the material attaining a bluish colour, which if the conditions are correct will give a degree of corrosion resistance.

Application range STEEL

Temper blueing is carried out by heating the clean steel in an oxidizing atmosphere at approx. 300°C–400°C. At this temperature the surface oxide formed will be blue in colour. The process requires that all surfaces are thoroughly cleaned and free of any existing oxide and presented to the oxidizing atmosphere. This means that small components will probably require to be mechanically moved during the process to ensure that all surfaces are oxidized. The oxide achieved is very thin and will have little or no corrosion resistance if presented to damp atmosphere.

Temper blueing is, however, commonly applied to materials used under indoor conditions, such as tooling, where there is contact with oil and the general atmosphere is low in corrosion potential. In order to obtain the same colour with every batch of components, standardized conditions are obviously necessary. Temper blueing is very often obtained during the normal *heat treatment* cycle when parts are *tempered* following the *hardening* operation. The hardness required will decide the choice of *tempering* temperature, and if this coincides with the temperatures used to achieve the blue colour, then temper blueing results as part of the *heat treatment* cycle.

Temper blueing will not have comparable corrosion resistance with '*black oxide*' but belongs to the same family of treatments. With '*black oxide*', the components are immersed in an oxidizing salt to achieve a thicker, more adherent surface. The *steam blueing* process uses a steam atmosphere for *tempering* of the parts. The result is a treatment better than temper blueing but inferior to '*black oxide*'. The technique of quenching steel parts in soluble oil from the *tempering* temperature gives a more homogeneous colour, and is comparable to temper blueing.

Temper Hardening

An alternative name for *precipitation hardening* or *ageing*. The process of precipitation or ageing was known before the reasons for its occurring were fully understood. This has resulted in a number of names being given to the process. Temper hardening is derived from the fact that the precipitation treatment is carried out at a temperature which would normally result in the *tempering* of steel, but in this instance the result is in an increase in hardness to the components in question. Described fully under *Precipitation Hardening* in Part Two **Heat Treatment**.

Temper Rolling

A technique used in the production of sheet metal.

Application range ALL METALS

The term is applied to the final rolling operation following *full annealing* which will produce a small amount of *work hardening* in the material. This work can generally be achieved without any appreciable reduction of ductility but does give a slight amount of stiffness to the material. The final sizing pass commonly given to sheet metal to produce a satisfactory finish and hold the final dimensions of the strip thickness is, in fact a form of temper rolling.

Tempering
The *heat treatment* which follows hardening.
Tempering must always follow some method of *hardening*, and results in an increase in ductility, generally with a decrease in hardness. Described fully in Part Two **Heat Treatment**.

Tensile Test
A destructive inspection test which supplies comprehensive information on the ultimate tensile strength, *yield* or *proof strength* and *ductility*. Described fully in Part Two **Mechanical Testing**.

'TFD'
An alternative name for *Toyota diffusion process*.
Further information from Degussa Ltd.

Thermal Treatment
An alternative term for *heat treatment*.

Thermit Welding
A very specialized form of *fusion welding*, where the heat source is obtained from the chemical reaction between powdered aluminium and iron oxide.
    Described fully in Part Two **Welding**.

Thickness Testing
An inspection method used to estimate the coating applied as an *electrodeposit*, by *painting* techniques, or by other deposit methods. Described fully under *Thickness Testing* in Part Two **Non-destructive Testing**.

Thuriting
A *softening* process.

Application range MILD STEEL

It is a form of *box annealing* where the sheet metal is placed in a box which is rapidly heated to above 900°C then cooled at a fairly rapid rate. The box and sheet metal is then heated to about 650°C and held for some time. This, then, to some extent is a form of *hardening* and *tempering*, where sheet metal is produced in a soft condition with controlled grain size. The rate of cooling from the high temperature will partially control the amount of *deep-drawing* which can be carried out on the sheet metal after Thuriting.

Hardening, Tempering and *Annealing* are described fully under these headings in Part Two **Heat Treatment**.

Tiduran

This is the trade name given to a salt-bath process used for titanium and its alloys where an improvement in wear properties is achieved on the surface without, it is claimed, any reduction in fatigue strength.

Treatment is carried out at approximately 800°C.

Further information can be obtained from Degussa Ltd.

TIG Welding

A method of *welding* using an inactive electrode and the *inert-gas shielded metal arc* technique. Described fully in Part Two **Welding**.

Tin Plating

One of the few non-toxic metals plated for corrosion resistance in applications in food processing. It is also commonly used for its 'solderability'. Described fully in Part Two **Electroplating**.

Tin–Copper Plating

Electroplated on electronic components to stop 'whisker growth'. The alloy has approximately 2 per cent copper, which does not affect 'solderability'. Described fully in Part Two **Electroplating**.

Tin–Lead Plating

Deposited to give a surface which can be readily soldered. The aim is to deposit an alloy with 60 per cent tin, 40 per cent lead which is even, free of 'treeing' and has no occluded material. This is necessary to ensure a reasonable shelf life and good 'solderability'. Described fully under *Solder Plating* in Part Two **Electroplating**.

Tin–Nickel Plating

Used on printed circuit boards for the electronics industry and also with some application for decorative purposes. This is a modern deposit which requires careful control to ensure 'solderability', etch resistance and good appearance. Some details are given in Part Two **Electroplating**, but users are advised to contact the Tin Research Institute for operational details.

Tin–Zinc Plating
Deposited for corrosion resistance or 'solderability', or where a combination of both is required. This is now seldom used as the cost of tin favours other methods. Described fully in Part Two **Electroplating**.

Tinning
The term applied to the specific process of dipping or wiping components with tin or tin alloy. It is occasionally, but incorrectly, used for *tin plating* by electrodeposition.

Tinning by the hot process involves *cleaning* the surface to be treated by any conventional means, removing any adherent oxide and applying a flux and subsequently dipping into the molten metal or applying heat by some other form such as a flame or conduction from a standard soldering bit. In the past tinning meant that the surface was, in fact, coated with a thin deposit of tin metal, but today it will quite often mean that the surface is coated with the alloy of tin and lead known as solder, and it is important that as soon as possible after the operation a subsequent *soldering*, or white metalling, is carried out. Tinning is now frequently carried out by the electrodeposition of a thin coating of pure tin metal, and for this reason the term is less specifically defined.

Tocco Process
An alternative term for *induction hardening* which at one time was reserved for the use of this process on crankshafts. However, it is now a general term. Described fully under *Induction Hardening* in Part Two **Heat Treatment**.

Tool Weld Process
The term applied to the weld deposition of hardened material which can be used for cutting. The use of Stellite-type, high-cobalt materials is an example of this technique.

Torch Brazing
The term used for the *brazing* process where the heat is supplied by gas flame. In general torch brazing is confined to the lower-melting-point alloys and can be used for aluminium in addition to copper alloy braze fillers. In skilled hands torch brazing results in high-quality joins as it is possible to use the torch to produce an atmosphere which is slightly reducing and thus prevents the formation of oxide. It will not, however, be possible to use this method without fluxes.

Torch Hardening
A term sometimes applied to the process described as *flame hardening*. Described fully under this heading in Part Two **Heat Treatment**.

Touch Welding
A term used for *tack welding*, and applied to *electric arc welding*, with or without shielded arc and also *oxyacetylene* and *resistance welding*. With touch or tack welding, the weld is very local and often designed to be readily broken after it has served its purpose. It can also be used to locate components prior to completion of the full weld.

Toyota Diffusion Process (TFD)
This is a technique which diffuses into the metal surface, carbides of vanadium, niobium or chromium.

Application range STEEL

This is a salt bath process where the components being treated are immersed after pre-heating in a specific salt depending whether the diffused layer is to be vanadium, niobium or chromium carbide.

Following the salt spray immersion process the components can either be oil quenched, water quenched, salt quenched or annealed and then re-hardened if necessary.

The process results in an increase in dimensions and a tightly bonded diffused surface layer of carbide.

This will invariably result in an increase in surface hardness. The adhesion is such that the component can be quenched and tempered to give a desirable, tough but hard core to support the very hard surface layer.

This layer, it is claimed, can be as hard as 3,000 DPN.

The corrosion resistance can also be improved where chromium carbides are involved.

The applications of this process are where punches or cutting tools are involved to improve the tool life. This will also apply to components such as dies for moulding, roller chains and any components which are not subjected to impact.

Toyota difusion process (TFD) is competing with processes such as chromium plating, boriding and certain types of metal spraying such as plasma coating and hard surfacing by welding.

Contact Degussa Ltd for further information.

'Tribomet'
This is the proprietary name for a technique where cobalt metal is co-deposited onto the surface being plated along with carbides, generally tungsten or

chromium. This technique is used for anti-frettage purposes. It is a composite technique discussed in Part Two **Electroplating**.

'Tri-Ni'

A multi-layer nickel *electrodeposit* where the nickel deposit is first dull or matt, followed by semi-bright with a third deposit of bright nickel. This will have a top deposit of chromium, which may also be multi-layer for top-quality corrosion resistance.

The purpose is to enhance the corrosion protection. Experience has shown that bright and semi-bright nickel has a lower corrosion resistance then the dull or matt *nickel plating*. This is a proprietary technique and further information can be obtained from Oxy Metal Industries (GB) Ltd.

'Trisec' Drying

A technique for the removal of water.

Application range ALL METALS

This is a proprietary process developed by ICI Ltd, which uses a standard vapour degreaser with trichloroethylene in the vapour phase in the same manner as for *vapour degreasing*. The equipment is modified and an additive made to the trichloroethylene having the effect of depressing the boiling point of water to below that of trichloroethylene at 87°C. Any water entering the degreaser immediately boils, but before reaching the condensation point of trichloroethylene it is condensed, together with the trichloroethylene, on the cooling coils. The condensed liquids are fed to a unit outside the degreaser instead of directly to the sump for re-use. This special unit has the ability to separate water efficiently from trichloroethylene. The water is rejected and fed to a drain while the trichloroethylene liquid is returned to the sump of the degreaser.

Using 'Trisec' drying it is, therefore, possible efficiently to remove water from the surface and the internal passages of components. The basic difference between 'Trisec' drying and a standard vapour degreaser is that, with the latter, water is retained as vapour with a boiling point of 100°C and thus will efficiently condense at a temperature of 87°C. It is possible, then, for vapour degreasers to build up a considerable quantity of water droplets in the form of mist. When the component is removed from the degreaser and the trichloroethylene evaporates it leaves a thin film of moisture on the surface of the cleaned component.

With the 'Trisec' additive the water is continuously removed from the vapour, and because of this additive has a lesser boiling point than trichloroethylene and consequently any water in contact with the component after removal from the 'Trisec' unit evaporates together with the trichloroethylene. This method of drying is listed and evaluated in the entry *Cleaning*.

'Tufftride' Process
A proprietary friction-reducing process.

Application range STEEL

The process results in the surface of the steel being impregnated with nitrides, giving an increase in hardness and a reduction in the friction characteristics. The process uses a salt-bath technique carried out by immersing the components in molten salt at a temperature of approximately 500°C. The time involved is approximately 1 h and results in little or no distortion, provided the components are submitted in a stress-free condition. It is, therefore, recommended that prior to final machining of fine limits a *stress-releasing* at a temperature of above 550°C is carried out.

The effective depth of the 'Tufftride' layer is very shallow, being less than 0.002 in (0.05 mm), with the total depth seldom above 0.005 in (0.12 mm). This means that any machining, or even abrasive *polishing*, following the process will remove the effective layer.

The colour of the treated components is a general matt grey with the possibility of some staining or mottling. If the components are polished to bright steel following the 'Tufftride' process, this will invariably remove the effective layer.

This is a process which is applied to the surface to improve friction characteristics. Friction is the result of cold *welding* of the components which are in contact with each other, and the 'Tufftride' layer, by improving the friction characteristics will reduce this tendency to a marked degree. The treatment achieves better characteristics than '*Sulphinuz*', but will not be as satisfactory as conventional *surface hardening* processes such as *carburizing*. It appears to have considerably better characteristics regarding austenitic stainless steel, and there is some evidence to show that an appreciable increase in life can be obtained from stainless steel components which show certain types of frictional wear.

There is some evidence that correctly carried out Tufftriding gives an increase in fatigue strength.

Further information from Degussa Ltd.

Tufftride QPQ
This is a proprietary process based on the standard Tufftride salt-bath nitriding which is then followed by salt-bath oxidation process with controlled cooling. The components are then subjected to some form of mechanical polishing or vibratory finishing and then given a further immersion in the oxidizing salt.

Tukon Hardness Test

It is claimed that this process gives improved corrosion resistance with better quality wear resistance because of the high standard mechanical finish.

Further information from Degussa Ltd.

Tukon Hardness Test
A *hardness testing* technique. Described fully in Part Two **Hardness Testing**.

Tumbling
An alternative name for the *deburring* process in which components are rotated or tumbled in contact with each other or, more generally, with abrasives and special compounds.

This is discussed in more detail under *Barrelling*.

Turner's Sclerometer
A form of *hardness testing* based on scratch resistance. Described fully in Part Two **Hardness Testing**.

Twisting Test
A form of torque testing which can be used to evaluate ductility.

Application range ALL METALS

This, as its name implies, is a test where the material is subjected to torsion or twisting and is usually applied to wire which is gripped in two vices, one of which is then turned. The number of turns, the length of the wire and whether or not reverse twisting is required will be specified as part of the test.

The twisting test can be applied to strip material when the same technique is applied and the number of turns and whether or not full recovery is expected is specified. With some tests twisting will be repeated until fracture occurs, with others the number of complete turns successfully achieved will be accepted.

Ultrasonic Cleaning
A sophisticated method of soil removal for tightly adherent soils and under some circumstances soil which is semi-adherent.

Application range ALL METALS

The user is referred to the heading *Cleaning* for information on the different types of soil which cause contamination and a general discussion on their methods of removal. Ultrasonic cleaning is necessary only where difficult contamination exists.

Ultrasonic energy can be applied to liquids and will pass through the liquid in the form of cycles of positive and negative energy. This means that as the positive cycle reaches any surface pressure is applied to this surface. As the positive cycle reduces the pressure is reduced to zero, and as the negative energy cycle builds up a vacuum is applied to the surface. Provided that liquid contact is maintained, the surface will be subjected to cycles of pushing and pulling at a high frequency measured in the kilocycle range. It will thus be seen that any loosely adherent material on the surface will be removed as the liquid gains access to the underside. Where the contamination is present as individual spots, as in the case of dust, ultrasonic cleaning rapidly loosens and removes the contamination. However, where contamination is in the form of a continuous film, or a film which has local porosity, the energy will be utilized only at the edges of pores. In these cases, then, ultrasonic cleaning is at the same disadvantage as *electrocleaning* and other methods which rely on the wetting of the contamination interface.

The liquid used in conjunction with the ultrasonic energy is of vital importance. It is obvious that the better the cleaning capability of this liquid, the better the action of the ultrasonics. However, it should be appreciated that, where the cleaning solution readily forms a vapour phase, the efficiency will to some extent be reduced as vapour will tend to build up during the vacuum cycle at the interface between the solution and the solid being cleaned. The vapour or gas will then act as a cushion and can seriously reduce the efficient action of the ultrasonic energy cycle.

A test piece in the form of a fine wire mesh which can be contaminated with the soil is recommended to evaluate the cleaning capabilities of different liquids of ultrasonic equipment. If the soil to be tested is removed from all the holes in the mesh instantaneously, then the equipment and cleaning solution will be satisfactory. If no action, or very little, occurs, either the solution or the equipment is not operating. The degree of efficiency can be quantified by examining the number of holes which have been efficiently cleaned.

When using this test piece, it is essential to examine the critical area being cleaned, and the angles which are inclined between the ultrasonic unit and the surface being cleaned. The reason for this is that ultrasonic energy can be extremely selective and can be produced in pencil-thin beams, which are extremely efficient provided that the component or surface is in the line of the beam of energy. It must be appreciated that outside this narrow beam the degree of cleaning will either be non-existent or at a low level, and it is difficult under normal industrial conditions to move components in such a manner that the complete surface comes within the area affected by this narrow beam.

Ultrasonic cleaning, then, is a cleaning technique which can be very efficient at removing difficult soils from simple surfaces. It has limited application where intricate components are involved as the energy cannot normally be passed through one solid to affect the liquid on the inner surface. Users are advised to carefully evaluate existing conventional techniques to ensure that they are incapable of solving the cleaning problem before embarking on the expensive ultrasonic method. Considerable care is also necessary in the choice of cleaning liquids. Simple, cheap water-based liquids should always be examined prior to trying the more expensive organic solvents.

Liquids which do not vaporize easily, such as water, will give better results from ultrasonic energy than liquids which rapidly vaporize. Many liquids which readily vaporize have good solvent cleaning properties.

With good-quality upset forging all evidence of the fused or cast metal is pushed into the 'upset' area which is removed thus leaving only a forged, high-quality weld.

Ultrasonic Crack Detection

A method of identifying defects such as porosity or cracking below the surface. This uses ultrasonic energy which is reflected from defects below the surface. Described fully in Part Two **Non-destructive Testing** under *Ultrasonic Flaw detection*.

Ultrasonic Soldering

This is used only when *soldering* aluminium and its alloys; it is briefly discussed under that heading in this book.

Unionmelt Welding

An alternative term for *submerged arc welding*. Immediately the arc is struck, it is covered with solid powdered flux and the weld continues under this layer of flux. Described fully under *Submerged Arc Welding* in Part Two **Welding**.

Unshielded Metal Arc Welding

A term used where any form of *welding* is applied without a shielding gas.

Application range ALL METALS WELDED

While in general unshielded welding is by far the most common, there are specific industries and materials where a considerable amount of *inert-gas shielded metal arc welding* is used and, therefore, this specialized technique becomes the standard procedure. Under these conditions the use of normal *welding* requires to be identified, hence this necessitates the use of the present term. There are now fluxes which produce a 'shielding' gas during welding, and this can thus add to the lack of precision of this terminology.

Upset Welding

A specific type of *forge welding*, where the join is achieved by 'upsetting'.

Application range ALL METALS WELDED

This upset is achieved where a local area is heated and formed into a mushroom by impact forces. The term comes from the conventional technique of upset-*forging* the end of a bar. With *welding*, both components are upset by the application of pressure to bring the areas being joined into contact with each other, and the upset forces out of the edge of the weld material in the form of a mushroom. This will normally be machined off subsequent to welding and, thus, all evidence of oxide or cast material produced during the joining operation is removed. Upset welds are most commonly applied in *flash butt*, *friction* and *inertia welding*. Some forms of *resistance* and *forge welding* also result in upset welding.

With good quality upset forging all evidence of the fused or cast metal is pushed into the 'upset' area which is removed thus leaving only a forged, high-quality weld.

Vacu-blasting

A specific form of portable blasting. Described fully under this heading in Part Two **Blasting**.

Vacuum Brazing

A specific form of brazing.

Application range ALL METALS BRAZED

With this technique the components are assembled, and held either by jigging or by an interference fit with the braze assemblies to hold the individual components in place. These are then placed in a vacuum chamber which is evacuated to give high vacuum and heat applied, generally by radiation, to melt the braze material. With this technique, fluxes are normally not possible as these would cause contamination of the vacuum pumps. It is, therefore, essential that the components to be brazed are chemically clean and the surfaces to be joined are free of oxide.

Vacuum Coating

This is a general term used when vacuum techniques are involved where a deposit is then produced on the surface of metals.

Further information is given under the heading *Vacuum-coating Deposition or Plating*.

The filler material, which can be any one of the common braze metals, is present either as a shim or a preformed ring or other shape, and commonly as an *electroplated* deposit. This latter technique is usual when copper *brazing* is used. The quality of the braze, as with other forms, relies on the clearances which exist prior to the operation.

Unlike other methods of *brazing*, where fluxes can be used, there is no possibility of any removal of contamination once the operation commences. In addition to the problem of a faulty braze caused by contamination, there is the possibility that certain types of contamination could damage the expensive pumps used to produce and hold the vacuum. Provided that the conditions are correct, vacuum brazing will give high-integrity joints for copper *brazing* and also for some of the higher-melting-point braze alloys based on nickel. If economic use can be made of the space within the vacuum chamber then vacuum brazing can be an economic method of producing high-quality brazed joints. In this way, it is competing with *hydrogen brazing* and to a lesser extent conventional *furnace brazing*, using fluxes, and *salt-bath brazing*. *Controlled atmosphere* and *inert-atmosphere furnace brazing* are also possible alternatives. Users are referred to the heading *Soldering* and to Part Two **Welding**, since very often the joint produced by vacuum brazing is an alternative, in particular, to welding.

Vacuum-coating Deposition or Plating

A process in which certain pure metals are deposited on a substrate.

Application range ALL METALS AND MANY NON-METALS

The technique uses the fact that, by reducing the pressure to produce a vacuum, pure metals can be vaporized at a low temperature inside a closed container. The metal vapour will then condense evenly on all cool surfaces to give a metallic coating. In practice, the components are held on simple jigs or fixtures which are then placed inside the vacuum chamber. When the chamber has been loaded, it is sealed and a very high vacuum is produced, which with modern equipment can be achieved in a matter of minutes. The chamber also contains a predetermined amount of the metal to be deposited. This is held in a container which is capable of heating the metal to above its vaporization point. Because the material is held in vacuum, the vaporization point will be relatively low compared with the boiling point of metals at normal pressures. As soon as the desired vacuum is achieved, the electric circuit heating the metal is closed. The metal will boil and the vapour entirely fills the chamber. The vapour then condenses and every surface within the vacuum chamber receives a uniform deposit of the metal. As the chamber surfaces, fixtures holding the components etc., will be coated with the condensed metal, the efficiency of this process particularly with expensive metals largely depends on the relative areas of component and equipment.

The only materials which can be successfully vacuum deposited are pure metals, since an instantaneous boiling point is necessary. Aluminium will probably be the most successful metal and the result is a very high reflective material with attractive visual appearance. Other pure metals such as gold, copper, etc. can be vacuum deposited but will not achieve the same degree of reflectivity without subsequent treatment.

The uses of vacuum deposition include headlight reflectors and mirrors of all types. Vacuum deposition is also used in the production of electronic devices, where thin films of metallic conducting material are required on dielectric materials for the production of condensers, etc. In this area it is used in conjunction with the *sputtering* technique for depositing thin metallic coatings. A considerable amount of vacuum deposition is carried out on relatively cheap plastic, in order to improve the visual appearance and many toys are vacuum plated for this reason.

Vacuum deposition, under controlled conditions, results in a homogeneous thin film, approximately 0.0005 in (0.01 mm) maximum, with excellent adhesion. Many instances of vacuum deposition, however, occur where substrate material is unsatisfactorily prepared resulting in failure of adhesion. The process is comparable to some extent to the plating of plastic, but will seldom achieve the adhesion possible with that process. As a mass-production technique it is more economical than conventional plastic plating and can result in an attractive finish difficult to reproduce by conventional *plating* or *painting*.

Vacuum Metallization/Vacuum Evaporation

A process which is similar to *vacuum coating* but having a varnish or lacquer applied to the components prior to coating.

Application range ALL METALS

This lacquer is designed to have good adhesion to the base material.

The components are subjected to the *vacuum coating* process in the vacuum chamber where the metal is evaporated and condensed on to the surfaces. On removal, the components are generally given a coating of clear lacquer to prevent tarnishing of the vacuum deposit. If necessary, this final lacquer can be dyed and, provided it is applied in a thin enough coating, the high reflection of the vacuum deposited metal will impart the necessary sheen to the lacquer retaining a metallic appearance. In this manner aluminium is commonly made to appear golden.

This is comparable to some forms of paint finish.

Vapour Blasting

A form of abrasive *blasting* where the particles are suspended as a slurry mixture, and air is used to inject this through a nozzle to blast the component. Further details will be found in the Part Two **Blasting.**

Vapour Degreasing

A form of *cleaning*. The process generally uses chlorinated solvents, which have excellent degreasing properties in their own right but also produce a vapour heavier than air.

Application range ALL METALS

The liquid is heated in an open-topped container and as it boils it produces a hot vapour which rises above the boiling liquid. The vapour is held within the container by means of a cooling coil which runs around the inside of the container a short distance below the rim. This produces a cold zone, resulting in the vapour condensing and being returned to the sump ready for re-boiling. This, therefore, is a form of continuous distillation.

When any cold component is placed in the area of the vapour, the liquid immediately condenses on the surface, producing a hot, clean solvent. The solvent dissolves any grease on the surface and as further chlorinated gas condenses it runs off again and is returned to the sump in a contaminated state. It will be appreciated that only clean solvent will condense on the surface of the component, since condensation is the result of the liquefaction of pure vapour. The liquid in the sump thus gradually becomes contaminated with the material removed from the surfaces of the component, but the vapour remains pure and uncontaminated.

Several aspects of vapour degreasing should be noted. First, the action will only proceed as long as the components being cleaned are maintained below the temperature of the gas. As soon as they reach this temperature, no further condensation can occur. There is, therefore, no point in leaving components for long periods in the vapour phase. Where difficult components require cleaning, and this consideration applies more to the intricacy of the geometry than the contaminant involved, then the parts should be removed, allowed to cool and then returned to the vapour phase. To some extent this disadvantage can be overcome by immersing in boiling liquid solvent.

Second, it should be appreciated that vapour degreasing efficiently removes only those contaminants which it is designed to dissolve. It is possible that the 'secondary' contaminants in intimate contact with the contaminants under attack by this method will also be washed off, but this will often be incompletely achieved and a thin film of dust will be retained which may, under some circumstances, become the more difficult to remove because the particles are no longer readily 'wetted' (see *Cleaning*).

Third, most if not all chlorinated solvents entail some health hazard. While this is not serious where the equipment is correctly used, it will always be present. The principal solvents are trichloroethylene and perchloroethylene, boiling points 87°C and 120°C respectively. The use of carbon tetrachloride is no longer permitted in the UK and trichloroethylene is not normally used in the USA.

A number of more complex materials are available which contain fluorine in addition to chlorine, these are known as fluorochlorinated hydrocarbons and generally have a low boiling point. These solvents are less active than chlorinated hydrocarbons and are finding considerable use in the electrical and electronics industries.

It should be noted that all chlorinated hydrocarbons when heated can be extremely reactive in contact with finely divided metals of the aluminium or magnesium type. While it is possible to stabilize the cleaners to reduce this danger, all users of light-alloy metals should take particular care that their vapour degreasing plant is kept in good condition and is scrupulously cleaned at regular, short intervals. Unless this is done, there will be a danger of an explosion from the reaction which can occur under certain conditions. The reaction can result in serious corrosion long before there is any danger of explosion.

Vapour degreasing is a useful method of removing oil and grease but has certain limitations. Prior to installing this type of equipment, the user is advised to give careful consideration to the type of soil removal and the standard of cleanliness required together with the subsequent processing carried out.

Vapour-phase Inhibitors (VPI)

A form of low level *corrosion protection*.

Application range STEEL

Use is made of chemicals in powder form which have a low vapour pressure and, thus, readily give off a vapour or gas. This will fill any confined space, the vapour either by condensing on the surface of components, or excluding water vapour, protects the articles from corrosion.

Vapour-phase inhibitors can be used in the form of a powder which is placed in a perforated container inside the package containing the components. More generally, the material is impregnated into paper or cardboard wrapping and the articles to be protected are either wrapped in paper, or covered above and below. Because the protection is achieved in the vapour phase, any draught or ventilation will result in protection being lost. It will thus be seen that vapour-phase inhibitors are of limited use in normal atmospheres and will not effectively protect articles packed in porous containers. The protection afforded is of a relatively low standard and will not, for instance, protect steel from corroding during a short sea passage even with the components packaged to normal industrial standards.

Vapour-phase inhibitors will, however, considerably help existing forms of *corrosion protection*. If oil or other forms of protection cannot be applied, it is essential that components are well wrapped in VPI-impregnated paper and that some free powder is liberally spread as near as possible to the airtight container.

Vapour-phase inhibitors do not give protection comparable with oil films and are certainly much less efficient than the grease used on engineering components. Where VPIs are found to be most useful is with delicate components such as mechanical or electronic instruments, which for obvious reasons cannot be coated with heavy corrosion-resisting oils or greases. These components are, therefore, frequently packed using inhibitors. These parts may also have silica gel packed in the airtight container to absorb any moisture built into the airtight container.

'Vaqua Blast'
A proprietary name for the form of *vapour blasting* which uses a high-pressure water slurry into which is injected high-pressure air. Further information can be obtained from Abrasive Developments Ltd.

Vibration Testing
A method of flaw detection. Described fully in Part Two **Non-destructive Testing**. See also *Reflectogage Testing*.

Vickers Hardness Test
The standard laboratory and reference type of *hardness testing*. Described fully in Part Two **Hardness Testing**.

Vitreous Enamel
A form of corrosion protection using a glass film.

Application range ALL METALS

This is a relatively ancient process which in essence is the application of a thin, non-porous coloured film of glass on the surface of metal. The quality of vitreous enamel varies from the poor, cheap product which readily cracks and crazes, either falling off directly or allowing corrosion to occur with subsequent failure, to the production of components capable of withstanding considerable shock loading, either of a mechanical or thermal nature, and in some cases having a high artistic content.

Vitreous enamelling was very popular in the late nineteenth and early twentieth centuries. With the advent of new materials, for example, aluminium in cooking ware, and better corrosion resistance from conventionally applied paints, the use of vitreous enamel declined and was confined very largely to articles such as baths where no ready alternative was available. Of recent years there has been a renewal of interest, with production of new enamels and better control at the production stages, giving superior bonding and more attractive colours.

The procedure is that the article to be treated is chemically cleaned and may have the surface prepared by *pickling* or chemical *etching*. More commonly, the component is *grit blasted* to give a controlled surface finish, with the surface being rough to give a good key. The material used is known as a 'frit' and consists of fine silver sand with borax, sodium carbonate and feldspar. Metal oxides are added for colouring. The frit is generally produced by ball-milling to give a fine dustlike material.

VITREOUS ENAMEL / 315

The methods of application are varied and are as follows:

1. Heating the article to above 900°C and either dipping the article into the powdered frit, or blowing or throwing the frit on to the heated article. Baths are produced in this manner with smaller components being dipped into the frit itself.

2. Mixing the frit with water and spraying this on to the prepared surface. The spraying can be of the conventional type and commonly, with modern equipment, uses electrostatic spraying to ensure good adhesion of the frit–water mixture. The components are then carefully dried to remove moisture and taken to a temperature generally above 900°C.

3. Using fluidized beds, where the dry frit is held in a container with the powder in suspension. This technique gives the dry powder the properties of a liquid and the heated components can then be immersed in the dry frit to give a very even coating. These components can be re-heated and re-dipped.

It is possible using standard *stop-off* procedures and frits of different melting points to produce multi-coloured designs, and frits are available which give a mottled effect as a single dip or spray. Glass frits with low melting points can be used to coat the lower-melting-point alloys such as aluminium or even magnesium. This new technique has increased the range of uses of vitreous enamel.

Provided the pre-treatment is correctly carried out and the frit is applied to a completely clean material with a sharp, roughened oxide-free surface, and provided the frit is applied as soon as possible after pre-treatment and successfully fused, then the result is a high standard material capable of withstanding considerable shock or abrasion. It will have good corrosion resistance even in very acid or very alkaline conditions, with very long life expected for normal atmosphere use. Where either acid or alkaline conditions will be severe, then it is advised that special frits are used as acid-resistant glass will probably be attacked by strong alkalis, and vice versa.

The uses of vitreous enamel in modern industry are generally where excellent corrosion resistance is needed. The agricultural industry is finding considerable use of vitreous enamel sheet metal in silos for storing highly corrosive materials such as sludge and manures. Considerable use is also made still in bathroom fittings, although plastics are making inroads. Vitreous enamel-ware is returning to the kitchen increasingly with sink units and pots and pans bearing colourful designs. The engineering industry too makes limited use of vitreous enamel, but more common is the area of office furniture and equipment.

The process is still used for decorative jewellery, and almost all high-quality metal badges will make use of vitreous enamel where bright colours are required.

It will have abrasion and corrosion resistance superior to most forms of *stove enamelling* but will not be able to withstand bending without cracking.

However, it should be considered when corrosion resistance of high standard is required, together with resistance to abrasion. The fact that Victorian mild-steel panels with advertising motifs can still be seen in industrial atmospheres, with no evidence of corrosion except where mechanical damage has occurred, is proof of the efficiency of the process.

Some modern electroplating and painting techniques have an equivalent corrosion and impact resistance.

Walnut Blasting

A technique very similar to, but not normally as efficient as, *plumstone blasting*. Described fully under *Plumstone Blasting* in Part Two **Blasting**.

'Walter Black'

A proprietary name for the *black oxide* treatment.

Detailed information on 'Walter Black' can be obtained from the Walterizing Company.

'Walterizing'

A proprietary form of *Phosphating*. Details of this process will be found under that heading.

Information can be obtained from Pyrene Chemical Services Ltd.

Warman Penetrascope

A portable form of *hardness testing* not commonly used. Described fully in Part Two **Hardness Testing**.

Water Gilding

A chemical procedure for gold deposition without the use of an electric current. It is described in more detail under the headings *Gilding* and *Electroless Plating*.

Water Hardening

The technique in the *hardening* of steel which uses water as the quenching medium. Described fully under *Hardening* in Part Two **Heat Treatment**.

Watt Nickel Process

A specific type of *electroplating* of nickel.

Application range ALL METALS

This is the traditional method of metal plating and uses the most common solution based on nickel chloride, nickel sulphate and horic acid. Described fully under *Nickel Plating* in Part Two **Electroplating**.

Wave Soldering
A mass-production technique of *soldering* simple components. More details will be found under *Soldering*.

Weathering
A term used to indicate that components have been stored for some time under specific conditions. It has been found to result in components being more stable dimensionally after machining. It is now known that a short-term *stress-releasing* operation at, for example, 150°C or 200°C for 2 h can achieve similar or superior stability. This use of weathering is, therefore, very seldom carried out in modern industry, but *stress-releasing* or preferably *normalizing* will be carried out prior to machining.

Weathering, known alternatively as *seasoning*, should not be confused with the *ageing* process. There are a number of alloys which increase in hardness when held at atmospheric temperatures, and this is a different process. *Stress-releasing (ageing)*, and *Precipitation Hardening* are described fully in Part Two **Heat Treatment**.

Weathering has been used to describe an atmosphere test to show the result of storage for long periods at temperature, or by temperature cycling.

Weibel Process
A form of *resistance welding* where carbon electrodes are used to supply the current. This has been replaced by the use of copper alloy and other metallic electrodes. Described fully in Part Two **Welding**.

Weld Deposition
The term applied to the method of applying metal normally to build up worn surfaces.

Application range ALL METALS

Weld deposition uses the standard methods of *welding*, although *oxyacetylene gas welding* is probably more common than *electric arc*. The technique is that the chosen metal is applied by *welding* to the worn surfaces. It is generally desirable that as little as possible mixing of the metal being deposited and the base metal occurs. Weld deposition is usually applied for corrosion resistance or *hard facing* purposes. It will, therefore, be appreciated that excess mixing of the base and deposited metals will result in dilution of the latter. This is sometimes known as 'bleeding'. It is for this reason that *gas welding* is more commonly used than *electric arc*, since the welder has more control over this technique regarding the mixing of the two metals, while the latter gives better weld penetration.

Weld deposition is in competition with *metal spraying* and some of the spraying techniques are very similar or identical to the present technique.

Electroplating of certain metals can also be an alternative to weld deposition.

Weld-decay Testing

An inspection technique used on samples to ascertain the susceptibility to weld decay, which is the result of carbides precipitating at the grain boundaries.

Application range AUSTENITIC STAINLESS STEEL

Precipitation results in a second phase appearing and, thus, reduces the stainless nature of austenitic stainless steel, which relies on the fact that it is a single-phase material for its corrosion resistance. The weld decay test takes a sample of the material to be tested and sensitizes this at 650°C for an agreed time, generally 0.25 or 0.5 h. This treatment ensures that if any precipitation of carbide occurs in the material it will be present in the test piece. The steel specimen is then immersed in a copper sulphate solution of sulphuric acid and boiled continuously for 72 h. At the end of this period the test pieces are removed, washed, dried and bent. If the test pieces can successfully withstand bending at 180°, then there is no susceptibility to weld decay. Any cracking indicates, of course, the opposite state. This is an approved test procedure and can be quantified by varying the techniques to include material in the 'as received' condition, and also by carrying out the bend test after 24 h and 48 h.

If the steel in the 'as received' condition fails by cracking after immersion in the copper sulphate solution for 24 h, then it will have a very serious weld decay problem. If, on the other hand, the 'as received' material is satisfactory after 72 h, and the sensitized test piece is satisfactory after 48 h but shows cracking after immersion for 72 h, the potential weld decay can be regarded as being limited.

This data can be used in conjunction with the known conditions of the treatment the steel is to have, together with those of service, to assess further likelihood of a weld decay problem.

This is a unique test for weld decay, but where it is known that stainless steel has been correctly stabilized with niobium (columbium), titanium or molybdenum it will be unnecesssary. Similarly where steel has been subjected to the correct *stabilized anneal*, and will not be heated subsequently, testing will not be required. It will often be more economic to carry out the weld decay test than to carry out chemical analysis or metallurgical investigation to find the specification of a stainless steel. The test is also known as the *Strauss test*.

Welding

This term covers the methods of joining using heat and/or pressure. All the techniques used and some general information have been collected in Part Two **Welding**. There is a discussion of the process and information on all the known methods of welding in this section.

Wet Nitriding
A form of *nitriding* using a cyanide salt bath. This results in the production of atomic nitrogen in the bath and the evolution of hydrogen. The atomic nitrogen combines with the steel resulting in *nitriding*. This process is also known as *Chapmanizing* and has largely been replaced by conventional *nitriding, carbonitriding*, or '*Tufftriding*'.

Wiped Joint
A term applied to the method of joining where a solder with a wide range of solidification is used, to produce a join by manipulating the molten metal.

Application range LEAD OR TIN

This can be either for a join of two pieces of lead pipe or for the repair of a hole or cut. The molten solder is poured on to a pad and this is used to build up a thickness of solder on the heated pipe. Considerable skill is required and the traditional material for wiping the joint is moleskin. With the gradual elimination of lead piping in favour of copper, stainless steel and plastic this process is rarely used nowadays. The process is a form of *fusion weld*.

Wire Brushing
A form of *cleaning* or *de-scaling* using either revolving or hand-held brushes with wire bristles.

Application range ALL METALS

Modern brushes generally use hard-drawn stainless steel wire but hard-drawn normal mild steel or low-alloy steel are also employed. Where materials other than steels are being brushed, it is often essential that stainless steel brushes are used to prevent corrosion staining on cleaned surfaces, which will be contaminated by ordinary steel.

Wire brushing is very often an alternative to *blasting* or *pickling* but only in exceptional cases will it be found to give comparable results, particularly with the former. However, where access for *blasting* is difficult or impossible, properly used wire brushing will be an acceptable alternative and is certainly superior to no treatment or the misuse of pickling.

Wohler Test
A destructive test carried out on prepared specimens to find the fatigue strength.

Application range ALL METALS

The round specimen has two diameters, having a large, carefully prepared radius between each diameter. It is rotated, held by the largest diameter and loaded at the end of the small diameter. It will be seen that this technique applies a load fluctuating between tension and compression on the test piece; this load will normally be concentrated at the change of section. With each revolution, the surface of the test piece at the change of section will vary from tension at the top, and compression at the bottom as the specimen revolves. Provided the specimen is prepared in such a manner that no stress raisers exist, then the true fatigue strength of the material undergoing the test will be obtained. It is now realized that the fatigue strength of a component is related to the tensile strength and is affected more by surface finish than by the basic fatigue strength of the material. The Wohler test is now accepted to be of limited application.

In general the useful fatigue strength of a metal is approximately 50 per cent of the ultimate tensile strength, but where stress raisers concentrate the stress applied at a local area the fatigue strength will vary considerably. It is now, therefore, advised that components or assemblies are fatigue tested, rather than producing carefully manufactured test pieces.

Work Hardening

The process of increasing the tensile strength by mechanical working.

Application range ALL METALS

Work hardening results in the distortion of the metallurgical grain structure, causing strain which increases the tensile strength of the material. When a material is hot worked, the grain structure is immediately recrystallized, thus removing grain distortion and strain. Many pure materials such as lead, gold and silver hot work at room temperature, and it is, therefore, necessary if work hardening is to be encountered that they are taken to below room temperature. As recrystallization will occur when the material is returned to room temperature, work hardening will obviously be of only theoretical interest.

Many materials cannot be *cold worked* because their metallurgical structure precludes the distortion of the grain and fracture at the grain boundary or of the grain itself will occur before any appreciable work hardening or grain distortion takes place. Under many circumstances the same materials cannot be successfully hot worked, and thus there are a range of materials which are not capable of being forged.

Work hardening is a means of increasing the tensile properties of many materials which cannot be subjected to *heat treatment*. This applies to austenitic stainless steel and many of the aluminium and copper alloys. The work is generally applied to metal while it is being produced. Thus, during rolling, a controlled amount of *cold working* can be successfully applied to sheet metal which will considerably increase the tensile properties, the 'yield' and 'ultimate tensile strength', with some reduction in the ductility. This will require to be within acceptable limits in those alloys which are used in the work hardened condition. The same conditions apply in the manufacture of tube and extruded sections where the amount of cold work being applied can be carefully controlled. The use of *forging* as a work hardening technique requires considerably more skill and care as there is generally a thicker section, or a greater mass, and thus predicting the amount of *cold work* being applied is more difficult.

It should be realized when work hardening is applied to any material, that subsequent heating of the component will result in a reduction in the tensile strength. Thus, *brazing* or *welding* of work hardened materials is only possible when this reduction in mechanical properties can be locally accepted in conjunction with the weld or braze. Work hardening can, however, be applied to those areas which have been heated in order to recover some or all of these mechanical properties. This technique is often referred to as *peening* and requires some skill.

Any of the materials listed above which can accept work hardening, and which are bent or hammered during manufacture, will be work hardened locally. Work hardening can also occur on components in service where they are subjected to bending stresses which result in distortion of the grain. Under these local conditions work hardening may be a defect which, because of accompanying local reduction in ductility, can cause cracking. If these bending stresses are continued, fatigue or brittle failure will result.

Different materials exhibit widely varying *cold working* characteristics. In general, with steel, it can be stated that work hardening characteristics are to some extent proportional to the hardenability of the material. There are, however, variations even in this in that high manganese materials will work harden to a greater extent than low manganese materials. Austenitic stainless steel, and other materials containing appreciable amounts of nickel and chromium, are susceptible to work hardening with an increase in tensile strength but decreased ductility. This decrease in ductility can, under some circumstances, be serious.

Further information on the work hardening characteristics of alloys should always be sought before components are used in conditions where bending stresses will occur, and where severe *cold working* is to be applied during the manufacturing process, without subsequent *stress-releasing*. Stress-releasing at a temperature of 650°C for 2 h will successfully eliminate all appreciable work hardening from ferrous alloys and there will be critical temperatures to remove the effect from other materials. Where materials are work hardened as part of the manufacturing process, for example, in wire-drawing, extrusion, rolling, etc., it is advisable that they should be tempered before going into service. This operation is carried out at the highest possible temperature, and should certainly be at least 20°C above the temperature to which the components will be subjected during manufacture, or in service. *Tempering* is described fully in Part Two **Heat Treatment**.

Wrapping Test

A destructive test used to ascertain the ability of wire to accept *work hardening* without fracture.

Application range ALL METALS USED IN WIRE FORM

The test has many variations and can be very specific when necessary. In essence, the wire is wrapped around a specified mandrel for a specified number of times and is either examined then for cracking in the 'as wrapped' condition, or after unwrapping and straightening. The test will often specify that the mandrel has the same or twice the diameter of the wire being tested, and that the minimum number of wraps is eight. During the actual wrapping, information can be obtained regarding the material being tested; if fracture occurs, or it is not possible to obtain a satisfactory wrap, then obviously the material is unsuitable. The tests should include an examination of the wire in the 'as wrapped' condition, when it will be possible to examine the tensile surface for fine cracks. This wrapping test can be used to test surface finishes and is common for wire which has been galvanized. On completion of wrapping, the wire should be carefully unwound, straightened and re-examined. This test is typical of many relatively simple, destructive inspection techniques, which clearly indicate whether or not a material is fit for a certain specified function. It is, in fact, a form of *Bend Testing*, described fully in Part Two **Mechanical Testing**.

It is necessary to have an agreed specification of the test parameters for the results to be meaningful.

X-ray

A form of *non-destructive testing*. Described fully under *Radiographic Test* in Part Two **Non-destructive Testing**. It is more generally known as *radiographic inspection*.

Xeroradiography

A specific form of *X-ray/radiography* which has certain special applications. It makes use of the fact that certain materials such as selenium are converted from an electrical insulator to become conductors when subjected to X-rays.

The technique is to coat the component, generally in plate form, with selenium, and this is then used instead of the photographic plate normally found in radiographical use.

The reader is advised to obtain specialized advice. With advances in *radiography* and other forms of flaw detection, this method is seldom used. The different types are listed and discussed in Part Two **Non-destructive Testing**.

Yield

This is a specific 'proof strength' identified visually during the tensile test.

Application range PLAIN CARBON AND LOW ALLOY STEEL AND SOME SIMPLE NON-FERROUS ALLOYS

The yield point is the load in tensile test where the straight line relationship between the load applied and stretch achieved alters.

It can be seen as a delay in the increase in load, while the specimen continues to stretch.

On the stress-strain graph the yield point is taken as 0.5 per cent proof strength. Described fully in Part Two **Mechanical Testing** under *Proof Test*.

Young's Modulus

A measure of the strength and elasticity of materials. Some details and values are given under this heading in Part Two **Mechanical Testing**.

'Zartan'

A proprietary tin deposit which is claimed to be an alternative to *chromium plating* as a decorative finish.

Application range NICKEL

The advantages are that current control is simpler as the current density is lower, thus allowing normal *barrel plating* in many instances which would not be possible with *chromium plating*. The throwing power is much better, allowing improved coverage and the equipment required is simpler. Further information can be obtained from M. L. Alkan Ltd.

Zerener Process

An alternative term to *carbon arc welding*. This is now seldom applied. It makes use of the arc struck between two carbon electrodes, or one electrode and the work piece. This arc is used as a source of heat to melt the filler rod and the component being welded. *Fusion Welding, Electric Arc Welding* and *Inert-gas Shielded Arc Welding* have replaced this technique. Described fully under these headings in Part Two **Welding**.

Zinc Coating

This term is used to cover several techniques where zinc is applied as a surface treatment.

Application range STEEL

These are almost invariably limited to the *corrosion protection* of steel, and include the basic hot-dip process known as *galvanizing* and also the electro-deposition of zinc (see *Zinc Plating* in Part Two **Electroplating**). There is also the process of *metal-spraying*, where zinc is deposited. The *sherardizing* technique diffuses zinc into the surface of steel and is, thus, a process similar to *aluminizing* or *chromizing* using zinc instead of aluminium or chromium.

These procedures all have their specific purpose. *Galvanizing* will produce a relatively thick deposit and, in most circumstances, is by far the most economical and technically superior method of coating steel with zinc. It has, however, the proven disadvantage that it is relatively unsightly, and cannot be applied where components have fine limit dimensions. It will be less economical and relatively difficult to apply to small components. It is because of these disadvantages that zinc plating is used. To some extent the process of *sherardizing* can be used on components which have relatively fine limits.

The *galvanizing* process finds another limitation on threaded parts, where in spite of the techniques of *hot wiping, spinning*, etc. devised to reduce the unevenness of the deposit, it is found that these components cannot normally be galvanized without subsequent machining. In addition to the obvious expense involved, there is the disadvantage that this operation will very often remove, either locally or completely, the galvanized corrosion-resistant film.

Zinc protects steel in direct proportion to the thickness of the deposit. The protection is achieved because the deposit of zinc prevents the atmosphere from contacting the steel. Oxygen, particularly when present with moisture and other elements such as chloride, sulphate, etc. results in the rusting of steel.

Zinc has a lower oxidation potential than iron, and while zinc is present the oxidation will be slower. Even when the zinc film is broken, allowing access of atmosphere to the steel, protection will continue. Since zinc is anodic to iron, an electrolytic cell is set up which produces minute quantities of electricity when an electrolyte is present where the zinc is the anode and corrodes, the iron being the cathode which is protected. Moisture forms the electrolyte, and it is obvious that the more electricity which is generated, the faster this reaction, and the more rapid the corrosion and removal of the zinc. Once the zinc is eliminated and the atmosphere is capable of attacking the steel surface, corrosion of steel commences.

Thus, *corrosion protection* of the steel will be a function, first, of the thickness of the zinc; second, of the aggressiveness of the atmosphere; and third, it is a measure of time. Therefore *galvanizing*, with which a zinc thickness in excess of 0.005 in (0.125 mm) is common, will have an advantage over electroplated zinc, with which a thickness of less than 0.0005 in (0.0125 mm) is acceptable and thicknesses over 0.001 in (0.025 mm) are undesirable.

The *sherardizing* process and *metal-spraying* are capable of giving zinc deposits between the two extremes, and it is the designer's problem that he must specify the thickest possible zinc capable of being used on his component. Zinc plating will generally be the most expensive method of depositing zinc with *sherardizing* and *metal-spraying* being intermediate, the expense varying considerably in the quantities involved and the geometry of the component. For quantity production, *galvanizing* will always be by far the most economical method and, at the same time, produces a zinc deposit which is technically superior to the other methods.

It is now possible to improve the corrosion resistance of zinc by treating the freshly prepared surface with a chromate solution. Painting, particularly with zinc or zinc chromate primers, can give equal protection but will often be difficult and expensive to apply and control at the necessary standard. An etch primer is essential when zinc is painted.

With electroplated zinc, it is generally agreed that the corrosion resistance obtained by *chromate* sealing is at least twice that of zinc alone, and the technique is now being applied to galvanized and metal-sprayed parts. *Corrosion protection* by the use of zinc is the cheapest and technically the best. If superior corrosion resistance is required, this will probably mean considering the use of stainless steel, aluminium or a copper alloy. *Chromizing* might be an alternative under certain circumstances. *Nickel plating* will give better corrosion protection in aggressive atmospheres, but does not have as good galvanic protection when damaged and is much more expensive.

Zinc Phosphatizing

A process applied only to electrodeposited zinc. The newly plated component is dipped in a zinc phosphate solution which has been acidified with phosphoric acid to convert the surface of the zinc to zinc phosphate. This zinc phosphate coating is then sealed by immersing in a dilute solution of chromic acid. This method of sealing zinc deposits to prevent the formation of unsightly zinc oxide will improve the corrosion resistance. For most purposes, *chromating*, or '*passivating*', is used and gives equal or better results.

Zinc Plating

A technique for the *corrosion protection* of steel. Described fully in Part Two **Electroplating**. Zinc is now the commonest metal deposited.

Zincate Treatment

The deposition of zinc by chemical displacement.

Application range ALUMINIUM

The process is only applied as a pre-treatment prior to electroplating.

The material is subjected to *cleaning* and is then *etched* in chromic acid or phosphoric acid to remove all evidence of the oxide film. The material is carefully rinsed then dipped in nitric acid to activate the surface, followed by rapid rinsing in clean, running water and immediate immersion in a sodium zincate solution.

The material, on removal from this solution after a few minutes' immersion, will have a dull grey appearance. This is caused by the reduction of metallic zinc on the surface of the aluminium. Immediately after the zincate treatment the components must be thoroughly swilled and rapidly as possible passed to the *electroplating* process. This will most commonly be the electrodeposition of copper which is then plated with any other metal. Aluminium components can be *chromium plated* using this method. The adhesion of the plated metal to aluminium and its alloys will not be as good as that of metal which can be plated directly. The problem with plating aluminium is the affinity of aluminium for oxygen, which means that immediately on chemical removal of the oxide coating and *swilling* of the component, an oxide film re-forms at once, and this acts as a barrier preventing adhesion of the plated metal. The chemical conversion used in the zincate process means that a conductive surface is produced which is capable of being *electroplated*. This is thus a form of *electroless* or *immersion plating*, where zinc displaces the aluminium.

Zincing
An alternative term to *galvanizing*, used with reference to wrought iron or mild steel sheets.

Zincote Process
A method of producing a coating of zinc chromate as a pre-treatment for *painting*.

Application range MILD STEEL

The component or material is dipped or sprayed after *cleaning* with a solution of zinc nitrate, which contains sodium acetate and bisulphite. This oxidizing solution activates the steel surface which is then converted to the chromate by dipping in a dichromate solution. The steel is then carefully dried and immediately painted. The process has now been replaced largely by *phosphating*. The phosphate treatment, when followed by a chromate seal, gives a superior corrosion-resistant finish to steel with better adhesion of paint.

'Zinkostar'
A bright acid zinc electroplated deposit.

Zinc Plating is described fully in Part Two **Electroplating** and further information on this proprietary process can be obtained from the Oxy Metal Industries (GB) Ltd.

'Zyphos'
The proprietary name of a range of phosphate solutions. Further information can be obtained from M. and T. Cruickshank Ltd.

PART TWO

Major groups of processes

Blasting

A term covering *Abrasive, Dry, Ferrous, Grit, Sand, Shot* and *Vapour Blasting*; *Shot Peening*, etc., all of which processes are listed under their separate headings in this book, but deserve fuller treatment in this section.

Application range ALL METALS

Blasting may be used on any material in addition to metals.

There are three distinct processes: (i) removal of scale, undesirable paint or other surface contaminants which are adhering to a surface; (ii) preparation of a surface for further surface treatment; and (iii) treatment of a surface for certain specific reasons such as *deburring, shot peening* or improvement of surface finish.

The process is carried out by three distinct techniques:

1. Using air as the means of adding velocity to the shot or grit.
2. Using an impeller to increase the velocity of the grit or shot.
3. Using a liquid, almost invariably water, as a means of carrying the grit or shot, this will normally be supplemented by air pressure.

The shot used will vary considerably, depending on the process itself, the industry and the articles being treated. Types of shot include cast iron and cast steel both of which can be controlled to very tight limits, depending on whether or not it is required that the shot will be very hard and sharp, and cut the components, or whether a softer more gentle action is needed. *Hard-drawn* steel wire cut to the desired size may be used for specific applications, when no cutting action is required.

Aluminium oxide, known as bauxite or alumina, is also used. This can be iron-free when used for non-ferrous blasting under highly controlled conditions, but is very often a crude form of bauxite containing a considerable quantity of impurities. Silicon carbide is a similar type of abrasive.

These materials are used for the production of many grinding wheels and are hard and abrasive. Also used are many other forms of abrasive but it is now illegal, in the UK at least, to use any form of material containing free silica (sand) for enclosed blasting, as it has been found that breathing in the dust formed during the blasting process can result in the lung disease silicosis. It is, therefore, a legal requirement that all blasting grits have a stated maximum quantity of free silica. Under some outdoor conditions the use of free silica (see

Sand Blasting) is permissible. Roughly sieved abrasive grit from slagheaps, etc. may also be used, usually for crude blasting, where the abrasive is employed once then rejected, and operators are protected with special clothing and breathing apparatus. These slag abrasives can, however, be of controlled grit size and homogeneous chemical composition, and under controlled conditions may be reused. Also in use are glass beads and materials such as crushed walnuts and plumstones. These processes are described below.

The parameters to be controlled during any blasting process are as follows.

(a) In many cases it is necessary to control the analysis of the grit to very high standards. This will apply to aircraft and nuclear-type work where any contamination from the blasting grit may cause problems. The size and shape of the grit particle may also be controlled. Again, the type of control will depend on the article being blasted, but generally the grit size is quoted according to mesh size.

Thus, 120–220 grit indicates the size of the particles which will be retained on a mesh containing 220 holes per square inch, but pass through a mesh having 120 holes per square inch. 30–60 indicates a larger, coarser grade, the grit passing through a 30 holes per square inch mesh, and being retained on a 60 holes per square inch mesh. Under some circumstances, as already stated, the particle shape will also be controlled. Thus, where cutting action is required, it will be necessary to specify that the grit has cutting edges, whereas in the case of *peening*, a rounded shot is necessary. Details of grit size testing are given under *Sieve Testing*.

(b) The method of imparting velocity to the shot also requires to be specified. In the case of *air blasting* the air pressure used, the method of inserting the grit into the air stream and the size of the nozzle will all be detailed.

(c) In the case of *airless blasting* the size of the impeller, the speed of rotation and the distance of the impeller from the work will require to be specified.

(d) In the case of *liquid blasting* the makeup of the slurry and the air pressure used will be detailed.

(e) The distance between the nozzle and the article being blasted may be specified.

(f) The intervals when the grit or shot requires to be changed will often be part of the specification.

Blasting equipment will generally require to be self-contained, that is, all the shot or grit, and the resulting debris produced during the blasting process, are retained within the blast area, or precautions must be taken to contain the dust produced.

There are available 'walk-in' types of blast cabinet, where the operator, wearing protective clothing and with his own air supply, carries out the blasting treatment in a confined space, thus containing all the blast media and the debris. This, obviously, has limited application and results in some problems regarding visibility because of dust and, in some cases, the subsequent separation of debris from shot for reuse.

More commonly blasting is carried out from outside the cabinet, the operator using a viewing window and wearing specially designed gloves which are inserted into the cabinet. There are many variations on this design, some of which have controlled dust extraction, shot separation and cleaning for recycling. This method, when correctly used, results in a minimum of dust escaping to the atmosphere.

Equipment using the *air blast system* is also available, where the nozzle has a concentric ring of flexible material with suction applied. This has the dual purpose of controlling the distance between the nozzle and the part being blasted, and also of collecting the spent grit for reuse. However, the method is only effective on reasonably flat surfaces and with equipment which is well maintained.

When blasting is carried out without precautions for collecting dust, or where poor maintenance allows dust to escape, there is a danger of dust contamination of surrounding equipment. There is the obvious problem of abrasive material entering moving parts, and the less obvious one, of dust contamination, resulting in serious corrosion where the active small particles contact painted, plated or any unprotected steel. Blasting under these conditions is often an environmental problem, in that the dust and debris produced are a health hazard, can cause corrosion and will result in an unsightly area associated with the process.

There may, however, be no technical or economical alternative to the blast process. Use must then be made of dust screens and water sprays, together with a general intention to reduce the hazard. Modern environmental awareness and laws are gradually ensuring that such precautions become standard.

One advantage of *vapour blasting* is that the grit, and the debris produced, are kept wet, thus reducing or eliminating the dust problem. Using this technique, it is possible to use higher velocities than with any form of *dry blasting*. This means that, using the same size and standard of grit, faster metal removal can be achieved. At the same time, because of the lubrication effect of the liquid, a superior surface finish results which gives better fatigue strength than that supplied by *polishing*, and is comparable to that of *barrelling*.

There is now a process where the grit is kept damp, allowing the use of higher air pressures and, in some circumstances, use of silica sand.

To summarize, blasting is used to remove metal and scale, to prepare or alter the surface for subsequent treatment, and to increase fatigue strength. The different blasting techniques are described in some detail in alphabetical order as follows:

Glass-bead Blasting   Shot Peening
Grit Blasting   Vacu-blasting
Plumstone Blasting   Vapour Blasting
Shot Blasting

The more general terms, *Abrasive Blasting, Air Blasting, Airless Blasting, 'Aquablast'* and *Dry Blasting* are listed and briefly described in the alphabetical section.

Glass-bead Blasting

A form of *blasting* which uses very small glass spheres.

Application range ALL METALS

It will result in the removal of brittle burrs, but very little surface material. Thus, the glass blasting process should not be used if a specific type of surface finish is necessary. There will, however, be a different visual appearance as the surface will become matt and this process can be used to produce a typical *frosted* finish which is visually attractive.

The fact that no surface material is removed, means that there is no improvement to the surface finish and no radii are produced. The process can be compared in some ways to *vapour blasting* and normal *grit blasting*, but both these processes will result in some metal removal which can be measured by ascertaining the loss of weight and measurable difference in surface finish.

It is used to *deburr* plastic components without affecting the soft plastic surface, and to give the necessary matt finish to some moulding dies.

Grit Blasting

Grit blasting is a relatively vague term where the material used for *blasting* is in the dry state, but uses no specific material such as cast-iron shot or aluminium oxide (alumina).

Application range ALL METALS

It is now illegal, in the UK at least, to use any free silica material such as sand for the purpose of blasting indoors, as this results in serious hazard to health.

It is, however, still commonly found that *sand* and *shot blasting* are equated with *grit blasting*. The term grit blasting should be taken to cover the blasting technique designed to remove scale, corrosion, paint and other surface films by a cutting action. It may also be required to produce a specific type of surface finish, again by a free cutting action. For descaling, *pickling* in acid can be alternative, but this will not produce the type of surface finish necessary for subsequent processes such as *painting* or *metal-spraying*, etc.

The grit can be used damp to reduce the dust problem, but must then be iron free to prevent severe corrosion staining.

Plumstone Blasting

A form of blasting where the 'grit' used is ground plumstones.

Application range ALL METALS

This is a material not unlike sawdust but with unique characteristics regarding its ability to first act as an abrasive and, secondly, an absorbent. These two characteristics allow the use of plumstones in a specialist field where relatively soft metals such as aluminium or magnesium have some contaminant either paint, or an adherent soil such as carbon or burnt oil from the products of combustion.

Typical components are diesel-engine pistons. Plumstone blasting is carried out after initial washing to remove excess oil and sludge. With the correct equipment, which will generally be as for *air blasting*, using pressures up to 15 lb/in^2 and a nozzle diameter of approximately 0.5 in (10 mm) and a distance of approximately 6 in (150 cm), then provided the carbon or paint is hard and brittle, plumstone blasting will efficiently remove the contaminant without in any way damaging the soft metal of the pistons.

The blasting technique appears to have characteristics which are not readily reproducible by any other materials. *Walnut blasting*, using the husks of walnuts, has been shown to give similar results but without the same degree of efficiency. Experiments with the various forms of sawdust and other abrasive absorbent materials have not as yet produced a material giving results comparable to plumstone blasting. Alternative methods of *cleaning* include those based on solvent action, such as paint strippers or organic solvents.

Shot Blasting

A general name used for a specialist process, but more often the generic name applied to any process where shot or rounded particles are impacted on to the surface of the material being blasted.

Application range ALL METALS

This should be distinguished from *grit blasting*, where the material being used is abrasive, and is designed to cut the surface.

Under many circumstances this process and *grit blasting* are designed to achieve the same end-effect, the removal of scale. *Grit blasting* achieves this by impacting plus cutting, whereas the present method is purely by impacting the brittle scale. As stated, there is a blurring of the two processes under many circumstances. *Shot peening* is a variation where specific properties are required as a result of the process, and this is described separately in this section.

Shot blasting should not be used for the pretreatment of surfaces for *painting, plating*, etc. This generally requires that the surface has a roughened type of finish and shot blasting (using rounded shot) will not normally achieve any cutting effect on the surface, and may result in debris being impacted into the surface. Shot blasting is, however, sometimes used for this purpose, when it will generally be found that to achieve a satisfactory finish, the shot is broken up, resulting in a sharp cutting action in addition to the ball-type material. The generic term 'blasting' covers this process and the reader is referred to the information given in the introduction to this section.

Shot blasting may be used as an alternative to *pickling* using acids, or paint removal by solvents, needle descaling or *wire-brushing*, but none of these gives the key which is necessary for good adhesion of *paint* or *metal-spray* deposit. Here the reader is referred to Part Two **Painting** for information on the use of blasting for paint preparation.

Shot blasting cannot be used damp as this will result in corrosion of the shot used.

Shot Peening
A method of locally strengthening the surface.

Application range STEEL AND ALUMINIUM

The process uses either rounded ball or cut wire which has been treated to round the ends. This shot or wire receives velocity, either in an air stream or by means of a centrifugal-type impeller, the distance between the shot leaving the nozzle or impeller and the component being treated being tightly controlled.

The purpose is to form a compressive skin on either the whole surface or, more generally, on a selected area of the component, so increasing fatigue strength. This is achieved because the presence of the compressive skin necessitates the application of a tensile load to this area before a null stress is achieved. Thus, it will be seen that an area of a component which has been correctly shot peened will require a higher tensile stress to produce a fatigue crack than when no such compressive stress exists.

The shot peening requirement will be laid down in a specification, produced as part of the design. Almost invariably the process is controlled by means of a trial peening carried out on a standard test piece. This is known as the *Almen gauge*, a thin strip of metal, either aluminium or mild steel, which is held – to prevent it stretching – by four screws on a hardened steel surface. This strip is then shot peened under identical conditions to those required for the component area. It will be found that some bending has occurred, and by the use of a height gauge the extent of the bend is indicated as a measure of the cold work applied to the surface.

The reason for this is that, since the strip is thin and only one side is shot peened, stress applied to this one surface is, therefore, sufficient in stretch only this side, and the reverse side being unaffected, the strip bends. The quoted *Almen number* is the extent of the bend measured in thousandths of an inch, thus an *Almen number* of 20 indicates that the gauge piece has bent to give an increase in height of 0.020 in (0.5 mm).

The design requirement should be that the size and type of shot particles is specified. In some cases hardened spherical balls are used, but the most common material is cold-drawn wire, commonly called 'piano wire', which has a specified diameter and is cut to the same length as its diameter; the cut ends are treated by a *tumbling process* to remove any sharp corners. The two most common specified wire diameters are 1/32 in and 1/16 in (0.03125 in–0.75 mm or 0.0625 in–1.5 mm). It is also necessary to specify the method of imparting energy to the shot particles. One of two methods is involved, first, using air pressure, where the shot is injected into an air stream, the air pressure and nozzle size being specified and tightly controlled; or, alternatively, using centrifugal force, where impeller diameter, type and speed must also be specified and controlled. In addition, it is necessary to define the distance between the nozzle or edge of the impeller and the area being peened. The final design requirement specifies the number of passes; a typical design specification would be that the specified area for shot peening using 1/16 in diameter piano-wire-type shot at a nozzle distance of 6 in (150 cm) and an air pressure of 15 lb/in^2 with a 0.750 in (20 mm) diameter nozzle, held at a distance of 6 in (150 cm) from the component, requires four passes to achieve an *Almen number* of 20 minimum.

Shot peening is a more controlled method of applying surface *work hardening* than *burnishing*, and there is no real alternative to this technique other than a major design change of the material or surface treatment. It must be emphasized that the increase in fatigue strength is on the existing surface. It is, therefore, necessary to produce, by mechanical means, a high-quality surface finish prior to peening.

Vacu-Blasting
This is a form of abrasive blasting.

Application range ALL METALS

This is a technique where the abrasive blast material is fed through a hose and impinges on the surface being treated. Concentric with and outside the hose there is a second hose larger in diameter with a flexible brush-type surface. This is approximately 0.5 in (10 mm) proud of the blast hose. The surface is flexible and very often consists of a brush-type component. This concentric hose has a vacuum or negative pressure applied such that it will collect the abrasive material and any of the contaminating surface which is removed.

Vacu-blasting is therefore a process which can abrasively treat a local surface with the size of the surface being treated dependent on the diameter of the hose. The difference between this and the normal abrasive blasting is that there is a serious attempt made to collect the abrasive grit and the debris produced and thus this does not either go to waste or contaminate the immediate atmosphere.

The problem with this technique is that it is extremely local and because of the necessity to have a tight control between the nozzle which applies the abrasive material and the concentric hose which collects the debris plus the abrasive material that the ability to sweep large areas is extremely limited.

This technique, while theoretically extremely efficient, has the problems that it is not very flexible and normally is only used where atmospheric controlled conditions are rigidly enforced.

It is normally more efficient and economic either to allow the blast grit and debris to go to waste or to protect the operator by efficient clothing and masking rather than to carry out this technique.

Users should also be warned that because of the relatively small distance between the nozzle and the material being treated that problems can exist regarding the quality of the blast finish.

Vapour Blasting

A process where the abrasive material is in suspension in water.

Application range ALL METALS

The slurry is pressurized and ejected through a nozzle in the conventional manner of blasting. It is found that, with vapour blasting, higher pressures can be used than with conventional *dry blasting*. This results in faster metal removal and at the same time a good surface finish can be obtained.

Two basic types of equipment are involved. First, is the more or less conventional blasting type of machine, where suspended material in water is fed by a pump at low pressure to the nozzle or, using the venturi principle, is sucked to the nozzle. At the nozzle, high-pressure air is injected to give the necessary velocity to the slurry. This air pressure can be 60–80 lb/in^2, and with a sharp grit, in the 200 to 80 mesh size, considerable metal removal can be obtained on mild or low alloy steel. With harder materials this metal removal will, of course, be less, but it will still be in excess of that achieved using the same grade of grit with *dry blasting*. In addition, surface finish will also be superior.

The second technique is a modification of the above, where the slurry is pumped through the nozzle at high pressure and air is then injected into the slurry which has already considerable velocity. It will be seen that this technique results in an even faster metal removal, and it has been shown that there is no reduction in the standard of the surface finish which can be achieved.

The method is more expensive regarding equipment than *dry blasting* but, since considerably higher rates of metal removal and improved surface finish can be achieved, it will generally be an economical *blasting* method when the purpose is to improve the surface finish. This technique can be used to increase the fatigue strength of high-integrity components by removing the stress-raisers produced during normal machining or polishing operations. Provided the grit is carefully controlled, vapour blasting results in a free cutting action with no danger of any *cold working* and, thus, this is desirable in areas where fatigue problems are known to exist.

Vapour blasting is used as a method of *deburring* where components, because of their geometry or the small numbers involved, are not suitable for *barrel deburring*; however, it will seldom be more economical where the latter process is technically possible.

Vapour blasting is comparable to *polishing*, but has the considerable advantage that the resultant surface will have no directional properties – *polishing* almost invariably results in some directional characteristics. The surface finish can be as low as 4–10 micro inches and using coarser grit can be up to 50 micro inches. This will be homogeneous in all directions.

The grit used for vapour blasting must be largely free of iron, otherwise corrosion staining will obviously occur on steel. The most common material is aluminium oxide (bauxite), but carborundum powder can also be used, and there is no reason why other suitable abrasive material should be excluded. As the abrasive particles break up, or are worn, a very fine powder is formed. An improvement of finish, then, is achieved, but the necessary metal removal will, of course, take longer. This aspect must be controlled by use of laboratory techniques or, more generally, the rejection and renewal of the slurry at stated intervals of time, or blasting of components. The choice will depend on the technical requirements of the finished components, and the economics involved.

Vapour blasting is seldom a technically or economically desirable method of paint preparation, but is often used as a method of producing an attractive visual finish before and after *electroplating*.

Where laboratory control of the grit is used, a sample is taken and allowed to settle in a Crow receiver. This allows the variation in grit size to be examined and quantified. Considerable care is required to ensure a good representative sample is obtained.

There is a technique where the abrasive grit is kept damp. This cannot be equated with vapour blasting, but is a form of dry blasting with the dust problem reduced.

Electroplating

Electroplating can, in theory at least, be applied to every metal, and the electrodeposition of every metal itself can theoretically be carried out. In practice, however, there are chemical, electrochemical and metallurgical considerations which considerably limit the number of metals which can be electrodeposited satisfactorily and economically. There are also a number of metals which, with the present state of knowledge, are extremely difficult or impossible to electroplate. This section examines the ease with which each metal can be electroplated with, as well as its suitability for electrodeposition onto, other metals. Each is briefly referred to under its appropriate heading in the alphabetical section of the book.

The process of electroplating or the electrodeposition of metals is comparatively recent. The theory was examined and postulated by Faraday who is known as the father of electroplating. Briefly, the process relies on the fact that when a direct current of electricity is passed between electrodes immersed in a solution containing a metallic salt, electrolysis will occur. This will result in all metals, as ions, migrating to the negative or cathodic electrodes. The non-metal part of the electrolyte, again in the form of ions, will migrate to the positive or anodic electrode. Water will be involved with all practical electrolysis as it is the solvent used for the metallic salts. There is, however, considerable research and development at present into the uses of non-aqueous, organic solutions and molten salts. Further information of the more technical type is also available through bodies such as the Institute of Metal Finishing, but it must be admitted that it is difficult to obtain practical information readily which is published and available under many headings. One reason for this may be the comparatively wide gulf which appears to exist at present between shopfloor technology and research, development and control laboratories.

The electroplating process itself should not be isolated from its accompanying processes. It is vitally important that components to be plated are first thoroughly *cleaned*. The standard of cleanliness necessary is very high, since the components must be chemically clean and this is difficult to achieve. Modern plating shops normally use a two-stage aqueous clean. The first stage involves a hot immersion soak, and this is followed immediately by a second clean generally of the same type but in many instances this will involve an electrolytic cleaner. (These processes are described under the appropriate heading.) While the standard of cleaning will never vary in that the components must be chemically clean, the method of cleaning will vary considerably, depending on the type and quantity of the soil and the shape of the component and, indeed, the kind of establishment involved. Thus, components which have been previously *cleaned* for some other processes or for inspection purposes will require relatively little prior cleaning, but components which have been subjected to *polishing* operations, press shop or heavy machining, or any of the multitude of mechanical processes, will require considerable efforts before they are in a satisfactory state. There is no doubt that unless this high standard of cleanliness is achieved problems will occur at plating: there is considerable evidence to show that an insufficient clean is the major cause of plating faults.

The components must then be thoroughly washed to ensure that all traces of the cleaning liquid are removed. At this stage it will be possible to check that a satisfactory clean has been achieved. Any water breaks, that is, discontinuity of the thin water film on the surface of the component, clearly indicate an unsatisfactory state of cleanliness. Any obvious soil discoloration or evidence of even loosely adherent solid particles must be returned for a further clean, if a satisfactory standard of plating is to be achieved.

Immediately after this swilling operation the components require to be *etched*. This process is sometimes termed *pickling* or *activation*. The purpose is to ensure that the cleaned surface is free of any remaining oxide. This allows the electrodeposition of the metallic ions and ensures a high-quality, intermetallic bond. This etching or activating process should not be used as a means of removing visible corrosion. While light fingerprint corrosion and a limited amount of light active corrosion will be successfully removed with the etch, it is not the prime purpose of the process to act as a *deruster* and, when necessary, this should be a separate controlled process. *Etching* must be carried out immediately prior to electroplating and no delays must occur between the commencement of the etch and the time the components enter the plating tank.

It is possible to hold in the cleaning-tanks or in the subsequent cold swill for periods of approx. 1–2 h without causing any damage. Any delay in excess of 5 min at the etch, however, will result in a plated deposit which will have less than perfect adhesion. The longer the delay, especially if the delay takes place subsequent to the *etch*, the greater will be the danger of faulty adhesion.

The actual etch solution will vary depending on the metal being plated and, to a lesser extent, on the metal being deposited. The common etches are as follows:

Steel 10–20 per cent hydrochloric acid at room temperature up to 2 min immersion. Sulphuric acid can also be used and a 50 per cent electrolytic sulphuric etch is commonly used where high-integrity components are involved or where good adhesion of *chromium plating* is essential.

Stainless steel and nickel alloys Ferric chloride used warm up to 30°C for periods of up to 5 min.

Copper alloys Sulphuric acid used cold at approximately 50 per cent strength. Etches with nitric and/or hydrochloric acid added to sulphuric acid are also used for copper alloys.

The user is advised to obtain technical advice on etches for other metals or special alloys.

Some indication of the solution makeup, the temperature range and current density requirements is given under separate headings for each metal in this section. Some plating solutions can stand considerable abuse and give reasonable deposits. Cyanide solutions in general are very tolerant to abuse but have effluent and safety problems. Other solutions will permit little or no deviation from specification. It is advisable to treat a plating specification as an important document, and ensure that its contents are followed.

The plating process itself relies on the two Faraday laws which state, first, that the metal will be deposited in proportion by weight to the current used. Thus, the higher the current density achieved, the thicker the deposit within any given time unit. Second, that the rate of deposition will be controlled by the chemical equivalent of the metal being plated. It is, thus, impossible to plate, for example, the same weight of silver under the same conditions of current density and the time as copper or any other metal having a different chemical equivalent. This second law has less significance than the first but both are of considerable importance to the practical plater.

The thickness of the deposit will be seen from the above to depend on the current used and the material being plated, other things being equal. It is, therefore, possible to predict accurately the thickness of the achieved deposit. This requires that the other parameters are within control. Finally, it must be appreciated that hydrogen behaves like a metal, and as soon as the conditions are suitable for the electrolysis of water, then some current will be used to produce hydrogen at the cathode together with, or in some cases instead of, the desired metal.

To summarize, the important stages of electroplating are:

1. Cleaning.
2. Etching.
3. Minimum delay between etching and plating.
4. Control of chemical analysis of solutions.
5. Temperature control.
6. Current density control.

If these six points are carefully controlled with efficient swilling between solutions, then good electroplated deposits will be obtained. It should be appreciated that a a detailed specification is essential, which must be written in a sensible manner and adhered to in the spirit rather than the letter. No basic deviation from the specification should be countenanced. In addition reasonable quality control must exist. This will monitor the six stages and measure thickness of deposit, adhesion and general appearance.

Excellent control of plating can be achieved by an initial estimation of the surface area to be plated. This figure can be used for control of current density, and also to make regular additions to the cleaning, etching and plating chemicals and to estimate accurately the effective life of solutions. Chemical analysis then becomes a matter of monitoring the solutions to ensure that they are in control, rather than of instructing the necessary additions.

The rest of this section is divided into the following headings. The material given under these headings deals with all the metals known to be deposited by electroplating, together with information on the techniques used.

Alloy Plating	Gold Plating	Rhodium Plating
Aluminium Plating	Indium Plating	Ruthenium Plating
Barrel Plating	Iron Plating	Silver Plating
Brass Plating	Lead Plating	Solder Plating
Bronze Plating	Lead-tin Plating	Speculum Plating
Cadmium Plating	Levelling	Tin Plating
Chromium Plating	Nickel Plating	Tin–copper Plating
Cobalt Plating	Palladium Plating	Tin–nickel Plating
Composite Coatings	Platinum Plating	Tin–zinc Plating
Copper Plating	Rhodanizing	Zinc Plating

Alloy Plating

The electroplating of any alloy on to any substrate. The process is used for decorative purposes and for the production of technical equipment. The term is used when two or more metals are codeposited from solution, and does not apply when two or more metals are deposited on top of each other. For example, nickel–chromium decorative plating is not alloy plating, whereas tin–lead solder plating is.

The process uses conventional equipment, with the electrolyte containing the metals being deposited. The anodes are either alloy, or the separate metals making up the deposit, or on rare occasions may be inert.

Examples of alloy plating commonly found are:

Brass plating – copper–zinc alloy – for decorative purposes.

Bronze plating – copper–tin alloy – for decorative purposes and for the buildup for salvage purposes, and sometimes used to *stop-off* at *nitriding*.

Solder plating – tin–lead alloy – *stop-off* at *nitriding* and for components to be *soldered*.

Speculum plating – copper–tin alloy – for decorative purposes.

Tin–nickel alloy – for electronic devices and for decorative purposes.

Tin–zinc alloy – alloy for electronic devices to give constant contact response.

Tin–copper alloy – for electronic devices to prevent whisker growth.

The reason for and some details of the solutions used for each are given under the alloy names in this section.

Considerable research has been and is being carried out into the plating of alloys. This is generally in the field of decorative, corrosion-resistant coatings, but also applies to other fields. This is, therefore, a field in which considerable advances are being made and users are advised to contact some competent technical authority for up-to-date information. The metal associations are listed in the Appendix; universities and private companies are also engaged in this area of research.

Aluminium Plating

Carried out almost solely on steel for *corrosion protection*. The electrolyte used is a fused mixture of aluminium chloride and alkali chlorides in a ratio of approx. 4:1, with the anode being pure aluminium. Before plating all impurities must be removed from the bath, usually by electrolysis. The need for a contaminant-free solution necessitates the use of containers made from aluminium.

Nominal operating conditions are to plate from a solution at 175°C using a current density of 15 A/ft^2 (1.6 A/dm^2). It will thus be seen that the process is highly specialized and requires sophisticated equipment; it is only used when alternative treatments are not suitable. Correctly carried out, the resultant coating is similar to but better than the standard aluminized coating and considerably more expensive. It is rarely used at present but is the subject of research and development because of the technical advantage of an aluminium deposit, and the fact that the relative cost of aluminium is reducing compared to other common metals.

Barrel Plating

An electroplating process for small, relatively uniform components where the parts are placed in a rotating perforated drum, which is then immersed in the plating solution. Electrical contact is made by dangling arms inside the barrel, with the normal anodes outside the barrel. The process requires higher voltage than jig plating to overcome relatively poor contact resistance. It has the advantage that the high current density areas are also the mechanical contact areas, thus, there is a tendency for the plating to be worn off; and as these are the points where generally excess plating is achieved a more even than normal metal deposit can be produced.

There are also barrels which contain the solution itself. These operate at an angle, with the anode just below the surface. The solution and plated parts are emptied out at the end of each cycle. The problems with barrel plating are that handling damage is more common and with many parts there is a danger of tangling, resulting in high handling costs for subsequent separation. It is generally unsatisfactory, but still quite common that only the *electroplating* stages in the plating sequence make use of a barrel, the remaining stages using baskets. The use of fully automatic barrel plating equipment, where the labour content of the operation is very low, is becoming increasingly popular, but has a high capital cost. With this, a controlled batch of parts is loaded in the barrel which then progresses through all the stages of the plating process. This technique can achieve excellent results.

Brass Plating

The *electrodeposition* of a copper–zinc alloy. The principal use is for decorative purposes, where mild-steel articles are plated to give an attractive brass finish. A limited amount of brass plating is carried out on steel which subsequently will be rubber-bonded.

For decorative purposes the alloy deposited is 60–40 copper–zinc, and for rubber-bonding 70–30. The solution for decorative purposes has the typical analysis of 9 g/l copper, 3 g/l zinc and 8 g/l sodium cyanide, used at room temperature at a current density of 4 A/ft^2 (0.4 A/dm^2). Some variation in this analysis will be found and variations in the current density, solution makeup and temperature of the solution will give a different alloy deposit. This will be readily seen as quite small variations in the copper–zinc ratio will result in different colours: the higher the copper the redder the brass, the higher the zinc the more yellow the deposit. There are available proprietary solutions with additives which give consistent alloy deposits and a high-lustre, as plated finish, can also be achieved.

Colour anodizing is now a serious competitor with brass plating for decorative purposes.

Brass colouring of steel is extremely thin and readily damaged. This is produced by chemical treatment and is not an effective alternative to brass plating.

The solution for depositing brass which is used to bond rubber will have the typical analysis of copper 12 g/l, zinc 4 g/l, and sodium cyanide 12 g/l. This is used at a current density of 10 A/ft^2 (1.25 A/dm^2) at a temperature of 35°C.

It will be appreciated that any bad adhesion can result in failure of the rubber bond, thus with this application it is essential that preplating procedures are correctly applied. As the bond of the rubber to the brass relies on a clean, oxide-free surface, the rubber bonding must be applied as soon as possible after plating, and care must be taken to prevent contamination of the plated surface. There have been recent advances in the technology of rubber bonding to steel and up-to-date information should be obtained where necessary.

Bronze Plating

A deposit of copper and tin, with the tin content generally held below 10 per cent but above 3 per cent. The result is a deposit with the pleasing appearance of bronze, which is darker than brass and does not have the red coloration of copper. The deposit retains its original colour better than copper or brass, but for permanent use a lacquer should be applied after plating.

The deposit is principally used for decorative purposes, but engineering uses include the *stopping-off* steel to prevent *nitriding*, where it is an alternative to tin, and for the buildup to replace wear under some circumstances.

Other uses are for *corrosion protection* under special circumstances, for example, marine conditions where appearance is important, and as a method of preventing slip. It is possible to deposit under 0.001 in (0.025 mm) on a shaft, for instance, which is then pressed into a bore to ensure an interference fit. The plated bronze is soft enough to be cut where necessary, and thus acts like a glue. Bronze presents no serious corrosion problems when in contact with steel.

The solution used is the cyanide type with a typical analysis giving copper 32 g/l, tin 16 g/l, sodium cyanide 15 g/l and sodium hydroxide 10 g/l. The plating can be carried out over a large range of temperatures up to 65°C at a current density of 24 A/ft^2 (2.7 A/dm^2). This should give a deposit of 0.001 in (0.025 mm) in approximately 30 min. Bronze plating is relatively easy to control and it is probably the tin alloy which gives the least trouble.

Alternatives would be brass or copper for decorative purposes, and it is possible to treat steel chemically to give it a bronze appearance, where this is required for short-term purposes. For buildup and similar engineering purposes tin and nickel could be considered, and comparative corrosion resistance can be achieved with these metals. It is now possible to produce plastic and *colour anodized* aluminium components which are identical in appearance to those achieved by the present method and, thus, are competitive where decorative purposes only are involved.

Cadmium Plating

This process is applied to steel to prevent corrosion. It is comparable in many ways to *zinc plating*, and under most circumstances there will be no visual difference between the two techniques.

Cadmium metal is more expensive than zinc and is much more toxic. For these two reasons, the use of cadmium is decreasing in relation to zinc for corrosion protection purposes. Cadmium metal, and many of its salts, present a serious health hazard at low levels.

Cadmium has some advantage over zinc in marine atmospheres, where corrosion is lessened, and for many purposes is still specified in place of zinc, but this is becoming less noticeable. Where steel requires 'galvanic' protection, and is in contact with aluminium, cadmium will generally be specified in place of zinc, because the corrosion resulting is much less and the corrosion products which can form are much less bulky. The aircraft industry has, thus, standardized on cadmium plating. Cadmium solders slightly better than zinc, and at one time a considerable amount of cadmium was deposited for this reason in the electronic industry, but because of price increases, and the toxicity problems, there is now comparatively little cadmium plated for this purpose.

A typical solution is cadmium 15 g/l, sodium cyanide 40 g/l, and sodium hydroxide 6 g/l. This is used at room temperature and a current density of 12 A/ft^2 (1.2 A/dm^2) to give a deposit of 0.0006 in (0.015 mm) in approximately 30 min.

Cyanide-free acid and low-cyanide solutions are now available which reduce or eliminate effluent problems. It is possible with additives to produce a bright deposit, but there is some indication that these reduce by some degree the corrosion resistance of the deposit. Most cadmium deposits are now '*passivated*' immediately after plating as it has been shown that this considerably increases the corrosion resistance.

Chromium Plating

The *electrodeposition* of chromium is generally applied to steel in all its forms. There are occasions when copper alloys are plated and metals such as nickel alloys may also be plated directly. Direct chromium plating of aluminium and its alloys on to the base metal is not possible at present but can be achieved with specific prior treatment. Because of the nature of the chemicals used during this plating process, most metals which have a tendency to rapidly form surface oxides are difficult or impossible to plate directly.

Chromium is deposited for two specific reasons. First, is the well-known decorative deposit. This is applied because of the bright, attractive appearance of chromium metal. This deposit is invariably on top of nickel, and it is the nickel deposit which supplies the necessary corrosion resistance. Details of the procedure are given below.

Second, chromium is deposited as a hard surface for engineering purposes. Use is made of a different type of solution and the plating conditions vary slightly but in essence both techniques are similar. While it is often desirable and always possible to plate nickel below the hard chromium deposit, this is not essential and in most cases the chromium deposit is applied directly on to steel.

Chromium plating is one of the metals which makes use of inert anodes during the operation. Apart from this it obeys all the other rules and requirements, details of which are given under the heading *Electroplating* at the start of this section.

When decorative chromium is required, the preliminary operation prior to *nickel plating* will very often include *levelling* using copper to give a smooth, bright surface. The type of plating and the thickness deposited will vary depending on the service requirement of the component. This can be as thin as 0.0002 in (0.005 mm) of bright nickel plating or, where very adverse conditions are present, as thick as 0.005 in (0.125 mm). Under certain circumstances two deposits of nickel, one of dull or semi-bright and the second of bright nickel,

CHROMIUM PLATING / 347

may be specified. The user is advised to obtain specialist advice on the type and thickness of metal depending on the service required. British Standards have now produced a specification which gives some assistance in this area.

Immediately after the *nickel plating* the components are swilled and then chromium plated. The solution used will be of chromic acid with some sulphuric acid present, the nominal strength being approx. 500 g/l of chromic acid with a ratio of chromic–sulphuric acid of 100:1. It is the ratio that is important rather than any specific strength of chemicals. Proprietary additives to solutions which are based on the above are now available. They improve the throwing power and control of the chromium plating.

The current density required is in the region of 350 A/ft^2 (38 A/dm^2) and the thickness of the deposit will seldom be above 0.0001 in (0.0025 mm); it will very often be considerably less. The solution is used at approx. 50°C. The purpose of this deposit is purely to give the necessary attractive surface, the substrate nickel will supply corrosion resistance. The thickness of this deposit and the quality of the nickel, then, will largely decide the resistance to corrosion of the finished component.

Hard chromium plating is also carried out using chromic acid as the solution, and again it is important that the ratio of chromic to sulphuric acid is maintained at 100–1. In this instance the strength of the chromic acid is nominally 250 g/l. The current required is in the region of 400 A/ft^2 (43 A/dm^2), and the temperature 50°C.

As with bright chromium, there are now additives available which help to regulate the solution and to help the distribution of current. With the hard chromium deposit, it is common practice for plating to be carried out directly on the substrate. This is readily possible with steel, nickel alloys, copper and its alloys, but seldom possible with other metals, most of which tend to oxidize too rapidly and, thus, form a barrier of metal oxide between the metal substrate and the chromium deposit.

Special *etching* techniques are required. A technically efficient method is to use *cathodic etching*, where the component to be plated is etched in 50 per cent sulphuric acid and is made the cathode in an electrical circuit at approx. 10 V. Other etches include the normal hydrochloric acid but very great care is essential as chloride will invariably contaminate the chromic acid plating solution and, thus, any dragout of the etch into the solution will give problems. Sulphuric acid *etching* alone, under some circumstances, will be satisfactory but again care is necessary to ensure that the component is free of all oxide and scale. With chromium plating, it is possible to *back-etch* in the plating solution. After cleaning the parts are placed directly into the plating solution and are made the anode for up to 1 min. By reversing the polarity, plating commences.

When being used, all chromium plating solutions will give off a spray unless a suppressant is added. This is the result of electrolysis, which causes the gases oxygen and hydrogen to be given off. In the UK it is necessary either to have efficient fume-extraction or to use an efficient proprietary spray suppressant to comply with Factory Act regulations.

Bright chromium plating is now competing with aluminium components which have been *electropolished* and *anodized* and, to a lesser extent, *silver plating*. In the motorcar industry chromium is in competition with stainless steel and plastic. There has recently been a return to bright chrome finish for cars, very often with plastic or zinc as the substrate material. *Bright zinc plating* is an alternative which can give comparable or better corrosion resistance and is much cheaper.

With *hard chromium*, there is serious competition from *electroless nickel* which is deposited homogeneously and can, thus, be plated in areas not reached by chrome, which has poor throwing power. This again has equal or better corrosion resistance than chrome plating. *Hard facing* either by *weld deposit* or *metal-spray* is also possible.

At the design stage the use of *carburizing* or other *surface hardening* techniques can be considered, but these will not give the corrosion resistance supplied by chrome plating.

Aluminium which has been *hard anodized* may be an alternative to be considered when the surface loads causing wear or abrasion are low. Under many circumstances the corrosion resistance of anodized aluminium is superb.

Cobalt Plating

A seldom-used specialist electrodeposit, as the characteristics of cobalt, while in many ways superior to nickel, do not justify the expense of the metal itself and the difficulties involved in plating.

Cobalt is a very stable metal, capable of resisting chemical attack over a wide range of conditions. As stated above, it is comparable in many ways to nickel but has certain advantages. One of these is in the stability of a relatively high reflective surface. When nickel is polished and maintained in a normal atmosphere, it will gradually produce a yellow surface patina. Cobalt has a lesser tendency for this to happen. It therefore finds some specialist application in this field. At one time it was plated on steel then polished as the reflector in carbon arc lamps.

Composite Coatings

These are bi-phase deposits which cover a range of materials.

Application range ALL METALS

The name covers basically any coating which has a matrix containing a suspension. While the name could in fact be applied to many coatings which are in fact paints, i.e. where there is a vehicle with a solid pigment, the term is confined in general to the co-deposition of a metal with some form of non-metal or metallic compound. The term 'composite' is also commonly applied to plastic, laminates or fibre implants. These are non-metallics and thus not a subject for this book.

Electroplating solutions have been devised which have a dispersion of the solid phase suspended in the solution. This can be a plastic, a metal oxide, or carbides. The process is in its relatively early development and is changing, thus the reader should seek further information. The process is at present finding some use where low friction properties are required and there are developments in this area. These appear to be in the low friction plastics such as PTFE, and also the very high hardness particles found with carbides.

Copper Plating

Copper is one of the earliest metals to be *electrodeposited* and there is a considerable history of its use. It is still used for decorative purposes, being one of the two truly 'coloured' metals, the other being gold. A considerable quantity of copper is still deposited on steel for decoration and there is a large industry using copper in various forms as foil on metallic and non-metallic substrates where chemical processes are used to manufacture decorative plaques and articles of various types. To a large extent, however, copper for decorative purposes is being replaced by techniques such as *colour anodizing* and to a lesser extent the use of modern paint finishes.

Copper is *electrodeposited* for conductivity reasons and the printed-circuit industry and the electrical industry in general are making increasing use of copper deposited in this way. With the increasing price of copper in relation to other metals, particularly steel, there is a tendency for copper plating to be used to create a surface conductor with steel supplying the mechanical strength of the component.

Copper is used as an *electrodeposit* to *'stop-off'* carburizing. It was found that the gases could not penetrate a thin, non-porous layer of copper and this was the traditional method of selectively *carburizing*. To a very large extent this technique has been replaced, firstly by the use of copper-rich paints which exactly reproduced the plating process and secondly by techniques such as *carburizing* all over, *annealing*, and selectively *hardening* the chosen area by *induction heating*. The use of copper as an *electrodeposit* is still found in the printing industry. It produces high-standard printing plates. This is now used much less frequently with the advent of photocopying and other new methods of printing.

Copper is also deposited in the manufacture of embossed rolls which are then used to produce cartons and various metal objects with raised decoration. This uses the same basic technique as the printing process, in which the steel roll substrate has a mask applied which will *'stop-off'* certain areas to give the desired design. The rolls are then *electrodeposited* with a thick coating of copper with the reverse of the design that will appear on the finished article.

Copper has been found to be a suitable material for this procedure. It has an efficient plating technique and at the end of each production run, which tends to be relatively short, the copper is stripped from the roll by chemical action ready for the next design.

By far the most common deposition of copper is for *levelling* purposes, in which a relatively poor surface finish can be improved using specially developed plating deposits. The original solution used was the copper cyanide solution but this is relatively inefficient and has effluent problems.

There are four basic types of copper plating solutions. These are: copper sulphate; copper cyanide; copper pyro-phosphate; and copper fluoborate.

1. *Copper sulphate*. This is an old technique; it is probably the simplest form of plating which unfortunately cannot be used directly on steel as the initial deposit will be of the electroless type, where the adhesion of the copper to steel is very poor. The electrolytic deposit will normally adhere in a satisfactory manner to this first deposit, but failure in service will almost invariably occur by peeling of the electroless deposit.

The traditional copper sulphate solution has a chemical analysis of copper sulphate 200 g/l, sulphuric acid 30 ml/l and potassium 12 g/l and is used at room temperature with a current density of 10 A/ft^2 (1.1 A/dm^2), giving a deposit of 0.001 in (0.025 mm) in 2 h.

With modern solutions containing proprietary additions, it is possible to plate at higher temperatures to produce a deposit with a high *levelling* action. These solutions can be used to deposit copper at the rate of 0.001 in (0.025 mm) in less than 10 min, using current densities of 140 A/ft^2 (15.5 A/dm^2).

2. *Copper cyanide*. This is the solution normally used when steel requires to be copper plated. The component is first plated in the copper cyanide solution to produce a thick, dense, non-porous film. As soon as this has been produced, to a thickness of approx. 0.0005 in (0.0125 mm), the components are removed, swilled and plated to the desired thickness using the copper sulphate procedure.

A typical copper cyanide solution used at room temperature would be copper 8 g/l, free sodium cyanide 5 g/l. Plating at a current density of 10 A/ft^2 (1.1 A/dm^2), a deposit of 0.001 in (0.025 mm) can be obtained in 1 h. Using a high-speed plating solution at a temperature of 50°C, the speed of deposit would be 0.001 in (0.025 mm) in 30 min at current desnity of 40 A/ft^2 (4.3 A/dm^2). There is no doubt, however, that these modern solutions require more laboratory control than is necessary with the old method.

3. *Copper pyrophosphate*. Like cyanide, this solution can be used for plating directly on to steel, but an initial 'strike' is necessary. This 'strike' solution will be copper 5 g/l, pyrophosphate 60 g/l and will contain oxalate at 5 g/l and chloride at 10 g/l. This can be used at 50°C, with a current density of 15 A/ft^2 (1.7 A/dm^2). The deposit required is very thin, and a normal plating time of 15

min is generally satisfactory. The parts are removed then placed directly in the standard solution without any necessity to swill.

This solution will have copper 20 g/l, pyrophosphate 160 g/l, with oxalate at 17 g/l and some ammonia at approx. 6 g/l. This solution requires constant control to maintain the correct pH value of 8.4. Additives are generally present to give good *levelling*. It is essential to maintain the ratio of pyrophosphate to copper between 7–8:1. The copper pyrophosphate solution requires careful control and is probably more expensive in chemicals than either sulphate or cyanide. It gives a bright, dense deposit with good 'throwing' power and an even film. When plated on zinc, a 'strike' of copper cyanide or nickel is necessary. To some extent the copper phosphate solution is being replaced by modern fast-*levelling* proprietary copper sulphate solutions.

4. *Copper fluoborate*. This solution is used when a rapid buildup of thick deposits is necessary. This will be required for *electroforming* and the production of rolls used for *embossing*. Like the pyrophosphate solution, the fluoborate system is finding competition from modern high-speed *levelling* copper sulphate solutions.

The copper fluoborate solution will only be used when high-speed plating is required and, thus, tight control of all plating parameters is necessary to ensure that optimum plating conditions exist. Laboratory control is generally required. The solution will contain copper 120 g/l, fluoboric acid 30 g/l and will be used at temperature 45°C. The current density will be up to 400 A/ft^2 (43 A/dm^2), when a deposit of 0.001 in (0.025 mm) will be plated in less than 5 min. This solution contains free fluoboric acid which is extremely active and requires to be handled with care.

Gold Plating

Gold has two specific attractions: first, it resists oxidation and corrosion to a very high degree, while second, it has an attractive colour, being with copper one of the two 'coloured' metals. Whereas copper in a normal atmosphere darkens, and in the presence of moisture becomes discoloured, gold retains its attractive appearance under these and more arduous conditions. Gold plating is, therefore, widely used where appearance and corrosion resistance are of paramount importance, and there is no doubt that but for the very high cost of this metal it would find considerably more application.

In industry gold is now finding considerable use in the electronics and electrical fields. While gold only has conductivity of approx. 60 per cent that of copper, it has the advantage that it will retain its contact resistance over a very wide range of conditions, unlike the oxides and other surface compounds formed by copper which seriously interfere with the contact resistance.

In addition to restricting the actual application of gold plating, the cost also contributes in some way to the difficulty of plating as there is emphasis on the application of very thin deposits which must be wherever possible completely free of pores. This means that control of the deposit must be at a high level. Traditionally gold was plated from high cyanide solutions very often produced by the simple expedient of immersing a gold component in cyanide, which dissolved the metal to form gold cyanide, and using this solution as the electrolyte with an inert anode. In general the most important factor of gold plating will be that the deposit is free of pores, since the purpose is generally to protect from corrosion in addition to other factors such as appearance or contact conductivity. Any pores will allow the atmosphere to attack the substrate and the corrosion product will then appear on the surface of the gold.

Because of this necessity, and the high cost inhibiting the use of thick deposits, it is generally essential to control the solution parameters particularly closely and to ensure that any contaminants which might be present are kept at a much lower level than is normally required in plating. For the same reason there is greater emphasis on an even deposit of gold than with most metals. Freedom from pores and thickness are strictly related, and while with other materials the minimum deposit might be accepted, since it would not be a serious disadvantage that some areas of the component had a thickness four or five times the minimum deposit, with gold this is not normally acceptable, for economic reasons.

In addition to the plating of pure gold, there are now alloys containing small amounts of nickel, platinum and sometimes copper which alter the characteristic of the gold deposit. As the alloy content is generally low, it has little impact on the parameters of gold plating. It will, however, be essential to maintain a more rigid control, as variation in current density can lead to selective plating out of the alloy content. A wide range of other conditions are possible where the control of thickness and porosity are not important.

Proprietary solutions are now available which range from the standard cyanide content to solutions with very low cyanide, and some solutions are cyanide-free. Some examples of these solutions and the current control and plating speeds are as follows. A typical high-cyanide solution contains approx. potassium cyanide 60 g/l and gold 19 g/l. It can be operated at a temperature up to 30°C but generally room temperature is used. At 3 A/ft^2 (0.3 A/dm^2) a deposit of 0.0001 in (0.0025 mm) is plated in approximately 15 min. With this solution additives can be used to give a bright deposit, and also increase the hardness. This solution is used for decorative deposits.

The lower-cyanide and acid solutions are more complex, and it is advised that use is made either of the proprietary solutions or specialist advice sought. These solutions operate at lower levels of gold, 2–8 g/l being usual concentration, and the cyanide is complexed. Cyanide-free acid solutions are readily available. These modern solutions give the same approximate rate of deposition, but a thinner pore-free deposit is possible and more homogeneous plating.

Indium Plating

Indium is a metal not unlike lead but with certain friction and corrosion resistant characteristics which appear to be unique, and is in fact plated for the sole purpose of improving the friction characteristic of high-rated bearings.

The technique is that the bearings have a shell of the required mechanical strength which supports a conventional high-quality bearing material such as tin–bronze. This is then *lead plated* to give improved friction characteristics, and the surface of the lead plated with a thin film of indium metal. Subsequent to this plating there is a *heat treatment* at 170°C for 2 h, when the indium diffuses into the lead, forming a lead–indium alloy which has been shown to have excellent wear and friction characteristics. This would appear to be related to the ability of the indium deposit to resist corrosion.

The solution used is an acid sulphate containing indium sulphate 15 g/l, generally with additives to prevent build up at high current density areas. Plating is carried out at room temperature using 10 A/ft^2 (1.1 A/dm^2). The speed of plating will be approx. 0.0001 in (0.0025 mm) in 10 min.

This metal is expensive and as it will be seen that the complete process is complicated, it will be appreciated that indium plating is only used where very high-integrity components are required. This was the case with the main bearings on aircraft engines of the reciprocating type.

Iron Plating

This is a seldom used deposit at present. Iron is the commonest metal used in engineering and has no obvious characteristic which would necessitate the plating of iron. There are, however, specific instances where iron is plated on top of copper soldering bits or bolts to prevent alloy of solder into the copper.

Iron is plated from acid solution, is relatively simple to control but readers are advised to obtain specialist advice before setting up any equipment.

Lead Plating

This is not a common metal for *electrodeposition*, for although relatively cheap and easy to plate, the deposit cannot be used for *corrosion protection* of steel as it is not galvanic and thus any 'pinholes' in the deposit or damage during service result in the active corrosion of the steel in an attempt to protect the lead film. Lead is an extremely toxic metal and this has also inhibited its use.

Lead plating is carried out to produce components such as the electrodes for lead acid batteries. Steel which has been plated with lead will be much stronger mechanically and lighter than the same thickness of pure lead. The fact that lead has excellent resistance to sulphuric acid gives it some limited application in this field, where components are manufactured from steel to give mechanical strength and then lead plated to prevent corrosion when in contact with sulphuric acid. Some types of stainless steel and titanium and its alloys are now competing with lead plating in this area.

Lead is still deposited as a bearing metal and the user is referred to the heading *Indium Plating* where reference is made to the lead–indium plating technique. With this, lead is deposited on the surface of the bearing, and on top of the lead a thin layer of indium is deposited which then diffuses into the lead.

Lead plating solution contains approx. lead 100 g/l, approx. fluoboric acid 40 g/l together with additives. The plating is normally carried out at room temperature with current density of approx. 20 A/ft^2 (2.2 A/dm^2). This gives a plating thickness of 0.001 in (0.025 mm) in approx. 20 min.

Lead–Tin Plating

This is an unusual alloy for plating, but it is possible to deposit a 10 per cent tin–lead alloy (approx.) which is used for bearing purposes. The solution is of the fluoborate type with a typical nominal analysis of total tin 7 g/l, tin in the stannous state 6 g/l, lead 88 g/l and fluoboric acid between 40 and 100 g/l.

Usually, plating is at room temperature but can be carried out at up to 35°C. A current density of 30 A/ft^2 (3.2 A/dm^2) will give a deposit of 0.001 in (0.025 mm) in 15 min. Unless the correct additives are present, only very thin granular deposits will be possible. With additives, much thicker deposits can be plated successfully.

An alternative to this alloy would be to manufacture bearings from solid material such as bronze or white metal. Cast white-metal lining is also possible. The choice will depend on the type of bearing, the quantities and loading involved. Other alternatives include a deposit of tin or lead, which would not be as satisfactory a bearing material. A deposit of lead, into which indium is diffused, would give superior bearing properties under most conditions, but is a more expensive process.

Consideration should also be given to coatings applied by other means, such as PTFE and graphite-impregnated films, applied for *dry-film lubrication*. *Metal-spraying* can also be used to apply coatings of bearing material.

Levelling

A term, in the metal-treatment context, confined to electroplating and to the metal being deposited.

It is a fact that *electrodeposited* metals will tend to become concenrated at sharp corners, peaks and ridges. The reason for this is that metals are deposited according to the current available, and it is also an electrical fact that the current distributed on any surface will tend to concentrate on sharp corners and be much less inside a concave surface and at the bottom of a valley. Thus, any object which is submitted for electroplating which has a rough, machined surface will tend to have this surface accentuated by the normal plating process.

Because of the expense of subsequent *polishing* operations, considerable research has been carried out in this area and the use is now made of certain constituents that can be added to an electroplating electrolyte which will concentrate at the high current-density areas. These substances produce a polarization effect which will prevent or reduce the *electrodeposition* of the metal involved. When correctly controlled additions are made, then high current-density areas, instead of having an excess of plating, will have less plating deposited, while the low current-density areas will continue to plate at their normal speed. This results in a 'levelling' of the surface being plated.

It must be appreciated that it is possible to produce a level deposited metal which is not bright and shiny. With modern techniques, the tendency is to ensure not only that the surface is level and smooth but also bright. This is achieved by the addition of brightening agents. Much of the economy and new technology offered by research and development in modern *electroplating* solutions is related to the use of 'brighteners', or additives which result in a level deposit which also has a good reflectivity.

Alternatives to levelling are *polishing*, *barrelling* and *electrolytic polishing*, all of which are much more expensive.

Nickel Plating

A very common form of electrolytic metal deposition, but seldom seen as an end-product. It is the nickel plating deposit which is applied prior to *chromium plating* which gives this process its corrosion resistant character, and it is also the latter deposit applied on top of the nickel which gives the bright, hard bluish finish.

There are two common forms of electrolytic nickel plating solutions used. These are the Watt's solution, and nickel sulphamate; nickel can also be deposited by the *electroless* process.

1. *Watt's solution.* This is the traditional nickel plating solution which has been in use for many years and is based on a formula containing nickel sulphate, nickel chloride and boric acid. The different constituents are necessary because nickel sulphate has a relatively low conductivity, while nickel chloride tends to deposit out rather than plate, thus by blending the two salts satisfactory plating can be achieved over a relatively wide range. The boric acid is necessary as a buffer to prevent the solution from becoming alkaline with use. Without the addition of boric acid, it would be necessary to feed an acid continuously to the nickel plating solution to prevent the plating conditions moving out of the specification range by going alkaline.

A typical Watt's solution would be nickel sulphate 300 g/l, nickel chloride 50 g/l and boric acid 35 g/l. Considerable variation on this will be found with the sulphate reducing as the chloride content increases. These solutions can be used over a range of current density, from 10 A/ft^2 (1.1 A/dm^2) to 80 A/ft^2

(8.5 A/dm^2). The temperature used is 40°C. Under these conditions the speed of the deposit will vary from 2 h per 0.001 in (0.025 mm) to 15 min. per 0.001 in (0.025 mm).

2. *Nickel sulphamate*. This is a relatively modern solution and is based on the single saltnickel sulphamate with certain additions. A typical solution would be nickel sulphamate 500 ml/l concentrate, boric acid 30 g/l and nickel chloride 5 g/l. At a current density of 400 A/ft^2 (43 dm^2), 0.025 in (0.64 mm) can be deposited in 1 h. Variations in hardness of the deposit can be achieved with current control, the lower the current density the harder the deposit.

Both Watt's and sulphamate solutions can be supplied ready mixed, and can have additives to give a bright, shiny deposit.

Nickel plating at present is seldom carried out to produce a dull, stress-free deposit but it is almost invariably plated in the bright condition. This achieves two purposes, first, the deposit is reflective, and is thus pleasing to the eye and can accept the *chromium plating* deposit without any further mechanical polishing; and, second, the bright solutions can be made to act as a *levelling* material. This will remove, or reduce, the surface imperfections which occur on normal sheet-metal components or after normal high-quality machining.

This *levelling* can be achieved without giving a high reflective surface, but for most purposes the nickel plating solutions are bright and generally known as *bright nickel*. Because of the high cost of nickel it is now common that the initial *levelling* is achieved by the use of a *bright copper* deposit.

The technology involved in producing these solutions is of a high order and more care is necessary in maintaining the correct level of brightener than any other single parameter in plating. Information on the technique of *levelling* is given under the appropriate heading in this book. Nickel plating is, as stated above, used invariably below the chromium deposit when this is applied for decorative purposes. Nickel itself is non-corrosive, and can be plated with thin deposits in a non-porous condition, thus preventing any attack of the substrate by the atmosphere. However, it is essential that the nickel deposit is of sufficient thickness to withstand the corrosive conditions which will be applied to the component in question.

This will vary considerably depending on the use of the component, and the British Standard on *chromium plating* now specifies the thickness of nickel which must exist below the chromium. This will vary with the substrate metal and the conditions of use. Some details are that the service conditions are graded into four types, with grade 1 being for adverse conditions such as those encountered on the exterior of motor cars, and grade 4 for indoor low humidity such as living rooms or offices. The minimum nickel deposit for each grade is shown in the following table.

Substrate metal	Grade 1	Grade 2	Grade 3	Grade 4
Steel	0.0012 in (0.03 mm)	0.0012 in (0.03 mm)	0.0008 in (0.02 mm)	0.0004 in (0.01 mm)
Zinc	0.0010 in (0.025 mm)	0.0010 in (0.025 mm)	0.0006 in (0.015 mm)	0.0003 in (0.008 mm)
Copper	0.0009 in (0.023 mm)	0.0009 in (0.023 mm)	0.0009 in (0.023 mm)	0.0003 in (0.008 mm)

In addition to this, nickel plating is still used occasionally in its own right as a means of *corrosion protection*. Under these circumstances it is advisable that dull nickel is used, as there is a considerable history to show that bright nickel deposits have less corrosion resistance than the dull nickel deposits. In addition, the dull nickel deposit will almost invariably be more ductile than the bright deposit, and if nickel plating thicknesses greater than 0.001 in (0.025 mm) are required, this higher ductility and better corrosion resistance can be a considerable advantage. This dull nickel, generally the sulphamate solution, is used for *electroforming*, where articles such as record masters and hollow-ware are manufactured. Further details of this are given under this heading.

The *electroless nickel plating* process is now commonly used. While this has less corrosion resistance, by a marked degree than the electrolytic deposit, particularly in the dull ductile deposit, the electroless deposit has the considerable advantage in that it can be hardened by a simple *heat treatment* after plating to a high degree of hardness, thus enhancing the wear-resistant properties. It is this high hardness of the electroless deposit which is finding considerable use in modern industry, but it is as an alternative to *chromium plating* rather than the conventional nickel plating that the electroless deposit is being used.

Electroless plating has the added advantage that the deposit will be homogeneous in thickness, whereas electrolytic deposits will be thicker at corners or points where the plating current concentrates. While this effect is reduced with bright or level solutions, it cannot be eliminated, and the solutions will not affect internal bores where thickness control is difficult with all electrolytic plating.

Palladium Plating

Palladium is a metal in the platinum group. It is, therefore, relatively inert, quite heavy and is resistant to most acids. Palladium is attacked by nitric acid and will oxidize when heated, but apart from this will resist tarnishing and discoloration. It is cheaper than platinum and is, thus, used as a substitute for this metal, particularly in jewellery.

The plating makes use of a solution of tetraamino palladium nitrate. This will contain approx. 10 g/l of palladium. The solution is used at an approx. temperature of 70°C at current densities of approx. 5 A/dm^2 (0.5 A/dm^2), a deposit of 0.000012 in (0.0003 mm) will be plated per minute. This solution will require to be held in a glass or plastic tank and will make use of platinum inert anodes.

Platinum Plating

Platinum is a precious metal which has excellent corrosion resistance, will not tarnish under the most severe corrosion conditions and is, thus, used for industrial purposes where these requirements are essential. It is also a popular metal for electrodeposition on other materials in jewellery and for decorative purposes where a white tarnish-resistance colour is required.

Under many circumstances it can be replaced by *rhodium plating* which is a harder material with comparable but not quite as good corrosion resistance. Rhodium is cheaper than platinum.

Platinum is soft and has reasonably good electrical conductivity and, because of its resistance to tarnishing, will retain its contact resistance over a wide range of atmospheric and temperature conditions.

The *electroplating* solution used is a more complex sulphatodinitrito platinous acid with a nominal metal concentration of 5 g/l. The solution is very acid and can only be retained in glass or plastic containers. It is used at a temperature of 30°C and current density of 5 A/ft^2 (0.5 A/dm^2). Under these conditions a deposit of 0.001 in (0.025 mm) is plated in approx. 2 h.

This metal is now finding some use in the printed-circuit industry where the stable contact conditions are of a considerable advantage. There is a proprietary platinum plating solution known as 'Pfanhauser's solution'. (See *Pfanhauser's Plating* in Part One.)

Rhodanizing

The name given to the technique of *electroplating* silver or gold components with a very thin coating of rhodium.

In the case of silver the purpose of the coating is to prevent tarnishing, and a large number of *silver-plated* objects which are not subjected to abrasion in their life are now *rhodium plated* during manufacture, and then do not require any cleaning or polishing throughout their life. The rhodium coating is extremely thin and cannot be visually identified but is sufficient to prevent oxidation or the fomation of sulphide on the silver surface. Rhodium is hard enough to withstand normal handling wear and wiping with soft cloths, but cleaning with abrasive polishes will eventually result in the removal of this very thin film. It is thus essential that silver components plated with rhodium are not cleaned using conventional methods.

In the case of gold components there is no need for rhodium to prevent tarnishing, as gold will not tarnish under normal atmospheric conditions, and conditions which would tarnish gold would probably also result in some tarnishing of the rhodium. Gold, however, is a soft metal and when used for contact switching purposes becomes seriously indented and malfunctioning can result. By applying a thin film of rhodium to the gold contact surface, better impact resistance is obtained. Further details are given under *Rhodium Plating* in this section.

Rhodium Plating

This is the most commonly plated metal of the platinum group. It is white in colour, has good tarnish resistance and can be plated at a high hardness. In addition it has a reasonably low contact resistance, which because of the excellent corrosion resistance, remains constant over a very wide range of atmospheric conditions. Rhodium plating, thus, finds considerable application in the electrical and electronics industries, where switches require make-and-break-type contact, and where the atmoshphere is either oxidizing or corrosive. Rhodium plating is commonly applied on top of silver for this purpose.

Silver for decorative purposes is very often plated with a very thin film of rhodium, which acts as a tarnish-resistant layer without detracting from the attractive visual appearance of silver. The same technique can be used for industrial mirrors and reflectors and similar types of equipment where high reflectivity is necessary under corrosive conditions.

The solution used is generally acid, containing approx. 2 g/l rhodium and 20 ml/l of concentrated sulphuric acid. The rhodium is generally present as a phosphate, and it is not advised that rhodium solutions are made from basic materials, but that they are purchased as ready-made solutions from a proprietary source. The solution requires inert anodes which are generally platinized titanium.

The conditions of plating are that the solution is used at temperatures up to a current density of 10 A/ft^2 (1 A/dm^2) where simple components are involved.

The coating thickness will be very thin with a satisfactory plate preventing tarnishing of silver being as low as 0.000005 in (0.000125 mm). Where resistance or contact problems are involved, a deposit of ten times that thickness will be normally used. For industrial use, the copper or copper alloy normally involved will very often have a nickel deposit prior to the rhodium.

Hollow articles are often rhodium plated by filling with solution and plating out all the rhodium, then rejecting the solution. The shapes require to be simple, but where qualitative reflectivity is required this can result in excellent quality control with considerable economy.

Ruthenium Plating

One of the metals of the platinum group which has properties very similar to those of rhodium. It is only recently that plating solutions have been devised which can allow *electrodeposition* of this metal. The solution involved is of a complex nature but several proprietary solutions are now available. They contain ruthenium at approx. 10 g/l and are used in the strongly acid condition at the comparatively high temperature of 70°C and a current density in the order of 10 A/ft^2 (1 A/dm^2). As with rhodium the anodes use a platinized titanium.

The deposit obtained is relatively bright and has a hardness estimated to be in excess of 700 DPN. The coating thickness is very thin again, being 0.000005 in (0.000125 mm). Thicker deposits are possible but the control becomes much more difficult. These solutions require agitation which is best achieved by mechanical movements of the parts being plated. Ruthenium plating is still being developed, but would appear to be a very useful competitor to *rhodium plating*.

Silver Plating

Silver and copper are the earliest metals to have been electroplated. Silver is deposited for decorative purposes and is still the most common electrodeposit for tableware and decorative household articles. It is generally deposited on a substrate of nickel silver, which is a copper–nickel–zinc alloy containing no silver but having an attractive silver-like appearance; thus, it backs up the silver deposit if this is polished away during service.

In addition, and on an increasing scale, silver is used in the electrical industry and to a lesser extent electronics for contact purposes. Copper is the normal conducting material and is considerably cheaper than silver, but in normal atmospheres an oxide is formed which has a high resistance. It is the most common reason for heating at contact points on switches and plugs. Silver deposited on the surface of a copper contact considerably improves contact resistance as silver oxide is not formed under most circumstances. The majority of industrial switchgear is now silver plated, as are quite commonly fuse contacts etc. in household electrical plugs. Silver is not suitable where sulphide atmospheres exist as black sulphide is readily formed, giving poor contact resistance.

Silver still finds a limited application for bearings but this is on a reduced scale with the advent of other materials such as lead, aluminium alloys and plastics such as PTFE. Finally, silver is now being used to prevent hot seizure: it was found that a thin deposit on either male or female components of fasteners required to operate in relatively high-temperature oxidizing conditions prevents the seizure of these components. This is a common problem on studs on the exhaust manifold of internal-combustion engines, and on many stud fixings on jet engines.

The silver-plated deposit will have a brilliant white finish but unless special brighteners are added it will have no lustre and requires polishing. It is now possible that an 'as plated' deposit is produced with high reflectivity, but this too will often require a final polishing to satisfy the high standards of eye-appeal necessary in many products.

Silver has probably the best 'throwing' power of any common electro-deposit. While this is a considerable advantage in that the deposit can find its way into deep recesses without the use of special anodes, it can under some circumstances cause problems where *stopping-off* is required as the silver deposit is notorious for finding its way behind conventional systems.

Almost without exception the plating solution will be of the cyanide type. Because of the high value of silver there is no 'drag out' problem of cyanide to the drain, thus the effluent problem is at a minimum.

Silver plating requires an underlay of copper or nickel between itself and steel to prevent electroless deposition and poor adhesion. A 'strike' with low current density, low stress plating is also advisable where thick deposits are required. Silver is a precious metal and deposits will seldom be thicker than the essential minimum. For uses on cutlery, for instance, a good-quality deposit would be 0.002 in (0.05 mm) and this would be expected to give up to twenty years' life under canteen, etc. conditions. Household cutlery is generally considered satisfactory with half this thickness. Thickness, then, will be a function of the service conditions, and in particular the number of times that a cleaning cycle is applied as each time the silver is cleaned the surface is abraded.

For engineering purposes, that is, where contact resistance is required, a deposit of 0.001 in (0.025 mm) is again generally considered satisfactory, but deposits up to 0.050 in (1.25 mm) will be required under certain conditions. For prevention of seizure, the thickness deposited is seldom above 0.0002 in (0.005 mm) and this thin deposit will result in a satisfactory coating.

Alternatives are *speculum* and *bronze plating* for household use and there is no doubt that austenitic stainless steel is competing successfully with silver plating in this area. On electrical contacts, gold and rhodium are finding application particularly where the conditions are unsuitable for silver. For seizure application, *dry-film lubricants* using molybdenum disulphide and graphite-impregnated paint have been found to be successful.

Decorative silverware is often given a very thin deposit of rhodium to prevent oxidation or the formation of the black sulphide. This in essence is a clear lacquer which is hard and becomes part of the electrodeposit film, successfully giving longer life without polishing, particularly for articles which are purely decorative (see *Rhodanizing* in this section).

Solder Plating

The term covers the deposition of tin–lead alloys over the range of 50–90 per cent tin, the remainder being lead. By far the most common alloy specified will be 60 per cent tin, 40 per cent lead, which is the solder generally used in the electrical and electronics industries. Where corrosion resistance is required together with 'solderability', this is also the most economical material. Tin–zinc deposits will be cheaper, but have much lower 'solderability'.

The control of solution analysis, additive and current density is important for a dense deposit free of 'trees' at corners and will control the alloy composition.

As the deposit is often plated for the dual purpose of acting as an *etch* resist (*stop-off* agent) and *soldering*, it will be appreciated that it is essential that it is free of pores and excess granular deposit, while remaining close to the desired alloy of 60–40 tin–lead. Good 'solderability' relies on a consistent 60–40 alloy, free of 'trees' and pores. The plating of pure tin, or tin–copper (2 per cent copper) and tin–nickel deposits are now alternatives to this solder alloy. These deposits, however, can present more problems regarding 'solderability' or as *stop-off* agents under many circumstances than the tin–lead alloy. At present, many advances in these deposits are being developed and the user is advised to contact the Tin Research Institute for up-to-date information.

The solutions most commonly used are fluoborates which are very corrosive and also a health hazard.

Speculum Plating

Speculum is an alloy of copper and tin, containing approx. 40 per cent tin. It is relatively hard and can be given a good lustre either by polishing, or by the use of additives to the plating solution.

The deposit is used for decorative purposes, usually as an economical competitor to silver. It is non-toxic and, thus, can be used for the same range of items as silver. The solution is based on a mixture of sodium stannate and copper cyanide and the anodes can be an alloy or a balance of pure copper and tin. Further details can be obtained from the Tin Research Institute.

Tin Plating

This is carried out for several distinct reasons. There is a considerable amount of tin plating on copper alloys which will be subsequently *soldered*, the purpose being to produce a 'solderable' surface. To a lesser extent tin plating is carried out on steel for the same reason.

Tin plating is employed on the most extensive scale on sheet metal to prevent corrosion in contact with foodstuffs. In tin-plate mills this is now conducted with highly sophisticated equipment where it is a continuous process with all plating carried out with the steel strip moving at speeds in excess of 30 mph. The end-product is used to manufacture the tin used in canning foodstuffs.

Tin plating is also carried out in order to 'tin' the base material prior to the *casting* of white metal bearings.

Finally, tin plating is still used, but on a limited scale, with modern metals, to prevent *nitriding* of steel components. In this way it is analogous to the use of copper at *carburizing*, and like *copper plating* is being replaced by alternative processing. In the case of tin this will be *bronze plating* or commonly the use of bronze-rich paints.

There are three types of tin plating solution. The different solutions all give a thin deposit which is capable of being used for any of the uses listed above. It should, in fact, be stated that all tin plating solutions are relatively difficult to control, and with high-speed deposits it is essential that laboratory control of all parameters is strictly carried out.

The following nominal thicknesses are listed for the uses listed above. For contact with foodstuffs or where first-class corrosion resistance of steel is necessary in contact with water, then 0.0012 in (0.030 mm) minimum is essential. For protection against normal atmospheric corrosion on a long-life basis on steel, the minimum requirements are 0.0008 in (0.020 mm). For corrosion conditions under moderate environments with occasional condensation, the minimum requirement on steel is 0.0004 in (0.010 mm).

For 'solderability' purposes on copper, bronze or brass underlays, where the storage conditions and shelf-life are moderate, a deposit of 0.002 in (0.05 mm) is considered sufficient. For coating on brass in particular, or steel where 'solderability' must be maintained on the base metal. Shelf-life is time-limited and, thus, storage time and conditions must also be controlled. For *stopping-off nitriding*, the minimum requirement is 0.002 in (0.05 mm).

It should be noted that where plated deposits are flowed by fusion in absence of air, generally by immersing in hot oil, then a lower thickness is acceptable. The reason for this is that the as-plated deposit will have pores of a certain depth, and so by plating to a certain thickness, it is possible to guarantee a pore-free surface. The pores are initiated in a random manner as plating proceeds and, thus, all pores do not penetrate to the substrate surface. The subsequent flowing by fusion eliminates the problem.

Without doubt, tin plating would be used on a wider scale for general *corrosion protection* if the metal were cheaper. With the continuing increase in the cost of tin relative to other metals, it is unlikely that its use will increase outside the fields of corrosion resistance where foodstuffs are involved and 'solderability' where high standards are necessary. Tin is one of the few metals which are non-toxic with good corrosion resistance. It is finding increasing competition in this field from aluminium, where *deep-drawn* cans are being used to replace the conventional tin-coated, mild-steel rolled and soldered ones. Plastics are also competing with tin in the packaging of food. For 'solderability', there are fewer economical alternatives but the use of tin–nickel and tin–zinc deposits are indications of the search for suitable alternatives.

Tin–Copper Plating

An electrodeposit confined to use in the electronics industry, where the components are subsequently to be *soldered*. The deposit is, thus, invariably on top of copper or copper alloy conductors; but it may have a thin layer of nickel, or pure copper, between the substrate and the tin–copper deposit to prevent diffusion of undesirable elements such as zinc into the tin, so reducing the 'solderability'.

The deposited alloy has between 1 and 3 per cent copper in the tin, and the purpose is to prevent the formation of 'tin whiskers' during service. Experience has shown that a pure tin deposit subjected to certain conditions of stress can result in the growth of a fine thread or whisker of pure tin resulting in current leakage between conductors. Over 1 per cent copper in the deposit will inhibit this whisker growth, but over approx. 5 per cent copper, while still inhibiting this growth, can give rise to 'solderability' problems. Other contaminants in tin also preclude whisker formation, but most of these such as zinc and aluminium can again give rise to 'solderability' problems. Tin–lead alloys are not prone to whisker growth. The solution used is a standard stannate type to which is added copper.

Tin–Nickel Plating

An alloy deposit which was developed by the Tin Research Institute. It is used on a limited scale by the electronics industry as an alternative to pure tin and tin–lead where consistent 'solderability' is of more importance than actual 'solderability'. The deposit requires active fluxes, but the 'solderability' remains constant over a wide range of storage conditions and time.

The alloy is also plated for decorative purposes, having an attractive high reflective sheen with a slightly reddish or pink colour. It retains its high lustre over a wide range of conditions.

The solution contains tin and nickel and is used at a high temperature. It is highly corrosive, thus presenting some difficulty in the equipment used, the choice of substrate metal, and when any selective plating necessitates *stopping-off*.

The anodes used are an alloy of tin and nickel, and the user is advised to contact the Tin Research Institute for up-to-date information.

Tin–Zinc Plating

An electrodeposit which is generally applied to steel components to supply corrosion resistance. The protection achieved is better than the normal zinc, or cadmium, thus a thinner deposit can be accepted for the same standard of protection. With the increase in cost of tin, however, this alloy is seldom used as it is more economical to plate slightly thicker zinc or cadmium than the tin–

zinc alloy. Where, for technical reasons, a very thin deposit is necessary, this alloy could be considered.

The tin–zinc alloy also has better 'solderability' than zinc, but it is not really comparable to pure tin or clean copper and will not be considered 'solderable' by electronics standards, where non-active fluxes are essential.

The alloy deposited is 75 per cent tin, 25 per cent zinc.

Technical difficulties may be encountered in controlling tin–zinc plating to give a consistent deposit, particularly when the solution is used with internal anodes. The deposit can be chromate *passivated* when the corrosion resistance is of a very high standard, but solderability is poor.

Where corrosion resistance is the prime consideration, zinc plating can be an economical, technical alternative. Where 'solderability' is a serious problem, and economics rule out the use of tin or tin–lead, then tin–zinc is unique.

Zinc Plating

This is now probably the most popular metal for *electrodeposition*, and the sole purpose is to improve the corrosion resistance of steel components.

For many years industry has successfully used *hot-dip galvanizing*, which coats the steel with zinc. The advantage of zinc as a material for *corrosion protection* is that it is 'galvanic', or anodic to the steel substrate. This means that when the coating is damaged for any reason, the zinc will sacrifice itself to protect the underlying steel. This 'galvanic' or 'sacrificial' corrosion gives excellent protection to steel and is the reason that zinc is popular.

The electrodeposition of zinc uses three basic types of solution, two of which are based on cyanide. The third solution is the more recently developed acid zinc, which is free of cyanide.

There has been considerable development work on the corrosion resistance achieved with zinc plating. This has shown that by *'passivation'* or *chromating* immediately after depositing the zinc, an increase in the corrosion resistance is achieved which almost equals that of the original deposit, and this technique is now almost universally applied to zinc-plated components. Developmental work has shown at the same time that, unless the various parameters which exist at zinc plating are correctly controlled, the deposit will have inferior corrosion resistance. These parameters are the chemical composition of the solution, current density to within reasonably tight limits and the thickness of the deposit. The thickness of deposit is a function of time and current density. Where high-speed plating is carried out, there is no doubt that with adequate control the deposit can equal the standard of the slower solution. Where quality plating is essential, then, laboratory-type control is necessary. Considering the high cost of electroplating and the disastrous consequences of faulty deposits, it is strongly recommended that zinc plating and *chromating* are always carried out under controlled conditions. The period of time over which zinc plating will provide protection is a function of the thickness, other things being equal: the thicker the deposit, the longer corrosion resistance will last.

Electrodeposited zinc will seldom have a thickness greater than 0.0015 in (0.038 mm) and will commonly be little in excess of 0.0003 in (0.0075 mm). For outdoor use *electrodeposited* zinc is not a serious competitor to *hot-dip galvanized* components, where the thickness will seldom be less than 0.003 in (0.075 mm) and is commonly in excess of 0.010 in (0.25 mm). There is now considerable data available on the life expectation of zinc coating and for up-to-date information users are advised to contact the Zinc Development Association.

Electrodeposition of zinc is commonly carried out on components which will be subjected to indoor conditions. The electrical and electronics industries, and the communications industry, are all large users of *electrodeposited* zinc and have been instigators in the improvement of control to guarantee thickness of deposit and quality.

The quality is generally assessed by a series of tests including *thickness tests, salt-mist accelerated corrosion tests, adhesion tests* and *fingerprint tests* to assess the ability of the coatings to withstand handling corrosion. All of these are discussed under their own headings in this book.

Because of the relatively even coating which can be applied as distinct from *hot-dip galvanizing*, zinc electroplating is now the most common method of *corrosion protection* applied to nuts and bolts. These will generally have a thicker coating than that for indoor purposes, and will almost invariably be subjected to post-plate chromate passivation.

Sherardizing is a means of evenly depositing zinc and this can be used for threaded components.

Electroplated zinc can be passivated and then painted using special paints, and this gives indefinite life, with positive protection when the paint film is damaged.

Hardness testing

An inspection procedure which is a measure of a metal's resistance to indentation or cutting abrasion and as such there is a relationship between hardness testing and the 'ultimate tensile strength' of the material.

Application range ALL METALS

There is no exact formula for the conversion of hardness testing to the ultimate tensile strength, but comparison of various hardness testing systems is given in the Appendix to this book and is related to the ultimate tensile strength of steel. The different hardness testing techniques can be divided into six types:

1. Using a machine with a penetrator which is loaded and impressed into the component being tested. The diameter of the impression is measured, and the larger the diameter, the softer the material. When the standard material used for testing approaches the hardness of the actual material being tested, then the size of the impression will be proportionally reduced.
2. Using a machine with a diamond (which is infinitely hard in relation to most materials). The diamond is loaded in a similar manner to the penetrator, described above, and cuts the material. The size of the diamond impression is then measured, and again the larger the impression, the softer the material.
3. Using a hardened indenter (diamond, etc.) to make an impression, and measuring the permanent depth of impression.
4. Using abrasion as a means of indenting or cutting. Since no material can be cut or scratched by a softer material than itself, it is possible to assess the hardness by noting whether or not the material is scratched by a series of standards.
5. Using vibrations, the difference between the material being tested and standard test pieces are assessed. These vibrations can be over a wide range of frequencies.
6. Using the reaction principle, that a weight of standard hardness will bounce higher from a hard surface than from a soft. The height obtained, or the speed of rebound is measured.

Hardness testing requires considerable care in order to ensure that a sensible result is obtained. It must, of course, be realized that hardness testing measures only the surface of the material or component being tested. It is possible by using heavier loads to average the hardness found over a thicker layer, but unless precautions are taken and any surface peculiarities removed prior to the test, unsatisfactory results will be obtained. Any removal must be by careful *grinding* or filing of surface layers which are not considered to be an essential part of the material. Where, of course, parts are *case hardened* or have some intended surface effect, then in general the test will be to find the hardness of the surface layer and no surface removal is possible. In these cases, it is necessary to be very careful regarding the depth of the impression and, thus, the load applied to the penetrator must be carefully chosen.

Again, the preparation of the surface must be carried out carefully, so that this does not affect the hardness to be measured; thus, *grinding* must not generate heat and any cutting method must not cause *cold working*.

The impression obtained with hardness testing is usually minute and demands painstaking measurement. It is, therefore, necessary that the finish obtained on the surface is relatively smooth otherwise the impression will be distorted. The surface must also be flat as elliptical impressions or elongated diamonds will not result in an accurate result. The material being tested should be well supported, otherwise when the load is applied, there will be bending or movement instead of cutting or impressing, and this will result in small impressions and a misleading appearance of high hardness.

The thickness of the material undergoing testing is also of critical importance. If this is too thin, the anvil on which it rests will be reflected through the component and will affect the hardness figures obtained. It is a general rule that the thickness of the material being tested should be at least five times the depth of the hardness impression.

The type of penetrator and load used will vary for different materials. The following points must be attended to, prior to attempting hardness testing. The material must have a smooth finish, must be held rigidly on a firm base, and the load chosen must be suitable for the material being tested and the type of indenter used and light enough not to penetrate deeply into a thin component.

It is generally not satisfactory to grip components on a vice without the base of the component also being supported. Minute slippage between the jaws can give wrong results.

The accuracy of hardness testing relies on the above factors and their control. It is essential that constant calibration of all testing machines is carried out by the operator. This is readily accomplished by the use of flat standard test pieces which have a guaranteed hardness with a high-quality surface finish and sufficient thickness. Calibration using standard test pieces is essential before setting up a hardness test. The calibration should be carried out at least once per day, or immediately prior to hardness testing where infrequent testing is used. The test piece used should be in the same range of hardness as the material being tested. It is recommended that test pieces should be employed which are below and above the hardness of the material, and for laboratory conditions this is essential. Periodically the hardness tester should be subjected to a series of standard test pieces covering the complete range of hardness carried out, and the results noted. It is recommended that test pieces should be available at approximately every 100–200 points of hardness on the *Diamond scale*, and where high-integrity results are required the daily calibration should ensure that test pieces are used below and above the hardness of the component being tested.

In addition to these relatively simple methods of checking the accuracy of the machine, it is recommended that production machines are compared to laboratory-type machines at frequent intervals and that the laboratory machine is either calibrated against a master machine or is checked at least twice per year by independent experts.

Hardness testing makes use of scientific equipment which is relatively delicate and, thus, can easily sustain damage. In many cases the result of the test will be of critical importance to the component being tested and if wrong can result in disastrous or expensive failure.

Where accurate results are essential, a number of tests on the one area should always be carried out. The first test is generally rejected as this will very often be affected by the bedding down of the component. An average over three or four impressions should be taken, and if this average varies more than approximately 5 per cent, then the situation should be investigated.

One of the more important points to realize with this test is that it is an accurate measure of the hardness and, as stated above, there is a relationship between hardness and the 'ultimate tensile strength'. Where hardness is required for wear resistance, the hardness test is the correct means of assessing this material property. However, where the test is used as a means of estimating the tensile properties of a material, it should be realized that the test does not measure the *'proof'* or *'yield strength'* of the material, which in general is the important factor associated with tensile properties. This is lower than the ultimate tensile strength, and can be affected by *heat treatment* in the case of steel and *cold working* in all metals.

Where relatively low-strength materials are involved as is the case with mild steels and most non-ferrous alloys, then the design safety-factor will generally be sufficient so that any variation in the elastic limit or proof strength will not result in failure. However, with many alloy steels which are heat treatable the wrong treatment can result in an appreciable lowering of the yield strength or elastic limit with no appreciable difference in the hardness obtained. The only accurate measure of the yield or elastic limit property is the *tensile test*. This is a destructive test and requires expensive machining of test pieces.

A number of textbooks have been written on the difficulties and control of hardness testing. The British Standards Institution supplies considerable information on this subject.

The following are the methods of hardness testing commonly used together with less common ones. The techniques used by industry are based on the diamond penetration test (Vickers), or the Rockwell or Brinell tests which are almost universally used by material laboratories and industry at present. Each type of test is listed under its own heading in the alphabetical section of this book, and are collated in this section where a brief description is given. The figure in brackets after the name refers to the six techniques referred to at the beginning of this section.

Brinell (1)	Jagger (4)	Pusey (1)
Diamond Hardness	Keeps (4)	Pyramid Diamond (2)
Edgewick (1+2)	Kirsch (1)	Rockwell (3)
Equi-Tip (6)	Knoop (2)	Sceleroscope (6)
File (4)	Ludwig (2)	Scratch (4)
'Firth Hardometer' (1+2)	Microhardness (2)	Shore (6)
	Mohs (4)	Tukon (2)
Foppl (1)	Monotron (2)	Turner's Sclerometer (4)
Fremont (2)	Muschenbrock	Vickers (2)
Haig Prism (2)	Pellin's (1)	'Warman Penetrascope' (2)
Herbert (6)	Poldi (1)	

Brinell Test

One of the original scientific methods of assessing hardness. It is applied to all metals but is limited to the relatively soft range. The reason for this is that the Brinell test uses a hardened steel ball of specified size. The load is applied directly to this ball and the diameter of the impression measured. It will be seen that the larger the diameter of the impression, the softer is the material being tested. It will also be appreciated that the hardened steel ball itself has a specific hardness, and this ball will distort as the hardness of the material being tested approaches the hardness of the ball itself.

The Brinell method together with the Diamond (Vickers) test and the Rockwell hardness test are the three most commonly specified in the UK. However, the Brinell test is not now commonly carried out as it is relatively slow and, as stated above, has limitations regarding the range of hardness which can be measured. It is usually replaced by the Vickers test, which uses a diamond and is based on basically the same principle as the Brinell – that is, measurement of the size of the impression. The larger the impression, the softer the material being tested.

There is a relationship between hardness testing and 'tensile strength' (see Appendix 5), but the relationship should be treated with caution as the ratio varies under certain circumstances. It is, however, a useful guide. The same comments apply to the relationship between the different methods of hardness testing, which again should always be approached with caution. A table showing the relationship between Brinell testing, 'ultimate tensile strength' of steel and various other types of hardness testing is given in Appendix 5.

Diamond Hardness Test
See Pyramid Diamond Test.

Edgewick Test
This test uses the Vickers principle, with a diamond or ball penetrator. This is held on a hinged device which, on completion of the load, is removed and the impression then projected on to a screen in front of the operator. This screen has graticules to allow rapid estimation of the impression size. With this tester, the load can be infinitely varied by the use of a lever, and it is for this reason that the device is not permitted by several inspection authorities, since it is considered too difficult – if not impossible – to ensure accurate application of the desired load. The technique can use a ball as an alternative to the diamond penetrator, when it is similar to the Brinell system.

Equi-Tip Test
This is a proprietary tester which uses the rebound principle. The indenter used is small, portable and spring loaded. When released it strikes the surface and rebounds. The speed of rebound is measured electronically and indicated on a meter. This is an extremely portable and rapid technique, but requires considerable care as the mass and surface finish are very critical, and extraneous vibrations can give wrong results.

File Test
A test based on the Mohs technique. This relies on the fact that a material cannot be scratched or filed by a softer material. With the Mohs test, a range of materials is used to classify the material being tested. With a file test, materials are tested in relation to the hardness of a sharp hardened file. The test is usually restricted to surface hardened components which have been subsequently machined or treated by some heat treatment, which can result in an extremely thin, relatively soft layer on the surface of the carburized component. Because the layer is so thin, it cannot normally be identified using conventional testing techniques.

With a file, in the hands of a skilled operator, it is possible to identify this thin, softened layer because the file will bite the surface, whereas with correctly hardened material, which will be as hard or harder than the file, it will skid off the surface. This is an extremely critical test, and there is no simple alternative method of identifying this very thin layer of soft material. During service, the softened layer can result in pickup, or galling, with subsequent pitting, and in serious cases, where high loads are involved, failure can occur.

'Firth Hardometer' Test

A machine based on the Brinell principle and using a ball or diamond. The head holding the indenter is borne on a swivel and, on completion of the load application, the head is swivelled out of position and a projection-type microscope replaces the indenter automatically. This means that the operator is not required to search for the impression projected onto the screen. This can have an engraved graticule which allows estimation of the impression size.

Foppl Test

This test is confined to round bars and is seldom used in modern industry. The technique involves placing two of the round components or short lengths of bar at right angles to each other and then applying a predetermined load which is held for a set time. On removal, a circular impression should appear on both components; the softer the material, the larger the impression. Provided both components have the same hardness, the impression produced will be completely circular and identical on each component. This test is now replaced by one of the standard production techniques discussed in this section.

Fremont Test

An obsolete test in which a weight with a pointed tip is allowed to fall from a known height on to the test piece. The size of the indentation produced was a measure of the hardness. The smaller the indentation, the harder the material.

Haig Prism Test

A test where it is necessary to prepare special specimens cut to form 90° prisms. These are held at right angles to their edges under a set load. The indentation thus formed is measured, and gives an indication of the hardness.

This hardness technique is now very seldom used, but has the advantage that no special equipment is necessary. The hardness value is obtained from the formula

$$\frac{\text{Load applied}}{\text{Width of indentation}} = \text{Prism hardness.}$$

There is a relationship between this and the *Brinell hardness* figure.

Herbert Test

A specialized hardness test using the pendulum principle: a ball indenter is allowed to swing. After it has made an impression under the specified load. The pendulum stroke is timed and measured for an indication of hardness. This is seldom, if ever, used in modern industry. The longer the pendulum swing the higher the hardness.

Jagger Test

A specialist, and seldom used, test using a diamond-pointed drill under a load of 10 g rotated in contact with the test piece. The number of rotations is counted to achieve a depth of 0.01 mm. The higher the number of revolutions, the harder the material.

Keeps Test

A test where use is made of a rotating drill with a standard bit revolving at a known speed with a standard load applied. A material of the same hardness as the drill will result in no cutting occurring, while the softest material will result in the drill reaching the maximum depth.

The Keeps system uses a diagram where the hardness of the drill is indicated by a vertical line, and the hardness of the soft material, for example, 50 DPN, is the horizontal line. The angle of the material being tested will indicate the relative hardness.

This test has certain advantageous factors where hardness is required to be known for assessing 'machinability' but has limited application with today's modern testing machines and knowledge of 'machinability' problems.

Kirsch Test

A hardness test which can be applied to any material. It makes use of a steel punch 5 mm in diameter to which a load is applied to cause a depression of 1 mm in the test material. It will be appreciated that, with a material of the same hardness or harder than the punch, it will not be possible to cause any depression and, thus, an infinite hardness will be recorded by this method.

This test has very limited application in modern engineering and has generally been replaced by the *Brinell*, *Vickers*, or *Rockwell* tests. For portable tests, the *Scleroscope* or *Poldi* are now commonly used.

Knoop Test

Based on the *Brinell* or *Vickers* principle, in which an indenter is loaded and cuts the test material. A diamond is used, but the indenter is rhombic in shape.

In general the Knoop test is confined to light loads and is usually applied to very thin sheets of material; it is also employed for *microhardness testing*.

Ludwig Test
In this test the indenter has a 90° cone. The load is applied to this to form a conical impression, and the hardness of the material is found by a formula where the load applied is divided by the area of the impression. This has now been replaced by the more conventional *Brinell*, *Vickers* or *Rockwell* tests.

Microhardness Test
In essence, this method is identical to the standard hardness testing technique with the Vickers principle. The difference here is that a metallurgical microscope is used and the object being tested is usually a microspecimen, specially prepared for metallurgical examination. The loads applied are very low, varying from 1 g to 200 g and, in some circumstances, 500 g.

Apart from the above modifications, the technique uses the normal hardness testing method, where an impression is made by applying a known load to a diamond indenter for a specified time, and the impression measured accurately. The larger the diamond impression, the softer is the material.

With this technique, it is possible to carry out comparative tests on actual grains within the metallurgical structure. With the lighter loads, below 100 or 50 g, the accuracy of the result obtained is reduced, but it is possible to obtain results which give a meaningful comparison.

Mohs Test
This testing technique is more commonly applied to rocks by geologists than as a treatment of metals. The technique uses the fact that a material can only be scored by a material harder than itself. Thus, a diamond can scratch glass, but glass cannot scratch the harder diamond. In metal testing this test is normally applied only to carburized or nitrided components.

Where the surface hardening technique results in a very thin case, or where in carburizing subsequent heat treatment processes result in an extremely thin surface or film of softer material on the hardened case, then conventional testing techniques will not be sufficiently accurate. Use is made of a new, sharp, hardened file of approximately the same hardness as the carburized case, and if the case is correctly treated, it will not be possible to file the surface. Any softening of the surface will be identified by the file scratching or holding rather than slipping off the surface.

This in skilled hands provides an inspection technique which will identify surface defects which cannot be found by conventional testing, and which are only identifiable by destructive metallurgical examination.

Using needle-pointed hardened steel scribes of varying known hardness it is possible to use this technique for assessing surface hardness. It is possible to use the technique in awkward corners where conventional systems would not be possible. While the degree of accuracy will not be high, when used with care and some skill sensible results are possible. The surface being scribed must be free of contamination, paint and scale.

The Mohs hardness scale used by geologists is as follows:

Mohs number	Material used as standard	Similar material
10	Diamond	
$9\frac{1}{2}$		Carborundum
9		Sapphire
8	Topaz	
7	Quartz	Carburized steel
$6\frac{1}{2}$		Garnet – Steel file
6	Feldspar	Orthoclase
$5\frac{1}{2}$		Pumice – common glass
5	Apatite	Opal
4	Fluorite	
3	Calcite	
2	Gypsum	Rock Salt
1	Talc	

Monotron Test

A test seldom used today which uses a diamond ball penetrator to a diameter of 0.75 mm. A load is applied to the penetrator by means of pressure, until the penetrator has reached a standard depth of 0.0018 in (0.028 mm). On achieving this, the pressure required indicates the hardness. This can be read directly from the calibrated gauge. It will be appreciated that the harder the material being tested, then the higher is the pressure required to achieve the necessary penetration.

Muschenbrock Test

A very early hardness test, no longer used.

A knife of standardized type was swung with a pendulum motion, not unlike that of a present-day *impact testing* machine. The number of blows required to cut a standard piece to failure was recorded and this gave the hardness number.

Pellin's Test

A form of hardness testing in which a ball in a specially designed frame is allowed to fall from a known height onto the specimen being tested. The hardness is then measured from the diameter of the impression obtained.

This test is no longer commonly used, being replaced by the more conventional *Rockwell*, *Vickers* or *Brinell* tests.

Poldi Test

A portable method of estimating hardness. With this, an impact is applied to a spring-loaded punch by a firm hammer blow. This results in an impression being produced on a test bar, at approximately the mid section of the punch, and at the same time the nose of the punch forms an impression on the material being tested. As the hammer blow is used to trigger the spring, it is the load on the spring which produces the impressions, they are not related to the energy of the hammer blow.

Two impressions are always formed, one on a test bar of known hardness simultaneously with the test impression, and it is possible by measuring the test impression to be certain that reasonable accuracy is obtained. Charts are available to convert the impression size into hardness figures, with suitable allowances being made depending on the hardness of the material being tested.

The accuracy of individual impressions, in skilled hands, can be of a high order, but the value of this lies more in the facility of carrying out a survey across a surface to identify any change in hardness, rather than to produce accurate figures. The technique is not normally accepted as a method of producing hardness figures for certification purposes.

It is also a slow, rather tedious technique, and will seldom be economical with the various alternatives now available, in particular the *'Equi-tip'*.

Pusey Test

A hardness test used on rubber. This requires that the part being tested is of a reasonably homogeneous shape and it is tested by applying a load of 1 kg to a ball of $\frac{1}{8}$ in (3 mm) diameter. The depth of penetration after 15 s is a measure of the hardness. The diameter of the ball may vary, depending on the type of rubber, and the time of application of the load will also vary.

It will be appreciated that the penetration must be measured while the load is applied, since, unlike metals, rubber recovers more or less completely on removal of the load.

This form of hardness test on rubbers has been largely replaced by the *Durometer* method, which, while basically the same process, is less complicated.

Pyramid Diamond Test

A term covering the various forms of hardness testing where a load is applied to a diamond to obtain an impression to indicate the metal's hardness. This can be applied to any metal and is probably the standard reference method of testing for hardness. It is also known as the *Vickers hardness test* as it most commonly uses the machine of that name.

The load is normally applied through a series of levers with an initial preload. This remains in contact for a specified time, usually approx. 15 s, and is removed automatically. The size of the resultant impression is measured using a movable graticule which will be calibrated. The apex of the pyramid shape is measured in each direction and it is required that the difference in measurement between the two opposite sides is small. This will otherwise be an indication that the test piece is not flat and the impression must be retaken. The load applied can be as low as 1 g with the *microhardness* technique and as high as 3000 kg using special machines.

The hardness value is arrived at by dividing the load applied by the area of the impression. This is generally achieved using supplied tables which give the results in the appropriate form. The value is known as the Vickers Pyramidal Number (VPN) or the Diamond Pyramidal Number (DPN) but there are other systems using this basic technique. While in theory the same results will be obtained using any load, there can in fact be slight variations and it is normal convention to report the load used when reporting the hardness obtained. For accurate results, it is essential that the surface finish being tested is of a reasonably high standard. The higher the load applied, the larger the impression and, generally, the more accurate the result obtained.

It is, however, important to realize that the diamond impression can be a serious stress-raiser, and it is general practice to make use of as small a diamond impression as is reasonable, where the impression must be left on stressed high-duty, high-integrity components. Probably the most commonly applied load for steel components is 30 kg, with 5–10 kg loads being usual with *surface hardened* components.

As with all hardness testing, it is essential to have sufficient material to support the load applied. If this is not the case, some reflection of the supporting base is obtained. It is generally accepted that the depth of the diamond impression must always be less than one-fifth of the thickness of the material being tested; this ruling also applies when a *surface hardened* case is tested, that is the case depth must be five times larger than the impression.

In general any error in hardness testing using the diamond-indenter method will result in a high reading. Any movement of the component, any wrong application of the load, will result either in a seriously distorted diamond impression, or one which has not achieved full depth. The only way in which a softer, that is a larger, impression will be achieved is by using the wrong load or impacting the diamond indenter into the surface of the test component.

The Pyramidal Diamond hardness test is relatively slow; unless specialized equipment is used, each hardness test will occupy approx. 0.75 to 1.5 min. It is, therefore, usually found that this test is the standard reference method or is applied only to high-integrity components. The *Rockwell* hardness test or one of the many modified versions of the *Brinell* are used for production purposes.

Rockwell Test

A hardness test which measures the depth of the impression formed.

The principle of the Rockwell test is that a shaped indenter, generally a diamond, but under some circumstances hardened steel is impressed into the surface of the metal being tested. This is carried out by applying first a preset load, and then a full load. It is the depth of the impression which is measured in the Rockwell test, and this is distinct and different from the generally applied laboratory type of test, where the diameter or the width of the impression is measured.

The Rockwell principle of measuring the depth of the impression allows the use of normal inspection equipment in the form of a depth gauge to give a more or less immediate result. This is shown as a reading on the dial which is directly calibrated in the scale being used. The Rockwell test is, therefore, commonly used for production-type inspection, where hardness figures of a reasonably high standard of accuracy can be obtained at a rate of up to 5 or 6 per minute, depending on the skill of the operator and the shape and size of the components being tested.

The Rockwell hardness test has a series of scales which employ different types of indenter and different loads. This allows the technique to be employed with relatively thin *carburized* components where a very light load (known as the 'Superficial Rockwell' tester is used) up to a 120 kg load, which gives accurate results on through-hardened components. As with other forms of hardness testing, there is a relationship between the different systems, but care must be taken in interpreting this and in general it is essential to quote the Rockwell system used together with the hardness obtained, otherwise the figure obtained will be meaningless. In general, the heavier the load used, the more accurate will be the figure. There are various tables for converting Rockwell to other hardness types, but these must be used with caution.

Scleroscope Test

A hardness test which can be applied to all metals. It uses the principle that the rebound of a hardened ball will give an indication of the hardness of the surface from which the ball rebounds. A variation is that a hardened steel hammer of constant weight can be used in place of balls. In both cases the principle of the test is the same: the harder the surface which is struck, the higher the rebound.

With the balls, a glass cylinder with the outside calibrated is used. By means of a simple hand-pump, the balls are dropped from the top of the cylinder and the height of the rebound is estimated visually.

The hammer is held, often by an electromagnet, released, strikes the surface and rebounds. The height of the rebound is measured. Depending on the weight of the hammer and the weight of the balls, the hardness is estimated using calibration charts. The Scleroscope test is of limited use and accuracy but, because of its extreme portability, has certain advantages over other more conventional tests.

The modern *Equi-tip* uses the same technique, but measures the speed of rebound.

Scratch Test

With this hardness test, by using a material of known hardness, it is possible to assess whether the material being tested is harder or softer, since it is impossible to cut a material harder than the cutting medium.

The test in its scientific form is known as the *Mohs test*. The *file test* is also based on this method. In engineering, it is commonly applied to *carburized* or *nitrided* components, where a hardened file is used, and provided the file is unable to mark or grip the component, then it is considered to be satisfactory. It is essential that the file has not been previously used for other purposes, otherwise the sharp cutting edge will be lost and this, of course will have a serious effect on the test itself.

If a component has been correctly surface *hardened*, a sharp file will slide off the surface. If there is any appreciable reduction in surface hardness, the file will grip the surface and, when further pressure is applied, fine scratches will be noted. The scratch hardness test can be modified to suit particular purposes, for example, the *anodized* coating of aluminium can be assessed with this technique. Using calibrated needles as described in the *Mohs test*, a wide range of material hardness can be estimated.

This, more than most commonly used hardness tests, requires considerable experience and some skill by the operator to ensure that sensible results are achieved.

Shore Test

This hardness test is similar in principle to the *Scleroscope*.

With this, a hammer with a hardened tip, either of diamond or tungsten carbide, is allowed to drop from a known height and the rebound is measured: the harder the metal being tested the higher the rebound, the softer the metal the lower the rebound.

It is doubtful if this technique results in accurate hardness figures being obtained in their own right. It is, however, a useful portable method of obtaining comparative hardness figures and, as it can be applied to massive objects which cannot be tested by conventional testers, it is useful in showing any variation in hardness across a surface. Meaningful results can only be obtained on a horizontal surface, but the test is extremely flexible and economical when used under these conditions. This test is used for rolls or other components where no impression on the finished surface can be accepted.

The Equi-tip equipment make use of this principle where the indenter is allowed to strike the surface being measured under carefully measured load, and the speed of rebound is accurately measured by electronic means. Further details are given in this section.

Tukon Test

A hardness test using the pyramid diamond technique in the microhardness region. It is commonly used for geological hardness testing but is also used metallurgically. It was developed by the US Bureau of Standards, and the test is expressed as the 'Knoop Hardness Number' which has a similarity with the Brinell technique. In the UK the Tukon test has been replaced by the standard microhardness technique, using the 'Pyramid Diamond Number' and the Vickers system.

Turner's Sclerometer Test

A test which, like the Mohs system, relies on the fact that no material can be scratched by a material softer than itself. The specimens must be reasonably flat and the test surface polished. A diamond is loaded with a known weight and is moved once backwards and forwards on the surface in a straight line. If no scratch is seen under conditions of side lighting, or felt with a fine point, the load on the diamond is increased and the test repeated. If, on the other hand, the surface has a well-defined scratch or score, the load on the diamond is decreased. The load is varied until a scratch is produced which can be seen only as a very slight marking with side illumination, or felt when a fine needle is drawn at right angles across the surface. The hardness is recorded as the weight required to produce this scratch. It will be seen that this method of testing is slow and there will be problems regarding standardization. It has been largely replaced by the standard Rockwell, diamond hardness and Brinell systems all of which are discussed in this section. This was an attempt to standardize the scratch or Mohs test.

Vickers Test

This shares the same basic principle as the Brinell hardness test.
   It was the first machine to make use of diamond penetrators, thus reducing the inaccuracy by the distortion of the ball penetrator.
   With the Vickers machine the load in the form of weights is placed on one arm of a lever. The material is held on a movable platform which is placed under the penetrator and carefully adjusted for height so that the diamond or ball is almost touching. The load is then applied in two stages and is held automatically for the required time. It is removed without any further action by the operator. The material is then moved on the platform under the microscope

and by means of an adjustable, calibrated graticule the size of the indent is measured.

The variables involved with the Vickers hardness tester are load and magnification, and some care is required when the ocular reading is translated, using the charts supplied, into the 'Vickers Hardness Number' (VHN) or the 'Vickers Pyramid Number' (VPN). It is now standard that the 'Diamond Pyramid Number' (DPN) is used in place of the older VPN.

This is identical to the Pyramid Diamond test or Diamond Hardness test.

'Warman Penetrascope' Test

A portable hardness tester where the machine is clamped to the component.

It is based on the *Diamond Pyramid* test, the load being applied hydraulically by means of a hand wheel. On removal of the load, the diamond is measured in the same manner as with the normal *Vickers* technique. This machine is not now commonly used, having been replaced by the portable *Rockwell* machines.

Heat treatment

The term used to cover all forms of heat or thermal treatment of metals. It is considerably abused in that it is very often loosely applied to any one specific and commonly employed form of treatment in any one industry, firm or geographical area: thus 'heat treatment' will be found to cover a wide range of techniques and care must be exercised in using the term.

A practical definition for the term would be: that process where the metal being treated is placed in a furnace or otherwise heated to a controlled temperature and subsequently cooled under known conditions.

The various types of heat treatment are listed alphabetically and very briefly described under their appropriate headings in this book. Each of these treatments appears in alphabetical order in this section with a fuller description. These can be conveniently divided into two types.

1. Treatments which are designed to increase the hardness of the material.
2. Treatments which are designed to reduce the hardness or remove stresses.

There are alternative names for some of the treatments, which will be found under these headings in the alphabetical section of this book.

The treatments in this section are listed below with the type number shown.

Age Hardening 1
Annealing 2
Austempering 1
Austenizing 1 & 2
Boriding 1
Bright annealing 2
Carbo-nitride 1
Carburizing 1
Classical Anneal 2
Controlled Anneal 2
Cyanide Hardening 1
Cyclic Anneal 2
Double Temper 2
Flame Hardening 1
Full Anneal 2
Gas Carburizing 1
Graphitizing 2

Hardening 1
Homogenizing 2
Hydrogen Annealing 2
Induction Hardening 2
Isothermal Annealing 2
Magnetic Anneal 2
Maraging 1
Marquenching 1
Martempering 2
Nitriding 1
Normalizing 2
Pack Anneal 2
Pack Carburize 1
Plasma nitride 1
Postheating 2
Precipitation Hardening 1

contd

Preheating 2
Quench and Temper 2
Quench Tempering 2
Refine 1
Secondary Hardening 1
Selective Annealing 2
Selective Carburizing 1
Selective Hardening 1
Self-annealing 2
Self-hardening 1
Skin Annealing 2
Softening 2
Solution Treatment 1 & 2
Spheroidal anneal/
 Spheroidization 2
Stabilizing 2
Stabilizing Anneal 2
Step Anneal 2
Stress-equalization Anneal 2
Stress-relieving/Stress-releasing 2
Subcritical Anneal 2
Subzero Treatment 1
Tempering 2

Age Hardening
This is an alternative name for *Precipitation Hardening*.

Annealing
An abused term. 'Annealing' covers a number of variations where in general the intention is to soften the material. Annealing can be carried out on all forms of metal. Descriptions of the various processes will be found under the headings *Classical Anneal, Cyclic Anneal, Full Anneal, Isothermal Annealing, Pack Anneal, Self-annealing, Skin Annealing, Softening, Spheroidal Annealing, Stabilizing Anneal, Step Anneal, Stress- equalization Anneal and Subcritical Anneal*.

Austempering
A specialized form of *hardening* and *tempering* which uses two furnaces at different temperatures to ensure the complete transformation of all the austenite to martensite, and which then has the temper characteristics required by the design engineer. It can be applied to a relatively small range of alloy steels. The term is also used to describe any form of delayed quench using liquid metals or salt baths, and can thus be synonymous with *martemper* and *marquench*. This means that instead of oil quenching to convert the austenite into martensite and then re-heating to the desired temperature to *temper* the martensite, the same end-product is obtained without cooling to room temperature.

With austempering, one furnace is at the *hardening* temperature and the parts are held at the correct temperature for sufficient time to ensure the complete formation of austenite. The temperature required will be in the range 720–900°C depending on the steel analysis. The time at temperature will be in the region 10–20 min per inch of ruling section. At the end of this time the parts are transferred to a second furnace which should be of the forced-air convection type to ensure that the parts attain the correct temperature rapidly. Molten salt or metal may be used instead of the furnace at this stage. The temperature will be in the range 250–500°C and the time at temperature is measured in minutes. It is this time–temperature range which must be tightly controlled, and makes the process quite difficult to carry out. The time or temperature requires to be varied as the size of the work load is varied, and the difference in time between good and bad quality austempering can be measured in minutes.

The process is therefore confined to high-quality mass production, where batches can be standardized. The result of good austempering is controlled heat treatment of a standard at least equal to that attained by *hardening* and *tempering* and, with certain materials, improved tensile-ductility characteristics. At the same time there is a reduction in distortion, as there is no drastic quench, and an improvement in heating economies can be achieved.

The *martemper* or *marquench* process is similar in many respects to austemper, with the parts being transferred to a liquid medium such as molten lead or more generally a conventional salt bath. The term austemper is sometimes used to describe a *hardening* and *tempering* process where the components receive a delayed quench. This can result in considerable reduction in distortion, while achieving the desired hardness range. However, unless the conditions are controlled to ensure the correct conversion of austenite to martensite, there will be a reduction in the yield strength and ductility of the components involved. The term is also sometimes applied to a form of *hardening* using a delayed quench, where adequate mechanical properties are obtained by the use of more sophisticated steel than is necessary, and then reducing the rate of quenching to control distortion. With modern knowledge and the correct use of equipment, it is doubtful if this procedure can now be justified.

Austenizing

An alternative term for the initial part of the *hardening, normalizing* or *full annealing* operation.

Application range STEEL

During any of these heat treatment operations, all of which are confined to steel, the components are taken to above the upper critical temperature and at this temperature the steel structure is fully austenitic. Austenizing, then, is the term sometimes used for this operation.

For *hardening*, the steel is rapidly cooled or quenched after austenizing; for *normalizing*, the steel is air cooled; and for *full annealing* the steel is cooled very slowly.

Further information on each of these treatments, and other forms of heat treating, will be found elsewhere in this section **Heat Treatment**.

Boriding

A form of surface hardening. Described fully in Part One.

Bright Annealing

A term covering softening whereby the component is removed from the furnace in an oxide-free condition.

Application range ALL METALS

While it is sometimes possible to produce a surface which is bright and shiny, this criterion is not essential for the correct use of the term. Provided the components are removed with no free scale and a limited amount of adherent oxide, then it can be stated that the parts have been bright annealed. This process can be applied to any material.

There is some difficulty in defining the term 'anneal', but any *softening* operation in which the components are placed in a furnace and removed in a softer condition with no evidence of scale can be said to be a bright anneal.

The process can be carried out by various means, either simply by placing in a box and ensuring that the amount of oxygen available is limited, by filling the box with components, or by placing in the box some other material which will absorb the oxygen faster than steel. Thus, a few chips of charcoal placed in a box filled with components and taken to above 900°C will result in a form of annealing for steel which, if the control is correct, will have little or no scale and will thus be a bright anneal.

There is a danger with incorrectly controlled atmospheres that the oxygen content during the *annealing* will not form a complete surface film, but will be sufficient to give intergranular oxidation. With this, it is possible to produce an oxide-free surface, but with the surface grains having oxidized boundaries. This intergranular oxidation can penetrate to a considerable depth, 0.025 in (0.63 mm) being quite common, and results in a low fatigue strength and relatively brittle surface layer. It can occur on all types of metal. This is a potentially much more dangerous defect than the formation of a thin visibly unsightly surface oxide, and can seriously reduce the mechanical properties of the components or material treated. If the correct control equipment is not available to monitor neutral or reducing furnace atmosphere, it is advised that bright annealing is not attempted on high-duty components.

Variations on bright annealing are *black* and *box annealing*, where a controlled adherent oxide is the desired effect.

It is advised that the technique required is correctly identified, and also that efforts are made to ensure that the required process is correctly controlled.

Carbo-nitride

A *surface hardening* technique.

Application range STEEL

The process in many respects is similar to *cyanide hardening* and results in a surface case containing carbon and nitrogen. The process is carried out in a sophisticated gas atmosphere in special-purpose furnaces.

The furnaces used can be relatively simple, but generally are two-stage, with a preheat area used to bring the components to temperature. This area is sealed from the active area and also from the air atmosphere. When the components are at a predetermined temperature, the inner door of the furnace is opened and the components are pushed into the controlled atmosphere of the furnace itself. This atmosphere will be manufactured from propane or butane or, in some circumstances, town or natural gas. Into this is injected ammonia, which results in the required atmosphere to produce the necessary amount of atomic carbon and nitrogen which will then react with the surface of the steel to produce iron carbides and iron nitrides.

One advantage of using carbo-nitriding is that the range of temperatures at which the process can occur is wider than with either the *nitriding* or the *carburizing* processes. This allows the use of a single temperature which can be used as the *hardening* temperature and, under many circumstances, the components are quenched directly from the carbo-nitriding atmosphere. This is known as the 'sealed quench' furnace.

The temperature used will range from 700°C–850°C with in some cases extension to these temperatures, but the most common will be in the 720°C–800°C range. The time of the process will vary, depending on the case depth required, the temperature chosen and to some extent the steel being treated. A case depth of approx. 0.005 in (0.125 mm) can be achieved in approx. 30 min.

In general, the process is used for shallow case depths on relatively small components where the use of a single temperature *sealed-quench* furnace shows considerable economy over all other forms of *surface hardening*. From the point of view of flexibility, the carbo-nitriding process can be compared to *cyanide hardening*. This has considerable dangers regarding toxicity, and with modern effluent control there is generally a requirement for treatment, and the economics of this process can then become such that carbo-nitriding is competitive. For case depth up to 0.010 in (0.25 mm) carbo-nitriding is an economical process when compared with *carburizing*, since this requires only single quenching under the majority of circumstances. Somewhere above 0.010 in (0.25 mm) and certainly below 0.050 in (1.2 mm), *carburizing* becomes economical when continuous processing is considered. The reason for this is the high cost of the ammonia required to generate the nitrogen. It will thus be seen that carbo-nitriding will almost invariably be more economical than the *nitriding process*, where ammonia is the sole source of the nitrogen atmosphere.

Technically the carbo-nitrided case is slightly inferior to the *carburized* regarding brittleness, but will be comparable regarding hardness. Except in instances where technical requirements of core hardness, case hardness, core ductility and case ductility are paramount, then carbo-nitriding will be technically comparable, and at low case depths has a considerable economic advantage. *Nitriding*, however, will have a technical advantage over the present process in that the case produced will be harder, will retain its hardness at slightly higher temperatures and will produce a superior fatigue-resistant surface. Nitriding has the additional advantage that it is carried out at a lower temperature, and requires no quench, thus distortion is minimal. It is, however, a more expensive method as being a lower-temperature process, it takes longer to produce the same case depth and uses more ammonia.

Carburizing

A technique in which a surface hardness of a very high order can be produced on a ductile, relatively cheap base metal. Surface hardness up to 700 DPN (Rockwell C 60–65) can be achieved, while the core will have a hardness in the range approx. 100–300 DPN (Rockwell C 10–30).

Application range STEEL

The process relies on the fact that the iron contained in steel will react with carbon in a controlled gaseous form, as carbon monoxide, and results in the production of iron carbide (Fe_3C), or cementite. This newly formed iron carbide or cementite on the surface of the component then diffuses from the surface towards the core. Provided the correct conditions exist, the result is a surface with a carbon content of approx. 1 per cent, giving a saturated iron carbide structure. As the 'hardenability' of steel is proportional to the iron carbide content, this means that the structure at the surface is capable of being hardened to a very high figure. The high surface hardness caused by the high carbon will drop to the hardness and carbon content of the material in the core, which is the basic steel used to manufacture the component.

There are two separate forms of this process. First, is the original *pack carburizing* method, whereby the components are placed in a heat-resistant box, covered with charcoal in the form of small pellets and activated by the addition of carbonates, generally of barium, sodium and calcium. When the box is filled to the brim with components and charcoal in the ratio of approx. 1:4, a lid is fitted and the box heated to 920°C.

Carburizing commences at approx. 830°C–850°C, and is ideally carried out in the range 900°C–920°C. Below 900°C carburizing will be slow, and as diffusion occurs at this temperature it is difficult, if not impossible, to obtain more than 0.5–0.7 per cent carbon at the surface. Above 920°C the carburizing process will continue at a faster rate; that is, the formation of the iron carbide at the surface will proceed at a more rapid rate. Unfortunately, the speed of diffusion of the carbides from the surface to the core will not increase at the same speed. Instead of obtaining the ideal condition in which the carbon content falls in a proportional manner from approx. 1 per cent at the surface, to the core carbon content, it will be found that the carbon content falls from 1 per cent to that of the core material over a very short distance. This means that a stress zone exists, with case and core being relatively easily parted at this sharp demarcation line.

Second, is *gas carburizing*, the more modern process, where in place of the box and charcoal, it is now common that a carbon-rich gas is used. This results in exactly the same process with the same conditions regarding temperature, but in this instance the gas is produced artificially, either town gas, boosted with an active gas, or more commonly at present propane or butane. There is also the *'Homocarb process'* where a liquid of controlled composition is allowed to drip into the hot furnace. The end-product is again a gas containing carbon monoxide, carbon dioxide, water vapour, methane, hydrogen and butane. Provided the ratios of the above gases are correct, then carburizing will take place.

At the end of the desired carburizing cycle, the components are removed from the furnace and allowed to cool slowly. In some cases cooling can be carried out in the furnace, but this ties up expensive capital equipment and is not normally an economical proposition. The components in the carburizing basket are generally placed in an insulated container which is sealed and allowed to cool relatively slowly. Under some circumstances where quality of the finished component is not of paramount importance, then quenching directly from the basket can be carried out. Considerable care is necessary as the carburizing temperature will not coincide with the ideal *hardening* temperature. This is known as *pot quenching* and is not recommended for quality components.

After cooling the *pack carburized* parts are removed from the box, cleaned of the excess compound and are then in the same condition as the components from the gas process, these latter are also cooled to room temperature. The metallurgical condition of the components will now be large grained, since they have been taken above the critical temperature for a considerable time; they will also be soft following slow cooling through the critical temperature. It is, therefore, necessary that the components now receive a specific type of heat treatment.

Firstly 'core-refining' is carried out at the temperature required for *hardening* the core. Depending on the alloy used, this will be in the range 840°C–900°C. This results in the core being 'refined', that is, the grain will be reformed to give a fine-grained structure, and as the heating is followed by quenching will result in complete *hardening* of the core. The case, however, will be taken above the ideal temperature for treating high-carbon steel and this will result in excessive grain growth and thus brittleness.

Secondly it is necessary thus to re-treat the carburized components, heating them to the ideal *hardening* temperature for high-carbon material, that is, in the range of 700°C–740°C, and when quenched the steel of the case will have an ideal fine-grained structure. The core material, which had been *hardened* and *refined* at the previous operation, will now be in the fully *tempered* condition, with a limited amount of the carbides being taken back into solution. This will have no appreciable effect on either grain size or 'hardenability'.

Finally, it is necessary to *temper* the case in order to obtain the maximum ductility. This will be carried out at a temperature of 150°C–250°C, depending on the application. This treatment will have no effect whatever on the core.

The post-carburizing treatments described above are necessary when a good-quality operation on high-duty components is required. Where, however, optimum properties are unnecessary, particularly in the core, the single-quench process can be satisfactory. This consists of cooling slowly from the carburizing temperature and then reheating to the temperature required for *hardening* of the case. After quenching a *tempering* operation is essential.

Adequate hardness of the case will be obtained by this single-quench treatment, but there will be reduction in ductility of the core as this will be in a relatively coarse-grained, thus brittle, condition. With the use of grain-controlled materials, the grain growth during carburizing can be reduced and thus the use of these materials will result in a more satisfactory core condition than with normal steels which do not contain grain-refining elements. It is not possible to obtain the same high-standard case and core properties as can be obtained using the refine and harden technique. Normally, it is not economically justified to use grain-controlled materials where *refining* is to be carried out.

The case properties achieved by carburizing are comparable to those produced by *nitriding*, *carbo-nitriding* and the *cyanide* processes. *Nitriding* results in a harder case with better fatigue resistance, and a greater ability to operate at higher temperatures without losing hardness. The carburized case is generally superior to *carbo-nitriding* and *cyaniding*, however, and certainly for a case depth greater than approx. 0.050 in (1.2 mm), the present process is more economical.

The purpose of carburizing is to produce a surface which is capable of withstanding wear and abrasion, and also accept a certain amount of deflection. The degree of deflection which the case will accept will depend on the thickness of the case itself and the toughness and hardness of the supporting core. It is essential that the core to case ratio is always greater than 3:1. That is, the core must be at least three times the case depth. Unless this minimum ratio exists, the case will be unable to 'breathe' in the correct manner and thus there will be little advantage over a through-hardened material, and brittle failure will result when any bending movement is applied.

It is the design engineer's responsibility to call for the correct core conditions, with sufficient case to withstand the action of the component in service, allowing for enough ductility for 'breathing' or movement to take place. These characteristics are particularly important with the design of gears where the teeth mesh with each other. As they are driven, movement will take place at the root of the gear teeth. With high-quality, high-duty gears, a sufficient case depth is necessary to ensure that cracking or failure by scuffing is avoided. It is further essential that the case depth is low enough to allow movement. The core, therefore, must be designed strong enough to support the case so that it will not crack by too much movement. The core must not be too brittle or of too high a hardness since this would prevent movement of the case.

It is essential to appreciate that the carburizing process results in a relatively thin, high-carbon surface. Even with a total case depth of 0.075 in (2 mm), the effective high-carbon case will be no greater than 0.02 in (0.5 mm). With a total case depth of 0.010 in (0.25 mm), the effective high-carbon case will not be any greater than 0.003–0.005 in (0.075–0.15 mm). It must, therefore, be realized that when distortion takes place, as it must during the carburizing and subsequent processes, this distortion will result in some areas having more metal machined off at any subsequent *grinding* operation. It is not good practice to allow for this distortion during carburizing by increasing the case depth as the result will be that the areas with the greatest amount of distortion will have the effective case removed in subsequent *grinding*. It is good engineering practice to ensure that distortion is held to an absolute minimum during the carburizing itself. The single, most important aspect is to ensure that components going to the carburizing operation are in a stress-free condition, and that distortion results only from the carburizing process. It is important, therefore, that the components are *normalized* in the as-received condition. This will ensure that all *forging* stresses are removed prior to carburizing. It has now been proved that the use of *normalizing* at the *forging* or rough-machining stage has done more to reduce distortion at carburizing than any other process.

It will be the planning engineer's responsibility to use his knowledge when necessary to call for some form of *stress-relieving* operation, at 650°C for at least 0.5 h prior to carburizing, if between it and *normalizing* there is a considerable amount of heavy machining giving rise to *work hardening*. This can result in distortion at the carburizing process. It should be appreciated that carburizing for several hours at 900°C is itself an excellent form of *stress-relieving* and any distortion which would result from a stress relief will be found here.

The planning engineer, in conjunction with the design engineer, can also help to reduce distortion by ensuring that no rapid change of section occurs at carburizing. Finally, the heat treatment operator will help to reduce distortion by holding or placing the components in a proper manner in the furnace such that stresses will not result. The tensile strength of steel at 900°C is very low, thus stress at a low level can cause distortion.

Classical Anneal

The technique of furnace heating from cold and furnace cooling to soften and remove all stresses.

Application range ALL METALS

The technique used is to load the furnace while cold with the components, and to heat the furnace and its load to the desired temperature. This will be at the same temperature as for *full anneal* of steel, or in the case of other metals the temperature required to remove all traces of *cold work* or previous forms of heat treatment. The furnace is held until all the components have reached the temperature and then for a further period which may be as short as 20 min or as long as 2 h. The heat source is then removed and furnace and load allowed to cool.

This technique will ensure that the components are in the fully softened, stress-free condition. There will, however, be considerable grain growth as the parts will invariably be held for a considerable time above the temperature at which this occurs. This can result in machining problems, since large grains result in a poor finish, and there will be a reduction in the ductility as a large-grained structure will generally be more brittle than a fine-grained structure of the same hardness.

The technique is expensive in capital equipment as the furnace will be occupied for long periods, and will have a shorter life because of the repeated heating and cooling stresses. Finally, this *annealing* method is expensive in fuel as the furnace has to be heated from cold with each load of work. Provided the rate of cooling is correctly controlled, and the heating rate adjusted when necessary, there is seldom any technical advantage over the *full anneal* with this method.

Controlled Anneal

The term generally applied to variations of the *full annealing* procedure, where either for the purpose of further *cold working*, but more generally where the magnetic properties of the material in question must be controlled, the rate of heating, and more particularly the rate of cooling, is carefully controlled.

Application range ALL METALS

This operation, very often in addition to requiring control of the speed of heating and cooling, will necessitate the use of some form of controlled atmosphere. Where iron, or very low carbon steel, is used in transformer manufacture, or where the steel in question is being used for armatures or cores of electrical relays, then the *annealing* will specify a rate of heating which is reasonable, and will tightly control the rate of cooling through various temperature ranges. For example, it is commonly required that the material be heated to 870°C plus or minus 10°C and held there for 3–5 h. This is to allow a certain grain growth which in some cases will be preferential, depending on the previous *work hardening* history. The components will then be cooled from 870°C to, for example, 500°C at a rate of 50°C per hour, and then continued to cool at a slightly faster rate to approximately 300°C. The components can then be cooled at any desired rate. As stated, this whole process will be carried out in a controlled atmosphere which may be of a neutral type such as produced by normal hydrocarbon gases, or contain nitrogen and hydrogen to give a specific amount of *decarburization* at the surface of the material being treated.

Many of the controlled annealing operations are proprietary, and are used to produce specific properties on highly controlled materials. The term can be applied to non-ferrous materials, but generally there is less scope for controlling grain size and producing directional properties.

Cyanide Hardening

A form of *surface hardening* which uses the cyanide salt bath as the active heating medium to impregnate the surface.

Application range STEEL – LOW CARBON

The process can be used with a wide range of plain-carbon and low-alloy steels, provided the carbon content is below approximately 0.25 per cent. The process is essentially a compromise between *carburizing* and *nitriding*. With the *carburizing* process, the iron on the surface combines with the carbon at a temperature in excess of 850°C, and the process will always result in excessive grain growth, requiring a subsequent *refine* treatment to recover the grain size and also to achieve the necessary *surface hardening*. The *nitriding* process is carried out at a temperature of approx. 500°C, where the surface iron combines with nitrogen to form the hard iron nitride. This process makes use of relatively expensive ammonia gas and is a long-term heat treatment commonly requiring a time in excess of 50 h at temperature. With this, there is no grain growth and the surface is inherently hard requiring no subsequent treatment.

The cyanide treatment makes use of salts of mixed sodium and potassium cyanide which are melted by heating, generally by gas. The cyanide compound consists of an atom of carbon combined with an atom of nitrogen, and a third atom will be present, such as sodium or potassium, to give the solid salt, or hydrogen to form the gas hydrocyanide, HCN. If the salt bath is used at a high temperature in excess of 800°C, the case produced will contain very little nitrogen and will be, in essence, a carburized case. If on the other hand the salt bath is used at a temperature as low as 500°C, the case produced will contain very little carbon, and will be almost equivalent to a nitrided case requiring the same long-term immersion to achieve any appreciable case depth. It is this ability to choose any temperature between approx. 650°C and 800°C which makes cyanide hardening an attractive method of *surface hardening*. This means that the method can be used as a flexible tool, with the temperature chosen to coincide with the *hardening* temperature of the alloy steel being *surface hardened* if desired. In the case of plain and low alloy carbon steel the chosen temperature can be low enough to prevent the core hardening and inhibit grain growth. Thus, direct quenching from the cyanide hardening temperature will result in a hard case with generally acceptable grain growth, with a core having no excessive grain growth and considerable ductility.

The technique is that the mixed cyanide salts are melted, the components to be treated are cleaned and thoroughly dried and immersed in the molten salt at the desired temperature which will generally be in the range of 700°C–780°C. The higher the temperature, the faster the rate of cyanide penetration; but the greater the grain growth, the lower the ductility of the case and the higher the brittleness of the case. Case depth of 0.005 in (0.125 mm) can be achieved in approx. 20–30 min. The components are removed from the salt at the cyanide hardening temperature and are quenched in water or oil to achieve full *surface hardening* and are then *tempered*. The case will have hardness comparable to that achieved with the *carburizing* but will be more brittle. Normally the core properties will not be of the same high standard as obtained with a fully treated carburized component.

Cyanide hardening is extremely convenient, with relatively low capital costs for equipment. It has a short process time and achieves a case of satisfactory hardness for most medium-duty purposes. This would not include components such as gears, where hardness plus ductility of the case and core are essential. The major disadvantage of the process is the high toxicity of the salt itself and the problem of treating the quench water. Most local authorities will not now accept cyanide at levels above five parts per million, and many authorities with sophisticated sewage plants will have limits of one part per million cyanide. These limits will generally preclude the use of a cyanide salt bath unless an effluent treatment plant is available. The cost of effluent treatment will be expensive, and if the plant is required for the sole purpose of treating the effluent from the cyanide hardening section, then it will generally be more economical to carry out some alternative treatment. The closest alternative is *carbo-nitriding* and *carburizing* being an obvious second if suitable plant is already available.

Cyclic Anneal

This is a two-stage softening process which is useful with materials which are difficult to soften.

Application range ALLOY STEEL – HIGH CARBON

If these steels are hardened by any previous treatment, it will be found either that they require excessively long times at temperature, or that an excessively high temperature is necessary in order to reduce the hardness to a satisfactory level for machining.

The cyclic annealing process takes the components to above the upper critical temperature, probably 850°C–950°C, holds them there for approx. 1 h and then transfers the components to a second furnace held at 650°C for 2 h and then to the original furnace above the upper critical temperature for a further period of approximately 1 h. The components are then re-transferred to the furnace at 650°C and held there for a further 2 h. With the majority of steels, the two cycles are sufficient and the components can then be cooled to room temperature. With some of the more complex high-tungsten, high-carbon materials, further cycles may be necessary.

This procedure results in *balling* or *spheroidizing* of the carbides involved, and *tempering* the matrix of the steel. This means that the cutting tools can more or less ignore the spheres or balls of hard carbide and cut their way through the tempered martensite. Cyclic annealing will almost always be used to improve 'machinability' and there will then be a final *hardening* and *tempering*. The fact that the cyclic anneal when correctly carried out results in fine, well-dispersed carbides, means that the subsequent *hardening* can be efficiently carried out.

The presence of the carbide spheres considerably enhances the ability of these steels to be abrasion resistant and they are commonly used in this condition for cutting blades, punches etc.

The treatment is known as *spheroidal anneal*.

Double Temper

A tempering technique used where sluggish metallurgical reactions can be expected. The treatment can result in an increase in hardness which is considered to result from the final transformation of any retained austenite to martensite.

Application range TOOL STEELS

The treatment is most important when the materials involved are used to manufacture any component which must be dimensionally stable throughout its life. The transformation of austenite to martensite results in dimensional changes.

It is considered that the double temper is required because the quench following *hardening* has been too severe and, thus, by controlling the quench rate, the second temper will not be necessary. There is, however, the possibility that without a rapid enough quench the best possible properties will not be obtained from the steel, and thus 'overhardening' followed by double tempering will be the most satisfactory procedure.

For some steels which are less highly alloyed the double temper is used to ensure that no stresses remain following the *hardening* operation.

The reader is advised to obtain further advice on the subject.

Flame Hardening

A form of *surface hardening*.

Application range MEDIUM-CARBON LOW-ALLOY STEELS

The steels involved are of the medium-carbon variety; but it is possible to use this process in a less efficient manner on low-carbon steels, where the degree of *hardening* will be relatively slight, and also on higher-carbon alloy steels, where there is some danger that their 'hardenability' will result in too rapid a demarcation between the case produced and the core.

The process uses a flame, either with oxyacetylene, or any of the other hydrocarbon gases, with oxygen, which gives a high enough temperature. The principle of the operation is that the flame temperature is considerably above the required *hardening* temperature and, thus, the surface of the component will reach the critical temperature before the core can be heated by conduction. As discussed in detail under *Hardening* in this section **Heat Treatment**, this requires that the steel in question has a certain critical amount of carbon and is taken to above a specific temperature, which depends on the analysis of the steel but will be in the region 750°C–900°C.

To achieve full hardening, it is then necessary to rapidly cool the steel. It will be seen that with flame hardening the time to bring the surface to the required temperature will be critical. The longer the time involved, the greater the thickness of steel which will be above the critical temperature as conduction will ensure that the heat is taken inwards. In addition, the surface of the steel will be taken well above the critical temperature, resulting in undesirable grain growth which can give excessive brittleness.

It will also be seen that if the correct time is chosen, then very efficient quenching can be achieved as the core of the material will be below the critical *hardening* temperature. This will dissipate the heat, ensuring a highly efficient quench with the heat being removed from the surface by the quench and drawn to the core by the mass of cold steel at the centre.

The advantage of this process is that components can be hardened locally in an economical manner. Of all methods of local surface hardening, flame hardening is the cheapest. Provided that the distance of the flame tip from the component and the speed of movement between the flame and the component relative to each other is controlled, and the flame temperature is maintained, then components will be hardened in a reasonably reproducible manner. It must, however, be appreciated that control of the flame to within the normal temperature limits used by heat treaters is impossible. When it is realized that the difference between good and poor-quality *hardening* can be to the order of ±20°C, it is seen that the range of temperature with the oxyacetylene flame is unacceptable for a high-quality process. The temperature of the flame, with the best possible control, can vary from 3000°C–3100°C and, with normal control, will vary between 3000°C–3350°C. When these figures are accepted, the difficulty of controlling flame hardening is appreciated.

The process can be compared with *carburizing, cyanide hardening* and *induction hardening*. It will always be much cheaper than any of the above, and particularly so regarding the *cyanide* process or *carburizing* for small-quantity production. Flame hardening is technically and economically comparable to *induction hardening*, where it is possible to achieve much tighter control of the temperature required on a mass-production basis; but because of the sophisticated equipment required, the capital cost will be more than with flame hardening.

Because of the relative inflexibility of induction-heating coils, there will always be a demand for the flame hardening process where components are large or awkward in shape.

Steels used are generally in the 0.3–0.5 per cent carbon range. With these, there is sufficient 'hardenability' and the reaction time is low. With alloy steels, particularly those containing appreciable amounts of chromium, it generally is found that they are too sluggish and, thus, are not ideal where this rapid process is desired.

Under some circumstances it can be technically and economically desirable to flame harden components which have been previously *carburized*. This is most often the case when an all-over process is economically desirable but for technical reasons only specific local areas are needed in the *hardened* condition.

Flame hardening, therefore, is a useful method for components of reasonably low mechanical duty which are not produced in high numbers and which have peculiar geometrical shapes or are large in size. Where high-production quantities are involved, then *induction hardening* will generally be found to be more economical, and permitting higher technical control. Where mechanical integrity is essential, components should be hardened using *carburizing* or *nitriding* processes. Flame hardening should never be used where any serious consequence may arise from failure of the hardened surface.

Full Anneal

The term implies that the material in question has had all *cold working* removed and where applicable has been fully recrystallized.

Application range ALL METALS

With steel, this means that the parts in question have been taken above the upper critical temperature and held there for approximately 45 min. to 1 h per inch (25 mm) of section and then cooled under conditions of equilibrium. This means that the cooling rate is such that no iron carbide is retained in solution and thus, the material is in its softest possible condition. It will generally be found that considerable grain growth occurs during this operation. This means that the material will be in a weaker condition with, in some instances, a reduction in ductility over the ideal.

The rate of heating will not in itself be critical, but with complex components and some of the more highly alloyed steels it will be necessary to heat at a reasonably slow rate in order to obviate any danger of cracking due to heat shock. The common methods of slow heating are either to heat in the furnace from room temperature, which is expensive and seldom essential or, more economical and satisfactory in the majority of cases, to preheat in a furnace in the range of 500°C–650°C and transfer the components from this furnace to the *annealing* temperature. These two stages will ensure that the rate of heating is slow enough to prevent cracking or excessive distortion.

The *annealing* range for steel will be between 700°C for high-carbon materials to approx. 950°C for low-carbon materials and highly alloyed materials, and is identical or slightly higher than the *hardening* and *normalizing* temperature range. The time at temperature will vary, again depending on the alloy content. For plain-carbon steels and simple alloys where no element such as chromium, molybdenum, vanadium, tungsten, etc. are present a time at temperature of 30 min per inch (25 mm) of ruling section will be sufficient. With more highly alloyed materials, this time may require to be considerably increased, and for certain complex tool-type steels times in excess of 24 h may be necessary. This time can be reduced by increasing the temperature where possible, and the skill of the heat treater is required in this instance to balance the undesirability of excessive time or excessively high temperature.

The rate of cooling will also depend on the alloy element present. This can vary from cooling the components in the furnace itself, which in general is uneconomical and comes within the heading *classical anneal*, to, with low-carbon mild steel containing no alloying elements, cooling in sand or ashes. With alloy steels containing small amounts of nickel or chromium, a satisfactory method is to heat the components in containers with sand, ashes, turnings or some other suitable inert material. This procedure regulates the rate of heating, and when the container is removed ensures that cooling is carried out at a reasonable speed to prevent the formation of any undesirable metallurgical structure.

This full annealing procedure will be required when components are subjected to excessive *cold work* such as *deep-drawing* operations, which then require to be further drawn or cold worked. A full anneal ensures complete recrystallization of the grain structure, but almost always will result in some grain growth.

Under many circumstances a subcritical anneal will result in acceptable machineability being attianed. This is especially so when machining hardness is the critical parameter.

Gas Carburizing

A surface hardening technique using controlled gases to increase the carbon content at the component surface.

Application range LOW-CARBON LOW-ALLOY STEEL

The components are loaded into a specially designed furnace into which is fed the gas used to produce the carbon-rich atmosphere. The furnace can be heated by any conventional method, but will of course be gas-tight and almost invariably will have forced circulation. The gas may be produced from town gas, natural gas or a manufactured gas, which will normally require to be enriched – it is more usual with modern equipment to use propane gas. The enriched gas in contact with the component will be a mixture containing carbon monoxide, carbon dioxide, hydrogen, water vapour, methane and butane. Unless these are present in controlled proportions, the *carburizing* potential will not be achieved. Described fully under *carburize* in this section **Heat Treatment**.

Graphitizing

A softening process which converts some carbide to graphite.

Application range STEEL AND CAST IRON

The process, in essence, is a form of *annealing* in which the carbon present as iron carbide, and therefore hard or hardenable, is converted to graphite which has certain 'machinability' characteristics and does not take part in any *hardening* operations either by heating or by *work hardening*. The difference between conventional *annealing* of steel and graphitizing is that with the former process the structure formed will generally be ferrite and pearlite, where the carbon remains as iron carbide in a soft condition, but graphitizing converts the iron carbide to ferrite and graphite and, thus, permanently softens the material. With normal *annealing* the structure can be converted into a harder material by some form of heat treatment.

Graphitizing requires that certain constituents are present in the material being treated, and that specific temperatures are used. It can thus be carried out only if the chemical analysis of the material is correct. It is a specialist treatment not commonly carried out with modern materials where the same effect can generally be obtained by alternative and more economical methods. An exception to this is spheroidal cast iron, which, provided the analysis is correct, can have iron carbides converted to graphite by controlled heat treating. With steel, an alternative would include the addition of free machining constituents such as sulphur or lead during steel manufacture. With cast iron, control of the analysis will result in the desired amount of graphite, but as stated above this may require a controlled *heat treatment* to ensure that the carbide breaks down without the graphite already present altering its form.

The process requires that the component is taken above the upper critical temperature of approx. 850°C–900°C and after a specified time transferred to a furnace at 650°C. Thus, it has some similarity to *cyclic annealing*. Because of the variety of materials involved, especially with cast iron, users are advised to contact the technical authorities for detailed information. The British Cast Iron Research Association (BCIRA) is suggested for cast irons.

Hardening
A process to increase the strength or abrasive resistance.

Application range STEEL

Under certain circumstances hardening can be achieved by *cold working*, and it is possible that the term 'hardening' is on occasion loosely applied in this context. The following information, however, is restricted to the thermal hardening of steel. Information on hardening by cold work is given under *Work Hardening* in Part One, and the hardening of non-ferrous materials under *Precipitation Hardening* in this section **Heat Treatment**.

Steel is an alloy of iron and iron carbide, and it is because iron exists in several states, one of which is capable of dissolving iron carbide, that hardening of steel is possible. It is this ability of steel to vary its mechanical properties through a wide range of conditions that allows engineers to design and manufacture the large range of products which exists at present.

The procedure for hardening requires that the steel being treated has a carbon content in excess of 0.05 per cent and that the carbon present in the steel exists as the compound iron carbide, commonly called cementite, with the chemical formula Fe_3C. When the steel is heated below 600°C no reactions occur, and the steel will consist of iron in the alpha or ferritic state and iron carbide. When the steel reaches a temperature of approx. 700°C, the iron starts to be converted to the state known as gamma or austenitic iron. This austenitic iron is capable of dissolving the iron carbide, and solution of iron carbide in the austenitic iron continues as the alpha iron is converted to the austenitic state. A second critical temperature exists above which no alpha iron can exist and all iron above this temperature will be in the austenitic state and is thus capable of dissolving all the iron carbide.

One of the complications of heat treatment of steel is that this second critical temperature is a variable depending on the carbon content of the steel and varies from approx. 900°C with 0.05 per cent carbon to approx. 700°C, where it meets the initial temperature of conversion of alpha iron to austenite at approx. 0.6 per cent carbon. This is shown in the well-known diagram for iron carbon (Fig. 1).

If the steel is held for the necessary amount of time in the austenitic state, then all the iron carbide present in the steel will be dissolved and held in solution in the austenite.

If now the steel is slowly cooled, the iron carbide will be allowed to drop out of solution and will return to its normal soft state, where alpha iron and iron carbide exist. If, however, the steel is cooled rapidly by quenching in water or oil, then the iron carbide will be trapped in solution in the alpha iron and this sets up a state of strain, resulting in an increase in hardness and tensile properties, and forms a new metallurgical condition known as martensite.

The skill of the hardener is employed to ensure that, first, all the available iron carbide is taken into solution in the austenitic iron. This requires that the material must be above the critical temperature where all the alpha iron is converted to the austenitic state. Secondly, he must ensure that the steel is held at this temperature for sufficient time to allow all the iron or alloy carbide which may be present to become dissolved. With some alloy steels this can be a relatively lengthy process. It is undesirable to hold the steel at too high a temperature, or for too long at the correct temperature, as grain growth will then occur, resulting in brittleness in the finished component.

The heat treater must, therefore, carefully judge the size of the component, the material being heat treated, and assess the dimensions to ensure that all these critical factors are looked after. In many cases the heat treater will be able to follow specific instructions supplied by the planning department or the laboratory, informing him of the temperatures and times necessary for each steel. While this reduces the responsibility of the heat-treatment operator to some extent, different sizes of component will of course require different hardening times and the handling of the components during quenching requires considerable skill.

The choice of quenching medium whether it is water or brine, or a slower type such as oil or even air, will probably be decided by the material being used. There is, however, some flexibility depending on the size of the component, and the operator should be skilled enough to realize, for example, whether the component is softer than desired and the correct material and temperature have been used. It will be possible to increase hardness by altering the quenching medium from, for example, oil to water. There are now polymer quenches which are faster than water, slower than oil.

Immediately after hardening the majority of steels should be in a brittle, hard condition. Many of the medium and high-carbon steel alloys will be so

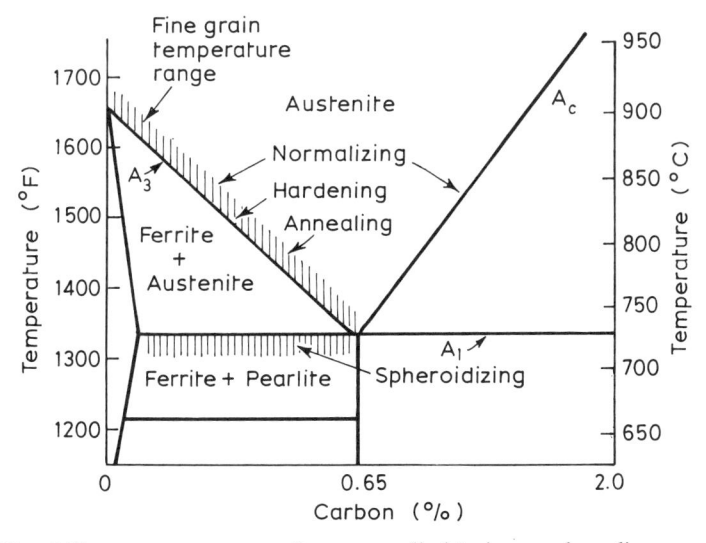

Fig. 1 Heat treatment operations as applied to iron-carbon diagram.

brittle that they will fracture with very little abuse and certain of the more exotic steels can in fact fracture from the internal stresses resulting from the hardening operation. It is, thus, essential that all components are *tempered* as soon as possible after hardening and certainly *tempering* of all medium and high-carbon steels with or without alloy elements should be carried out within hours of quenching.

It is good practice that one component in a heat-treatment batch should be hardness tested is the as-quenched condition, to test whether or not good hardening has been achieved. The hardness will vary within relatively wide limits for any one material, depending on the geometry and size of the component. It should, however, be possible to obtain the same hardness value within very narrow limits from all components of the same shape in the same material at any time and, it is desirable that the ideal figure should be ascertained during the planning stage, and this should be the lowest acceptable figure throughout. This practice will ensure that good-quality hardening is carried out at all times. Any reduction in hardness from the standard would indicate faulty quenching, incorrect material or temperature, and should result in the parts being rehardened or the material queried. Any increase in hardness from the norm will indicate problems and should be investigated as it could result in brittle or too hard components, or that the wrong material has been used.

The importance of making sure that the component is in the hardest possible condition is not always appreciated as the *temper* which should invariably follow the hardening operation results in a reduction in the hardness of the component. It must, however, be appreciated that the hardness value measured is some indication of the *'ultimate tensile strength'* of the steel. It is not a measurement of the *'proof'* or *'yield stress'* of the material, and this property can vary considerably depending on the efficiency of the hardening operation. It is not possible to measure the *'proof'* or *'yield strength'* of any material by any means except *tensile testing*. This is expensive and there can be considerable difficulty in making certain that the tensile test piece does in fact represent the component being tested.

Hardening of steel is a unique metallurgical process of considerable importance. If surface hardening is all that is required, then alternative methods are *chromium plating*, various forms of *metal-spray* or *electroless nickel plating*. Hardening by *work hardening* with materials or components of simple shape is possible but will never be as flexible, and seldom as satisfactory, as thermal hardening.

Hardening has the unique ability to increase the *'elastic limit'*, *'yield point'* or *'proof strength'* of steel and also to increase the hardness or *'ultimate tensile strength'* of the material. Correct choice of material affects the possibility of increasing tensile properties, without at the same time reducing the ductility of the steel involved. No other process can effect this improvement in the range of steel properties.

It is possible with good-quality hardening, and *tempering*, to obtain mechanical properties with mild or low-alloy steels identical to those obtained with poor-quality heat treatment or no heat treatment at all on more expensive higher-alloy steels. Good design, planning and quality control ensure that the correct quality steel is chosen and then correctly treated.

Homogenizing

An alternative term to *normalizing*.

Application range ALL METALS

When steel is taken to above the upper critical point, causing recrystallization, and held at that temperature for sufficient time to ensure complete recrystallization, then cooled at a reasonable rate in air, a fine grain will be produced.

The term homogenizing can be applied to any material, in fact, which can be recrystallized and, thus, the effect of variations in *cold work* and heat treating are eliminated. Homogenizing should not be used as an alternative term for *annealing*, which, while usually resulting in homogenizing, is intended as a softening process. Homogenizing may not result in complete softening since cooling in air under many circumstances will result in some degree of *hardening* being achieved with steel.

Hydrogen Annealing

A softening treatment in a hydrogen atmosphere.

Application range STEEL, COPPER AND NICKEL ALLOYS. ALUMINIUM. NOT TITANIUM

The process results in an identical metallurgical condition to that obtained by standard *annealing* but is carried out in a hydrogen atmosphere. Thus, no air is in contact with the material and oxidation is eliminated. Given the correct conditions of time and temperature, the components will be removed from the hydrogen annealing process in the soft condition, the surface being either free of oxide, or in the same condition as when the components entered the furnace. It is possible, however, with this process for oxide on the surface to be removed by reduction and, thus, the material can be removed from the furnace in a brighter condition than it entered. This is not a recommended practice as peculiar conditions of intergranular attack may occur under certain circumstances. It is therefore advisable that components are thoroughly cleaned, with particular emphasis on the removal of oxides and any oil. The components are then placed in the hydrogen furnace, taken to the required temperature and then slowly cooled to comply with normal *annealing* conditions.

The annealing furnace can be of the 'hump-back' type, whereby a conveyor enters the furnace and moves up a slope into the hydrogen atmosphere as it enters the heated zone. Hydrogen is fed in at the bottom of the hill and leaves at the peak, being burned at the exit. The components increase in temperature to the maximum at the peak of the slope and are reduced in temperature as they move downwards, leaving the hydrogen atmosphere. This equipment is ideal for small mass-production-type components. Alternatively, the components can be installed in a heat-resistant box with the lid either welded in position, or clamped. Considerable care must be taken to ensure gas-tight seals, as hydrogen and air form an explosive mixture. The use of specially designed furnaces is recommended where high production and efficiency are required.

If steel is maintained for any length of time above approximately 800°C–900°C in dry hydrogen, then a limited amount of *decarburization* will occur. This will be to a lesser extent than in an oxidizing atmosphere, but it is a defect which must be accepted if long-time hydrogen annealing is essential. With high-duty components subsequent machining of the surface is essential, thus reducing some of the advantages of hydrogen annealing.

Hydrogen annealing is comparable with *annealing* in inert atmospheres, such as nitrogen or argon, where there is no danger of *decarburization*, but at the same time positive reduction of oxides can be achieved using hydrogen. *Vacuum annealing* is also comparable, the same remarks applying as to inert gas annealing. *Nitrogen annealing* is probably the cheapest form with the hydrogen process next cheapest, *vacuum* and *argon annealing* are the most expensive.

Induction Hardening

A form of local hardening to produce a case using the induction technique for heating.

Application range MEDIUM CARBON, LOW-ALLOY STEEL

It makes use of the technique of induction heating and will be technically similar or identical to *flame hardening*. While the method can be used to harden steel throughout its section, this will seldom or never be economical, and in the majority of cases the result of induction hardening is that only the surface of the steel is treated.

As stated under *Induction Heating* in Part One, where the frequency used is high, then this will have the effect of heating only the skin or surface of the steel. As the frequency is reduced, so this skin effect becomes less. It will, therefore, be seen that where *case hardening* is required, high-frequency induction heating is the more useful and, as with normal steels, it is a requirement that this takes the surface of the steel to above the necessary critical temperature. The steel, then, must be heated to above the critical temperature to ensure that all the carbides are in solution, and is cooled rapidly enough to form the unstable, hard constituent known as martensite.

Because induction heating is a very rapid method of heating steel, the surface temperature is rapidly increased to above the critical point, while the core or centre of the steel is still cold. Thus, when the source of heating is removed, by switching off the heater or removing the steel from the induction coil and the metal surface is quenched, the actual rate of cooling will be very rapid as the cold core material will also remove heat from the surface metal. Thus, induction heating can be used as a means of *surface hardening* where the surface is very efficiently quenched. This can only be effectively applied to a relatively narrow range of steels. With the carbon content below 0.2 per cent, the effective 'hardenability' is too low, and with carbon above 0.7 per cent, the 'hardenability' is such that cracking is a high probability, exfoliation of the case is also possible, and the core ductility will be too low to support the brittle case.

When any alloy content, such as chromium, molybdenum and to a lesser extent vanadium and nickel are present, then there can be a sharp demarcation between the hard case and the soft core: with the increase in 'hardenability', there is a smaller area of steel which will be incorrectly hardened. Without a slow change from case to core there will be a plane of weakness, and this can result in flaking of the hard case, called exfoliation, along this plane of weakness which is parallel with the surface. This is particularly dangerous with shallow case depths.

Induction hardening is, therefore, of most use with medium-carbon steels with little or no alloy content. Steels in the range 0.3–0.5 per cent carbon are those most popularly used. Case depths from 0.005 in (0.125 mm) up to 0.1 in (2.5 mm) are commonly found and, in some circumstances, cases of greater depth are used.

Where alloy steels with high 'hardenability' are required to have a high surface hardness, it is recommended that consideration is given to the double-hardening technique, in which two successive heating cycles are employed. The first cycle will be slightly longer, and on removal from the heat source, the steel is air cooled. This is followed by the second, shorter cycle and quenching. The controlled technique ensures that a good demarcation is in fact produced.

In addition to induction hardening of plain carbon steels, this method can be used to locally harden components which have been carburized all over. This is a form of *selective hardening*. Using the same idea, induction heating can also be used for *selective annealing*.

The technique, then, is similar to *flame hardening* as a method of heating, but much greater control of the variable parameters is possible.

Isothermal Annealing
A method of reducing hardness.

Application range STEEL

With this, the material is taken to above the upper critical temperature, and held at this temperature for between 20 and 30 min per inch of section. The furnace is then cooled to just below the lower critical temperature, or the parts transferred to a second furnace at this temperature. They are held at this temperature for sufficient time to ensure the complete transformation of all the austenitic phase to pearlite.

This *annealing* form is usually applied to those steels which are difficult to soften. It has the advantage over the classical method, where the parts are cooled to ambient inside the furnace, that expensive furnace equipment is not tied up for long periods. There is also greater control of grain size and less danger of producing undesirably large grains. Superior results can generally be achieved more economically using *cyclic* or *spheroidal annealing*.

Magnetic Anneal

A treatment designed to produce specific magnetic properties.

Application range IRON NICKEL ALLOYS

The procedure is that the material is heated to above a certain critical temperature which will vary depending on the alloy used. It will be above 600°C and can be as high as 800°C.

It is then cooled to a temperature of approximately 500°C and is held at that temperature in a magnetic field, then slowly cooled to room temperature while held in this magnetic field. This results in the production of a magnetic effect where the hysteresis loop is almost rectangular.

The result is that the material has greater permeability, that is, it will be a stronger magnet, than if the treatment had not been applied, but of equal if not greater importance, is the fact that the material will then have a lower coercivity, that is, the magnetism remaining when the magnetic source is removed will be lower. It is these two properties which result in the rectangular hysteresis loop. It is also of importance that the magnetic field drops rapidly to zero on removal of the magnetic force.

The material is then more valuable for components such as the core in transformer windings which require a high permeability (magnetic strength) for a very short period and as the magnetic force is reversed or removed must immediately lose their magnetism. The same property is required for the armature of a contact type of mechanism which operates by magnetic attraction, but which must be released either to open or close a circuit immediately the magnetic field is removed.

Materials and processes are now available which do not require that the material being treated is held in a magnetic force during the cooling cycle. These can be of the very low-carbon-steel type and the process applied is again referred to as magnetic annealing. In this case the material will be taken to above the Curie point (the temperature where iron becomes non-magnetic). This will be approximately 800°C and is then subjected to a controlled cooling cycle. The details of this will vary depending on the material involved and the magnetic requirements, but can involve controlled cooling over a 24-h period. This information will be contained in the material specification.

Maraging
A specific form of *ageing*.

Application range MARAGING STEEL

The term is applied to a very narrow range of ferrous alloys known as maraging steel. These contain approx. 20 per cent nickel, are very low in carbon, in the range of 0.01–0.05 per cent, with other alloying elements such as cobalt being essential. This is a heat treating term for these specific steels. When these steels are *hardened* using quenching techniques, they form martensite in the normal manner, but because of the very small amount of carbon present, this is of a relatively tough, ductile nature rather than hard. This means that it can be machined and shaped using the normal techniques available for production engineering.

When this fully *hardened* material is taken and held at a temperature in the region 350°C–450°C, then *precipitation* occurs within the martensite, and will be similar in most respects to that achieved in the *precipitation hardening* or *ageing* process, but in this instance achieves a dramatic increase in mechanical properties. The steel in the 'as quenched' condition will be in the 40–60 ton f/in^2 (620–930 N/mm^2) tensile range (350–500 DPN) and by the maraging process these mechanical properties are increased to 120–140 ton f/in^2 (1860–2170 N/mm^2) with 0.2 per cent proof in excess of 110 ton f/in^2 (1710 N/mm^2), but with ductility measured as elongation in the 5–10 per cent range. This results in the strongest metal available for engineering use which has any measurable ductility.

These mechanical properties are comparable with many *surface hardening* processes, or materials in the through-hardened condition which would normally result in a very low ductility, of less than 1 per cent elongation.

The materials are expensive to produce and require carefully controlled heat treating, with attention paid to the temperatures and time involved. Care is also necessary to ensure that only a limited amount of *cold work* is applied between the quench and the maraging operation. Controlled *cold work* at this stage can improve the mechanical properties of the fully maraged steel.

When correctly carried out, there is no comparable material which has the strength and ductility approaching that achieved by maraging steels.

Marquenching
A form of delayed quench hardening.

Application range HARDENABLE STEEL

408 / HEAT TREATMENT

With marquenching, the material is taken to above the upper critical temperature at which all the structure is transformed to austenite, and then cooled to a temperature which allows the transformation of austenite to the desired martensitic type of structure. This very often involves quenching in hot oil or molten metals such as lead, or using modern techniques such as some form of salt bath.

It is similar in many respects to *austempering* and *martempering*, and in some circumstances the terms are interchangeable.

As with *austempering*, the problem is that unless tight control is maintained over the mass of material being treated at any one time, then a very large range of metallurgical structures will be obtained and many of these either will be undesirably hard, or too soft and ductile. Again, as with *austempering* and *martempering*, the present process should only be used after considerable laboratory experiments have been carried out and the conditions required carefully recorded and a procedure instituted to ensure that these are maintained at all times.

This is a sophisticated form of *hardening* and *tempering* and, when correctly carried out, will result in less distortion, slightly or considerably better mechanical properties, depending on the alloys involved and the dimensions of the component. It is an economical means of achieving the same end-product but, as stated above, considerable control is necessary. The technique is not advised for the jobbing type of heat treatments or where batch-size control is not possible.

Martempering
A form of delayed hardening.

Application range HARDENABLE STEEL

The components are taken to above their upper critical temperature at which the structure is converted to austenite. The material is then quenched in a liquid salt bath.

There is a convention that the term 'Austemper' indicates an air quench, while the name 'Martemper' is used where a liquid quench is involved. This can be oil, salt or molten metals.

As with *austempering* and *marquenching*, the problem is that the mass of material being treated is very critical and, thus, specific conditions must be found and held within very fine limits with different masses, and for most types of heat treatment, these conditions preclude the use of martempering as an economic proposition. Where it is economically and technically possible to have tight control over the size of batches, then the process can result in an economical method of heat treatment which will result in improved mechanical conditions when compared with conventional forms of *hardening* and *tempering*.

It can also be considerably more economical as the energy involved is less than with conventional hardening and tempering.

Nitriding

A *surface hardening* process that results in one of the hardest surfaces to which steel can be taken by heat treatment.

Application range CERTAIN ALLOY STEELS

The process consists of holding the components in an atmosphere rich in atomic nitrogen, this being obtained from ammonia at a temperature of approximately 500°C. The process is lengthy, taking anything up to 100 h of continuous processing at the correct temperature.

In theory the nitriding process can be applied to any steel, but in practice it will be found that a large number of steels either will not nitride at all, or the result will be a patchy surface with soft areas.

The theory is very similar to that which applies to *carburizing*, where by enriching the atmosphere at the steel surface with active carbon the compound iron carbide is formed and diffuses into the steel. With the nitriding process the atmosphere at the surface is enriched with atomic nitrogen and, provided the conditions of the steel surface are correct, this nitrogen will combine with the surface iron to form iron nitride, which then diffuses away from the surface. This iron nitride has the chemical formula Fe_3N, is a very hard substance and once formed requires no further heat treating to achieve a high degree of hardness, this hardness being a function of the amount of iron nitride which has formed at the surface. Assuming the correct steel has been used, the extent of the *hardening*, that is, the case depth, will depend to a limited extent on the temperature used, but will be highly dependent on the length of time that the components have been held at the correct temperature and proper control of the ammonia gas.

The nitriding process is expensive and, thus, must be carefully controlled at all stages. As stated, only a limited number of steels can be guaranteed to nitride correctly. These steels are generally in the medium-carbon range, 0.25–0.36 per cent carbon, and must contain some vanadium or aluminium to ensure that correct nitriding occurs. In general the steels will also contain chromium, molybdenum and very often nickel. British Standard BS 970 now specifies those steels which are suitable for nitriding.

While other steels, particularly plain carbon steel, are often used for nitriding, it will be found that where careful checking is carried out subsequent to nitriding, soft spots are commonly produced where the iron nitride has not formed correctly.

The nitriding process is occasionally used to produce a very thin case, and under these circumstances a satisfactory though patchy case can be achieved with steels other than the alloy steels mentioned above.

When the process is applied, the chosen steel, generally as bar or forging, should be *normalized*, then *hardened* and *tempered* to give the correct properties required for the core. These will generally be in the range 250–350 DPN (60–80 ton f/in^2 – 930–1230 N/mm^2) tensile strength. The chosen steel must be *tempered* at a temperature not lower than approx. 550°C.

The *hardened* and *tempered* components are then machined to within 0.010 in (0.25 mm) of the finished size on all areas where these final dimensions will be held to a tight tolerance. At this stage the components are then *stabilized*. This is a form of *stress-releasing*, carried out for 2 h at a temperature of 520°C–530°C. As this is below the *tempering* temperature, it will not affect the final mechanical properties of the core, but as it is above the nitriding temperature, it will ensure that this does remove any of the stresses imposed during the machining cycle. The components are then returned to the machine shop and finish-machined, with great care being taken that no further stress is imposed. Wherever possible this final machining either should be of a grinding type, or high-quality turning, using carefully controlled feeds and speeds to ensure an absence of surface *work hardening*.

The components should be machined to the finished dimensions, and it must be appreciated that the nitriding process will result in some increase in size. This increase in size is of the order of 0.00075 in (0.02 mm) in diameter and, since it is caused by the formation of the iron nitride during the process, it will be realized that the growth is a surface phenomenon and the overall dimension of the component has no effect on this growth. If no machining subsequent to nitriding is to be carried out, then allowances must be made for this growth and where necessary the component must be machined to dimensions smaller than the drawing requirements.

The components in the *stabilized*, fully machined condition must then be *cleaned*. There is considerable experience to show that oily surfaces can inhibit the process, and it would appear that when an oxide film of a certain type forms on the component surface at an early stage of the process, then this will prevent nitriding taking place. This oxide film need not be visible to the eye.

The *cleaned* and *dried* components are then placed in the furnace, and it will be appreciated that they must be carefully supported to prevent distortion occurring. The tensile strength of steel at 500°C is relatively low, and thus unless correctly supported, it is possible that sagging or twisting can occur. There are several types of nitriding furnace, the most common being the 'top hat' type. With this system, the components are placed on a flat hearth and built up using supports to produce the furnace load. When this has been completed the 'top hat' or bell is then lowered to cover all the components and the seal at the rim is made. The seal can be one of a number of types, but must be an excellent gas seal.

The sealed furnace is then heated and ammonia gas is passed into the chamber. The critical temperature for the process is approx. 470°C, below which no appreciable nitriding occurs. Very little nitriding will, in fact, take place until the components are at a temperature of 490°C, and most good-quality nitriding is carried out between 490°C and 500°C; it is possible to use temperatures up to 520°C but there is some indication that above this temperature the case is less than satisfactory.

The control of the process is by temperature and the dissociation of the ammonia gas. Ammonia has the formula NH_3, and when an ammonia atmosphere is heated 'dissociation' occurs. This means that a reaction which is reversible under normal circumstances occurs, to give atomic nitrogen in constant contact with the other constituents. It is the presence of this atomic nitrogen in the atmosphere in contact with the *cleaned* and active surface of the component which allows nitriding to proceed. As stated above, this results in the formation at the surface of iron nitride and this compound of iron and nitrogen can then diffuse towards the core of the component.

At the end of the desired time the temperature is reduced while the gas is still in contact, and only when the temperature has dropped to below approx. 200–300°C should the gas be stopped and the furnace opened. Any earlier opening or reduction of the gas will result in the components being blued.

On removal of the components after correct nitriding, it will be seen that they are bright in appearance and have little surface difference from the components which entered the furnace.

Subsequent processing of a nitrided case must be carefully carried out. The case produced is extremely hard, being in the region 800–900 DPN as nitrided. This very high hardness means that the surface is extremely brittle, and on high-integrity components at least, no attempt should be made to grind without removing by chemical means the extremely brittle surface layer. Lapping is a satisfactory method, and it is common practice for components to be machined to finished size prior to nitriding and for this lapping method to be used to bring the nitrided component to the finished size, that is, to remove by lapping the 0.00075 in (0.02 mm) of growth which should have occurred during the process.

Lapping is the technique using hard abrasive particles in a lubricating medium with either no or very low loading applied.

The nitrided case in addition to being very hard is also compressive, and it is this high compressive stress which gives the correctly nitrided component the improvement in fatigue characteristics. This also makes the finding of grinding cracks on a nitrided surface extremely difficult.

Any grinding carried out on the as-nitrided surface will almost invariably result in grinding cracking. The surface is so brittle that, even with the greatest care, it is virtually impossible not to produce grinding cracks. This cracking may initially be only in the very hard, very thin, very brittle surface layer but will invariably propagate from this area into the underlying layer which will not have sufficient ductility to absorb these cracks. The fact that the nitrided case is so compressive, means that the cracks will be held shut and, thus, will be extremely difficult to identify by all the normal means of *crack detection*. It will also mean, however, that the advantages regarding fatigue strength have been eliminated.

Where it is essential to grind subsequent to nitriding, the very brittle thin layer must first be removed. This can be achieved by the lapping process described above, but can be more economically removed by chemical *etching*. This may be one of several processes, *sulphuric acid etching* being the most commonly used. This is a form of *electropolishing* whereby the surface layer of 0.0005 in (0.012 mm) is removed.

The advantages of nitriding over other forms of *surface hardening* are that the hard surface is produced with no subsequent heat treatment operations, thus there is a minimum of distortion. This is the principal reason why the process is used at present.

It is a fact, however, that the nitrided surface is harder than with conventional *case hardened* processes such as *carburizing* or *carbo-nitriding* and has a higher compressive stress. In addition, the softening temperature of a nitride case is considerably above that of these two alternatives. This softening temperature is in the region of 300°C or even up to 400°C for short periods.

Nitriding, being a lengthy process using ammonia, which is relatively costly, is an expensive *case hardening* method and, wherever possible, alternatives should be considered. Some care is required in that the process of *carbo-nitriding* is sometimes referred to as 'nitriding', and although this process can produce a satisfactory case for many conditions, it will never produce a case that is comparable in all respects to the nitrided case.

Normalizing

A treatment designed to remove evidence of previous thermal treatments and produce a material in a homogeneous condition.

Application range STEEL, but the term is sometimes used for other metals

The necessity of normalizing resulting from the different treatments applied to steel during the *forging* operations, or the varying rate of cooling after *casting*. This will vary from batch to batch and, in many instances, within the same batch. These variations include the temperature used at the forge operation, dependent on the temperature of the forge hammer, and the temperature of the billet at the beginning and end of the operation. This again will vary and in some cases can result in *work hardening*.

When the operation is completed, the forgings are allowed to air cool. This also causes considerable variation between components. Some are rapidly cooled, for a variety of reasons, others are held at the temperature for a considerable time and cooled very slowly because of the large mass of material in a pile of hot forgings held in a still atmosphere. The forged structure then can contain a considerable amount of the hard, brittle martensite, which gives steel its hardness, or have very large grain structures of ferrite and pearlite, comparable to that obtained in the *fully annealed* condition. There can also be a considerable amount of *cold working*, resulting in distorted grains.

Castings also vary, depending on the temperature at which the casting is knocked out of the mould and the method of cooling to room temperature. These heterogeneous structures will exhibit different 'machinability' characteristics and can result in distortion occurring during service and have been blamed as the reason for excessive distortion during subsequent heat treatment operations, in particular *carburizing*.

The normalizing treatment consists of taking the components to above the upper critical temperature, that is, the same temperature used for *annealing* and *hardening* in the range 700°C–900°C, depending on the composition of the steel. The components must be carefully placed in the furnace so that they heat at an even rate. They are held at temperature for approximately 20–30 min per inch of ruling section and are then removed from the furnace and allowed to cool in still air.

This treatment results in the complete recrystallization of the grain structure, and as the time at temperature is controlled, and the rate of cooling from above the crystallization point is quite rapid, the grain growth is held to a reasonable amount. As the components cool in air, the rate of cooling will decrease by a measurable amount at temperature approx. 650°C and, thus, there is little danger of cooling stresses remaining. These stresses will have time to relieve themselves during the relatively slow cooling from red heat to room temperature.

Normalizing is a valuable and economical treatment which should be carried out by the forging company or at the foundry. With many steels containing small amounts of chromium and nickel at the 1 per cent level, the operation may result in the formation of some martensite to increase the hardness by an undesirable amount. In these instances the normalize should be followed by a *subcritical anneal*, or *temper* at 650°C for 2 h.

The full annealing operation will also result in recrystallization and removal of all stresses. There is, however, a much slower cooling rate with full annealing resulting very often in undesirably large grains giving brittle components.

Pack Anneal

A softening process where the components are kept out of contact with air by packing.

Application range ALL METALS

The term covers two types of processes.

First, there is the specialist process where sheet material is packed together so that air is excluded between the separate sheets and, where necessary, the sheets are bound together to ensure the absence of air. The pack of sheets is then heated in a furnace to the necessary temperature, and because little or no

HEAT TREATMENT

air is in contact with the surface, no oxidation will take place. This *annealing* method is commonly applied to mass-production processes when sheet metal is involved.

Second, there is the process where individual components are packed in a heat-resistant box together with cast-iron cuttings, sand or some other inert material. The box, with a close-fitting lid, is then fed into a furnace and *annealing* is carried out at the appropriate temperature. Again, the result will be that the components are softened with little or no oxidation at the surface.

Care must be taken to ensure that the components are clean and dry prior to insertion, since any contamination of the components' surface in the sealed box may result in serious surface damage such as corrosion pitting. The choice of the inert material requires some care and this material must be clean and dry.

The pack annealing process is analagous in all ways to other types of *annealing* with the exception that an effort is made without using sophisticated equipment to exclude air from contact with the surfaces during the heating cycle.

Pack Carburize

The process of case hardening in a box with activated charcoal.

Application range STEEL

This is a traditional method of *carburizing*, where the components are packed in a heat-resisting box together with compounds, consisting of small pieces of charcoal, which have been surface-activated with carbonates. The loading in the box is to the ratio of approximately four charcoal to one of components, and care is taken to ensure that a layer of charcoal separates the components. The box must be completely filled and fitted with a lid.

The furnace is heated to a temperature between 900°C and 920°C, the *carburizing* of the steel is the result of reaction between the activated charcoal and the air in the box to produce the essential gases which induce the formation of iron carbides at the surface of the steel.

Pack carburizing has now largely been superseded by *gas carburizing*.

Plasma Nitride

A sophisticated technique using high vacuum and the 'plasma' technique to nitride harden.

Application range ALLOY STEEL

The technique uses a vacuum chamber to treat the components. Hydrogen and nitrogen gases are subjected to the 'plasma' treatment to alter their molecular structure to the atomic state, and then to interfere with their atomic configuration.

This is achieved by high voltages applied for some time to the gases to interfere with the electron configuration of the outer layer of the atomic structure.

This considerably increases the energy of the gas involved, and can be recovered as heat when the gas contacts the surface of the component. The component temperature will rise, and as the nitrogen gas is present in the atomic state nitriding will take place, provided the correct steel has been chosen, and the surface is in a satisfactory state.

Once the steel is in the correct condition, and there is atomic nitrogen present above 450°C then the process will proceed as described under nitriding in this section.

Plasma nitriding has the advantage that the process time will be reduced as the surface temperature will reach the reaction temperature very rapidly.

It is also claimed that the nitride surface achieved is superior in all respects to the conventional nitride layer, and that the super-hard brittle surface formed by normal nitriding is not produced thus lapping or etching subsequent to nitriding is not necessary. There is also some indication that nitriding can be produced as low as 400°C, and that alloys such as high chromium steels can be successfully nitrided more easily than with the conventional process.

In general it is claimed that the process can have lower maintenance, production and energy costs, with a higher capital investment.

Further details can be obtained from the Wolfssen Plasma Processing Unit, University of Liverpool, or J. J. Casting Investments (Heat Treatment) Ltd.

The plasma nitride is in direct competition with conventional nitriding, and the reader is advised to obtain further technical information before considering this process.

Postheating

A general term which can be applied to any form of heat treatment following on a previous process. In the main, however, the term is confined to that treatment carried out on many components and assemblies following *welding*.

Application range STEEL – MAY APPLY TO OTHER METALS

This postheating will very often be referred to as *stress-relieving* and, in effect, results in the removal of all expansion and contraction stresses, or at least the majority of these stresses which occur during the *welding* process. This relieving of stresses is brought about by the fact that as the temperature of the part being heated is increased, so its yield strength is decreased. Any stress within the assembly greater than the yield strength, will result in 'yielding' or movement of the assembly. As this is a plastic deformation the movement will be permanent, but the stress will be removed. Considerable care is required in the temperature chosen, which, whenever possible, must be above that at which the assembly will operate. It is also essential that the assembly is allowed to yield and that no mechanical restriction is placed on the assembly during the *stress-relieving* operation.

Postheating can also act as a *tempering* operation. If during *welding*, the rate of cooling is faster than a certain critical rate, which will vary with the carbon content and alloying elements, then the *hardening* achieved can be considerable. Unless this hardness is removed, the assembly will have a brittle zone associated with this hardening, which can result in failure during service.

The *tempering* operation must be carried out at a higher temperature than that at which the material will be used during service, and with alloy steel some care must be taken in the choice of temperature, particularly if previous *heat treatment* has been carried out for mechanical properties. Users are advised to obtain specialist advice on important components. Further information is given under *Stress-relieving* in this section.

Precipitation Hardening

The process whereby hardening takes place at relatively low temperatures by precipitation following solution treatment. This is a modern term covering the process of *age hardening*.

Application range STEEL, ALUMINIUM, COPPER, TITANIUM, MAGNESIUM AND NICKEL ALLOYS

It is necessary that certain metallurgical conditions exist before an alloy can take part in the precipitation hardening process. Briefly, there must form within the alloy certain compounds which can be taken into solution in the matrix of the alloy. These compounds are known as 'intermetallics'. The resultant solution must be unstable, and allow the intermetallic compounds to precipitate from the solution in such a manner that the lattice of the alloy is strained, and it is this straining which results in the increase in hardness.

The precipitation hardening process invariably has a prior *solution treatment*. This means that the alloy must be taken to above a certain critical temperature and must be held for sufficient time for the intermetallic compound to be taken into solution. This is a solid–solid reaction, and in many cases the *solution treatment* is achieved during cooling after *casting* or following *welding*, but the actual dissolving of the intermetallic compound is achieved while the material is in the solid state.

When any of the materials which take part in this process are in the fully *solution treated* condition, the alloy is at its softest and most ductile state. It is, therefore, the ideal condition for any *deep-drawing* operation or the application of other heavy *cold work*.

The precipitation hardening operation in some cases, particularly with certain aluminium alloys, takes place at room temperature, and it is this increase in tensile properties with time at room temperature which gives rise to the term *ageing* or *age hardening*. With the majority of modern alloys, it is necessary that they are heated to a certain tightly controlled temperature for a specified time for the *precipitation* to occur, which results in the *hardening*.

Only by holding the material at the correct temperature for the correct time will the maximum hardness or 'tensile strength' be obtained. The difficulty of controlling precipitation hardening is that the time and temperature are interdependent in that the higher the temperature, within a reasonably wide range, the shorter is the time required to achieve the same mechanical properties. Thus, an aluminium alloy which can be precipitation hardened following *solution treatment* at a temperature of 120°C for 8 h could achieve the same mechanical properties by, for example, heating to 160°C for 2 h, and the same properties obtained by heating for a very short time above 250°C.

All precipitation hardening alloys have the common property that holding at temperature for too long a time, or heating to too high a temperature, will result in a decrease in the ultimate hardness and tensile strength. Errors in the choice of temperatures or time result in softening, which makes the choice of reasonably low temperatures for reasonably long periods of time preferable to higher temperatures for shorter times. Without further information it will not be possible to ascertain whether this low hardness is the result of under- or over-ageing.

Details of the alloys which take part in this process are as follows:

Aluminium	(a) 4% copper, with other additions
	(b) 1% silicon, 1% magnesium
	(c) 4% zinc with additions of silicon, magnesium and chromium
Copper	1.5% beryllium with nickel or cobalt. This alloy can result in the strongest non-ferrous material, with tensile strengths up to 100 tonf/in^2 (1550 N/mm^2).
Magnesium	(a) 6–10% aluminium, zinc, manganese
	(b) Rare earths+zinc and manganese
Nickel	15% chromium, with additions of titanium, vanadium and cobalt.
Stainless steel	(a) By adding copper and nitrogen to austenitic stainless steel, precipitation hardening is made possible.
	(b) 17% chromium with 4 or 7% nickel and other additions.
Titanium	(a) 4% aluminium, 3% molybednum, 1% vanadium with other additions
	(b) 4% aluminium, 3% manganese
	(c) 6% aluminium, 4% vanadium

Preheating

This term may be used to cover any heat treatment which is applied prior to further processing, but is usually confined to that prior to further heat treatment processes, and to heating prior to *welding* or *brazing* operations.

Application range MAINLY STEEL

Preheating when carried out in conjunction with other heat treatments is generally carried out in a furnace which will be at a temperature lower than the critical temperature for the subsequent operation. This is only necessary for those materials capable of a rapid change in metallurgical structure as this results in a volume change. These are generally high-chrome, high-carbon or tool steel materials. If the components are placed directly in the furnace at a temperature in the region of 900°C or higher, and the components have considerable change in section, then there is a possibility of cracking as certain sections go through the change point, because of stresses set up in these very brittle materials.

To prevent this, use is made of a second furnace held in the range 500–650°C as a preheat. Because this temperature is below the critical change points, the components can be placed in the furnace and when they reach this range are transferred to the higher temperature. The speed at which the metal rises in temperature is a function of its surface area and the temperature of the furnace. Thus, with the lower furnace temperature, the rate of heating is reduced. The parts must be held at this lower temperature for sufficient time to stabilize at the desired temperature and are then transferred to the furnace, and held at the final temperature.

For *welding*, or in some cases *brazing*, the preheating operation is necessary to produce a sufficient heat sink to ensure that when these operations have been completed, the rate of cooling of the area adjacent to the weld or braze will be sufficiently slow to ensure that no brittle structure is produced. This preheating can be carried out in a furnace, but will very often be by the use of a gas flame, or resistance heaters in contact with the area to be treated.

The temperature required must be specified and controlled. Control will most commonly be achieved with heat-sensitive materials, either a paint which changes colour at the specific temperature, or materials which have some other property, such as a softening or melting point.

The use of pyrometers spot welded or 'glued' in position is becoming quite common. It is essential that this equipment is correctly calibrated, and also that it is correctly measuring the surface temperature. It is good practice to check the surface temperature approximately one minute after the heat source has been removed or switched off.

It is quite common to find that the technique used to monitor the surface temperature can be influenced by the method used to apply the heat. If there is any appreciable difference in the temperature recorded before and after the heat source is removed, this must be investigated. A difference of 1–2°C could be significant and must be explained.

It is important that the *welding* or *brazing* operations are carried out while the components are at the specified temperature, and no delay is permitted between removing the source of heat and commencing of these processes.

Preheating prior to *flame cutting* will only be necessary for materials which have air-hardening characteristics, or where a very large mass of metal is cut while cold, for example, under outdoor winter conditions. These comments also apply to arc air gouging. Preheat must be considered when welding or brazing is carried out on steel above a carbon equivalent of 0.5 per cent approximately.

There is no real alternative to preheating, but where a controlled heat treatment operation is carried out immediately after *welding, brazing*, or *flame cutting*, it will not normally be essential except with steels of a high hardenability, above 0.7 per cent carbon equivalent. When preheating is required prior to *hardening* there is no real alternative for satisfactory results.

As a very rough guide the following information is supplied:

(a) Below a carbon equivalent (CE) of 0.3 per cent no preheat.
(b) Between 0.25–0.4 per cent CE preheat required on high-integrity components.
(c) Above 0.5 per cent CE preheat should be called for when high excess heat is applied, for example arc air gouging or *flame cutting*.
(d) Above 0.7 per cent CE any process such as *flame cutting*, arc air gouging, and some forms of arc welding should not be considered. Technical advice from an unbiased source should always be obtained before using any form of high-energy heating with these materials.

Quench and Temper

This is an alternative term for harden and temper.

Application range STEEL

This is the specific technique to ensure the best mechanical properties for steel regarding the ratio of yield to ultimate tensile strength with good ductility. There is no alternative, and high-quality control is essential if adequate results are required.

Quench Tempering

A form of delayed or slow quenching.

Application range STEEL

With this, the steel is quenched from the required *hardening* temperature into oil, or some other liquid such as molten salt or liquid metal held at the required temperature. As this always will be a reasonably low temperature, this will in addition give a form of *tempering*. This technique is not advised when high-

420 / HEAT TREATMENT

quality heat treating is necessary, and is not possible when different sized batches are involved, since the best mechanical properties are not achieved.

The term is sometimes applied to the technique of quenching in oil and removing the components while still hot, thus allowing them to cool from the *tempering* temperature at a reasonably slow rate. This again is not advised where superior heat treating is required as it will be impossible to control accurately the temperature at which the components are removed from the quench. The method, then, is an uncontrolled form of *hardening* and *tempering*.

Refine

A term applied to the core hardening process after carburizing. The full term is 'refining the core'.

Application range STEEL

Unless special grain-controlled steels are used in *carburizing*, then considerable grain growth will occur and this will affect the mechanical properties of the finished components, particularly the ductility of the core, if the defect is not rectified.

Refining is, therefore, carried out after the components have been removed from the *carburizing* atmosphere. This consists of taking the components to the correct temperature for *hardening* the core material or base metal and the component must be left at this temperature for sufficient time to allow recrystallization to occur but insufficient time for grain growth to occur. Refining of the core is identical to *hardening* of the material in the uncarburized condition.

The term refining has a much wider application in extraction and production metallurgy, where various techniques are applied to many of the smelting processes for different metals for removal of undesirable elements, oxides or gases. These, of course, are outside the scope of this book.

Secondary Hardening

The process where an increase in hardness is achieved at a second temper.

Application range ALLOY STEEL

This is now considered to be caused by the retention of austenite at room temperature after quenching. This austenite is not fully destroyed during quenching or the first *tempering* operation, and on the subsequent operation at the same temperature, the remaining austenite decomposes to form martensite. There is an appreciable increase in hardness as martensite is considerably harder than austenite.

It is sometimes considered when this *secondary hardening* occurs that there is an indication of over-quenching; had the hardening operation been correctly controlled, no austenite would have been retained, with no secondary hardening possible. Whether or not an ideal quench rate is economically possible is open to debate.

Where components have fine limit tolerances which must remain stable during service, *secondary hardening* is essential as the change from austenite to martensite also results in a volume change. This means that if the breakdown of austenite occurs during service, as is quite likely to happen, then there will be a change in dimensions. As the steels commonly used for gauges are among those most likely to have retained austenite, the importance either of controlling correctly the *hardening* operation, or having a second *temper* will be appreciated. In most instances the use of a double temper will be the most economical and technically satisfactory method of ensuring a fully *hardened* and dimensionally stable component. Deep freezing between the two tempers is sometimes advised to ensure the changes take place.

Selective Annealing
The local softening of previously hardened surfaces.

Application range STEEL

Use is made either of a *welding* type flame or *induction heating* to achieve the local effect. With both techniques some care is necessary to ensure sufficient heat is applied only to cause softening, and that no area of the component is taken above the critical temperature and then rapidly cooled. This *tempering* method is commonly applied to mass-production components. The operation is identical to annealing and involves the same wide variations covered by that term. Any *softening process* involving only specific areas on components is said to be selective annealing.

Selective Carburizing
This term is applied when a component is partially *carburized*.

Application range STEEL

1. *Using a stopping-off medium* This is either by electrolytic copper plating or by the application of copper-rich paint or some other proprietary medium to the area where *carburizing* is not required. This must be carried out immediately prior to the process, and it is necessary to ensure that the *stop-off* medium is correctly applied, particularly in areas where stress concentrations are likely as any local hardened spots in these areas could result in fatigue failure.

2. *By machining* With this, the entire component is carburized, followed

by *refining* to harden the core. The component is then softened by holding at 650°C for at least 2 h. The surfaces required to be soft are then machined to remove surface metal and ensure that no high carbon remains. This usually means that twice the designed case depth is removed, that is with a 0.050 in (1.25 mm) case the diameter will be reduced by 0.20 in (5 mm). This is much more expensive than *stopping-off* as an oversized forging is usually required. The increase in machining costs can be considerable but this does ensure the elimination of local hard spots on the finished component. Because of distortion problems and the importance of ensuring even removal of metal, this technique often demands extremely accurate machining to very tight tolerances. It is usually applied to aircraft components and other high-integrity parts.

3. *Selective hardening*   With this the entire component is *carburized* and, where necessary, *refined* in the normal manner, but instead of *hardening* the whole surface, only the selected areas are re-heated. This makes use of *flame* or *induction hardening*, and can be the cheapest method of selective carburizing for high-production parts.

Selective carburizing is required when wear resistance is necessary on certain areas, with high ductility on other areas.

Alternative methods of achieving a similar local hardening involve the higher carbon steels again by *flame* or *induction hardening*. *Chromium plating* is possible, but this will not normally permit the gradual change from hard to soft, it thus can present design problems. It is also possible to carburize and harden components in entirety then locally soften specified areas by *selective annealing*. *Metal-spraying* and *weld deposition* of hard metal can be employed under some circumstances. Selective carburizing, however, probably achieves a better compromise of a hard abrasive-resistant surface, associated with high ductility areas where high integrity is necessary.

Selective Hardening

The technique of locally hardening specified surfaces.

Application range STEEL

In general the term is not applied to *surface hardened* components but, under some circumstances, this may occur. Components which have been locally *carburized, carbonitrided, cyanided* or *nitrided* will be included under this heading. It is, however, also used when a through-*hardening* type of steel is treated locally. The process obeys all the laws of conventional *hardening* of steel. Alternatives are not possible where through-hardness is required, but *chromium plating* and *selective carburizing* are alternatives for surface treatment. *Metal-spraying* and *weld deposit* of hard metal are also possible, if the design parameters permit.

Local application of cold work is also a possibility, and this would apply to all metals.

Self-annealing

This term is applied to the *softening* process when there is a sufficient mass of material to ensure that the rate of cooling will result in the components being reduced in hardness.

Application range ALL METALS

As only steel is *hardened* by the use of a critical cooling rate, it is obvious that steel components must be handled carefully when removed from a *normalizing* or *annealing* furnace. The *full anneal* uses the furnace, and the heat sink of the furnace bricks, etc. to achieve the required slow rate of cooling. When the components reach the desired temperature, the furnace is switched off and the components and furnace are cooled together. With self-annealing the components are removed from the furnace and it is then essential that a large enough mass of material is present to ensure a sufficiently slow rate of cooling.

Another use of this term is where electrical conductors are *softened* in service because of the resistance heating which may occur with overloads. Thus, the conductor is heated, any *work hardening* is removed and the conducting material is softened. Electrical conductivity is improved as any *work hardening* increases the electrical resistance.

Self-hardening

An alternative term for *air hardening*.

Application range STEEL

With this, it is necessary that certain constituents are present in the steel, these in the main being a minimum carbon content of at least 0.3–0.4 per cent together with hardening elements, particularly chromium, and also nickel, molybdenum and vanadium. When these are present in a significant amount, and the mass of the material is low enough, then on removal of the steel from the hardening temperature, the cooling rate of air is sufficient to cause *hardening*. Described fully under *Hardening* in this section **Heat Treatment**.

The concept of 'carbon equivalent' (CE) is important in this area. This assumes that certain elements have an influence on the hardenability of steel, and that these can be related to carbon. There are several formulae in use; the British Standard stated in BS 4360 is:

$$CE = C + \frac{Mn}{6} + \frac{Cr + Mo + V}{5} + \frac{Ni + Cu}{15}$$

With a carbon equivalent (CE) above about 0.7 per cent self-hardening can be expected. Where a large mass is heated locally hardening can occur with the CE as low as 0.4 per cent.

Skin Annealing

A term used to describe the process of removal of surface *cold work* often produced during the manufacture of high nickel components. The process is, then, a short-term *solution treatment*.

Application range NICKEL/CHROMIUM ALLOYS

The components in the heavily *work hardened* condition will not be capable of accepting further *cold work* without causing damage to the tooling, or embrittlement or cracking of the components themselves. In addition, because of the very local *work hardened* skin, there is some possibility that surface cracking can occur during the final machining operation.

The skin anneal operation itself is carried out at approximately the *solution treatment* temperature for the particular alloy. With the nickel–chromium alloys this will be in the region of 1050°C–1200°C. A salt bath is usually employed but any means of achieving relatively rapid heating will be successful. The time at temperature will be as short as 10 min and will never be in excess of 30 min. If conventional furnaces are used, it is difficult when a number of components are involved to control this short time at temperature as those at the outer edge of the furnace will reach temperature before those at the centre. The rate of cooling is important. If it is too slow, some *precipitation hardening* will occur, and apart from being undesirable where further machining is necessary, this can also interfere with the efficiency of the final treatment.

Skin annealing is, then, a form of *annealing* specifically applied to a range of nickel–chromium alloys which are subjected to relatively heavy machining. The process is carried out between the *solution treatment* operation and final *precipitation ageing treatment*. With some of the nickel–chromum alloys, better 'machinability' has been identified when they are in the partially precipitated or aged condition. This, however, will not eliminate the *work hardening* and skin annealing will still be necessary. With these alloys, the necessity to control the time and temperature under very rigid conditions is essential otherwise the final component will be over-aged.

The term can be applied to other techniques where only the surface is affected, though this is more generally referred to as local *softening* or *annealing*.

Softening

Described fully under *Annealing* in this section **Heat Treatment**.

Solution Treatment

The first of two stages in the *precipitation hardening* process. This results in the intermetallic compounds being dissolved and when correctly carried out is the softest, most ductile condition of the metal involved.

Application range ALLOYS OF ALUMINIUM, MAGNESIUM, TITANIUM, NICKEL, COPPER AND SOME SPECIAL STAINLESS STEEL MATERIALS

The process requires that the constitutional diagram of the material has certain characteristics in that the material contains compounds, known as 'intermetallics', which can be dissolved in the matrix of the alloy.

The solution treatment will be the first of a two-part process. The solution treatment itself will invariably result in the alloy achieving a soft condition, and if the treatment is correctly carried out, then it will be in its softest possible condition. The second part of the treatment is called *ageing*, or *precipitation treatment*.

Any *deep-drawing* or other forms of *cold-forging* operations where maximum ductility is required should be carried out in this condition. A satisfactorily treated material will be softer and more ductile than in the *annealed* condition, although the difference might not be to any great extent.

The process requires that the material is taken to a specified temperature and held for sufficient time for the necessary metallurgical reactions to take place. These reactions are the taking into solution of the intermetallic compounds which affect the treatment. The condition achieved will be relatively unstable, and this will result in the precipitation from solution of these compounds: it is this which results in an increase in the mechanical properties of the material for which the treatment is carried out.

Under some circumstances the term solution treatment is used to describe the metallurgical reaction which occurs when other compounds, such as iron carbide, are taken into and retained in solution. These include the *hardening* of steel, where this is occasionally referred to as 'solutionizing' or 'austenizing', and some of the *annealing* operations. The common definition, however, will be that given above, where compounds are taken into solution and this solution is unstable.

The taking of intermetallic compounds into solution occurs in the solid state and is a solid–solid reaction but otherwise identical in all other respects to the more conventional meaning of solution as dissolving, as when, for example, solid sugar is dissolved in tea or coffee. The process uses conventional furnaces generally with good temperature control but controlled atmospheres are seldom required.

The temperature, of course, will depend on the alloy, and in most cases will be critical. If too low a temperature is used, the compounds will not all be taken into solution resulting in an unsatisfactory treatment. If too high a temperature is used, grain growth at the least will occur and this results always in a more brittle material. There is also the possibility of some liquefaction at the grain boundaries or actual melting of the components as it is quite common for the solution treatment temperature to be relatively close to the melting range. As stated, the ideal temperature varies to a considerable extent particularly, for example, between copper alloys and aluminium alloys. Even within the latter, wide variations exist and can result in the defects described unless the correct temperature is employed.

The material must be held at temperature for a specific time. Again, if insufficient time is used, the compounds will not all be taken into solution; and holding for an excess time, while not resulting in any melting problems, can cause excessive grain growth.

On completion of heating, the components are quenched. This, in high-quality processes, will generally be in water or oil, but air-quenching either by blasting with cold air or even the use of still air is commonly applied. The choice of quenching speed will depend on requirements and to a large extent on the geometry of the component or assembly. This may often preclude the correct and efficient use of a quench.

Only by ensuring that all the intermetallic compounds are retained in solution will the full effects of solution treatment be realized. The more the quench is delayed, the fewer compounds will be held in solution and, if the quench is slow enough, an *annealing* type of operation will result with a considerable reduction in the hardness achieved by *ageing*. This, however, will often be acceptable, since the alternative is to have no form of solution treatment at all.

It must, however, be realized that the yield or proof strength will also be lower, and if this is critical to the design it may need to be specified, and then checked.

This will be the case when the solution treatment is only part of another process. For example, during *casting* the metal will be at a temperature in excess of that required for the present process, but if the component can be cooled rapidly enough, a reasonably efficient 'solution treatment' will result. This is commonly practised, particularly with aluminium and magnesium alloys, where by careful design of the moulds relative to the size of the casting, and with careful placing of the castings within one mould, satisfactory results are obtained. This, of course, will not be as satisfactory as removing the casting from the mould, reheating to the correct temperature and then efficiently *quenching* in water. However, the difference in the mechanical properties achieved may not be sufficient to justify the expense involved.

The same comments apply to *welding*, for components which have been correctly solution treated and *aged* are very often joined by welding. *Ageing* is destroyed locally but, provided the weld is correctly designed, the heat of the process and the rapid cooling achieved by the adjacent metal can give a successful treatment. This means that the weld area will be in a soft, ductile condition and thus capable of accepting the stresses imposed during cooling.

Here the *welded* components are a more attractive proposition than when the process is carried out where rapid cooling results in *hardening* of the steel and the danger of brittle cracking. It will, however, be realized that somewhere in the weld zone there will be an area lacking the full mechanical properties, even allowing for subsequent *ageing* of the welded assembly, because this has been overheated to soften previously aged material but insufficiently heated for solution treatment to occur.

The solution treatment process will not be carried out in isolation and there will always be subsequent *ageing* or *precipitation* treatment. This might be a suitable room-temperature age or a furnace or oven treatment. The process is further described under the heading *Precipitation Hardening* in this section, where the different alloys which can be subjected to this form of *hardening* are listed together with information on the temperature range for solution treatment and *precipitation hardening*.

Spheroidal Anneal/Spheroidization
A form of softening with the formation of a spheroidized metallurgical structure, that is, there appear well-defined spheres in the micro specimens.

Application range STEEL

This results in the material being softened and made more machinable than with the conventional operation. It can also result in an abrasive resistant material with considerable ductility.

The technique results in the carbides present in steel, either as a separate phase or in solution, being altered to round, small, discrete particles which are held in the matrix, this being pearlite, or tempered martensite.

The treatment consists of heating the components to above their upper critical temperature for a period of approx. 20 min to 1 h. The components are then transferred to a second furnace and held at 650°C. On removal after approx. 2 h, they are transferred to a furnace at the original temperature. After a period of up to 1 h, the components are returned to the lower-temperature furnace at 650°C. This cycling, also referred to as *cyclic anneal*, results in full softening of the material's matrix by the formation of pearlite or tempered martensite, and as the time held above the upper critical temperature is reasonably short, there will be much less tendency for excessive grain growth. Holding the component at 650°C will temper the martensite without resulting in any grain growth, and with high-carbon steel will result in spheroidization of the carbides present.

With lower-alloy steels, satisfactory spheroidization can be produced either with a single high temperature–low temperature cycle or, under some circumstances, with a longer treatment at 650°C. With high-carbon, high-chromium steels, particularly when other alloying elements such as nickel are also present, more than two cycles may be necessary to produce the desired degree of softening.

This *annealing* operation is useful, first, for softening steel in order that machining can be carried out, but the technique has also been found useful with high-chromium steels in that the resulting material has considerable advantages in ductility and hardness over the conventional methods. Second, it also enhances the 'hardenability' of the material as the fine spheroidal carbides will go into solution more easily at the *hardening* operation and will, thus, be available for the production of the hard martensite. Where a cutting edge is necessary, the high-chromium, high-carbon steels in the spheroidal condition have been found to have certain advantages over other forms of treatment regarding ductility, hardness and wear resistance.

This spheroidization as an *annealing* operation has considerable advantages over conventional high-temperature methods, where in order to achieve the same standard of softness there will be very considerable grain growth. It is doubtful whether holding at 650°C alone will achieve the same degree of softness as the spheroidization process.

Stabilizing

A form of stress release to give dimensional stability.

Application range STEEL

This usually indicates the treatment of components which are to be *nitrided*, where sufficient machining allowances remain to remove any distortion which might occur during this subsequent process. It is obvious that the term stabilize, or stabilizing treatment, can be used to define other forms of processing where the article is more stable after the treatment than before. These, however, will be local definitions.

Components which are to be *nitrided* are *normalized* then *hardened* and *tempered* at a temperature in the region 550°C–600°C at an early stage in their manufacture. The components are then machined to within 0.010 in (0.25 mm) of finished dimensions, and at this stage they will be stabilized at a temperature in the region 520°C–550°C, which is above the *nitriding* temperature but below that of *tempering*. This treatment will, therefore, effectively remove all stresses without affecting the hardness.

The components are then finish-machined with care being taken that no stresses are imposed, particularly on fine-limit diameters. The components are then *nitrided*, and as the temperature will be in the range 490°C–510°C, it will be seen that no stresses will exist which can be removed during this process, thus no dimensional changes should occur.

The stabilizing treatment is, then, a specific form of *stress-releasing*. When the process is used in conjunction with low-stress machining between stabilizing and *nitriding*, no movement or distortion of the components will occur at the latter operation.

Stabilizing Anneal

The term applied to the specific treatment to take carbides into solution and to hold them in this condition. The term can also be applied to any other *softening* treatment where metallurgical or dimensional stabilization is achieved, but it is most commonly used for stainless steels.

Application range STAINLESS STEEL

This eliminates the danger of intergranular corrosion which is known as *weld decay* and is caused by the precipitation of chromium carbides at the grain boundaries. When straightforward 18–8 chrome–nickel steels are held at temperatures between 300° and 800°C for any length of time, the chromium carbides will precipitate from solution and appear as a separate phase at grain boundaries. The most critical temperature for this reaction to occur is 650°C. The appearance of this second phase reduces the corrosion resistance at grain boundaries thus intergranular corrosion will occur. This is known as *weld decay*.

The stabilize anneal is carried out at 1000°C–1050°C at which temperature the carbides are re-dissolved and held in solution. Provided the cooling rate is above a specified minimum, no re-precipitation will occur, thus the material will remain as a single phase metal with excellent corrosion resistance. The rate of cooling is not critical, and normal air cooling is satisfactory, provided the components are correctly spaced to allow air movement and are of a reasonable mass.

Two other techniques are available as alternatives to the stabilize anneal, both of which involve the material itself rather than being treatments. By controlling the carbon content of the stainless steel to a low level, fewer carbides are formed and thus the problem of carbide precipitation can be controlled. These steels are often classified with an 'L' suffix, denoting low carbon. This is generally at the level of 0.03–0.06 per cent, and it will be obvious that at the higher carbon levels more carbide precipitation will be possible than at the lower levels. This means that some precipitation can occur if the component is held for sufficient time at the critical temperature of 650°C. This precludes the use of these steels for *brazing* where a large mass is involved, if they are to be used under aggressive corrosion conditions.

The second method is to stabilize the carbides in the steel by the addition of elements such as niobium (columbium), titanium or molybdenum. These elements form stable carbides which do not precipitate and, provided that all the carbon is combined with the stabilizing element, no precipitation can occur; thus *weld decay* is eliminated.

To summarize, stabilize annealing is used to remove precipitated carbides in stainless steel. This treatment is necessary to prevent *weld decay* which can also

430 / HEAT TREATMENT

be prevented by reducing the carbon content, or adding a stabilizing element. Details of a testing technique to evaluate this problem appear under *Weld-decay Testing* or *Strauss Test* in Part One.

Step Anneal

A form of annealing which can more correctly be called *stress-releasing*.

Application range STEEL

It is required for components which generally have an intricate shape, and reasonably thin sections and which have had *work hardening* applied. These parts therefore are capable of distorting at reasonably low temperature.

In order to reduce the distortion which might occur at the *stress-releasing* operation, the anneal is carried out in steps or stages with the temperature being raised between each step. Depending on the material involved, the components, their shape and dimension, etc., these temperature steps will vary and can in fact be of different durations as the temperature increases. Commonly the stages will be between 50°C and 100°C with the components being held at each stage for sufficient time to equalize the temperature throughout the component. This technique ensures that the component is taken gradually to temperature, and that the stresses which might be imposed owing to change of section are kept to a minimum. With this process, it is generally advisable that cooling should either be carried out by steps, or at least should be controlled to a slow, steady rate. The temperature involved, times at each stage or step, and total heating and cooling time, will all depend on the component's shape and size and the known history of distortion.

Stress-equalization Anneal

This is a low-temperature *stress-releasing* operation which ensures minimum distortion at any subsequent machining.

Application range NICKEL AND ITS ALLOYS

The treatment is intended to develop some mechanical properties by a form of *ageing*. It is assumed that the components have previously been cooled at a reasonably rapid rate from a high temperature and are, thus, either correctly *solution treated* or that the intermetallic compounds which might be present are mostly in solution.

The mechanical properties of nickel alloys at room temperature are not markedly increased by the conventional *solution treat* and *ageing* processes but some of the high-temperature properties, particularly creep strength, are improved. If this property is of importance, then stress-equalization annealing will not normally be carried out.

This is a compromise process designed to give some improvement in mechanical properties and simultaneously to reduce the tendency to distortion during service. It is a form of *stress-releasing* but, depending on the alloy involved, some improvement in mechanical properties can be obtained.

Stress-relieving/Stress-releasing
A form of *heat treatment* to remove mechanical stress.

Application range ALL METALS

The purpose is to relieve stress which, for whatever reason, may be locked up in the component or material. These stresses can be caused by *work hardening*, by *welding* resulting in expansion and contraction stresses or by any other form of heating or treatment.

Strictly speaking, a stress-relieving operation must not affect the metallurgical structure of the component; (where this is not the case, the term *tempering*, *ageing* etc. should be applied). The stress-relieving operation chooses a temperature which is above that at which the component will be used in service, but below the temperature of previous metallurgical operations. With high-quality engineering components, the *tempering* or *ageing* carried out after *hardening* or *solution treatment* will generally be the limiting temperature regarding the metallurgical structure which has been produced. If possible, the stress-relieving operation should use a temperature between 10 and 20°C lower than the *tempering* or *ageing* temperature for a period of approximately 2 h.

The term stress-relieving is now commonly applied where components have been *welded* and where it is known that the operation has resulted in the formation of certain specific metallurgical structures, for example, in the case of steel, the formation of martensite. The postweld stress-relieving operation is then designed to alter this metallurgical structure and so is, in fact, a form of *tempering*. It is probably more vital that this should be carried out than a normal stress-relieving operation, which will merely remove the stresses imposed by contraction and expansion during heating and cooling at *welding*, while the formation of the brittle martensite can result in disastrous failure from fracture.

With steels, there is the general rule that when in doubt stress-relieving should be carried out at 600°C–650°C for 2 h. This may result in the softening of alloy steels which have been *tempered* at a lower temperature, but failure for this reason will be by ductile fracture and this will be less dangerous than brittle fracture arising from the brittle martensitic condition. Stress-relieving should always be treated as an important metallurgical process and as much information as possible must be obtained regarding the previous history of the component and the operating conditions under which it will be used.

Stress-relieving is advisable on all engineering components after finish machining or at a very late stage in machining if the components are going to be used at above room temperature. It is generally accepted that it is the manufacturer's responsibility to ensure that machining stresses are not removed during service running as this can cause distortion, resulting at the worst in engine seizure and at best undue wear.

During the stress-releasing operation it is important that the components are taken to temperature at a slow enough rate to ensure that equal heating occurs and consequently that there will be no danger of additional stresses being imposed owing to different expansion rates from different dimensions. It is also important that the component or assembly should be supported, since a stress-relieving temperature in the region of 600°C will result in a serious reduction in the mechanical strength with consequent distortion.

In practice it is this reduction in the tensile strength of the material at the stress-relieving temperature which allows the stresses to relieve themselves. As the temperature is increased so the mechanical strength of the material being treated is decreased, and if any stresses are present in the component or assembly which are greater than the yield strength at the stress-relieving temperature, then these stresses will result in the material yielding, thus the stresses are removed.

It is, therefore, true to state that if there is no measurable movement in components during routine stress-relieving, then the operation is not achieving any purpose and can be removed. It is unfortunate that there is no simple method at present of measuring the internal stress on a component and, thus, there is little doubt that a very large number of components and assemblies are subjected to unnecessary operations. There is, however, no argument that this stress-relieving is necessary in order to protect components or assemblies during service, and it is much cheaper to carry out a purposeless operation than that one component should fail in service.

The only alternative to stress-relieving would be to treat components during manufacture with such care that no stresses are imposed. Normally this will never be economically possible in engineering and so the alternative is not valid.

Subcritical Anneal

A *softening* process.

Application range STEELS WHICH CAN BE HARDENED

The term is specific to the *annealing* carried out at a temperature of approximately 650°C. Normally the time at temperature will be 2 h.

The term is derived from the fact that the temperature used is slightly lower than the first critical temperature used in the heat treating of steel. This is the temperature at which the necessary reactions required to harden steel commence. Below this no solution of iron carbide in the iron can take place as it is at this temperature that iron begins to be converted into that material which can successfully dissolve iron carbide. At the temperature of 650°C all the martensite – which is the structure giving steel its hardness – is fully tempered. This means that no hardening remains. At the same time any *work hardening* will be removed, as *cold work* is removed by increase of temperature. This can be noted in the microstructure by the removal of the grain distortion caused by the *work hardening*.

Subcritical annealing, then, is the treatment generally used when materials are required in the soft condition for 'machinability'. It has the considerable advantage over the *full anneal*, which is carried out at above the upper critical temperature, in that no grain growth can occur and, thus, no undesirable increase in the grain size results in undesirable brittleness. However, where sheet-metal components have been subjected to *deep-drawing* operations, and further drawing is required, then the present operation will not result in a material capable of accepting further *work hardening* on a considerable scale. Apart from this exception, the process results in a fully *softened* material.

In general the process is referred to simply as 'annealing'. The *stress-relieving* operation, commonly applied at this temperature following *welding*, is also in effect a subcritical anneal. It is further true that the present process is a full *tempering*. The term *stress-relieving*, under many circumstances, will be taken to mean subcritical anneal. It must be appreciated that in the metallurgical sense they are quite distinct processes. *Stress-relieving*, in theory at least, should never result in any metallurgical changes to the structure, while the subcritical anneal always results in metallurgical changes and resultant structure will be fully pearlitic in most cases. Provided it is understood what the process should achieve, the nomenclature applied is irrelevant.

Subzero Treatment

A form of heat treatment which is applied to certain alloy steels, where, because of the alloying elements present, there is a possibility of retaining austenite after quenching. Austenite is the phase which is obtained when iron is heated to above the upper critical temperature, and is the material which is capable of dissolving and holding carbides in solution. The *hardening* process requires that all the austenite is converted into the metallurgical structure known as martensite in order to achieve full 'hardenability'. If for any reason some austenite remains, unconverted to martensite, then this will have a double disadvantage in that the steel will not be in the fully *hardened* condition, and the components will be dimensionally unstable as there will be a tendency for the austenite to convert into martensite during service with a resultant change in dimensions.

The subzero treatment is carried out after *hardening* and *tempering* at temperatures below zero, and solid carbon dioxide in a chlorinated hydrocarbon liquid is the most common method, but deep-freeze units can be used. A time of approximately 2 h is required. The treatment must be followed by a further *tempering* operation.

This operation is necessary when alloy steels with a high or medium-high carbon content are used. It is relatively difficult to retain austenite at room temperature, and is only possible with steels which have elements such as chromium, since these result in reducing the critical quenching rate necessary to convert austenite to martensite.

It is particularly important to carry out the subzero treatment on components such as gauges or similar pieces which require to be dimensionally stable over a range of very fine tolerances. These components are commonly manufactured in high-chromium, high-carbon materials which are most prone to retain austenite after *hardening*.

There is some evidence to indicate that, provided the correct quenching rate is chosen for the steel in question, no retained austenite is possible. Thus, for high-production components it may be more economical to control the quenching rate rather than carry out this relatively expensive treatment. Many steels are double *tempered* to eliminate retained austenite, this being known as *secondary hardening*. All steels which are subjected to subzero treatment must be double *tempered*, and generally the low-temperature treatment is only carried out on very important components in highly alloyed steels where maximum stability is essential.

Tempering

The treatment which follows *hardening*, and one of the most important metallurgical treatments.

Application range STEEL

It is not possible to discuss this operation without describing the *hardening* process, and vice versa.

When steel is *hardened*, it is taken to above the upper critical temperature and quenched when the austenite produced will be converted into martensite. This is the hard, unstable metallurgical structure which ensures that the steel is in the fully hardened condition.

Steel can also be *work hardened*. In this case the grain structure is distorted, straining the lattice, resulting in an increase in hardness. For all practical purposes it can be assumed that the result of *work hardening* is identical to *thermal hardening*, but seldom achieves the dramatic increase in properties, and is certainly a much less flexible process.

The martensite which is produced by *hardening* is a very brittle constituent. When proper *hardening* has been achieved and the carbon and alloy content are high enough, then the metallurgical constituent will be alpha-martensite. This, in addition to being hard, is also brittle, thus failure can occur with only slight movement, which will result in cracking. By heating to approx. 150°C, this alpha-martensite is converted into beta-martensite, again hard and brittle but having a measurable reduction in brittleness over the former constituent.

This treatment of heating hardened steel, to convert alpha to beta martensite, is tempering and will be the lowest temperature at which the operation is carried out. It will not cause any measurable reduction in the hardness of the steel.

As the temperature is increased, the metallurgical structure will be altered. The martensitic structure is essentially an extremely hard and brittle needle-like material. The size of the needles is controlled at the *hardening* operation. The purpose of tempering is to produce a structure giving the maximum ductility together with the desired degree of hardness. It cannot be over-emphasized that unless the steel has been correctly *hardened*, and is fully martensitic, then the ideal mechanical properties cannot be obtained by subsequent tempering.

It is the design engineer's responsibility to specify the mechanical conditions, normally given as the desired hardness range. The assumption is made that this specific range indicates the 'ultimate tensile strength' and that this in turn is related to the 'proof' or 'yield strength' of the steel in question. It must be appreciated, then, particularly by the heat-treatment personnel, that while tempering can be carried out to achieve the specified hardness figures, the maximum possible 'proof' or 'yield strength', which is the design criterion for failure, may not have been achieved because of incorrect *hardening*. Thus, both *hardening* and tempering operations require to be rigidly controlled to ensure, first, that steel is in the fully hard condition and, second, that tempering achieves the desired mechanical properties.

As the temperature is increased, the metallurgical structure is gradually altered from the needle-like appearance of martensite, first, to an indeterminate structure, then by further gradual changes in which the needles form into spheres. This structure indicates the full temper and is achieved at a temperature of approx. 650°C–750°C; it is thus identical to the process known as *subcritical annealing*. The final hardness achieved at this temperature will be a function of the steel analysis. The various structures produced at different temperatures are generally named after the metallurgist who first identified them. It should be realized that no dramatic demarcation exists between these structures, a new structure gradually forms as the original structure disappears. The modern tendency in metallurgy is not to use these names, but to describe the different structures as temper products of martensite.

Fully tempered martensite will be spheroidal in structure and will have properties not unlike that obtained from the pearlite structure achieved by *normalizing* but with a better yield strength.

One important aspect of tempering is that the process produces a certain structure and, provided the steel is not heated to above the tempering temperature, no alteration to the structure takes place no matter how long the steel is held at or below that temperature. This makes steel treatment markedly different from the ageing of alloys, where that process is time as well as temperature dependent. Once a steel has been held at temperature for approx.

2 h, the metallurgical condition for that temperature will have been stabilized, and the component can be held at the temperature for an infinite time without any further changes in the metallurgical condition.

Where incorrect hardening exists, there can be variation in hardening by prolonged heating. With complex steels long-term heating at, or about, the tempering temperature may result in some alterations in the hardness but this will never be of any significant extent.

The tempering process is carried out below red heat, and this means that the furnaces will have some difficulty in controlling the temperature as only convection heating is possible. Tempering furnaces should, therefore, be designed with some mechanical agitation of the air, or use should be made of liquid-heat transfer. The most common furnace employed is the forced-air circulation furnace, which is circular in shape with a fan in the base or lid which forces the heated air through the mass of material then across the heating elements. In this way economical heating is achieved in that the component can be relatively rapidly brought to temperature, and also a temperature gradient between the heating element and the cold component is eliminated. Liquid heating using salt baths, molten metal or other heat-transfer liquids such as oil and stable chemicals are also successful, but these are normally usable only at the lower tempering temperatures.

The use of the *fluidized bed* technique can ensure excellent heat transfer in an economical manner. This is basically hot gases agitating and heating fine inert powder.

The speed of cooling following the operation is not generally critical, unless too rapid cooling is attempted when some distortion will be found on components which have a rapid change in section, or other geometrical peculiarities. Invariably no metallurgical changes take place between room temperature and 650°C which makes this aspect of cooling of no importance.

There are, however, a series of steels which require to be rapidly cooled after tempering. These are the chromium and nickel–chromium steels, particularly those with more than 2–3 per cent chromium. When these steels are held at approx. 300°C–400°C for any length of time, there is considerable danger of precipitation at the grain boundaries of chromium carbides. This precipitation of carbides is known as 'temper brittleness' and can produce a material which is extremely brittle indeed and can fail in service below the design limits. With these steels it is therefore essential to temper below this critical range, and where high hardness is necessary the operation generally takes place at the range 150°C–250°C to avoid problems. Where maximum ductility is essential, the range will be 550°C–650°C and it is then desirable that the components are oil or water quenched to ensure that they pass through the critical temperature range as rapidly as possible. Tempering of these steels should never be attempted at 300°C–450°C. Most nickel–chromium and many chromium steels now have molybdenum added during manufacture. This element stabilizes the carbides and prevents their precipitation at the critical temperature, thus obviating the necessity for any rapid cooling after the operation.

It is advisable that the 400 series stainless steel, that is, 12 per cent chromium steels, are quenched in water following any heating operations in excess of 500°C.

Tempering of steel will always be preceded by a *hardening* operation. In the vast majority of instances this will be carried out under carefully controlled conditions, with the correct quenching rate specifically chosen. With *welding*, however, this is not possible and this process will under many circumstances result in the very local production of a martensitic structure. This local area of hard, brittle material has in the past resulted in many disastrous failures, where stresses have produced relatively slight movement causing the martensite to crack.

It is now completely accepted that, under many circumstances particularly where high-tensile steels are *welded*, and this includes any steel with a tensile strength greater than approx. 30 tonf/in^2 (50 kg/mm^2), a subsequent heat treatment is necessary. This operation is commonly referred to as *stress-relieving*, but where any martensite has been formed during *welding* this will, in fact, be a tempering process. It is generally carried out at 600°C–650°C, but with certain types of specific high-tensile, high-integrity components, the *welding* will be carried out after *hardening* and tempering and the subsequent *stress-relieving* or tempering operation will then be at the temperature for the latter process.

From the above, it will be seen that the process of tempering is of vital importance to the engineering industry. There are many statistics that prove conclusively that mishandling of the operation results in failure of components in service. However, it is seldom possible to single out this process for blame as it is generally the combination of *hardening* and lack of tempering which causes the problems.

No alternative exists for this operation.

Mechanical testing

The general term applied to a large variety of tests, the majority of which are *destructive*, used to evaluate the properties of a material. Once the properties have been successfully proven, and the necessary specification prepared, mechanical testing is used to ensure that the chosen material complies with the specification.

With certain exceptions, mechanical testing is a relatively expensive procedure which requires the preparation, very often by accurate sophisticated machines, of standard test pieces. These are then destroyed by the application of the stresses necessary to measure one or more of the parameters required.

In engineering, mechanical testing is usually considered to take the form of *tensile testing*, *impact testing* and *bend testing* and various forms of *ductility testing*. Mechanical testing can, however, in addition to the above include the measurement of other parameters such as *torsion*, *fatigue*, *creep*, etc.

The choice of the mechanical test required and its frequency should be given considerable thought. For example, if the important parameter is that components are capable of long life in service without fatigue failure, then it is possible that the actual tensile strength of the material is of less relevance than the state of the surface to ensure that this is free of stress-raisers. Ensuring that a material has an ultimate tensile strength of, for example, 500 N/mm^2 (35 tonf/in^2) with a yield of 450 N/mm^2 (28 tonf/in^2) and appropriate elongation, is largely irrelevant if the component eventually fails because a stress-raiser has concentrated the stresses by a factor of 10 or more. The important parameter in this case is the absence of stress-raisers.

Many similar examples exist and the user is advised, wherever possible, to carry out sensible quality control in conjunction with, or in some cases instead of, expensive mechanical testing. The various mechanical tests applied to metals are briefly defined in the alphabetical section of this book. All these tests are now grouped together alphabetically in the present section with fuller descriptions. These are:

Bend Test – including	Crack Opening Displacement (COD)
(Full bend test)	Crack Tip Opening Displacement (CTOD)
(Repeated bend test)	Creep Test
(Reverse bend test)	Elasticity Test
Compression Test	Elongation Test

contd

Fatigue Test
Fracture Test
Impact Test
Nick Break Test
Nicked Fracture Test
Notch Bar Test
Proof Stress/Proof Strength

Proof Test
Shear Test
Tensile Test
Torsion Test
Young's Modulus of Elasticity

Bend Test

A test involving the permanent deformation of the test piece by bending.

Application range ALL METALS

This is a general term covering a wide variety of tests. It is not possible to carry out the test without a particular specification being supplied. The point of the test generally is to indicate the degree of ductility by bending.

Originally the bend test was used to prove the ductility of 'malleable iron' but has since been extended to cover all types of material. The different test types are as follows:

1. *Beams* Strictly, a deflection test carried out on beams. The beam is loaded with a specified weight, or one which varies according to the beam's length, and the amount by which the beam deflects is measured. Thus, for this test to be meaningful the load applied and deflection allowed must be specified. The test should never result in permanent deflection, it ensures that the beam will be loaded within its 'elastic limit' rather than being a ductility test. This is a non-destructive test.

2. *Plate or sheet material* The bend test is applied to test pieces cut at right angles to each other if the direction of rolling is unknown. These test pieces are 125–250 mm (6–10 in) long, the edges are radiused to reduce stress-raisers and the bend will be 180 degrees round a former. Where it is only necessary to ascertain the direction of rolling, the diameter of the former is not important. Generally the thickness, or twice the thickness, of the material is specified as the diameter of the former. There are, however, specified tests where the shape and size of the former are defined.

With certain types of sheet and plate the method of rolling and impurity content will result in directional properties such that greater ductility exists along the grain, that is, in the direction of rolling, than across the grain. The bend test will indicate whether or not such a difference exists because of earlier failure by cracking or fracture at the bend on the cross-grain specimen or the specimen representing the direction of rolling.

When a sheet or plate is required for components to be used under high-integrity conditions, the bend test may be specified in its own right. In this case the direction of the test and the force applied to cause fracture is specified. With this, the angle of the bend or force required to cause fracture will be measured, and the fracture face assessed for porosity, lack of fusion, etc. The shape of the former and method of applying the load will generally be part of the bend test specification.

These are all destructive tests.

3. *Welding* A bend test is now a routine method specified for testing welds. A test piece is produced using actual *welding* conditions, which may be rigidly specified. Depending on the final use of the weld, specific bend testing will be applied. This may be a face, side or root bend. Thus the test involves bending to place the surface or the root of the weld in tension.

In addition to the manner of bending the angle of bend will be specified, and very often the size of the test piece. Whether or not machining of the tension side of the bend is permitted, and the amount or degree of cracking which is acceptable for any specified degree of bend, will be part of the specification. It should always be clear before testing if the weld bead has to remain in position or be machined flush. This is a destructive test.

4. *Tubes and pipes* With these, the bend test can be better described as a 'flattening test'. A prescribed length of tube will be squeezed between parallel surfaces and examined for cracking after an agreed degree of permanent distortion. A standard degree would be one-third of the total diameter, but considerable variation on this exists.

Tubes and pipes may also require testing to comply with longitudinal bending. These tests will also specify whether or not the tube has to be filled, the angle and degree of bending and, under some circumstances, the speed of bending.

These are destructive tests.

5. *Bars* Bend testing on all forms of bars is commonly required as an economical method of rapidly assessing the ductility of the material. The test will vary from a single 90-degree bend with unspecified radius of bending, to repeated bending through 180 degrees with the angle of bend specified.

Each test should be designed with the end use of the material in mind. The parameters for specification will be the number of bends achieved without cracking, the angle of bending, and the radius of bending. In some cases the temperature of bending will be specified together with the dimensions of the test piece and the speed of bending.

Full bend test This is a form of crack opening displacement (COD) test using the full size component. A fatigue crack is produced from a machined slot, to leave a predetermined cross-section area which is then bent to open the crack. The load to fracture or bend to the specified angle is recorded. See *Crack Opening Displacement* in this section for further details.

Repeated bend test A destructive inspection process in which the metal being tested is held rigidly at one end but repeatedly bent. The test will vary widely and in order to be meaningful the requirements must be strictly specified. It is common that wire is tested in this manner, where a length 150 mm (6 in) is clamped in a vice and bent 90 degrees through a specified radius. The bend is then reversed in the opposite direction, giving a total bend of 180 degrees. The specification will state the number of complete bends which must occur without any fracture or cracking.

Points which must be stressed are that the specimen must be homogeneous, must be rigidly held and must be bent across a well-formed carefully designed radius. Bending must always occur in the same manner for the test to *be meaningful*. It will be seen that the test is a measure of ductility or ability to accept cold work without fracture, and will generally be specified only when the material is to be subjected to severe bending moment in service or during manufacture.

Reverse bend test A destructive mechanical test which has several variations. In essence, it is bending either sheet or wire through a stated angle, generally 90 degrees, over a specified radius. This bend is then reversed generally through 180 degrees and inspected for cracking. The second bend is again across a specified radius and the variation in the test is in the angles prescribed and the amount of cracking permitted.

The term is also, on occasion, used to define the testing of a weld where the root of the weld is placed in tension. This test is more correctly specified as the root bend test.

Bend testing, then, takes several forms, and the instigator of the test should be careful to specify the requirements: they should be based on some relevant feature regarding the method of manufacture, or the stresses imposed during service. Bend testing can be a very severe test which finds obvious defects cheaply. Some care is required in the interpretation of the results obtained.

Compression Test
A test to measure the strength in compression, that is the opposite to tension.

Application range ALL METALS

The test is carried out by compressing a cylinder with a length twice the diameter along its length, generally in a standard tensile test machine. With brittle materials, failure will be dramatic and will occur by shattering or collapse of the cylinder. With ductile materials, failure is difficult to define as with a fully plastic material, such as lead, where no *cold working* occurs, the cylinder will deform and eventually flatten. This method of failure will occur to a lesser extent with different degrees of ductility and the ability to accept cold work.

Agreement, therefore, is required to define failure. This may result from the load required to produce a certain degree of distortion. Under some circumstances, the load is recorded at the first evidence of bursting or cracks on the outer diameter which is taken as the point of failure.

In engineering pure compression failure is extremely rare as generally the load applied to metals results in some form of tension load. The tensile strength of metals is considerably less than the compression strength, thus failure will almost invariably occur because of the tensile load rather than any compression loading. The compression test is commonly applied to concrete, and special cubes are cast and tested under compression loading as part of routine quality control.

Crack Opening Displacement (COD)
A test to identify the ability of a test piece to resist crack propagation.

Application range ALL MATERIALS – VERY OFTEN WELD METAL

The test uses a test piece, or in some circumstances a component, when it is often referred to as a full bend test. A slot is machined, and a fatigue crack produced from the root of the slot. The fatigue crack is controlled to produce a known accurate cross-section from its root to the rear face. The load required to open the crack, either by tensile pull, or by bending, is recorded.

This is a relatively recent form of testing developed to give stress and design engineers some indication of the safety factor which could be expected when cracks of a known size exist in components subjected to tensile or impact load.

Care is required in using the information obtained, as it is possible to interpret the result to allow the acceptance of readily identified cracks, rather than to strive for crack-free components.

The test is obviously related to good-quality impact resistant, ductile material and tensile plus impact testing will generally be more economical.

It must be realized by the design engineer that material variations, particularly heat treatment and welding, can affect the material properties, and thus the results of this test. The test is unlikely to be economical for routine quality assurance checking.

Crack Tip Opening Displacement (CTOD)
Alternative name for *crack opening displacement (COD)* testing.

Creep Test
A test to find the amount by which a metal will stretch under its own weight, or with low loading at any given temperature.

Application range ALL METALS

The characteristic of creep is that a metal test piece suspended in air will stretch when loaded within its 'elastic limit'. Certain metals, notably lead, are subject to creep at room temperature, but on the whole metals do not normally creep at room temperature. Until the requirement existed for higher temperatures for steam generation and, in particular, the gas turbine or jet engine, the phenomenon of creep did not normally enter the engineer's calculations. With use of higher temperatures, however, it was found that components, particularly rotating components subjected to centrifugal force, and also those subjected to either no stress or very little stress, could fail when at high temperatures.

Creep testing is carried out on test pieces of a specified shape. The test apparatus consists of a furnace, designed for the individual test pieces, and measuring equipment which can accurately measure any increase in length of the components. The test piece is inserted into the furnace which is heated to the specified test temperature and controlled to tight limits. Measuring equipment is used to ascertain any increase in length greater than that caused by normal linear expansion.

A creep test curve is produced by plotting a known stated amount of extension against increase in temperature and time. It will be found that, below a certain temperature, creep does not occur at infinite time. At just above this temperature, creep will occur only after a very long interval, and as the temperature increases so does the creep. However, experience has shown that it is not possible to predict accurately the amount of creep which will occur above the measured temperature, thus, there is no alternative to long-term creep testing.

Some indication of the creep or high-temperature strength of metals can be found by adding a weight to the creep specimen. Technically this is a form of high-temperature *tensile test* and some care must be taken when interpreting the results. It does, however, give much more economically and rapidly than the present test, an indication of the high-temperature tensile properties of a material. Again, it must be pointed out that this high-temperature method and the true creep are not necessarily related.

Elasticity Test

This is a method of finding the elastic properties of materials, and is also termed the *proof* or *yield* test.

Application range ALL METALS

This test can be applied but is very seldom carried out in its own right, being generally part of the *tensile* test.

The elastic limit, or elasticity, of a material is the maximum load which can be applied without causing measurable permanent stretch or increase in length of the test piece. All materials are elastic in that they will stretch and return to their original shape or size on removal of the load. For engineering purposes, it is important that designers ensure that their components remain within this elastic area during normal use.

The *tensile* test is used to identify the 'yield' or '*proof stress*' which is accepted as the elastic limit. Any increase in load beyond this will result in permanent stretch. Some materials, notably mild and low-alloy steel, have a well-defined 'yield' point readily identified as a drop in the rate of application of load. With other materials this is not well defined and the elastic limit is taken as a 'proof' stress. This will generally be 0.2 per cent proof but some specifications use other figures. This is the load which will cause a permanent stretch of 0.2 per cent of the gauge length of the test piece. It is accepted that 0.5 per cent proof stress equals the yield point. This can be found by repeated application of load at increasing loads with the test piece gauge length measured after each removal of load. The load to cause the specified percentage permanent increase after removal of the load is the '*proof strength*'.

It is more conveniently found from a curve drawn, generally automatically, during the *tensile test*. By drawing a line parallel to this curve from the agreed permanent increase in length, the load required to produce this increase can be found. The increase in length as the load is applied is readily found, using an extensometer which is clamped to the test piece and records the increase in length with increase in load applied.

There is no alternative method of determining the elastic limit other than with *tensile testing*. As the elastic limit can vary independently of other mechanical properties, the importance of controlling any metallurgical process which affects this parameter must be appreciated. The elastic limit, found by elasticity testing during the *tensile test*, is related to *Young's modulus* of elasticity, and Hooke's law.

As the elastic limit, yield stress, proof stress, etc. can vary independently of the ultimate 'tensile strength' in steels, depending on the prior treatment, the importance of controlling the treatment carried out cannot be overstated.

The ultimate tensile strength is related to the hardness figure, but there is no non-destructive test available to evaluate a material's elasticity.

Elongation Test

A method of evaluating ductility.

Application range ALL METALS

This test procedure will very seldom be called up by itself, but is part of the normal *tensile test*. It is a destructive inspection requirement.

The method is that prior to *tensile testing* the specimen is carefully marked. One common procedure is to use a 50 mm (2 in) gauge length and to mark this on the parallel section of the test piece. A number of different formulae are used for elongation, and the gauge length is generally related to the diameter of the tensile test piece.

For sheet metal the elongation can be of considerable importance where *deep-drawing* is involved and there is a method whereby a parallel specimen has every 25 mm (1 in) along its 250 mm (10 in) length identified. The marked specimen is then submitted to the standard *tensile test* and pulled to destruction. The broken test piece is placed on a level, smooth surface and carefully reassembled. With normal test pieces the two marks, originally 50 mm (2 in) apart, are remeasured. They will now show a distance greater than 50 mm (2 in), and it is this increase over the original length which is shown as a percentage and called the 'elongation'. The different methods of measuring elongation on cylindrical test pieces are described below.

With the sheet metal test pieces described above, which are scribed along the 250 mm (10 in) length at intervals of 25 mm (1 in), the 50 mm (2 in) section involved in the actual fracture is ignored. The increase in length of the remaining 200 mm (8 in) is taken and used to assess the percentage elongation. It will be seen that with this the figure obtained is very much less than with the conventional test piece as the part of the specimen with the greatest elongation will be that area where failure occurred. This method indicates the maximum permanent deformation which can be expected without severe local thinning or necking of the material taking place. It is, therefore, of particular interest to the press-tool engineer involved in *deep-drawing* operations.

The elongation figure gives an accurate indication of the ductility of the material being tested. It is, in fact, related to the stress–strain diagram, and the ratio of the yield (or proof) strength and the ultimate tensile strength is generally in proportion to the elongation: the higher this ratio, the higher the figure.

Care is required in comparing elongation figures as the method of measuring can give considerable variation. Thus, where an accurate figure is required, the specification must define the method to be used, and where actual elongation figures are close to the specification requirement, some care is required in interpretation. It must also be realized that the technique of measurement is relatively crude, and it should not be expected that figures are reported to second decimal places or that duplicate tests will give identical figures.

Elongation is, however, an important mechanical property. There are a number of simple tests, based on bending, which are cheaper to carry out than the *tensile test*, and for specific components or materials these can be very useful. These include *bend* and *wrapping tests* which are described under the appropriate headings in Part Two and Part One respectively.

The elongation figures obtained will be proportional to the reduction in area which is also measured on the tensile test piece after fracture.

As a rough guide materials with an elongation of less than 5 per cent are brittle, those with figures above 25 per cent are ductile and those with figures above 50 per cent are very ductile, almost plastic materials. The different formulae for measuring elongation on circular test pieces are 5.65 times the square root of the cross-sectional area, $5.65\sqrt{So}$; and 4 times the square root of the cross-sectional area, $4\times\sqrt{So}$. The conversion is:

Elongation (%)

$4\sqrt{So}$	$5.65\sqrt{So}$
10	7
12	8
14	10
15	11
17	12
18	13
20	15
22	17
25	20
26	21
28	23

In some specifications elongation is taken as the 'stretch' which occurs when a load is applied, whether or not this 'stretch' remains when the load is removed. Thus the term elongation must sometimes be carefully defined.

Fatigue Test

This is a fluctuating or cyclic *tensile test*.

Application range ALL METALS

In general it is possible to obtain considerable technical information on the fatigue strength of any metal or alloy from one of several sources, and it will normally be found to be approximately half the ultimate tensile strength of the metal.

When it is realized that the fatigue stength of any material is found by revolving or reciprocating a test piece, and thus changing with each revolution the tensile load applied at any one point on the surface to a load in compression rather than tension, it will be appreciated that given perfect conditions, the fatigue strength will be approximately half the ultimate tensile strength. This is found by applying a longitudinal load to the material, that is, all the surfaces will be stressed in tension.

Where engineering components have relatively simple stress patterns, or where the stress can be accurately predicted, then the necessity to carry out testing will be limited, as the design engineer can with safety assume that the fatigue strength of his material will be half that of the ultimate tensile strength, and that the stresses imposed during service will be evenly distributed across the stressed surfaces.

Materials such as steel have what is termed a 'fatigue endurance limit'. This is the stress level at which ten million cycles (10^7 cycles) of stress will not result in failure. It has been found that any stress lower than this figure will not result in fatigue failure, no matter how many cycles are applied.

With most other materials, however, it is not possible to produce an endurance limit, as although the general stress pattern is similar to that of ferrous materials, there is no level below which it can be categorically stated that fatigue failure will not occur. Thus, with ferrous materials reference will be made to an endurance limit of, for example, ± 400 N/mm^2 (25 tonf/in^2) fatigue strength, with other materials such as aluminium alloys their fatigue strength will be recorded as 170 N/mm^2 (11 tonf/in^2) at 10^5 cycles. This means that the stress below 170 N/mm^2 (11 tonf/in^2) will have very little effect on the fatigue life of the components, and for normal engineering purposes this can be used in the same manner as the endurance limit for steel. The reason for this is that the normal safety factor built in to this type of component will result in considerably lower fatigue stresses being applied and, thus, within periods of 20–50 years fatigue failure should not occur, obviously depending on the usage, and design accuracy.

With aircraft components, which are used closer to their ultimate tensile strength and yield strength because of the importance of weight, it is necessary to carry out routine *crack tests* at regular intervals. This testing can be on a simple time basis but is more often related to the stress level and stress cycles. Thus, in aircraft structures, etc. the test schedule will depend on a complicated formula between the number of takeoffs, hours of flying and the passage of time.

As stated above, the information on the fatigue strength of metals is now considerable and it will seldom be necessary for any design engineer to require fatigue testing of a material itself. What must be appreciated is that the geometry of the component, the method of machining, and the standard of machine finish will have a considerable effect on the fatigue strength, and for this reason it is often necessary for high-integrity components to be tested. In this case it is not the material as such that is being assessed but the component itself.

It is important when fatigue testing is carried out that the component should be typical, or of a lower standard and quality than the components used in service. Under modern conditions it is often the final assembly which is subjected to fatigue testing. This may require the use of expensive rigs with complicated stressing equipment to simulate the stresses imposed during service.

When it is realized that a radius at a change of section differing between 1 mm (0.050 in) and 0.1 mm (0.005 in) can be the difference between the fatigue strength of a component being 450/650 N/mm^2 (30–40 tonf/in^2) and

10 N/mm^2 (0.5 tonf/in^2), then the importance of having the material in the correct condition will be appreciated. With modern engineering, fatigue failure is almost invariably from a change of section, such as described above, or from poor machine finishing or corrosion pitting. *Welding* is also a common cause of fatigue failure and thus must be carefully controlled. It will seldom be satisfactory to base design or quality standards on one fatigue stress cycle. As fatigue testing can be a relatively lengthy process using sophisticated equipment it is, therefore, expensive, and the more information that can be supplied to the fatigue testing laboratory the better. Thus, there is little point in producing fatigue strength information in the 450–650 N/mm^2 (30–40 tonf/in^2) range, if it is known that in service the component will never be used above the 75 N/mm^2 (5 tonf/in^2) level. The reverse is obviously of equal importance.

It is also of importance that the component is stressed in a manner as close as possible to that which will occur during service. This is often impossible within a development budget, but with modern types of machines variations of stress patterns can be achieved. Advice from stress engineers is invaluable at an early stage, before components are produced. They will often be able to advise on economical methods of reducing stress-raisers.

Once the fatigue strength of the component has been measured, then the approximate safety factor involved during normal service running will be known. If this is less than 100 per cent, Quality Assurance and Inspection should be advised of this fact to enable them to take the necessary action during production to ensure that no rogue parts are supplied.

Fatigue testing of components and assemblies is normally carried out only on high-stress, high-integrity parts. It should not be embarked upon until there is full knowledge on the fatigue strength of the materials involved and stress-raisers present have been controlled. This control must then be rigidly applied to production pieces.

Fracture Test

This term applies to a number of tests resulting in the test pieces being broken.

Application range ALL METALS

The term is not normally used for either the tensile or impact test, both of which cause fracture.

In skilled hands this test gives some indication of notch brittleness and it can be used to show ductility. Unlike *tensile* and *impact testing*, specific information in the form of figures which can be compared with a specification is not supplied. However, it supplies enough general information to the experienced investigator to obviate the need to carry out further expensive *mechanical testing*.

Fracture testing takes a number of forms: it can simply involve the specimen being held in a vice and struck sharply with a hammer. More detailed information is supplied by cutting a notch in the specimen and assessing the strength of the blow in relation to the depth required to cause fracture. The amount of bending prior to fracture is also useful information.

The most useful evidence will generally be supplied by visual examination of the fracture surface. If this is fine grained and homogeneous across the surface with ears at the edge of the fracture, this will be an indication that the material in question is ductile and, thus, probably satisfactory. If, on the other hand, the fracture occurs at low loads, is flat with no evidence of bending or ears, and with a large-grained crystalline fracture face, then the material is clearly of a brittle nature, and reference should be made to the specification.

Gross porosity can also be identified by examination of the fractured surface, and any prior cracking or surface defects will also be shown. The macro test and examination is in some ways an alternative test, but more generally will be carried out in conjunction with the fracture test, particularly when failure investigation is involved.

Impact Test

This test is an attempt to evaluate the ability of a material to withstand impact forces.

Application range ALL METALS

As the presence of stress-raisers or notches has considerable influence on the ability of any material to withstand impact, the test is generally based on a test piece which has a standard notch machined on one face.

The test piece, which is usually square in section, is inserted in a vice and a pendulum hammer is allowed to fall or swing in such a manner that it strikes the test piece, causing fracture. The energy absorbed by the hammer retards the distance of the hammer swing. Thus, materials such as glass or very hard, brittle steel will have very little resistance to the swing of the hammer and will not seriously interfere with its passage, whereas a heat-treated alloy steel, designed for toughness, will result in a much reduced distance of swing.

The length of the hammer or pendulum swing after striking the test piece is calibrated such that the greater the swing the lower the impact strength, and the lesser the swing the higher the impact strength of the material being tested.

It is difficult to calibrate this equipment, thus care is required to ensure that it is always in good condition. The pendulum must be free swinging and zero adjusted regularly.

There are a number of impact test systems but, in practice, only the *Charpy* and *Izod* are commonly used for metals. The *Charpy* is now the more common and is to some extent superseding the *Izod* test.

The difference between the tests arises basically from the notch machined in the test piece, and in the manner in which the test piece is held; details of the tests are given under the appropriate headings in Part One. Because of the sensitivity of the notch, it will always be found that at least two impact tests must be carried out, with three tests being commonly specified. Different specifications, then, will have different allowances for results being below requirement, in some cases a variation of a certain percentage means that the test must be repeated or the material rejected. In practice it is often found that these variations are a result of the machining of the notch rather than any variation of the material being tested. The importance of careful, consistent machining, and careful inspection of the notch, cannot be over-emphasized. The use of a standard gauge or comparator is essential for inspecting the notch.

Broaching with standard equipment generally gives more consistent results than milling.

Impact testing is often required over a range of temperatures. This applies particularly to steel, where it has been found that there is a dip in the impact strength at a specific temperature. This temperature varies from approximately $-50°C$ to room temperature and this is within the temperature range commonly found in use with components in many industries. The drop in impact values can be dramatic, being in the order of 100 per cent over a range of 10–20°C. It will, thus, be seen that a material having a satisfactory impact strength at room temperature may be found to have a low impact strength, and have notch brittleness, at 0°C. It is now known that this drop in impact value with temperature has been the cause of several disastrous failures and thus the necessity to carry out impact tests over the complete range of temperatures to which engineering components will be put is essential. Once the lowest operating temperature is known, impact tests at this temperature, or 20°C lower, should be specified. Industries such as petrochemicals, gas and oil, and processes where low temperatures are liable to be encountered are those most affected. Low temperatures can result from rapid decompression of gases in addition to static conditions.

The various types of impact test are:

Impact Test	Test Piece	Notch
Seaton and Judy	4 in×0.5 in×0.5 in	V $\frac{1}{8}$ in deep
Charpy	60 mm×10 mm×10 mm	keyhole or V notch
Fremont	30 mm×10 mm×8 mm	Square 1 mm
Izod	75 mm	V or none
Mesnager	60 mm×10 mm×10 mm	1 mm wide, 2 mm deep with radius.

Nick Break Test

A quality control test of a destructive nature; it is used during the assessment of *welding*.

Application range ALL METALS WELDED

A cut or nick is made along the centre of the weld and parallel with the weld deposit. The welded specimen or test piece is then fractured in such a way that the nick is opened. It is then possible to examine the root of the weld and assess the quality of the weld metal itself. The test will identify lack of fusion and will also show porosity.

This is one of the routine tests carried out, particularly on *fillet welds*, for welder approval and the procedural testing of welds. Any lack of fusion or gross porosity can be identified. If the same operator makes use of the technique, it is possible over a period to obtain some 'feel' regarding the ductility of the weld. The nick break test is generally confined to fillet welds as the equivalent to *bend testing* of butt welds.

Nicked Fracture Test

A destructive quality control test for assessing bar or sheet material where gross manufacturing defects can be identified.

Application range ALL METALS – GENERALLY STEEL

The test piece is cut from the bar or sheet and is 'nicked' at the centre. Impact loading to cause failure is applied in a direction to open the nick and, given a certain amount of training and some experience, considerable information can be gained. This can be improved on by increasing the size of the nick in order to ascertain the ductility or brittleness of the material. The fracture obtained is examined and will show any central weaknesses, such as piping or lamination, and closer examination shows undesirably large grains or fibrous-type fractures which might indicate a low standard of cleanliness.

This is a relatively crude method of assessing metal quality but will identify any serious defect. It is commonly applied to each bar end before high-integrity components are manufactured. It is to some extent a cheap *impact testing* method but cannot be said to be its equivalent.

Notch Bar Test

A general term covering the various forms of *fracture testing* where a notch is involved.

Application range ALL METALS

The test is carried out to assess the ability to withstand shock impact loading and, in addition, whether a material is more susceptible to impact when a notch is present. The test is seldom carried out in isolation and is usually part of a series including the *tensile test*, where the ductility of the material can be measured by elongation. There are various types including the *Charpy*, *Izod*, and *Mesnager tests*.

Invariably the impact test is carried out in duplicate, at least. Most specifications insist that a minimum of three test pieces are produced and tested under controlled conditions, and indicate the degree of scatter which is acceptable.

The importance of ensuring that the notch produced is of a high standard carried out under first-class conditions cannot be over-emphasized, otherwise the results will be meaningless.

Proof Stress/Proof Strength

This is the part of the destructive tensile test where the load required to produce a permanent deformation is known as the proof load.

Application range ALL METALS

With normal testing this is generally 0.1 or 0.2 per cent of the specified gauge length. That is, the proof strength is the stress required to produce a permanent extension of 0.1 or 0.2 per cent of the gauge length.

The yield stress is very similar and it is generally accepted that this will be 0.5 per cent proof stress. There is a visible indication of yield during the tensile test as the increase in load applied slows or stops as the material stretches or 'yields'. This is readily seen with mild steel and certain other materials but is much more difficult to identify with the majority of materials. Thus, it is preferable that a proof strength is identified.

The technique for carrying this out is that the load is recorded at standard increments and the degree of stretch achieved for each load is also recorded.

Where a 'yield' is identified it will be readily seen by either the instrument recording the load and amount of stretch applied, or the operator taking the readings that no, or very slight, increase in load, is being applied for a considerable increase in length.

Where this is not readily seen the graph produced following the completion of the tensile test will be prepared. This will show initially a straight line where the extension and load applied are proportional. At a certain point it will be noted that the load applied will give an increase in length to the test piece greater than the previous increase for any unit load. This is the elastic limit and the proof load is then measured either 0.1, 0.2 or 0.5 per cent of this figure.

With modern tensile machines the load and stretch are recorded automatically and the proof load identified.

The proof stress is then worked out from the cross-sectional area of the test piece and the load required to give the desired stretch.

Proof Test

This is a term generally applied to materials or engineering components where a known load is applied which must result in either no permanent movement or a well-defined and controlled permanent movement.

Application range ALL METALS

It must be appreciated that the term 'proof test' can also be applied to all manner of other pieces of equipment and materials, for example, there is a proof strength applied to alcoholic drinks, which indicates the alcohol content.

Where proof testing or proof loading is applied, the specification will give the load which is required. This should invariably be above the load which will be applied in service and will very often be 100 per cent greater than the normal working loads. It is common that the proof load is applied immediately after manufacture and that a safe working load is then specified which is the maximum which can be used during service. This load will be 75 per cent or thereabouts of the proof load or less.

It is also common that proof loading is applied at regular intervals during the service life of equipment such as lifting tackle. Again the load will be specified and it is common that this will be below the original proof load. Sometimes the proof load applied at regular intervals will equal the safe working load or a specified amount above that.

Proof loading requires either a standard tensile machine if the equipment is small enough or may require special equipment. In the case of chains, lifting tackle and wire ropes etc. specialized equipment is essential. This must be calibrated at regular intervals, at least once per year, and the results of the calibration recorded.

Proof loading requires that the load is applied to a known length. This will be measured immediately prior to the load being applied and care is essential that this measurement is accurate and that a load is applied to ensure that no kinks or deformation exist. Commonly, a 10 per cent load is applied during this measurement.

The proof load is then applied, sometimes the length at proof load is measured but this is not common. The load is then removed to approximately 10 per cent of the total proof load and again the component or equipment is measured. With many specifications no increase in length whatever is permitted at proof loading. With other equipment, such as chain, then some slight stretch will be expected and accepted. This will be specified before testing commences.

The purpose of proof loading is to ensure that the equipment itself is correct and will not fail when a tensile load is applied below the proof load. Failure in this context would not of necessity mean that fracture had occurred but would mean that stretching of a permanent nature had taken place, above the agreed limit.

There is no alternative to proof loading. While specimens can be taken and tested for ultimate tensile test, this will indicate only that the test piece itself has been satisfactory. The purpose of proof testing is to ensure that the component itself is satisfactory.

It must, however, be appreciated that the proof load will almost always be applied in a slow, gentle manner, whereas service conditions will often impose bending, impact or torsion.

As stated above, there can be other interpretations of proof testing and it is essential that the specification is obtained before proof testing is carried out.

Shear Test

Shear is a form of compressive failure where the load is applied at right angles to the normal axis of the material. Shear is thus the pure guillotine action.

Pure shear is compression only. It must be appreciated, however, that it is very difficult to load components in pure shear under test conditions.

Where, for any reason, the pure guillotine action does not exist and where any space is present between the loading plattens, then some bending will take place.

The most common production technique using shear is where a punch is used to produce a hole. This will invariably be pure shear provided there is sufficient metal around the hole to prevent any bending action.

It is very seldom that shear testing is carried out in its own right, and generally the *torsion test* is used to identify the shear strength of a material.

With pure shear, that is the guillotine action, then the surface of the component at each side of the platten loading the surface will be in compression, with the compression load being reduced proportionally to the opposite surface where on the same side of the platten the load would be zero.

With *torsion testing* or loading then the load at the surface is in compression which decreases to the centre where the load is zero, increasing to compressive loading of an equal amount on the other surface.

Tensile Test

A destructive inspection process to find the strength of the material being tested.

Application range ALL METALS

In general, use is made of specially machined test pieces but tensile testing can in fact be successfully carried out on finished machined components and raw material such as bars or tubes.

The purpose of tensile testing is to ascertain the mechanical strength of the material when a tensile load is applied along the length of the component or test piece with the purpose of stretching or pulling apart the material. The test piece used can take various forms but certain characteristics are common. The central portion must be parallel. Any lack of parallelism will result in failure occurring at an area which might not be representative of the material as a whole, and makes the estimation of the cross-sectional area very difficult, if not impossible.

The portion of the test piece where failure is expected to occur, in addition to being parallel, must also be smooth and free of all notches, machining defects or any other surface imperfections. The length of this parallel section should be as long as possible within reason, 25 mm (1 in) being the generally accepted minimum, and must gently merge with the larger diameter which is used to grip the test piece in the tensile machine.

Tensile testing as well as identifying the ultimate tensile strength, that is, the strength of the material when fracture actually occurs, is commonly used to identify the 'yield strength' or '*proof strength*' of the material. This for engineering purposes is much more important than assessing the 'ultimate tensile strength', which in the majority of cases is only of academic interest, as the designer will not consider the fracture point as being of importance to the engineering design. The 'yield strength', '*proof strength*' or 'elastic limit' is the strength at which the material being tested ceases to be wholly elastic and becomes a plastic material. It is this point of plasticity which is of vital interest to the engineer, since any permanent stretching or deformation on engineering components is generally accepted as being failure.

In addition to these two parameters it is also possible to use the tensile test to give an indication of the ductility of the material. This is either by measuring the length of the test piece accurately on its parallel length before and after application of the tensile load to cause fracture and also the smallest diameter of the test piece after fracture has occurred. The results are reported as elongation, and reduction in area.

Information on '*proof strength*', 'yield' and 'elongation' will be found in this section **Mechanical Testing**.

The tensile strength should be assessed along with ductility, impact and hardness. There is a direct relationship between the ultimate tensile strength, and hardness, and as the hardness test is much cheaper this can be used to find the approximate tensile strength. Appendix 5 gives the conversion. It is not possible to use hardness to find the yield, but if sufficient information and experience on the material involved is available then an intelligent guess can be made regarding the 'yield strength'.

There is no alternative to tensile testing when full specification information is required. Tests such as *hardness*, *fracture* and *bend* all supply limited information which can be useful.

Torsion Test

A destructive test to identify the shear and twisting characteristics of a metal or component.

Application range·ALL METALS

A load is used which results in twisting of the test piece and, thus, involves shear loading. The test places a cylindrical shape into a machine which causes twisting under a known load and measures the angle of twist thus caused.

Where torsion load is applied to a round component then the surface will be in shear which is a compressive load which will decrease towards the centre where the load will be zero, increasing again to the opposite surface.

Where, however, a flat component is subjected to shear then the compressive load will be applied on the surface and will proportionally decrease towards the opposite surface where a zero load will be applied.

Because of the extreme difficulty in applying a pure shear load under test conditions, the torsion test is more commonly applied.

Provided sufficient load is applied, three characteristics are found by this test:

1. The elastic limit for shear.
2. The ultimate shear strength.
3. The shear modulus, or the modulus of rigidity.

The elastic limit for the test is the load applied to the test piece which is the maximum which can be accepted without any permanent twist occurring. Any load less than this would result in the specimen returning to its original condition on removal of the load. Any further loading beyond the elastic limit will cause an increase in the amount of permanent twisting, until a maximum load known as the 'ultimate shear stress' is reached.

Hence the 'ultimate shear or torsion stress' differs from the ultimate strength of the *tensile test*, where the load decreases as the specimen elongates after the 'ultimate tensile stress' has been reached. With 'shear stress' when materials are cold worked the load will increase with the angle of twisting by a considerable amount. Those metals which do not cold work will show very little increase in load between the elastic limit and the ultimate shear load.

The modulus of rigidity or the shear modulus is found from the slope of the load–twist graph as the load is imposed. Thus, with materials with a low modulus of shear, a large angle of twist will result from a relatively low stress. That is, the slope on the graph will be relatively low, whereas a material with a high modulus of rigidity will require the application of considerable stress in order to produce any appreciable angle of twist, and the slope in the graph will be steep.

Young's Modulus of Elasticity

The modulus is a function which is evaluated using the standard *tensile test*. It is a measure of the stretch achieved for load applied.

Application range ALL MATERIALS

When a tensile stress is applied, then the material acts in an elastic manner.

This means that as the load is applied the material stretches, and on removal of the load it returns to its original size. All engineering materials when correctly used are stressed within this elastic limit.

Young's modulus is a measure of this load. Thus, a material with a low modulus will be considerably stretched by a relatively low load, whereas any material with a high modulus requires considerable stress to be applied in order to cause appreciable stretch. Approximate figures for Young's modulus for some materials are:

	(tons per in^2)	*(pounds per in^2)*	*(Newtons per m^2)*
Cast iron	9000	20×10^6	140×10^9
Steel	13200	30×10^6	206×10^9
Copper	7000	16×10^6	110×10^9
Aluminium	4600	10×10^6	68×10^9
Brass	6600	15×10^6	103×10^9
Phosphor bronze	7000	16×10^6	110×10^9
Wood	650	1.5×10^6	10×10^9
Rubber	32	0.7×10^6	0.5×10^9
Glass	4000	9×10^6	6×10^9

By use of this figure, engineering designers are enabled to predict the amount of movement which will occur when a known load is applied or, alternatively, to load the material when a known movement is required.

In order to find Young's modulus, it is necessary to carry out *tensile testing* and to accurately plot the increase in length with increase of load. This test is also applied to find the 'yield' or '*proof strength*' of the material. The 'yield strength' is the maximum stress which can be applied without causing permanent deformation, while Young's modulus is a measure of the slope of the line produced when stress is plotted against elongation.

Non-destructive testing

A series of inspection techniques designed to identify any flaw in a component or material without destroying or damaging the article being inspected. It can be applied to any material.

In this section are listed those techniques most commonly used by industry under the heading of Non-destructive Testing (NDT). It should be obvious that a most important method will be that of visual inspection, and it cannot be emphasized too strongly that the critical eye to ensure that no obvious faults are present should always be encouraged. Too often sophisticated methods are used to identify minor defects when a cursory visual examination will show any serious problems such as severe corrosion or surface porosity. The processes in this section are:

- 'Acoustic Testing'
- Air-pressure Test/Gas Pressure Test
- Crack-detection Testing (general)
- Dye-penetrant Crack Test
- Eddy Current Test
- Etching (general)
- Gas Pressure Test
- Holiday Test
- Hydraulic Test/Hydraulic Pressure Test
- Magnetic Crack Test/Magnetic Particle Inspection/ Magnetic Flaw Detection
- Porosity Test
- Pressure Test (general)
- Radiographic Test
- Sonic Test
- Spark Test
- Strain Gauging
- Thickness Testing
- Ultrasonic Flaw Detection
- Vibration Testing
- Visual Inspection

'Acoustic testing'

A recently developed technique of non-destructive testing of large assemblies.

Application range ALL METALS

The technique is based on the fact that all structures give off signals when stress is applied. This is the method which was formerly applied to wheel-tapping on railway wheels, where the stress was applied by tapping with a metallic hammer and this resulted in an audible sound signal which could be interpreted by the tapper.

The acoustic testing system uses extremely sensitive transducers which are placed on the equipment under examination, or can be placed some distance away from the defect. Unlike the tapping method involving sound vibration mentioned above, here the transducers pick up signals from the natural stresses in the equipment and thus no sound producing equipment is necessary. It is claimed that defects at a distance of 1000 ft from the transducer can be identified using this technique. The method has only recently been developed, and the basic information was obtained from an abandoned pipeline which was thoroughly examined to produce the necessary background information. The method works on a similar principle to a microphone which is in contact with components subjected to *pressure testing*, and allows any sudden release of stress to be detected and investigated.

Air-pressure Test (Gas-pressure Test)

A specific form of *pressure testing*, using gas as the medium to apply the pressure. All types of hollow assemblies can be subjected to this test.

Application range ALL METALS

The component or assembly must be 'blanked off', leaving one inlet pressure point. The pressure is applied slowly and wherever possible the component should be immersed in water during the application of pressure. Any leakage will be shown by the appearance of bubbles. The sensitivity of the test can sometimes be increased by using surface tension-reducing liquids. It is also possible to apply this test by wetting the surface of the component, or listening for leakage or looking for a reduction in a constantly applied pressure.

Gas-pressure testing is a design requirement and should never exceed 50 lb/in^2 (0.35 N/mm^2). Air is a compressive gas and testing at higher pressures under normal circumstances may result in explosive failure in the case of any components with a serious defect. Depending on the size of the component and the quantity of gas supplying the pressure, the use of lower pressures than the above can under some circumstances also be dangerous. If testing is necessary under these conditions, the operator should be protected by some form of screening.

Air is by far the most common gas used for pressure testing. Where pressures above 50 lb/in^2 (0.35 N/mm^2) are necessary, great care is required firstly to prevent decompression bursting, and secondly to eliminate explosive reactions when an oil and an oxidizing agent are present. Gas testing above 200 lb/in^2 (1.50 N/mm^2) should use nitrogen.

For higher pressures, consideration should be given to the use of *hydraulic pressure testing*, where by employing liquids such as water it is possible to apply extremely high pressures with complete safety. Liquids do not compress, thus release of the pressure by failure does not result in an explosion. The test, wherever possible, should be used subsequent to other forms of non-destructive testing, such as the *crack test*, which will identify any serious defects.

Crack Detection Testing
An inspection process for surface defects.

Application range ALL METALS

The various forms of crack test, or non-destructive testing, are designed to detect cracks or other defects. The term without further qualification should be viewed with suspicion. The general term non-destructive testing (NDT) can also be taken to include other inspection techniques such as *hardness testing* and *strain gauging*.

There are two basic methods of crack testing each of which has variations:

1. Using physical means to highlight the defect at or near the surface.
2. Using instruments which show, by various means, any differences within the component. This instrumental method can sometimes allow complete examination of the component, unlike the first system, which examines only the surface and the material immediately under.

Physical crack detection offers three methods: (i) *magnetic crack test (MCT)* or magnetic particle inspection *(MPI)*; (ii) *dye-penetrant test*; and (iii) *etching*. Each of these methods is described in this section.

Instrumental crack detection offers five tests: (i) *radiographic examination*; (ii) *ultrasonic testing*; (iii) *eddy current flaw detection*; (iv) *sonic testing*; and (v) *vibration testing*. Again each of these methods is described in this section.

The complete section is in alphabetical order.

Dye-penetrant Crack Test
An inspection technique to identify surface defects.

Application range ALL METALS

The principle is that a dye is allowed to penetrate any surface defect, highlighting it for examination in detail. *Cleaning* is important as it will be obvious that a surface defect which is filled with dirt will be more difficult to find than a clean, 'empty' crack. The method of cleaning is not important but *vapour degreasing* will serve the dual purpose of removing any oil and grease from the defect and heating of the component, thus causing expansion of any cracks present. The cleaned component – if desired it may be heated – is immersed in the liquid, or has the liquid applied by brushing or spraying. The time of liquid contact with the surface is not critical but generally a period of not less than 2 minutes and not more than 1 h is allowed.

The liquid itself, in its simplest form, consists of paraffin or a paraffin oil to which is added a small percentage of lubricating oil and a soluble dye. The more sophisticated proprietary liquids are again oil-based but include materials with extremely low surface tension. A common constituent of this type is methyl salicylate. The use of fluorescent materials in place of, or in addition to, a normal dye is not relatively common. It is essential that the choice of liquid ensures complete coverage of all critical areas and that it has the ability to penetrate the most critical cracks it is required to find.

After the chosen period of contact, the excess liquid must be removed. There are various techniques for this, varying from lightly tumbling in warm sawdust or wiping with rags, to the more sophisticated controlled cleaning with water or other soluble liquids. There is no doubt whatever that the method of washing, and the control of the washing technique, is by far the most important single operation of dye-penetrant crack detection. Since the purpose of washing is to remove the excess liquid and dye from all surfaces, it will be readily appreciated that too much effort put into this operation may remove the dye from some of the cracks. Cracks, then, which are relatively open or wide mouthed, or cracks having a relatively large radius at their bottom extremity, will be more readily cleaned of the dye than the tight cracks which are small, relatively compressed, and have no wide base radius. As this washing operation is generally by hand it will be obvious that the necessary control will not be easy to achieve.

Following the washing operation, the surface is treated generally with some form of chalk. In essence this acts in a similar manner to blotting paper in that it will absorb into itself any dye at the surface, simultaneously drawing out from cracks any dye which remains after washing. It will be appreciated that inadequate washing of the surface will be immediately highlighted by the complete surface showing evidence of the dye. The difficulty, if no evidence of cracking is found, is that this may result from no crack actually being present, or all evidence of the dye being washed out of the crack. The chalk, commonly called the 'developer', is applied by lightly tumbling the components in a bed of French chalk, or more generally by spraying either with the dry powder or with powder suspended in a volatile liquid, or as an aerospray. It is generally advisable to allow the components to stand for a short period of up to 1 h before carrying out detailed inspection. This inspection should use the dye as a means of highlighting the area to be examined. The chalk and dye should then be removed by lightly rubbing with a clean duster and the area examined at higher magnifications.

Dye-penetrant crack testing is the most universal means of non-destructive crack testing for surface defects. Kits are available which make the process extremely portable in three small aerospray containers. The technique can be applied locally, and to any material. It is necessary to ensure that the operator is skilled in cleaning the component prior to application of the dye, and that he is trained in the correct removal of the excess dye prior to applying the 'developer'.

The sensitivity of this crack detection will be a function of the dye used, the efficiency of the washing and the state of the surface finish. With modern dye-penetrants, minute cracks and porosity can be highlighted, which will not be found using a paraffin–oil mixture. By careful washing to remove only the surface dye, then very shallow cracks can be found which will be lost with excessive washing. Too little washing results in the complete surface having background dye, thus small cracks cannot be seen. If the surface finish of the component being inspected is porous or rough, then again the sensitivity will be reduced as the washing necessary to remove excess dye can also remove the dye from fine cracks. Different coloured dyes, or fluorescent dyes, can be used to overcome the problem of coloured components having the same colour as the dye. This crack detection method cannot be used for defects which are not at the surface. Thus, painted or plated components which have the base metal cracked, but not the paint or plating deposit, are not available to inspection by this technique. It is less sensitive than *magnetic crack testing* correctly carried out, but is much more flexible and can be applied to all materials. It is more economical than any other method of crack detection in most cases.

Eddy Current Test (for defects)

Like the *metal-sorting eddy current* method, this is based on the principle that any metallic material of a specific shape will give a pattern on an oscilloscope when eddy currents are produced.

Application range ALL METALS

These eddy currents are the result of an electric current affecting the material. The technique is that the component or material is contacted by a coil through which an electric current is passed. This is often achieved by placing the component within the coil, but there are techniques using a probe scanning the surface of metals, and it is this probe scanning which is most commonly used for the detection of flaws. The result is that the eddy currents are disturbed and the disturbance can be measured by feeding the results to an oscilloscope.

With flaw detection, it is possible to use two coils which are perfectly balanced, and will thus produce a true circle when the output of the coils is passed to the *x* and *y* plates of an oscilloscope. When two identical components are placed, one in each coil, then the result will again be a perfect circle. However, if one of the components is different to any degree, then the result will be a distortion of the oscilloscope trace. This distortion can be caused by the metallurgical structure of the metal and, thus, can be used to detect materials which are different because of heat treatment. This is commonly used to ensure that small components have all been subjected to the correct treatment and is much cheaper than hardness testing.

The problem is that any dimensional difference in the component will also show as a difference on the oscilloscope trace, and unless the components being tested are known to be dimensionally identical, the technique is not possible.

Using the probe system, it is possible to scan a surface which is homogeneous. Any difference such as corners, drilled holes or cracks will be shown as a distortion on the oscilloscope screen. This technique is, therefore, similar in many ways to *ultrasonic flaw detection* but can be affected by the mass of the material. This will be of no significance where the material is homogeneous as, for example, a pipe, but for conventional engineering components it is unlikely that eddy current testing will be a serious competitor to other forms of flaw detection such as the dye-penetrant, *magnetic crack testing* or *ultrasonic flaw detection* methods. There have been recent advances in the sensitivity and scope of this technique and the interested reader is advised to obtain specialist advice.

Etching

This form of flaw detection makes use of the fact that any chemical attack, whether electrolytically or by straightforward chemical action, will be concentrated on edges or any sharp corner. This principle is used to widen the actual crack itself and thus make it more easily visible either to the naked eye or optical equipment. The technique is occasionally used prior to the dye-penetrant crack test.

In addition it is sometimes possible to identify surface metallurgical effects using etching techniques. This is the *macroetch* effect described under *Macroetch Test* in Part One.

Application range ALL METALS

The process can use a number of chemical solutions, which will vary depending on the material being used or on the surface condition of the material. In addition to straightforward chemical attack, use can be made of electrolytic etching, where the component being examined is made the electrode in an electrolytic circuit with a carefully chosen acid as the electrolyte, and a second, generally inert, electrode. The choice of the procedure will depend, first, on the material, and second, on the critical nature of the inspection.

The simplest etching technique, generally confined to steel, is that the material is first cleaned and then either immersed or swabbed using a nitric acid solution. This nitric acid will be a 1–5 per cent solution in water or alcohol. The strength of the acid and the solvent, as stated, will vary depending on the critical nature of the inspection and the standard of the surface finish on the component: the rougher the finish and the less critical the inspection, the more the tendency will be to use nitric acid in water at a higher concentration. Alternative etches include hydrochloric acid, sulphuric acid, used electrolytically with the steel component being made the anode, or ferric chloride in water with hydrochloric acid also present. Where non-ferrous materials are involved, ferric chloride is a general etchant but other solutions such as ammonia alone or with ammonium persulphate are also commonly used. The variety of solutions and the concentrations which are available for this technique are such that it is not readily possible to produce a meaningful table, and it is recommended that technical advice is sought where this crack detection technique is considered as a means of aiding production.

It should be understood that this technique is used only as a means of highlighting surface defects: the etch will enlarge surface defects which can then more readily be identified by visual examination. It will, therefore, be possible using a magnification of, say, ×5 to ×30 to identify and eliminate defects without resorting to any other form of *crack detection*. By using an acid etch, defects can be identified with the unaided eye or at less than ×10 magnification. The economy of visual examination of relatively complex components at low magnification as distinct from the use of magnification higher than ×10 makes etching a viable proposition. Further details of *etching* are given in Part One.

Gas-pressure Test
See *Air-pressure Test* in this section **Non-destructive Testing**.

Holiday Test
An alternative name for porosity tests on paint films. See *Spark Test* in this section.

Hydraulic Test/Hydraulic Pressure Test
A form of *pressure testing*.

Application range ALL METALS

The component is filled with fluid, commonly water or oil, but in some cases a special liquid is employed which is capable of frothing under pressure so enabling ready identification of minute leaks. The test is frequently applied to assemblies to prove their mechanical integrity, and that they are leakproof.

The procedure is that the component or assembly to be tested is fitted with plugs at all outlets. One plug has a bleed valve, and a second plug is the inlet. With the bleed valve open, the hydraulic fluid is pumped into the component. The location of the bleed valve and the inlet must be carefully chosen to ensure that all air is eliminated by the entry of the test fluid. This is important with high-pressure hydraulic testing, where the compression of air could result in explosive failure. With liquids which do not compress there is no risk of explosion even when serious defects are present.

When the liquid is seen to flow from the bleed valve, and the component is known to be filled with liquid, the bleed valve is closed and the specified pressure is applied to the liquid. There are two basic methods of inspection. First, and most commonly, is visual inspection for leakage. This might be in the form of a fine spray where serious porosity or cracking is present, or may depend simply on the evidence of hydraulic fluid on the surface. It will, therefore, be seen that cleanliness of the outer surface of the tested component is essential if fine leaks are to be found. Access to all surfaces is also important.

Additives have been designed to cause frothing of the hydraulic liquid and thus highlight the escape areas.

Second is the more sophisticated method based on the principle that if any leakage has occurred, pressure cannot be maintained. In this case the specification in addition to calling for a certain pressure will also include the duration for which this pressure must be maintained and the maximum pressure drop permissible. With this sytem, a minute leak will be shown by loss of pressure and it is, thus, essential that the complete hydraulic system, including the pump, is of high standard, with all fitments well maintained. In high-integrity components only a minute pressure drop is permitted over a long period of time. Under some test schedules, the pressure will be increased at a controlled rate over a period of time, with stated intervals to allow the test fluid to degassify and the assembly to stabilize. With pipelines, test schedules can cover a period of time in excess of 24 h.

Air-pressure testing at 50 lb/in^2 (0.35 N/mm^2) probably has the same sensitivity as hydraulic pressure testing at approx. 500 lb/in^2 (3.5 N/mm^2). The reason for this is that gas has greater ability to penetrate porosity and minute cracks than even the most mobile of liquids.

It is perhaps unfortunate that *pressure testing* using gas is too dangerous to be carried out at high pressure where, in addition to searching for minor defects of porosity, the test must also be able to prove the ability of the component to withstand high pressures in service. Hydraulic pressure testing is often used in conjunction with *crack testing* and, under some circumstances, *X-ray examination*. The sequence of testing will be, first, to *crack test* to identify serious flaws, followed by *gas-pressure testing*, followed by hydraulic testing, followed by *X-ray* or *ultrasonic examination* when high-integrity components are involved.

Magnetic Crack Test/Magnetic Particle Inspection/Magnetic Flaw Detection (MCT-MPI-MFD)

A crack detection technique using magnetic principles.

Application range STEEL

The principle is that when any magnetic material is magnetized, each extremity will be a pole and each pole will attract and hold iron particles. If the magnet is cut in two, equal and opposite poles will exist at the central join and again these poles will attract the fine iron particles. If instead of a complete break, there is some discontinuity at right angles to the direction of the magnetic flux, this will appear as equal and opposite poles and again will attract the magnetic particles.

It is important to realize that this effect will occur only when the lines of magnetic flux are at right angles to the cut or crack. If, for example, a bar is magnetized along its length by inducing magnetism by any conventional means, then any cracks or defects which run circumferentially on the bar will be identified but cracks or defects which are longitudinal will not so readily be found. It will, therefore, be seen that some care is required at the magnetizing stage.

Methods of magnetizing include the use of a direct electric current passing along the component to produce a magnetic flux in the direction of the electric current. There are now sophisticated alternating-current magnetizing machines which can be used to induce the magnetic flux and again this flux will be in the direction of the electric current. It is also possible to produce the necessary magnetism by using powerful magnets, by placing one north and one south pole in contact with the components the areas between the two poles will be magnetized, and again the magnetic flux will be along the line between these two magnets. A large horseshoe magnet will have the same effect on the area between the poles. This last method is probably the safest, where any doubt exists regarding the flux lines, as it is possible to prove conclusively by the simple act of placing the magnet that the area between the poles is correctly magnetized. The use of electric current is not as positive but will generally result in more rapid magnetization. It is, however, more difficult to prove that the material has in fact been magnetized. Yet another method is the use of a direct or alternating current solenoid coil, where the component is placed within the coil. This results in a magnetic flux at right angles to the solenoid and, thus, it is possible to identify cracks along the length of, for instance, a bar. The magnetized component is then either immersed in a light oil such as paraffin with fine iron particles in suspension or sprayed with the iron particles in the dry or suspended state. Where it is possible, immersion is generally accepted as being the best method. The use of dry powder is not recommended, except under controlled conditions.

The cracks or defects are indicated by a concentration of the magnetic particles at the crack. The particles will be attracted to any discontinuity such as the ends of a bar or a change in section, and if cracking is suspected very close to these areas, then particular attention must be paid to the subsequent visual examination as these cracks can be hidden by the general concentration of particles in the area.

Where components are complicated in shape, careful planning of the magnetic crack test is essential and more than one inspection will be necessary. With complex components it is often necessary to magnetize local areas and inspect these in detail before removing the residual magnetism and re-magnetizing adjacent areas. This will require that the magnetic flux is applied at an angle to the original magnetism. When using portable magnets this is simple, but with electric means using straightforward induction or the solenoid coil, then the de-magnetizing technique is to pass the component through a coil with alternating currents, and then to re-magnetize in the desired direction and repeat the inspection.

On completion of magnetic crack testing it is generally important that all magnetism is removed. Unless this is correctly carried out the residual magnetism can affect cleaning, welding, thickness measurement or electronic equipment. It is also essential to *de-magnetize* between different magnetic operations carried out on one component. *De-magnetizing* is achieved by passing the component through an alternating current coil.

The suspension of iron or magnetic iron-oxide dust particles in a paraffin liquid is generally termed 'magnetic ink'. This is normally seen as dark particles in the form of a line where a crack or defect exists. Where the component is relatively dark, for example, when castings or forgings or components immediately after *heat treatment* are inspected, the contrast between material and magnetic ink is relatively small, and fine cracks can be missed. It is, therefore, advisable under these circumstances to increase the contrast either by whitewashing or white painting the component, using a very thin layer in order not to interfere with the magnetic flux pattern, or as an alternative to use coloured magnetic ink or modern fluorescent magnetic ink and examining under ultraviolet light. This latter technique is now accepted as the best method for the large variety of surfaces and components inspected by this technique.

Considerable care is necessary when using a background paint, that this is extremely thin, otherwise the crack detection sensitivity will be affected. Where bright metal exists there is no need for these paints.

Magnetic crack testing, in addition to showing surface defects of an extremely low order, can be used to identify defects below the surface. These are indicated by an interference in the flux patterns, but will not be the well-defined defects shown by surface cracking. It will be appreciated that a large defect immediately below the surface will be better defined than defects of smaller size at greater distances below. It is possible with experience to grade these defects as, in general, defects which do not appear at the surface are less dangerous than defects at the surface.

Magnetic crack testing is a common method of inspecting ferrous components. It is not always appreciated that the process will be complex with components which have a complicated geometry. This inspection technique is competitive with *crack testing* using the dye-penetrant fluids, but it is generally accepted that the magnetic crack test will identify surface defects of a finer nature than those found using the dye penetrants. That is, MPI is more sensitive than dye penetrant. Neither *ultrasonic* nor *radiological* methods of non-destructive testing will find surface defects, since the surface of the component will be clearly identified and, thus, this positive identification will mask any fine surface defect which exists. As magnetic crack testing is much cheaper than either of these instrument methods, then it is usually planned that the surface is first tested by this or the *dye-penetrant* method before more expensive *ultrasonic* or *radiological* examination is carried out. In this way components or assemblies may be rejected or rectified prior to carrying out expensive techniques.

Surface defects will almost always be potentially much more dangerous than defects below the surface.

Porosity Test
An inspection method used to identify pores or breaks in a coating.

Application range ALL METALS

The term can be used for metal porosity itself, which is identified by the *pressure crack test*, *X-rays* or *ultrasonics*. Porosity testing can be applied to any type of coating on any metal, and some techniques are available for non-metallic substrates. There are four basic methods. These are: visual examination; chemical methods; electrochemical methods; electrical methods. Where porosity of the metal of the component is involved then X-ray or ultrasonic inspection will be necessary.

1. *Visual examination* This can be simply visual or at magnifications up to ×40, with up to ×10 being common. This inspection will seldom be for porosity alone, but can be used to identify such defects.

It is common that visual inspection and thickness measurement are the general methods of coating inspection. Other methods of porosity testing are applied only when high-integrity components are involved.

2. *Chemical methods* These rely on a colour change which takes place where chemicals react with the substrate metal but does not occur with the coating. For steel substrate, the most common is the use of potassium ferri-cyanide solution in very dilute acid or in a solution of sodium chloride. This turns deep blue (Prussian blue) when in contact with iron. The test is very sensitive and the parts being tested must be thoroughly cleaned. The user is referred to the *feroxil test*, described in this book, for further information as this uses the same technique.

Where the base metal is not steel, or is austenitic stainless steel, other chemicals or tests must be used: the less the reactivity of the base metal, the less the sensitivity of this method. Thus, stainless steel, titanium alloys and nickel alloys are not capable of being readily tested, but will seldom have simple coatings applied.

A substrate of copper or a copper alloy can be tested using a sulphide after *cleaning* and activation. This is not as sensitive or rapid as the *feroxil test* and is seldom applied. There is, however, a method where the components after *cleaning* are held in a sulphur dioxide atmosphere for 24 h, followed by a further 24 h in a hydrogen sulphide atmosphere. Any porosity is shown by black spots.

Nickel and nickel alloys can also be tested with this technique.

Salt-mist and *humidity tests* can be used to identify porosity, but in general are used to supply additional information. These rely on the corrosion of the substrate metal.

3. *Electrochemical methods* These use the same principle as above, but apply a direct current and, thus, in effect, produce a plating cell. With the substrate metal as anode, the second electrode as cathode, and the sensitized

paper as electrolyte, any attack on the substrate metal will cause discoloration of the paper. An absence of pores will result in no reaction. The test is described in more detail under *Electrography* in Part One.

A modification of this is the *Holiday test* where, instead of reacting with a sensitized paper, the current produced at the pore is used to light a neon lamp or sound a buzzer.

4. *Electrical methods* These methods are based on the fact that paint coatings are generally non-conductors. Any porosity to base metal can then be identified by the passage of an electric current, which can be made to sound a buzzer, light a neon lamp or produce high-frequency sparks. Details of this test are given under *Spark Test* in this section **Non-destructive Testing**.

Pressure Test

A non-destructive test carried out on hollow-type components or assemblies.

Application range ALL METALS

Generally these components will be used under pressure in service, but there can be occasions where pressure testing will be used as a general inspection method to prove the integrity of an assembly or components, including testing for mechanical strength.

The choice of pressure and the method of pressure testing requires some care and should be a design requirement. It is obvious that the pressure for testing must be above that of the component in service, but it is often difficult, if not impossible, to reproduce by a static type of pressure test the pressures to which the assembly will be subjected in service.

There are two basic types of pressure testing, the most common being the use of air or some other gas, with the component being inspected under water or some other liquid. Any leakage will be shown by fine bubbles which appear on the surface. This test can be made more critical by the judicious choice of liquid and gas. A variation on this theme is that the assembly is taken to the desired pressure and that this pressure must then be held for a specific time. Any drop in pressure greater than a specified amount is reason for rejection. This test is carried out using the gas and temperature at which the component will operate in service or higher and is an extremely critical test. It is described under *Air-pressure Testing* in this section **Non-destructive Testing**.

The second method of testing is the *hydraulic pressure test*, where the component is filled with a liquid and leakage of the liquid is an indication that the component is porous. This test can again be made more critical by the careful choice of liquids. For example, by adding some detergent or foaming agent to water, then very fine cracks or porosity can be readily identified, where with water alone this might be missed.

As with *air-pressure testing*, it is possible to take the assembly to the specified pressure and to hold, with any drop in pressure above a specified limit being the reason for failure. As liquids are non-compressible, this is a less critical test than with the *air-pressure test* at the same pressure.

The fact that gases such as air can be compressed means that considerable care must be taken when components are being pressure tested, and this is particularly so with pressure in excess of 50 lb/in² (0.35 N/mm²). Above this pressure any failure can result in an explosive type of bursting and with large vessels this can be disastrous.

With *hydraulic testing*, where the liquids cannot be compressed, this danger does not exist as any release in pressure by bursting will result in the pressure dropping immediately to zero without any explosive forces. It must, however, be pointed out that very great care is needed when high-pressure *hydraulic pressure testing* is carried out to make sure that the vessel is correctly bled of air, otherwise air or gas testing will be carried out, and this can result in explosive failure.

Pressure testing is a form of non-destructive testing which is particularly valuable to identify the porosity in components and to show any weaknesses in *brazing* or *welding*. Except in very specific instances, pressure testing will seldom be required on wrought components which have not been welded and which will not be subjected to pressure during normal service. It is, however, quite common to carry out pressure testing on assemblies, particularly pipe assemblies. It is generally accepted that pressure testing should be preceded by some form of *crack testing* or inspection which will identify any gross defects. High-pressure *hydraulic pressure testing* can be used on high-integrity components, to prove that the component or assembly under test has the mechanical strength and sufficient safety factor to withstand the pressures applied during normal service.

Radiographic Test

This makes use of the X-ray sources used in medicine, of which there are now a number available. These include the standard radiographical X-ray machine, which produces the necessary energy waves or signals by conventional oscillators, and portable sources using isotopes of metal which produce rays. The different techniques, and sources, are important in the extent and sensitivity of the resulting photographic film but are not of vital importance. The convention is that the term 'X-ray' is used for machine-generated rays, while 'radiological examination' implies the use of radioactive metal isotopes.

Application range ALL METALS

The principle in all cases is identical in that the signal or source of energy is passed through the object being examined and allowed to strike a photographic plate or film which is located on the reverse side. These rays have exactly the same effect on the photographic film as does light. That is, the denser the material through which the rays pass the less the effect on the photographic film will be, and the less the density the darker the area on the film will appear. This means, then, that differences of thickness or dimension on the examined component will be shown as different intensities on the film. It also means that any porosity, cracks or other defects will be shown relatively as darker areas on the film. It is obvious that the extent of the defect will affect its radiographic intensity and, thus, pinpricks of local porosity will appear as very slight areas of local darkening on the film which require some skill, first, to ensure that they are picked up on the film and, second, that they are identified in any subsequent visual examination. The operator, then, must first choose the correct exposure or power for the initial radiographic shot (various techniques are available but knowledge of component dimensions is generally advisable), and second, he must have the technique to develop the film correctly and report adequately on the findings.

The advantage of X-rays over other forms of fault detection, is that a permanent film record of the complete cross-section of the component is available for detailed discussion. It will be appreciated that the direction in which the radiographic examination is carried out can have considerable bearing on the sensitivity of the result. A thin crack which is radiographed in the same plane as a larger defect might not show up on the film, or appear as a very slight darkening, since the difference between the surrounding metal and the crack will be only the width of the crack itelf and this may be less than 0.0005 in (0.01 mm). If the crack were radiographed at right angles to the defect, then the difference will be not the width but the length of the crack and this will generally be of considerable significance.

A disadvantage of radiographic examination is that sophisticated equipment is necessary, and while the use of metal isotopes has considerably increased portability over conventional valve equipment, there still remains the necessity for darkroom and viewing equipment. Portable caravan-type equipment is now available which can be moved to the site and with the metal isotopes complete portability is possible. A second disadvantage is the danger involved in working with X-rays or radiographic waves and considerable precautions are required by law with the use of this equipment. Thus, during radiographic examination a specified distance of operators and, in fact, all living creatures from the equipment must be maintained, and this clearing of the area involved can result in considerable upset to production. In addition, the radiographic examination will generally be more expensive, because of the capital outlay involved and the cost of photographic film, than, for instance, *ultrasonic examination*. While both machine-generated (X-ray) and radioactive isotope (radiography) sources are similar, the X-rays will produce the more sensitive and precise image. Because of the high cost of this method of flaw detection it is quite common that *dye-penetrant* or *magnetic crack detection* tests are used initially, while *ultrasonic examination* is used as a routine technique with the present methods used for cross-checking. Methods, and the degree of testing, will however vary, depending on the importance and cost of failure.

Sonic Testing

This is a crack detection technique using sound.

Application range ALL METALS

The traditional method of checking the wheels of railway rolling stock by wheel-tapping was a method of sonic testing, thus this technique uses variations in the vibration pattern to detect flaws in a component.

The theory is that a satisfactory component or assembly will have a well-defined sound pattern, generally resulting in a ringing sound. Any defect such as a crack or the loosening of an assembly will result in an interruption in this pattern. This can be detected by the trained ear and further investigation carried out. Sonic testing is not now normally carried out in this simple form as with modern equipment other forms of crack detection are more efficient, such as the *ultrasonic*, *dye-penetration* or *magnetic crack tests*. The modern techniques are based on the same principle, where by vibrating the assembly a pattern will be obtained and can be recorded. This recorded pattern is stored either in the form of a sound on tape or as a vibration pattern traced on paper. A periodic inspection can then be carried out with the identical vibration being submitted to the test points. The resultant sound pattern is compared with the original and any alterations are examined. This is a modern technique still in its development stages but shows promise for large intricate assemblies subjected to variable stresses in service. It has the obvious advantage that such structures can be monitored more cheaply than with existing methods. Any structural alteration carried out between sonic tests will affect the results, thus modifications should have the structure tested before and after the modification.

Spark Test/(Porosity Testing)

An inspection process applied to metal components which require complete insulation films, for example, paint. The test can also be used to ensure that tanks are completely plastic-covered with no porosity. The process is sometimes known as '*Holiday testing*'.

The test makes use of a high-frequency or high-voltage spark discharge. One end of the electrical circuit is connected to the metal of the component, while the other end of the circuit is in the form of a metal probe. The hand-held probe can be used to scan entire surfaces, or in the case of welded joints to concentrate on these significant areas. Provided the insulated film is complete and intact, no action takes place. Any conducting path between the probe and the metal will result in a high-frequency discharge. In addition there will be a buzzing sound which will aid inspection. Certain equipment makes use of the discharge to produce an audible whistle when porosity is present, and the passage of low-current, high-voltage electricity can be used to light a neon lamp.

It is possible to vary the sensitivity of this inspection process by increasing the applied voltage. Below a certain voltage the high-frequency discharge will have little energy, but as voltage and energy are increased it will be possible to break down a non-conductive film. Thus, the technique in skilled hands can additionally be used as a form of *thickness testing*.

This inspection technique is comparable in some ways to the *dye-penetrant crack test*, but where thin insulating films are applied it is doubtful whether a dye-penetrant would successfully identify incorrectly treated areas. Where tanks or vessels are plastic-lined to protect the metal from attack or to protect the contents of the tank from contamination, visual inspection will be insufficient. It is doubtful too whether dye-penetrants will have sufficient sensitivity. Where paint films require a continuous, non-porous coating, this technique using a brush or sponge electrode can be the most economical method of ensuring that no porosity is present. The test operator must be supplied with detailed information on the voltage to be applied and the type of equipment. The results obtained will vary for different types of materials and for different thicknesses and the method of application. It is thus essential that a detailed specification is available regarding the coating itself and testing technique when this system is used.

Strain Gauging

This form of non-destructive testing involves the application of strain gauges to the components being tested.

Application range ALL METALS

These strain gauges are small devices, the conventional ones producing an electric current when tensile stress is applied resulting in elongation. These devices are, therefore, glued or strapped to the component being tested and the gauges wired to electronic equipment capable of identifying and recording minute current or voltage changes.

The same idea is employed with the use of glass, having almost zero ductility, which is glued to the structure or component being tested. A relatively small elongation of the component will result in fracture of the glass, thus indicating the presence of strain. This technique can be modified to some extent by the use of controlled ductility materials, which will fracture when a certain elongation is applied. There are now brittle lacquers, applied as paint, with the pattern of cracking caused by stress indicating the type and location of the strain. These require considerable skill to apply and interpret the results.

Another variation on the strain gauge uses a known pattern, generally of minute holes in the form of a fine mesh, which is glued to the area being examined. The reflection from this pattern, either by the use of optics or laser, or by taking a mould from the mesh and examining in the laboratory, shows any changes in the formation which indicates that stress has been applied. This last technique has the advantage over the others that no extraneous wires are involved. In addition, it will not be necessary to destroy the gauge to find that stress has been applied and skilled interpretation of the results can be done from the examined structure. Since strain gauging is frequently required under arduous conditions, this last factor can be a considerable advantage. This technique will also find strains applied in any direction, while the other methods require individual gauges. The choice and application of the different forms of strain gauging, and the interpretation of the signals, require care and users are advised to seek specialist advice on the use of strain gauges as a method of testing.

Thickness Testing

Techniques used to measure the thickness of deposits.

Application range ALL METALS

The majority of coating thickness and measurements, however, are related to steel and to paint or plating deposits. There are six basic methods of thickness testing:

1. *Mechanical measuring technique*   This uses standard measuring equipment to find the exact size prior to the deposit being applied and compare this to the size on completion.

This method has several limitations, first, that in general coating deposits are not homogeneous and therefore, unless careful measurement is carried out on exactly the same spot before and after, the results will be meaningless as the variation in thickness will generally be at least equal to the minimum thickness required. Second, there is the problem that the method is time-consuming and, unless carried out on a number of points and a number of components, the results will again have limited application. Finally, because the component is a test piece, and this will often be known to the operators depositing the coating, tests will not be fully representative of conditions where normal techniques are applied.

It is also possible to use this method by measuring the coating and stripping. This overcomes the last objection above, but has the problem that, under some circumstances, the stripping will attack the substrate metal and will invariably result in the scrapping or re-treating of the component. Thus, it is expensive.

The technique known as the *British non-ferrous jet test* (BNF jet test) works on the same principle but uses a fine spray to determine the thickness of the deposit of plating at non-specific points. The time taken to remove the deposit is a measure of its thickness.

There are now available more sophisticated electrochemical systems using the same principle which should be evaluated by the reader. See *British non-ferrous Jet Test* in Part One.

2. *Strip and weigh method*   This is still looked on as a reference method for

plated deposits. The component is weighed in the as-plated, dry condition, is carefully stripped to remove all evidence of the deposit, and then washed and dried and re-weighed. The difference in weight will be the weight of the deposit applied. Formulae which can be used to estimate the deposited thickness are available.

A problem involved with this technique is the danger of assuming that a deposit has been homogeneously applied. This will seldom, if ever, be the case and, while the method can be useful as a control for routine plating, it will not in fact give sensible results in many cases.

3. *Magnetic method*   Variations of this method are based on the principle that magnetic pull will be a function of the material involved, the strength of the magnet and the distance of the magnet from the magnetic material. The method, then, can be applied only to materials which are magnetic. While iron, nickel, and cobalt are ferro-magnetic materials, only iron (and steel which is iron-based) will in fact have sufficient magnetic pull to be used with this method.

There are various relatively simple pocket-sized instruments using either a calibrated spring or a weight moving on a lever arm which rely on the magnetic pull achieved and can be surprisingly accurate. One approximately pencil-shaped type is placed at right angles to the point being measured then simply pulled off. This acts upon the magnet which is connected through a calibrated spring to a pointer. The point at which the magnet is released from the surface can be shown by movement of the pointer on a scale calibrated in the desired units. This very simple instrument can be used repeatedly on the same spots to ensure that reproducible results are being obtained, and because of the simple operation a large number of readings can be rapidly taken to obtain an indication of thickness. This instrument can be used for paint or plated deposits but may require re-calibration when used on nickel plating of appreciable thickness.

A similar instrument which uses the same principle has a weight which slides along an arm as the instrument is operated. This arm acts as a lever and the distance of the weight from the fulcrum point is calibrated as an indication of thickness. Again this device is simple and readily used. Both instruments can be used in any reasonable position and will give reproducible results.

A more complex instrument uses the magnet as the fixed point in a circuit, similar in all respects to the Wheatstone bridge. With this, a magnet of known magnetic strength is balanced with a magnet of variable strength resulting from the thickness of the material. This instrument gives a reading on the dial and is generally accepted as being more reliable than the devices described above but is much less portable.

All three instruments require to be calibrated at regular intervals. This should be at least twice daily when used regularly, and re-calibration should be carried out immediately prior to use if the instruments are not in constant use. This calibration should be on the same material as that being tested, that is mild or low-alloy steel, and making use of shims of plastic or, if desired, metal which are themselves calibrated at frequent intervals. It is essential that these cover the range of thickness measurements being carried out as the scales will not, of necessity, be linear.

Where routine thickness testing is carried out on components, it is advisable that the components themselves are used as test pieces. This will obviate any possibility of spurious effects. These instruments will not normally be upset by changes in mass or geometry but will be affected by any residual magnetism remaining in the component.

In general these instruments will be the ones most commonly used in industry, and are the first choice for the complete range of materials applied, but have the disadvantage that they are only applicable to the testing of steel components.

4. *Eddy current measurement* This is the same technique used for *metal-sorting* and *crack detection*. With this, an electric current is passed through a coil which is used as the probe. The result of bringing any metal within the range of this coil will be that the eddy currents which surround the coil will be affected, and the degree of this will be proportional to the materials involved.

Probes can, therefore, be produced which are calibrated to show the thickness of any material on any metallic substrate. This makes these instruments potentially more useful and universal than the magnetic measurement instruments.

The problem with eddy current measurements is that extraneous matter, for example the mass of metal below the coating, can affect the readings obtained. Thus, it is essential that the calibration test pieces are of the identical, or at least similar, shape and mass to the inspected components. There is an additional problem in that different substrate metals or the same metal deposited from different solutions will give different results. This is generally overcome by either a complete re-calibration or, more usually, a separate pre-calibrated probe is supplied.

There is some indication that the technique used to apply the coating can affect the readings obtained. Thus the analysis of the solution used for electrodeposition, or type of paint applied could affect the thickness results obtained.

A number of these instruments are available, ranging from battery-operated, hand-held pocket devices to sophisticated bench equipment.

5. *Radiation back-scatter* This is the most sophisticated method of *thickness testing*, and as would be expected the most expensive. It can be used on all materials and has the advantage that it determines the thickness of deposits of very thin films, much thinner than any of the conventional methods described above. In addition it can be calibrated against standards to determine the thickness of undercoat layers. Thus it will be possible, for instance, not only to measure the thickness of gold plate on nickel but also the thickness of the nickel deposit itself on the copper substrate.

The principle is that a very low radiation source is used and this radiation will be absorbed proportionally by the metal of the component itself and also by the thickness or amount of metal left. Non-absorbed radiation will be reflected, and it is this back-scattered radiation which is measured on a standard Geiger counter.

This instrument uses very low power sources of radiation which are decaying rapidly and, thus, must be constantly calibrated and adjusted to allow for this decay. It must be appreciated that although very slight, unless protected, the source will be giving off radiation. The source should not, therefore, be carelessly handled. The more radiation which is measured, the thinner the coating.

6. *Jet test*   This is a technique which can be applied only to metal deposits and uses the fact that different metals dissolve at different rates, and that the rate of solution on a local area (using a jet or drops) will be proportional to the thickness. There are two basic techniques, the best known being the *British non-ferrous jet test* (BNF jet test), described under this heading in Part One. The second instrument is more sophisticated and makes use of the fact that metals have different electrode potentials. This instrument employs an electric current which is carefully adjusted for the metal deposited. Immediately the solution dissolves away the top coating, contact is made with the base metal and the voltage and current conditions are altered. This alteration is used to sound an alarm or stop the time clock.

Both this and the BNF method in skilled hands are excellent means of accurately measuring thickness but they have the disadvantages, first, that they are destructive, and second, that the test applies only to a single spot.

Ultrasonic Flaw Detection

This is basically an 'echo sounder' using frequencies above that of sound.

This technique uses relatively sophisticated electronic equipment developed in recent years. While the operation appears essentially simple, it should be appreciated that considerable skill in operation and interpretation is necessary if the results are to be meaningful. The technique can be used on all metals and non-metals, and has been applied to living matter for medical purposes.

Application range ALL METALS

The principle is identical to that of radar. With this, a pulse of high-frequency electromagnetic energy is sent into the atmosphere. Any solid object in the path of this pulse of radiowaves will cause some of the energy to be reflected along the same path as it was originally sent. As the pulse can be sent in the form of a thin, pencil-type beam, it will be seen that the direction of the object can be accurately pinpointed, depending on the direction of the returned pulse. The fact that the speed of the radiowave is also accurately known, means that using a time-base the distance of the object from the transmitter/receiver can be accurately found. As stated, exactly the same principle is used with the present method of flaw detection, where a pencil beam of ultrasound is used, at a much higher frequency than can be heard by the human ear. This is transferred to the component being examined through a probe. Various techniques are available, but in general modern equipment uses a single probe containing the transmitter and receiver. Much of the sensitivity and accuracy of the modern machines relies on the type of probe employed. The beam of ultrasound will be reflected from the surface of the component on which the probe rests. It will also be reflected from the other side of the component.

Both these surfaces are shown on an oscilloscope in the form of pips or bleeps: the operator will recognize each surface and will, therefore, ignore them. Any indication of a defect such as internal cracks and blowholes or porosity or, in the case of some materials, a difference in metallurgical structure will again be shown as a pip or bleep between these two markers.

With the original ultrasonic equipment it was not possible to identify surface defects, or defects lying very close to the surface. However, development of probes, and also liquids to produce an interface between the probe and the inspected surface, has made it possible to separate the entry pip from the surface and, thus, this technique can successfully be used for defects quite close to the surface. Except with sophisticated, highly specialized equipment which is suitable only for mass production of identical components, it is not readily possible to obtain any record of defects identified by ultrasonic inspection. The procedure, therefore, will rely on capable operators correctly identifying problems. With the correct equipment and skilled operators, there is no doubt that ultrasonic examination is capable of identifying defects below the surface of the metal which cannot be found by any other means at the same cost. These defects can be extremely small, to the order of 0.005 in (0.1 mm) long, 0.002 in (0.05 mm) deep and 0.0001 in (0.0025 mm) wide. It is also possible to show by ultrasonic means relatively slight differences in metallurgical structure. The different structures obtaining, for example, between spheroidal graphite cast iron and normal grey cast iron can be identified with the correct equipment in skilled hands.

Ultrasonic crack detection differs from the surface crack detection of the *dye-penetrant* and *magnetic crack tests* in that it is capable of identifying below-surface defects. Surface defects are technically, and certainly much more economically, found using dye-penetrant or MPI techniques. It is, however, possible to use ultrasonic flaw detection at an earlier stage in manufacture. Below-surface flaws will be identified without the cost of machining. The disadvantage of this method is that no permanent record of this test will be kept and, thus, any dispute will be difficult to prove one way or the other. It also relies on the integrity of the operator regarding the area being scanned, and the interpretation of the results.

This technique cannot find surface flaws, but with a skilled operator and sensitive equipment quite small sub-surface defects can be identified. This is more expensive than, and not as sensitive as, either *dye penetrant* or *magnetic particle inspection*.

Vibration Testing

The sound or vibration pattern produced when a simple metallic component or assembly is vibrated by being struck a blow is a function of the shape of the component. Any alteration, particularly at the surface, will affect this pattern and can be identified by a different tone.

This is the basis of the wheel-tapping which was practised to identify any cracks in the wheels of trains. The modern system feeds the vibration via a microphone to an oscilloscope, where the pattern can be examined, measured and photographed if necessary. Any cracks or other defects which subsequently appear will alter the pattern and thus are identifiable. More sophisticated techniques feed vibrations of known amplitude and frequency into complex structures to give considerable information. At present the technique is in its infancy but considerable development work is being carried out, and this with *acoustic testing* could become the standard flaw-detection methods on complex static structures in the future. The *Reflectogage Test*, briefly described under that heading in Part One, uses high-frequency vibrations.

Visual Inspection

The use of eyes and magnification up to about ×40 to examine all surfaces.

Application range ALL METALS

This is probably the most important of all non-destructive types of inspection. It uses firstly the unaided eye to examine all surfaces for surface texture, contamination, damage, and other aspects such as burrs, incorrect radii, incorrect machining. Use should then be made of some form of magnification, generally ×5 or 10 to concentrate on critical areas.

This should find defects such as porosity, cracks, material defects and machining errors much more economically than any other technique. It should therefore be the first form of non-destructive testing.

Painting

The general name given to a large number of basically similar but variable methods of coating the surface of a material.

Application range ALL METALS AND MOST OTHER MATERIALS

All true paints have one characteristic in common, in that they are applied as liquids and harden or are cured into solids. The conversion may be achieved by two distinct means: natural oxidation or chemical reaction occurring at room temperature; or chemical reaction requiring stoving at elevated temperature or oxidation taking place at elevated temperatures.

A film produced by fusing dry powder is now also regarded as a paint film.

Painting is seldom required for a single reason, but is more often carried out for both *corrosion protection* and appearance. Other reasons may be for identification purposes, or to produce certain specific surface conditions such as electrical conduction, electrical non-conduction, resistance to moisture penetration, lubrication, anti-galling etc.

While there are exceptions, in the main paints consist of three components:

1. *Pigment* This is the solid medium which can be either a metal, a metal oxide or, very often with modern types of paint, a stable organic compound. Pigment gives body to the paint, supplies the colour and, where *corrosion protection* is essential, may be the positive protective medium.

2. *Vehicle* This is the liquid phase which carries the pigment and will eventually be converted to a solid. Traditionally the vehicle was an oxidizing oil, which on exposure to the atmosphere was converted by oxidation from a liquid to a solid or a plastic-like solid. The vehicle in modern paints generally hardens or solidifies by a more complex process. This may be a simple chemical reaction not unlike oxidation or, as with the two-pack types of paint, a more complex reaction, resulting in a very hard solid. Many modern paints require temperatures in the region of 100°C–250°C for the necessary reactions to take place, and these are generally known as stoving paints or stoving enamels.

The traditional enamel was quite different, where, in fact, the vehicle was driven off during the heating cycle, and the pigment (usually metal or glass) then fused to give the enamel.

Some modern plastic materials in powder form can be applied to metal and fused in position. These materials are known as paint, but have no vehicle, and in some respects are similar to 'enamels'.

3. *Thinners or solvent*   This is the component which is used to produce the correct viscosity for the best application; it is not present in all paints. Thinners or solvent have no significant reaction in paint. After application either by brush or spray the thinners will evaporate, the vehicle solidifies and the pigment is held in suspension.

With celluloid paints only two of the above three components are involved. In this case the pigment which is held in solution by the thinners is celluloid, which is dissolved in the solvent. The liquid is sprayed or brushed on to the component, the solvent then evaporates leaving a thin film of celluloid behind. Other variations are clear lacquers or varnishes. Here there may only be a vehicle (to which some thinners may be added for easier application). The vehicle cures after evaporation of the thinners, giving a clear, hard film.

At present it is possible to spray a fine dry powder on to metal which is then cured by heating. This technically is plastic coating but is known as *dry-film painting*.

The methods of application of painting are many and varied and the more common are briefly described in this section, together with some information on the preparation for painting. The processes are:

Paint preparation	Enamelling
Brush painting	Lacquering
Curtain painting	Powder painting
Dip painting	Roller coating
Electrophoresis/Electropainting	Spray painting
Electrostatic painting	Stove enamelling

It is not within the scope of this book to supply technical details of available paints, as this information can best be evaluated for a given metal under specified conditions.

Paint Preparation

The most common purpose of painting is for *corrosion protection* coupled with decorative colouring. It is, therefore, essential that the paint chosen and the method of application should result in a paint film which will succeed in the chosen purpose. A statistical analysis of paint failures in a large number of instances has shown that the reasons for paint failure are as follows:

1. Approx. 75 per cent of failures are the result of faulty preparation.

2. Approx. 20 per cent of failures are the result of improper application of the chosen paint film.

3. Less than 10 per cent of paint failures can be attributed to the choice of the paint itself.

PAINT PREPARATION / 483

From the above it will be seen that it is essential that the correct method of paint preparation is chosen and that its quality is controlled. In this book, under their separate headings, are given the various methods which can be used and these will be put into context in this section.

Cleaning is essential as the first stage of paint preparation. This applies irrespective of the material being painted and the type of paint applied. *Cleaning* requires that all grease, oil or moisture is eliminated where high-quality painting is required. Paint failure commonly results from lack of appreciation of the importance of cleaning. Adequate cleaning is never achieved by wiping a component prior to painting. At best this will result in any soil being evenly spread across the component, rather than concentrated in specific areas. Thus, wiping can result in a complete paint failure, whereas no attempt to clean might have confined failure to local areas.

In the majority of components, the presence of a fine dust film is not particularly damaging. This does not mean that contaminated parts can be successfully painted, but it should be appreciated that oil, grease and moisture are more serious with regard to painting (see *Cleaning* in Part One).

Surface preparation   The necessary preparation will depend to a large extent, of course, on the material being painted. Treatments for common materials are as follows:

Application range WOOD

Usually this will need to be scuffed, either by hand or mechanically, resulting in a slight roughening of the surface and removal from the surface of any impregnated contaminants. With already painted wooden surfaces, this preparation is also usually required. Today, modern paints are formulated so that they will key with each other and, thus, intercoat preparation is not as important as previously. All evidence of loose or flaking paint must be removed.

The adhesion of wood to paint is improved if some paint is absorbed by the wood, and intercoat adhesion will always be better when the two films of paint are compatible and diffuse. It will, thus, be seen that the *primer* should have low viscosity and good penetrating ability along with other factors, the most important being compatibility with the next coat.

Application range PLASTIC

Again this can be prepared by abrasive means but plastics are often 'keyed' by the choice of the correct solvent. This can be as preparation prior to the application of paint but is also commonly a component within the paint. This will 'key' by solvent action, which results in some dissolving of the existing surface with the paint being applied.

Application range STEEL

Mild steel is the most commonly painted material in engineering. This material is very often received for painting in the scaled or oxidized condition. Oxide must be completely removed with the exception of the fully adherent millscale which is rolled into the surface of the steel plate. This is fully oxidized iron and, provided it has good adhesion, will make an excellent 'key' for the paint.

It is advisable that the surface has all evidence of loose scale, oxide or any other form of surface corrosion removed. There are two methods by which this can be achieved, either blasting or pickling.

1. *Blasting* This is the most common method of pre-treating steel components for painting, particularly components involved in heavy or structural engineering. With this, components are blasted with grit or shot, which has the dual purpose of removing undesirable surface oxide and contamination and at the same time producing a roughened surface. It is this roughened surface in which the shape or contour consists of rounded humps, with a controlled ratio of height to width, which gives good paint adhesion. Various specifications are available which lay out the contour requirement in considerable detail; probably the best and most commonly used is the Swedish Standard.

Where the contours achieved are too sharp, the paint will run, leaving the sharp peaks bare with relatively thin paint coating, and thus liable to corrode. Where the contours are too shallow, the paint will lack sufficient 'key', thus adhesion will be poor. The standard of *grit blast* finish required is an important part of any specification, and with high-integrity parts must be carefully controlled.

2. *Pickling* This is the chemical process using a specified acid to dissolve oxide or scale. With this treatment, it is essential that correct control of the acid and subsequent washing is ensured. Subsequent processes must be carried out immediately, as the pickling will leave a highly reactive surface, and unless all traces of the acid are removed, oxidation will be rapid. Removal of acid is difficult, and special equipment and a copious water supply are necessary. Acid pickling, therefore, is generally inadvisable as a means of pre-treatment for paint unless it is immediately followed by a process such as *phosphating*. Technical advice and excellent control are essential when acid pickling is used.

3. *Chemical preparation and phosphating* Chemical treatments are sometimes applied, which are essential where high-integrity corrosion resistance is necessary. Such treatments are applied to the cleaned, descaled surface; some include chemicals which can cope with light rusting or staining. It is not always necessary to have grit-roughened surfaces, since pre-treatment systems are designed to supply the necessary adhesive key, but wherever possible blasting should be carried out.

By far the most common chemical preparation for painting is *phosphating*. This makes use of soluble phosphates in an acid medium which converts the surface of the steel to iron phosphate with or without other phosphates also being involved. The phosphate coating is adherent to the steel and is porous, so that the liquid paint is absorbed into the phosphate film. As this film is highly adhesive to the steel surface, excellent paint adhesion is obtained (see *Phosphating* in Part One).

In addition to paint adhesion the phosphate film contributes considerably to the corrosion resistance of the steel. A phosphated component which is painted and subsequently damaged will be much less liable to under-film corrosion than components where the paint is applied directly to the steel surface. Where the phosphate has been sealed with chromate, corrosion resistance is considerably enhanced. There are a large number of proprietary phosphate coatings, some of which are listed under *Phosphating* in Part One.

It is also possible to include additives in the paint primer. These are generally based on phosphoric acid or phosphates often with the addition of chromate. There is a wide range of these paints which are known as 'etch primers', many having metal or metallic oxides added to improve their long-term corrosion resistance. Some of these are two-pack systems and have a short shelf-life. Other pre-treatment methods for steel include *chromating* and various solutions which form iron oxide and attempt to ensure that this adheres to the steel surface. In general these are less popular and efficient than phosphating.

Application range ALLOY STEELS

These in the main are treated in a similar or identical manner to mild steel. The fact that they are often harder means that *grit blasting* will probably require more care and will be less effective. It must also be appreciated that where a high-tensile steel is used and protected by painting, then paint failure can be much more serious because local paint failure will result in corrosion pitting which, acting as a stress concentration, can be disastrous. It is, therefore, most important to ensure that the chosen system is satisfactory and correctly applied.

Application range ALUMINIUM AND ITS ALLOYS

The necessity to paint aluminium and its alloys is much less than with steel, since corrosion resistance is generally superior. There are, however, many occasions when painting is required. Paint will not adhere successfully to aluminium, and it is necessary for one of three preparation methods to be used when painting aluminium.

The simplest method is to use a specially formulated paint having an *etching* constituent within its make-up. This will commonly be based on phosphoric acid, with or without chromate, and the result will be the *etching* of the surface, giving a key for the paint film; the preparations are known as 'etch primers'. These paints are expensive and many of them have a short shelf-life and thus are not economical for high-production use.

The second method, by far the most common, is to treat the aluminium or alloy surfaces with a *chromating* solution. This results in the conversion of the aluminium surface to an oxide which then absorbs into itself some of the chromate. The process is applied by brushing, spraying or dipping.

The third method consists in *anodizing* the aluminium surface. If the subsequent anodic film is not immediately sealed, but sent for painting, this will result in the paint being absorbed into this anodic surface which, having an excellent corrosion resistance, enhances the paint film giving a very good corrosion-resistance material.

Application range MAGNESIUM AND ITS ALLOYS

Because of the chemical activity of magnesium, it is imperative that this surface is protected when in service. When any contact with atmospheric conditions is liable to occur, and particularly where these are moist and oxidizing, or there is a marine connotation, then considerable care must be taken in the preparation prior to painting. There is a magnesium *anodizing* process but this is expensive, requiring specialized equipment, and it is seldom used as a paint preparation.

By far the most common method of preparation is by *chromating*, where the components are immersed in a hot oxidizing solution of chromate which results in the formation of a chromate-absorbing oxide. Correctly chromated surfaces are dark blue or black in colour. This results in a key for paint, but it is important for best results that the component is coated with a protective paint or varnish such as 'Phenolic', within hours of *chromating*. The initial paint film must then be coated with a sealing top coat (or coats) which protects the primer and helps to prevent mechanical damage.

Because of the danger of corrosion with magnesium, any break in the paint film must be immediately and carefully treated. Where the chromate film is also damaged, it is advisable to treat the damaged area with *selenious acid* prior to touch-up painting.

Application range OTHER METALS

Most metals, apart from those outlined above, either do not require painting, or such requirement is very limited. The most common reason for painting these materials will be for decorative or identification purposes. Where a material has good corrosion resistance, then it is necessary only to roughen the surface to provide a key for the paint film. This is most commonly achieved by *blasting*, but under certain conditions the material can be roughened by chemical *etching*; the chemicals used and technique will vary considerably and specific advice should be sought in each case.

It should be realized that the reason that most metals do not corrode is because they have a non-reactive surface. This means that paint adherence will often be poor and, thus, only a roughened surface provides a key.

To summarize, paint preparation is of vital importance and paint failure can be expected if preparation is poor. Whether the paint is applied for *corrosion protection*, decoration, etc., it should be obvious that, if failure is unimportant, no painting need have been done in the first place. The problems involved in removing faulty paint for salvage purposes can be serious, and the more nearly satisfactory the preparation, the greater the problems involved in removing the paint. It is essential, then, that all preparation for painting is carried out in a responsible manner, the specification agreed upon and an inspection or quality control system employed to ensure specified parameters are achieved. See also under *Blasting*, *Cleaning*, *Corrosion Protection* and *Phosphating* in Part One, all of which have relevance here.

Brush Painting

This is undoubtedly the best known method of paint application, where all the transferring of paint is by a hand-held brush.

Application range ALL METALS

A certain viscosity over a fairly narrow range is obviously required, as if the paint is too thin it will not be held on the brush, and if too thick cannot be transferred to the surface. Thixotropic paints are now available which are in the form of gels in the static condition, thus ensuring easy retention on the brush but when pressure is applied they alter their viscosity and act like liquids. These paints have an obvious advantage for amateurs doing overhead painting.

In general brush painting is labour expensive and thus is not generally used in industry. It is extremely flexible and in skilled, controlled hands can achieve very high quality results. One not often appreciated advantage is that a skilled painter can continuously inspect the result of his paint film, and also the state of the component being painted.

Curtain Painting

With this, a curtain of paint is pumped or allowed to fall through slots, producing a thin wall of liquid paint. Through this wall is passed the component to be treated. By judicious control of the quantity of paint in the wall, and the speed at which the components pass through the wall, the thickness of the applied coating can be controlled. This method has a limited application to certain types of simple shaped components. (The remarks related to *spray painting* regarding solvent quantity, type of vehicle and pigment are relevant here also as curtain coating is a variation of this technique.) It will be appreciated that only upper horizontal surfaces can be coated in one pass.

The condition of the surface to which the paint is being applied is also quite critical.

Dip Painting

This, as the name implies, requires that the components are dipped into the liquid paint and then removed for drying in air or subsequent stoving.

Application range ALL METALS

In the vast majority of instances this technique will result in drops or tears on the bottom edges of the components being painted and, unless very careful control of paint viscosity and dipping techniques is achieved, the upper surfaces will have considerably less paint than the lower surfaces.

Dip painting is very often followed by some form of spinning or centrifugal action whereby the excess paint is removed, and the disadvantages described above are eliminated or considerably reduced.

This technique can be used for single articles but is also commonly used for low-cost, simple shaped articles in high-production batches. The technique of dipping heated articles into a fluidized bed of epoxy, or other plastic powder to result in a plastic coating, is discussed under *Powder Painting* in this section **Painting**, and is a variation of 'dip painting'.

Electrophoresis/Electropainting

A painting technique using electrolysis and osmosis to apply a finely divided colloidal solid in water to the chosen surface.

Application range ALL METALS

This is a physical electrochemical system which makes use of electronically charged colloidal particles which attract or repel according to normal electromagnetic laws. This technique can be used in several forms, the most commonly applied being *electro-osmosis*, which is finding considerable application in medicine and in the production of potable water from brackish supplies; this involves true osmosis and no solid particles are present. In industry electrophoresis is confined, on a production scale, to painting. This is now more commonly called electropainting and with this particles in colloidal form are present. (*Electro-osmosis* is described separately in Part One.)

Electropainting can be applied to any metal, but in general is applied only to steel. By osmosis, it is possible to separate liquids from each other, and to remove from liquids certain solids in solution. With this process, which has been known in laboratory conditions for some time, it is possible to separate miscible liquids from each other. A membrane is used to keep the liquids apart and through this membrane the liquids or solute pass. Osmosis is a complex and very slow process, but by the application of controlled direct current electricity, much faster results are obtained as the charged ions in solution are attracted by the opposite charge of electricity. With electropainting, colloids are present instead of, or in addition to, ions in solution, and are attracted by the opposite charge to that applied to the article being painted. The skill of electropainting is in keeping the colloidal particles in the correct condition and within the specified concentration.

The paint itself will be water 'soluble', and this has the considerable advantage over the majority of paints in that it is non-flammable. As stated, the ions or colloids are attracted to the article being painted, which has the opposite electric charge, and as with all electrochemical processes, like attracts unlike. Once there, the different charges match up and the paint particles adhere to the surface. Since the painted surface no longer has an electric charge, further particles will not be attracted to any part of it. The result is that the paint particles will tend to search out, and become attracted to, areas of the component where no paint already exists, and thus this process is excellent for ensuring complete coverage.

It is possible to include an ammeter in the circuit and to show that at the start of the paint cycle a current is flowing, but as the process proceeds, this current will fall off and gradually return to zero. When this has occurred, no further purpose is served in continuing, and the component is removed. The surface is then sprayed lightly with water to remove all excess paint, while those particles adhering by the electrical attraction are retained. Components must then be stoved at a temperature of approximately 150°C for approximately 1 h, resulting in the fusing and curing of the paint film.

It will be seen that electrophoresis or electropainting provides an excellent undercoat. For best *corrosion protection*, it should be applied to a *phosphated* steel. Almost invariably a conventional spray paint is applied to the surface as the electrophoretic coating has limited mechanical strength, and is intended as a primer paint which will search out and cover all intricate corners and crevices.

This paint system has no real equivalent, but for 'covering power' it may be equated to *dip-painting*, without the disadvantages of the latter process, where runs and tears are part of the expected finish. There is also no danger of air pockets.

Electrophoresis painting has achieved a considerable advance in paint technology, first, in giving a much more uniform and comprehensive covering and, secondly, being a water-based paint, in reducing fire hazard. It has been suggested that the reduction in insurance possible by the use of non-flammable paint, may almost pay for the capital cost of the equipment. The technical control required is greater than with normal paints, and unless this is of a high standard considerable technical trouble can be expected.

Electrostatic Painting

This is a variation of *spray painting*, using specially formulated paints where the pigment and vehicle are controlled such that the pigment particles will accept an electrostatic charge. Any thinners used must also be non-polar to prevent the discharge of the static electrical charge held by the pigment particles.

Application range ALL METALS

The technique uses painting guns, modern ones having the application of a small amount of air pressure so that the paint is ejected at some slight velocity.

The particles leaving the gun are given an electrostatic charge of anything up to 30 000 V. As the current involved is negligible, there is little safety hazard. The component to be painted is at earth potential relative to the paint particles, and thus there is considerable attraction between paint and component. As the paint particles arrive at the component, they are attracted and adhere. This results in the neutralization of the static charge, and added attraction of the paint particles to any areas which have not been painted.

Electrostatic painting, therefore, can be successfully used to paint round articles and is particularly useful in painting tubular components, for example stacking chairs, where, by conventional painting techniques, the operator requires considerable skill and a large number of passes to ensure covering all the tubular surfaces. With electrostatic painting, one pass in the general direction of the tube ensures complete coverage. Provided the correct paint is chosen, with the necessary control of voltage and distance between nozzle and article, then it is possible to have little or no overspray, and effect a considerable saving in paint usage.

Since this technique uses a very high attraction of particles, it is most useful for articles which do not have large, plane surfaces. Where flat panels are involved, the economy is much less significant. This method of painting initially had the disadvantage that the electrostatic charge on the paint particles, and the equal and opposite charge on the components, would be concentrated on edges, and thus these would receive more than their fair share of the paint film. However, since this is the opposite to conventional painting, where sharp edges have a tendency for the paint film to be thinned, there is in fact an obvious advantage.

Against this considerable advantage may be placed the disadvantage that re-entrant corners (the opposite of edges), either receive very little or no paint at all, since no electrical charges in these areas exist. However, as these areas do not normally have the same corrosion potential as sharp edges, the danger is not as much as might be expected. With modern equipment, to a large extent, this is overcome by the slight air velocity given to the paint particles as they leave the gun, ensuring that some paint will be driven into the re-entrant angle, provided the operator uses some skill.

Electrostatic painting is now commonly applied both as a hand-painting technique, with the operator holding and controlling the electrostatic gun, and in fully automatic painting equipment, where a number of guns reciprocate to give complete coverage. Because of the attraction, it will be readily seen that components of a relatively intricate shape which might require complicated automatic equipment will be successfully coated using electrostatic means. The method is an economical alternative to *spray painting*, because overspray is much reduced. The presence of more paint on edges tends to be a technical bonus.

Enamelling

The application of fused glass with pigmentation. The term, strictly speaking, has very limited application, but is commonly used to mean *stove enamelling* which is the modern use of paints which are cured by heating or stoving, described in this section.

Application range ALL METALS

Traditional enamelling is an ancient process, used by the Chinese and Japanese centuries ago. It uses glass in ground form to which are added various metal pigments for coloration purposes. Components are generally enamelled for purely aesthetic reasons. They are first thoroughly cleaned and the metal is treated by *scouring*, *wire-brushing*, *blasting*, *etching* or some other method to roughen its surfaces to give a key for adhesion. The powdered enamel is then applied to the area to be treated. Traditionally this will be done using a very fine brush, but there is no technical reason why *spraying*, *silk-screen printing* using masks, *dipping* or any other method whereby powdered enamel can be applied in the required thickness, cannot be used. The component is then heated to above the fusion point of the powdered glass. This can be as low as 400°C or as high as 1000°C, depending on the type of glass and the pigment used.

It will thus be seen that, although in theory enamelling can be applied to any metal, its use will be restricted to those metals which either do not readily oxidize, or are not melted or *softened* by the temperatures used during the process.

It is common that, for ornamental purposes, a number of colours are used during enamelling to produce the attractive designs required. These different colours will be applied successively to build up the pattern. It will be appreciated that, if each enamel has exactly the same melting point, or if enamels already applied melt at lower temperatures to those applied later, it will not be possible to prevent the colours blending. For some purposes this blending of adjacent colours is desirable, but where this is not the case a range of melting points is necessary, with the highest melting-point enamel applied first and the lower ones successively applied.

The process described above is still used for the production of ornaments such as brooches, badges, household vases etc., together with certain high-quality, semi-industrial articles such as name-plates, lapel or car badges etc. However, it is an expensive method requiring considerable skill, and for many purposes has been replaced by *silk-screen printing*, using normal modern paints which are then cured by stoving. These paints do not have the high durability and scratch resistance which is obtained with the pigmented glass, but may have better ductility and the ability to withstand temperature changes and shock loads without cracking.

Lacquering

A term used in the painting process which at one time had the very specific meaning of applying a clear varnish to protect an existing finish. While this is probably still the most common meaning, there is little doubt that the term 'lacquering' is now commonly used as a synonym for 'Painting'.

Application range ALL METALS

In modern industry lacquer is used for a variety of purposes but the most common will be as a clear, non-porous film for protection. Lacquer types range from the cellulose-based materials, where cellulose is dissolved in a solvent and applied either by brush or spray: when the solvent evaporates, a thin film of cellulose is left on the surface of the component. This has limited adhesion and is not used to any great extent in modern metal finishing.

Any modern paints, e.g. the phenolics, acrylics and polyurethane types, can be produced to give a clear film. These paints vary in their ability to achieve good adhesion and also (this might be very important) in stability and the retention of clear opaque properties.

Among the prime uses of lacquer is the protection of polished metal surfaces where oxidation results in tarnishing. The advantages of clear lacquer for this purpose is that it is cheap and easy to apply. The obvious disadvantage is that it will not withstand wear to any great extent and unless good adhesion is present, peeling will occur. It will also be appreciated that for this purpose the lacquer must be lightfast and not subject to changes of colour during normal service. Lacquers are also used for electrical insulation purposes, again it will be necessary that the adhesion is adequate; but for this purpose lightfastness is not important.

One of the original purposes of lacquering was to produce a lightfast, clear, non-porous film for application on top of other paints. Until comparatively recently the ability to produce stable colours over a range of conditions was very limited and the theory was that unstable undercoats would thus be protected. To some extent this use still remains and a lacquer film is applied to some modern paints to prevent deterioration. Here the lacquer is almost identical to a clear varnish.

During manufacturing processing lacquers are used to protect materials from damage. The lacquer will be of a specific type, often incorporating a resin, for example, 'Bedacryl resin' in solution, and it is this resin which acts as a dry lubricant on the surface, thus protecting it from minor damage. Materials which damage easily such as aluminium are often lacquered in this way and can then be handled without the need for great care, if necessary. Instances are known where the material has been trodden on and has apparently shown severe surface scoring, but when the finished article has been cleaned, using a solvent to remove the resin, it was found that no scratching damage was found to exist on the metal surface.

Lacquering is also used in *deep-drawing* operations. These lacquers are sometimes applied directly to the metal surface but more often in conjunction with some other treatment, for example, *phosphating*, where the lacquer is absorbed into the phosphate layer and, thus, more intimately adheres to the substrate. Again these lacquers have special characteristics, designed to improve the friction coefficient, but it will be seen that material similar to that described above will also be suitable. In addition, where higher temperatures are involved, substances such as powdered glass can be used. This, then, is an instance where the term lacquering is not the best nomenclature.

Finally, during *electroplating*, lacquers are often used to prevent or *stop-off* plating of local areas. The components are cleaned and dried and the lacquer is then applied either by *brushing*, *silk-screen printing* or, under some circumstances, photographic processing. In this instance the main characteristic of the lacquer must be its ability to withstand the chemical attack of the plating solutions involved. Under many circumstances a dye is added so that the presence of the lacquers on the surface of components is readily seen. Again it will be appreciated that adhesion to the substrate must be of high quality. The procedure is discussed under the heading *Stopping-off* in Part One.

It will be seen from the above that the term 'lacquer' is not easily defined. In many cases it will imply that a surface film is applied for protection, but it is also commonly used today to indicate the application of any surface film to components.

Powder Painting

This is a recently developed technique where the paint has no vehicle or thinner.

Application range ALL METALS

The material is in the form of a dry powder which is applied either by electrostatic means exactly as discussed under *Electrostatic Painting* in this section **Painting**, or very commonly by heating the component and plunging it into the powder. Under many circumstances use is made of a fluidized bed for the latter technique. The thickness of the coating can be controlled in the first case by the number of applications, but with *electrostatic painting* this will never be very great, and a thickness in excess of 0.005 in (0.15 mm) is unusual. Where necessary, further coats can be applied after stoving, but as the thickness of the paint film increases, so also does the electrical insulation property, thus the adhesion of the powder particles is reduced.

With *dip-coating*, thickness can be controlled within wide limits by adjusting the temperature of the component and the time of immersion. It is also possible using this technique for more than one coat to be applied, and no problems arise because of reduced thickness with subsequent applications, since no electrostatic charge is involved. The *dip-coating* technique can be a single process, where the temperature of the component results in the powder curing.

This will depend initially on the temperature of the component prior to dipping, but also to a large extent on the mass of the component, and for high-quality components, a subsequent stoving is advisable to ensure complete curing.

With electrostatic deposition, subsequent stoving is essential. The most commonly applied powder paints are the epoxy powders and these require curing or stoving at a temperature of 150°C–200°C. The necessity for roughening or phosphate coating prior to *dry-film painting* is much less than with any other method. This is because of the excellent adhesion and high impact resistance of the correctly applied and cured plastic film of paint. It is essential that the material being painted is dry, clean and free of corrosion.

If *phosphating* is used, it must be a thin film, otherwise adhesion will be reduced. This is because the powder paint is not absorbed into the phosphate as occurs with normal wet paint. Abrasion resistance and adhesion characteristics of dry-powder painting render this technique of considerable use for high-quality industrial and commercial articles. The economies are also favourable, since there is little overspray with electrostatic application and no waste with the fluidized bed. The fact that paint preparation here consists not in *phosphating* or *blasting*, but high-quality *cleaning*, also adds to the economic advantage. This does not mean that rusted or scaled components can be treated. Because curing is carried out above 150°C, use of this method of *corrosion protection* is restricted to those articles which can be oven-stoved, or to sophisticated equipment or large components.

Where there is any possibility of the film being damaged on steel components, it is advisable that a thin phosphate film is applied immediately prior to painting. This will reduce the danger of corrosion creeping below the paint at the damaged area.

Roller Coating

A technique where the paint is applied by rollers.

Application range ALL METALS

The technique, with modifications, can be used for mass production of relatively simple flat surfaces. In mass production, the paint is applied automatically to rollers which are then rolled across the surfaces. With this, both sides of flat panels can be painted, and with sophisticated apparatus ends and edges can also be coated. This term is also applied to the popular sheepskin, lambswool or absorbent plastic rollers used on occasions by professional house painters and very commonly by amateurs.

Spray Painting
The application of paint as a spray.

Application range ALL METALS

After brush painting, spray painting is the most popular and, without doubt, the largest volume of paint is applied by this method. There are various techniques involved, all of which require that the liquid paint is ejected from a nozzle in the form of atomized droplets. These droplets impinge on the surface and a film of liquid paint is achieved.

This film will have a specified thickness. If it is too thin, complete coverage will not exist or the colour intensity will not be correct, or in the case of corrosion resistance, there will be insufficient pigment to ensure adequate protection. If the film achieved is too thick, this will result in runs if the surface is anything but horizontal. Other defects may include the surface curing prior to the main body and, if thinners are applied, these will evaporate and break down the cured surface film to give the defect known as 'orange peel'.

The techniques used for spray painting can include the use of a steam of air under pressure into which is injected the liquid paint. A variation of this is where the air stream is used to produce a vortex which sucks the liquid paint into the air stream. Other systems make use of pressure containers, which pressurize the liquid paint and this is then ejected through a nozzle in the atomized condition. This airless spraying can be used as a basic method to feed liquid paint at low pressure to the nozzle where it is atomized and ejected by air pressure. It is also possible to use a spinning disc to give the necessary velocity to the paint particles.

It will be appreciated that the correct viscosity of the paint is essential and much of the quality control of spray painting is associated with viscosity. The most popular method is to use a thinner or solvent to reduce the viscosity to the desired value. These thinners in the ideal paint would evaporate prior to the vehicle and pigment reaching the surface being painted, but this is seldom possible except in high-production, single-item components. In the main the paint which reaches the component has some thinners remaining, and this must be allowed to evaporate before the surface cures trapping the thinners which then burst out leaving an uneven surface, and the defect known as 'orange peeling'. With high-production, high-quality spray painting this can be achieved by spraying thinners into the atmosphere at the entry of the stoving oven. The quantity of the thinners in the atmosphere is then progressively reduced, and with this technique it is possible to ensure that the thinners are evaporated from the paint film in a controlled manner before any curing of the surface occurs. These thinners are condensed and recycled, thus there is little increase in running costs.

Another method of controlling viscosity is the 'hot cup', where the container and paint are electrically heated to reduce viscosity. By controlling the temperature and the choice of the original paint, it is often possible to obtain the correct viscosity with the use either of no thinners or a very small quantity. The dual economy of using little or no thinners is achieved together with the ability often to apply a thicker film of paint in a single operation.

A common technique of spray painting is to use the 'wet on wet' method in which a thin film of paint is sprayed and allowed to 'flash off', that is, the components are allowed to stand at atmospheric temperatures to allow the thinners to evaporate. A second coat of paint is then sprayed on to the first wet coat, and again this may be left for the thinners to evaporate prior to the stoving operation. This overcomes the problem of runs, slips and drag when too thick a wet film is applied.

There are a large variety of paint types suitable for spraying, and any liquid paint may be used. These can vary from the standard paint which dries under normal atmospheric conditions within 24 h, but there is an increase in spray painting used prior to stoving. It is generally advisable, as with other painting techniques, to make certain that when different films of paint are to be applied, these are compatible with each other. This is best achieved by the use of a complete paint system recommended by one supplier.

Stove Enamelling

An operation where the paint is subsequently subjected to a heating operation.

Application range ALL METALS

The term 'enamel' was traditionally used for the application of metals or metal oxides which were mixed with powdered glass and fused in position. This technique is still used for ornamental purposes and has some limited industrial application. As long as painting operations used ambient temperatures to effect curing there was no problem in terminology, but with the advent of paints requiring curing at above room temperature the terms 'stoving' and 'stove enamelling' have become commonplace. It should be remembered, then, that these paints have no relation to the original enamels (see *Enamelling*, also *Spray Painting* in this section **Painting**).

Welding

A method of joining where heat and/or pressure is used and where the joint involves 'massive mingling' of the metals joined.

Application range ALL METALS

The term welding now covers a range of procedures, and many non-metallic materials can be joined using one of the many welding processes. It must be appreciated that the expression 'massive mingling' is used metallurgically and indicates 'macro' rather than 'micro'.

The *Encyclopaedia Britannica* defines welding as a method of joining using heat and/or pressure.

The welding processes appear in Part One under their indvidual heading but each is discussed and related in alphabetical order in this section.

The original welding process was carried out by the blacksmith where steel or wrought iron was heated to above red heat and a 'weld' or 'join' was then produced by hammering. This process is a form of *forging* and, when correctly carried out, results in a join which cannot be identified by *non-destructive testing* or metallurgical examination.

When modern techniques were introduced using a gas flame, it was possible actually to melt the edges to be joined and, with or without the use of a filler rod, a join was produced. This technique was given the same name as the join produced by the blacksmith, but in the case of *fusion welding* it is always possible to identify the weld itself, sometimes by non-destructive testing and certainly by metallurgical examination. The fact that two relatively distinct and separate processes are referred to by the same term, can cause some confusion when welding is discussed.

Welding can be carried out over a very wide range of temperatures. When it is appreciated that plastic materials can be successfully welded using hot air at a temperature very little in excess of 100°C, and that this will be a true *fusion welding*, the difficulties of defining processes by temperature alone are realized. It must then be appreciated that materials such as gold, lead, silver and many other similar materials can be successfully welded by *forging* at room temperature.

Welding requires that the material is joined by mingling with itself or the filler material or the material to which it is being joined: this is perhaps the most satisfactory definition of welding as distinct from *brazing* with which on occasion there may be some confusion.

When welding is involved, at least three of the following five characteristics will be present. These are: (1) forging, (2) casting, (3) heat treatment, (4) chemical changes, and (5) physical changes. Each is described below.

1. *Forging* The *hot working* of the material so that, as energy is given to the material, it will immediately recrystallize, eliminating all evidence of *work hardening*. Thus, many materials which recrystallize at or below room temperature can be *forge welded* at room temperature.

2. *Casting* All materials involved in *fusion welding*, where they are taken above the melting point, come within this category. Where *fusion welding* is involved, then the materials being joined, in addition to the filler metal, must be taken above their melting point. It is this melting of the metals to be joined that differentiates welding from *brazing* and *soldering*. The resulting cast structure is weaker than a forged structure of the same material, and there are the additional problems of differential cooling causing excessive grain size, and shrinkage problems resulting in porosity or stressing, and the trapping of gases and oxides in the molten metal.

3. *Heat treatment* It is impossible to carry out the welding of metals without affecting any *heat treatment* already carried out, and without producing some effect on metals which are heat-treatable. It is this characteristic of the present process which has produced the greatest problems, in that *heat treatment* carried out by welding was, until relatively recently, largely uncontrolled resulting in metallurgically undesirable end-products.

Tack welding or 'stray arcing' on cold metal are examples of uncontrolled welding with local undesirable hardening and brittle metal being produced.

4. *Chemical effects* All metals are more reactive at higher temperatures than at lower temperatures. This activity results in a chemical combination with surrounding materials, and as the most common material is air, the result is that oxidation in particular occurs, with other side effects such as the formation of nitrides. Metallic oxides are invariably brittle, refractory materials and if they become trapped the mechanical strength and ductility of the weld is reduced.

The use of fluxes to prevent this is one example of the application of chemistry in the welding process. The flux is designed to prevent the chemical combination of the hot metal with air, and at the same time it can be used to make additions to the weld pool. It is now quite common to achieve alloying of the weld material by adding the necessary elements to the flux. This will better control the alloy content of the weld, and it is a more economical method of alloying than the use of alloy welding rods, which may also be more difficult to weld. Care must then be taken to prevent damage or loss of the flux.

5. *Physical effects* When metal is hot it expands, and when it is cool it contracts. The higher the expansion rate, the greater will be the stress during and after welding. This is because as welding proceeds more and more metal

becomes hot and expands, but it is seldom that the complete assembly will be taken to the welding temperature. Cooling takes place at different rates, depending on the amount of heat which has been applied, and the geometry of the components, and the larger the mass the slower the cooling.

It will, therefore, be seen that considerable stresses can be set up particularly at a change of section, or with structural welds, where a series of local welds are produced, with relatively thin members between. These can be subjected to tensile and compressive stresses. It will be seen that the larger the heat sink produced during welding, the less local will be the stresses involved.

When any welding operation is being considered, each of the above five characteristics should be looked at and its effects during welding given consideration. It is also necessary to consider the effect each may have on the completed component.

The following are the welding processes described in this section:

Argon Arc	Inert Gas Shielded Metal Arc/
'Argonox'	Inert Gas Shielded/Inert Gas/
'Argonshield'	Metal Inert Gas (MIG)
Atomic Arc	Percussion
Carbon Dioxide (CO_2)	Plasma
Electric Arc Welding	Pressure
Electron Beam	Projection
Electropercussion	Resistance
Electroslag	Roller Spot
Explosive	Seam
Flash Butt	Spot
Forge	Stitch
Friction	Submerged Arc
Fusion	Tack
Heliarc	Thermit
High-frequency Induction	Tungsten Inert Gas (TIG)
Hydrogen	

Argon Arc Welding

The method of *fusion welding* where the weld and adjacent area are shrouded by argon gas and, thus, protected from oxidation. This also means that it is the argon which is ionized by the electric arc to give the welding heat. As this is a lesser source of energy than air, it is a slower process than *carbon dioxide welding* or normal electric welding; further details will be found under *Inert Gas Shielded Metal Arc Welding* in this section **Welding**.

'Argonox' Welding

The technique of shielded arc welding where argon is used with a percentage of oxygen present. Argon, being an inert gas, prevents oxidation while the oxygen supplies considerably more energy of ionization than argon or any other inert gas. This is another technique where attempts are made to control the chemistry, that is, the oxidation of the molten weld pool and, at the same time, improve the economy of welding. Costs are reduced, since argon is more expensive than oxygen, and in addition the production rate will be considerably faster because of the higher energy of ionization of oxygen than argon. '*Argonox*' is similar to '*Argonshield*' but more expensive; it is used to stabilize the arc rather than for purely economic reasons.

'Argonshield' Welding

A form of shielded gas welding where the gas shield is air to which a proportion of argon has been added. There are various grades of 'Argonshield', generally with the number indicating the percentage of argon present in the gas.

This process, together with 'Argonox' where the gas is oxygen with argon, is an attempt to give welds of a technical standard close to that achieved by *inert-gas welding*, but at a more economical rate. Economy is improved in two areas. First, costs are much reduced since air is considerably cheaper than argon and, secondly, probably more important, the energy achieved with the arc when air plus argon is ionized is much greater than that achieved with argon or other inert gases. There is evidence that the 'Argonshield' technique is easier to use than *carbon dioxide welding* and under many circumstances is comparable.

Considerable advances are being made in this field and the reader is advised to obtain further information from a reputable supplier. The fact that an oxidizing gas is present will always result in some danger of oxide forming, and this will be in relationship to the tendency of the metal being welded to oxidize. Thus metals such as titanium cannot normally be joined using this technique, and stainless steels, aluminium and similar metals will tend to have some oxide or slag produced during welding.

Atomic Arc Welding

A specialized welding process, where very high but localized heat sources can be obtained using the energy of an electric arc to break down the molecular structure of a gas into the atomic state.

Application range ALL METALS

The gas used is usually hydrogen. This is an unnatural, unstable condition, and reversion to the natural, stable molecular state, releases the collected energy which can be used for *fusion welding* or other processes requiring a very concentrated high temperature. With the use of hydrogen gas, it is possible to obtain additional characteristics such as reduction of oxides.

The temperature attained by this process is extremely high and very localized and, therefore, it finds very limited use. The technique has been largely supplanted by *inert-gas shielded metal arc welding* for joining, or *plasma heating* for cutting. The arc is generally struck between two tungsten electrodes, but it can be struck between one tungsten electrode and the component being welded.

Carbon Dioxide (CO_2) Welding
An *electric arc* method of welding.

Application range STEEL

The arc is struck between a consumable electrode and the material being welded in an atmosphere of carbon dioxide. The welding in some ways can be compared to *inert-gas shielded metal arc welding*, but does not have the same flexibility, and cannot be used on materials such as stainless steel, aluminium, nickel alloys, titanium etc.

Inert-gas shielded metal arc welding can be used on any material, and the inert gas will under no circumstances react with the material being welded. Carbon dioxide welding is useful with mild steel and low-alloy steel, but the carbon dioxide atmosphere will be found to react disastrously with many materials which can be successfully welded using the *inert-gas shielded process*.

Carbon dioxide welding has been developed for high production welding of mild steel. Usually, but not always, it is an automatic type weld, whereby the filler metal, sometimes fluxed but generally bare metal, is fed automatically through the welding torch. Over the surface of the welding wire is passed the carbon dioxide gas, which is supplied at a controlled flow rate from the solid in pressure vessels.

By this method of welding, higher production can be achieved than using a true inert gas. The method is more rapid and economical than inert-gas welding using argon or helium but, as stated above, it is less flexible. In general the process is confined to high-production welds using mild steel of relatively thin gauges up to approximately 0.5 in (10 mm). The method is comparable to but more flexible than *submerged arc welding*, although this latter method is faster and will generally give superior welds, particularly with thicker gauges of metal.

It is possible using carbon dioxide welding to have several welding runs, with a minimum of interpass de-scaling, and thus to produce satisfactory welds on thick mild steel. The energy produced by the carbon dioxide arc is greater than that of argon or helium arcs, but less than that of the arcs produced in air or with flux.

Unless care is taken there can be problems with less than perfect fusion of the weld metal to the base metal giving rise to lack of side wall fusion, which is difficult to identify.

There are now available mixed gases of carbon dioxide and argon; carbon dioxide, argon and air, which attempt to achieve the maximum heat energy from ionization of the gases with some blanket effect to present serious oxidation of the molten weld pool. These gas mixtures are compromises and cannot eliminate oxidation, with production speeds comparable to air or flux ionization. Advances in carbon dioxide techniques are continually being made; the reader is advised to obtain up-to-date information.

Electric Arc Welding
See under this heading in Part One.

Electron Beam Welding
A method of *fusion welding* using high vacuum and a thin beam of electrons.

Application range ALL METALS

In this fusion form of welding, melting of the joint faces occurs; filler metal is seldom used since considerable manipulation problems would arise.

The energy source is an electron gun, similar in many respects to that used in television sets but with a considerably higher energy output. The parts to be welded are placed in a chamber which is then evacuated to very high vacuum, thus energy when released from the electron beam is not dissipated in the intervening space between it and the components being welded. The electron beam width will be in the order of 0.002–0.005 in (0.05–0.1 mm), thus the area affected by the weld is considerably smaller than with conventional welding. The energy used is controllable to a fine degree and can, when necessary, be of a high order. This means that considerable penetration is achieved, greater than 2 in (50 mm) being common.

Thus, electron beam welding is unique in achieving high penetration welding, having an insignificant cast area and an extremely small heat-affected zone. This means that the effect of *casting* and *heat treatment* are virtually unimportant, while the chemical and physical effects are practically non-existent. As stated, electron beam is a relatively new technique, and gives a new dimension to welding design. With this, it is possible for example to use plate and tube material to achieve economically an alternative to *forging*. It is also possible to repair major components using full penetration welding.

It is possible to programme the beam itself to follow a predetermined path, thus making possible repeated intricate welds.

It is also possible to mechanically move the component, or components, under the static beam. A combination of both the above is also possible.

Electropercussion Welding

A specific form of *resistance spot welding*.

Application range METALS WITH GOOD ELECTRICAL CONDUCTIVITY

With these metals, normal *resistance welding* is difficult as the electrical conductivity of the electrode will be close to that of the metal being welded, resulting in non-existent or poor-quality welding of the components. Percussion welding is used to overcome this where the welding current is condenser-discharged at a high voltage. At a predetermined setting, a pulse of electrical energy is discharged to the welding electrodes.

There are other techniques using direct current pulses fed to transformers which achieve the same effect, producing a short-time cycle pulse of high electrical energy. These pulses are repeated at frequent intervals, for example 50 times per second using normal mains frequency. The energy level can be controlled depending on the electrodes used and the metals being joined. Between each electrical energy pulse, mechanical energy is applied as an impact load which *forge welds* the plastic metal in exactly the same manner as conventional *resistance welding*. Electropercussion welding is, therefore, a specialized technique for the *spot resistance welding* of materials such as aluminium, copper, silver and magnesium alloys which have good electrical and heat conductivity, and thus cannot be joined by normal *resistance welding*, information on which will be found in this section.

Electroslag Welding

A form of high-production welding where the weld pool is covered during the cooling cycle to prevent contamination, and the heat input is achieved by resistance heating of the weld flux.

Application range STEEL

This has some similarity to the *submerged arc process* but is possible only on vertical welds. The weld commences on special starting plates placed at the bottom of the pieces being joined. At either side of the vertical gap, which may be up to 1.0 in (25 mm) and can be as much as 1.25 in (30 mm), there are two movable plates or 'shoes' which form a mould with the sides of the plate being welded. As soon as the weld is struck it is covered with powdered flux, and the welding continues by raising the movable shoes at the same rate as the weld metal is produced. The weld metal is then melted by resistance heating between the wire and the molten flux, the heavier molten metal dropping through the flux to produce the fusion weld. The flux can contain active ingredients, capable of carrying out all the necessary chemical functions of a flux, but will seldom be used to control the metallurgical condition of the weld metal.

This form of welding is not unlike *submerged arc welding* with the weld proceeding in a vertical manner but there is no continuous submerged arc with this technique. High-productivity welding of a high integrity can be achieved, provided the control is correctly maintained. The equipment for this is more sophisticated and the control more difficult than with *submerged arc welding*, and the process, as stated, is usually confined to relatively simple, high-productivity, high-integrity welding. This technique is, in fact, casting, and as the energy input is very high but the steel sections involved are generally relatively thick, then the rate of cooling is quite slow. This may result in undesirably high grain size, which can be controlled by correct use of grain-controlled steels and flux additives. The slow rate of cooling is advantageous to *heat treatment* and reducing the stresses imposed by cooling contraction. The fluxes can very successfuly eliminate the atmosphere and any trapping of slag.

Explosive Welding

A relatively new procedure using explosives to give a forge weld.

Application range HIGH-DUCTILITY METALS – GENERALLY MILD STEEL

This, therefore, limits the method to materials such as aluminium, copper and very low carbon mild steel. It is not a satisfactory welding method where materials are hardened by *work hardening*, since a layer is formed which is hard, brittle and does not diffuse to give a satisfactory bond. In many respects the method is similar to *forge welding*, with the application of high impact force.

Briefly, the procedure is that the parts to be joined are placed in a female mould. Either using contact explosives or, under some circumstances, normal explosives, the mould is used as the back-up to withstand the pressure wave caused by the explosion. This results in heavy impact forces which cause *forge welding*. The mould can be immersed in water where the explosive can be made to react against the head of the water. For obvious reasons, the user is advised to obtain specialist advise before attempting explosive welding; further general information will be found under *Forge Welding* in this section **Welding**.

Flash Butt Welding

A form of forge welding which can only be applied to relatively simple shapes.

Application range MOST METALS

It uses the electric arc as a means of heating, but the heat involved will be conducted into the area of the components which subsequently will be welded.

The technique is that the two components are securely clamped, the clamps being of high-conductive material which, in addition to holding them securely, conduct electricity to each of the components. The clamps are positioned so that the surfaces to be joined are in light contact with each other. One clamp is rigidly fixed, while the other is movable. A high electrical current is passed to each component, resulting in an 'arc' where they meet and heating by electrical resistance. The arc will cause some burning and melting of the material at the interface. As this cycle proceeds, one clamp is moved forward at a pre-determined rate to feed and maintain the necessary pressure for resistance heating. This results in heating of the metal between the clamps and this will eventually become plastic.

When materials of different heat conductivity and specific heat are to be joined, the amount of metal protruding from each clamp will be different, to ensure that both pieces reach the desired forging temperature at the same instant. When this predetermined temperature has been reached, the current is then stopped. An upset force is immediately applied to ram the movable clamp against the fixed clamp. This results in upset forging at the interface, and when the correct conditions have been achieved, all the cast and oxidized material produced by arcing will be pushed out of the weld and appear in the weld flash. This flash is removed during subsequent machining.

It will be seen, therefore, that flash butt welding is a form of *forge welding*, where some casting and oxidation is involved at a preliminary stage of the process. Any evidence of cast structure or trapped oxide which appears in the finished weld is indicative of a faulty set-up and the weld will be less than perfect.

One of the advantages of flash butt welding is the possibility of welding different materials resulting in high-integrity joints. This is difficult or impossible with conventional methods of *fusion welding*.

With flash butt welding, all interface oxide is eliminated at the upset part of the cycle and thus no fluxes are necessary. As a reasonable sized area of heating is involved, the problem of quench cracking on steel will seldom be encountered; but when high hardenability steels are to be welded by this technique, it is then desirable to have a pre-welding cycle. Here resistance heating of a much lower order than arc heating can be used to pre-heat the complete component and clamp, thus reducing the danger of quench cracking. The same comments apply to contraction stresses caused by the expansion and contraction of the material during welding. Because the components joined by this technique are relatively small and of simple shape, it is generally possible to carry out controlled *heat treatment* subsequent to welding. This can be used to eliminate any heat treatment problems.

Flash butt welding can produce the type of *forge weld* where the mechanical properties of the weld itself and the heat-affected zone are identical to the basic material. When materials having no strong directional properties shown by flow lines are flash butt welded, then it is possible to produce a weld which is undetectable by metallurgical inspection techniques.

Friction and *inertia welding* are now to some extent replacing flash butt welding as they are more flexible and have a lower capital cost. (This comment applies only to smaller diameters and the simple shapes applicable to the newer techniques.)

Forge Welding
A method of welding which uses heat and pressure only.

Application range ALL METALS WHICH CAN BE FORGED

Cobalt alloys and similar refractory metals which have an impractical, limited *forging* range, are not joined by this method.

This is the original form of welding, used during the Industrial Revolution by craftsmen such as the village blacksmith, and still used in many companies by craftsmen blacksmiths.

Today, the term 'welding' is generally taken to mean *fusion welding*, where casting or melting of the pieces achieves the join. With forge welding, no liquid phase enters into the process.

The parts to be joined must be heated to the *forging* range, which will vary, depending on the materials being joined. For example, with aluminium alloys the range is 400°C–550°C, with copper alloys it is usually 700°C–950°C, and with normal steel it is from 1000°C–1250°C, with some variation at either end of the temperature range depending on the material used. The majority of pure metals, such as aluminium, gold, lead, tin, iron, etc. which do not work harden, can be forge welded at room temperature. Room temperature forge welding of other materials can be carried out to a limited extent, but this will depend on the amount of *cold working* which the material can accept without grain distortion becoming such that diffusion of the grains is not possible.

Forge welding requires that when force is applied to the interface of the two metals, this results in the micro grains joining together and bonding in such a manner that any metallurgical examination will not detect the line of the forge join. This does not preclude the occurrence of some *work hardening* but the cold work will be of a limited extent. Once *work hardening* affects the surface grain to a sufficient extent, the grains will no longer be capable of mingling with those of the other material being joined. In general forge welding is restricted to the joining of similar or identical metals, but there is no technical reason, provided a careful choice is made, why dissimilar metals cannot be joined using this process.

With metals that do not show cold work, or metals which are forge welded in the correct range of *hot working*, then it is impossible to identify metallurgically the resultant join. This, then, is a perfect join which does not affect the mechanical properties of the component. When a limited amount of cold work exists, this can be eliminated by subsequent *annealing, normalizing* or any *heat treatment* where the material involved is recrystallized. All evidence of grain distortion is also eliminated. *Heat treatment* cannot accomplish this, if there has been sufficient *work hardening* to prevent the correct technical forge. It will remain evident and will be a plane of weakness on the completed component, usually showing itself as a crack.

Forge welding is a relatively expensive method of joining, because of the labour involved. Under some circumstances the method can be improved from a production engineering point of view, but it is then generally renamed to indicate the method of heating. Thus, processes such as *resistance welding, flash butt welding, friction welding*, etc. are all examples of forge welding. The term forge welding is, therefore, generally confined to the method of heating, using conventional means such as gas, furnace, induction heating, etc. Force is then applied either with a hand-held hammer, or in a normal forge. For production purposes, forge welding has now largely been replaced by other methods, the most common being *fusion welding*. As stated, forge welding is more expensive, but when correctly controlled will generally result in a high-quality join.

Friction Welding

A form of *forge welding* where friction supplies the necessary heat.

Application range METALS WHICH CAN BE FORGED

This process cannot be applied to materials such as cast iron or cobalt alloys or other materials where the *hot working* range is too small to be practical.

The process makes use of the fact that, if friction is applied to any metal, this will result in the release of energy in the form of heat. This heat is used to increase the temperature of the component to the *forge welding* range. This is achieved, as stated, by friction at the rubbing interface, heat being conducted away from the interface to produce a heat sink. The amount of pressure and time will be related to the temperature at which *hot working* can be carried out.

The procedure is that one of the components is held rigid and static. The second component is revolved at a comparatively high speed in contact with the first. The degree of contact pressure and the speed of revolution will vary with the components being welded and the materials being used. The friction at the interface results in heat being conducted away from this area, thus achieving a heat sink. When the temperature achieved is within the *hot working* range, the rotational motion is stopped and the component which has been revolving is rammed against the static component. This is the *forge welding* part of the procedure and results in oxidized metal, caused during the friction heating cycle, to be rejected in the forge flash.

This form of welding has the considerable advantage that it can be used with completely dissimilar materials, the heating cycle, pressure applied and speed of revolution being adjusted to suit the materials in question. As stated, this is a *forge welding* method, the heating technique allowing use of high-speed machines.

It is comparable to *flash butt welding*, and utilizes less sophisticated equipment with less power and the output can be considerably greater. It is also possible by careful design to use a variety of shapes, giving high-quality joins. Thus, it is no longer necessary to have equal sized bars, but hollow tubes and shapes can be joined with this method, and studs can be successfully welded to plates or larger components. It is also possible to weld three components by rotating two outside components against a static central component.

A similar process is *inertia welding*, for less high-integrity joins; *stud welding* and *resistance welding* can also be used. *Flash butt welding* will generally be more expensive, but where large or awkward parts are involved it will be a more practical proposition. The various types of *fusion weld* will always be possible alternatives, but seldom give the same quality of join.

Fusion Welding

The general name given to all forms of welding, in all materials where the materials being joined are taken above their melting points and two or more areas of molten metal are allowed to fuse together.

Application range ALL METALS WHICH CAN BE CAST

Fusion welding, it will be seen, involves casting. In general the more rapidly a casting is cooled, the finer will be the grain structure, and this normally is a desirable characteristic. At the same time, however, rapid cooling can have serious effects on the structure if the material being welded is *heat treatable*, and also on the stresses imposed because of differential expansion. When a casting is allowed to cool slowly, relatively large grains will be produced, often with strong directional properties, and this will reduce the ductility of the weld but could increase the ductility of the heat affected zone (HAZ).

With fusion welding no *forging* is involved, but it must be appreciated that there are welding techniques, described separately, which make use of fusion and *forging* in the welding process. *Heat treatment* will occur, if the alloy is heat treatable. The same comments apply here as to *forge welding* in that the slower the cooling rate, the less serious will be *heat treatment* effects.

In fusion welding there are a number of techniques for supplying the necessary energy. Each is described under its own heading, and some indication is given regarding possible *heat treatment* effects. Some methods achieve high heat input which affects local areas, resulting in a brittle zone with possible cracking during cooling or in subsequent service. Regarding the chemistry of the components this will invariably have a more serious effect than *forge welding* as *casting* will always occur at a higher temperature than *forging* for any specific material.

During the fusion welding process fluxes are used to a much greater extent than with any other type of welding. The purpose of the flux is generally to remove surface oxide, and to act as a blanket on the weld pool to prevent further oxidation, and also prevent oxidation of metal during arc transfer. Fluxes can also be used to maintain a fluid weld, thus allowing trapped solids to float to the surface. Finally, fluxes may be used to add alloy metals to the weld.

The physical changes caused by fusion welding vary considerably depending on each type of welding. Where high-energy-production welding is involved, very local stresses may be produced, in a similar manner to those produced in *heat treatment*, because of the relatively small heat sink which results. This can under some circumstances cause a buildup of local stresses. The extent and seriousness of this will again vary, depending on the design, sequence of welding and size of the components, but with the lower productive methods, such as *oxyacetylene welding*, there will be a larger heat sink. The affected area is, therefore, larger and the stresses are distributed over a greater surface, resulting in lower stress-raisers. Some designs exist, however, where because of the sequence of welding, a large heat sink results in special stress problems and this must be assessed for all welds.

To summarize, fusion welding is a general name for a number of welding processes, each of which is described in this section under its separate heading. It was the advent of fusion welding which brought the welding process into common use as a method of fabrication, and in its absence the welding industry would not have advanced to its present state.

Heliarc Welding

A method of *fusion welding* using the inert gas helium as a shroud.

Application range ALL METALS WHICH CAN BE WELDED

The term heliarc is sometimes used as an alternative to *argon arc*. The present process is unusual in the UK where argon gas is used; details of this, and *argon arc welding*, will be found under *Inert-gas Shielded Metal Arc welding* in this section **Welding**.

Helium gas is available in some countries, notably the USA.

High-frequency Induction Welding

The use of high-frequency induction to supply the heat source for a weld join.

Application range ALL METALS WHICH CAN BE WELDED

The method of welding varies considerably but heating invariably uses high-frequency techniques. The procedure is that the components are fed into an induction heating coil and the energy is used to heat the material to within the *forging* range. Upset pressure is then applied and, thus, it is seen that high-frequency induction welding is a method of *forge welding*.

Theoretically, there is no reason why this process should not be used in *fusion welding* as it employs sufficient energy to cause melting of most metals. Practical and economic considerations, however, generally prevent *fusion welding* making use of this heating method.

Because of the technical difficulties involved in using this technique, it is generally confined to smaller components in the electronics industry or where high-production welding under carefully controlled temperature conditions is required.

By careful design and control induction heating can be used to ensure that the same heat is used to produce identical temperatures over a large number of components, and under conditions where high-integrity *forge welds* are required, this method is commonly applied. It is comparable in many ways with *friction* and *flash butt welding* regarding the control available, and the quality of the final assembly.

Hydrogen Welding

A form of fusion welding where the hydrogen acts as a shield to prevent oxidation or even reduce surface oxidation already present, as well as being the source of heat.

Application range ALL METALS WHICH CAN BE WELDED

The procedure is that an arc is struck between two electrodes, generally tungsten, in a hydrogen atmosphere. It is the ionization of the hydrogen molecule which results in high temperatures. Because of this, and the obvious danger of hydrogen contamination, this weld procedure is not now commonly used, and *inert-gas shielded metal arc welding* (either MIG or TIG welding) has replaced the majority of its uses. However, where very high local heat, with a strongly reducing atmosphere, is required hydrogen welding may still be used, provided there is no danger of the trapping of atomic hydrogen causing *hydrogen embrittlement*.

Inert-gas Shielded Metal Arc Welding/Inert-gas Shielded Welding/ Inert-gas Welding

Methods used in *fusion welding* where the weld itself is shrouded by an inert gas, namely, argon, helium, krypton, neon, radon or xenon.

Application range ALL METALS WHICH CAN BE WELDED

All these gases have identical chemical characteristics in that they cannot form chemical compounds under any circumstances, hence the term 'inert'. They are used to prevent the chemical reactions which normally occur between hot metal and the atmosphere. In practice only argon and helium are used, and in the UK, argon alone.

Inert-gas welding may be used to join any metal, and is commonly employed for welding more sophisticated metals having an affinity for oxygen. It is seldom economical for use with mild and low alloy steels.

In this method the electrode has a ceramic or metal shroud through which it projects to produce the arc with the metal to be joined. The shroud is used to pass a stream of inert gas around the electrode, and is designed to ensure that the weld pool and adjacent metal are also covered or shrouded. The arc itself, therefore, will be struck in this gas. As the heat of ionization of argon or helium is considerably less than that of air, the energy available for inert-gas welding is lower than for *electric arc welding*. The method will not normally be used where *forge welding* is involved, although there is no reason why argon or helium gas may not be used to prevent oxidation of surfaces during that process, where no weld.

Because of the lower heat input resulting from the poor heat ionization of the arc, it is not possible to produce the large weld pool associated with conventional *arc welding*. This means, then, much lower production speeds. Since rapid production of heavy weld deposits cannot be achieved, the present process is not an economical alternative to *electric arc welding*. This is balanced, however, from a technical viewpoint, in that there is sufficient heat to ensure correct fusion, while insufficient to cause problems from burning, gouging, etc. In addition, it does not permit the production of very local hot areas surrounded by cold metal which cause hardening and brittleness.

Thus, there are fewer *casting* problems and fewer dangers from *heat treatment* using the inert gas process than with normal *arc welding*. The gas blanket precludes air and so the chemistry of the weld is much simplified; fluxes are, therefore, never used with this method. Problems of expansion and contraction are reduced with inert-gas welding as compared with *electric arc*, as the differential between the hottest zone and the cold area will be less dramatic.

In many respects inert-gas welding is comparable to *oxyacetylene welding* and this similarity is considerable when *TIG (tungsten electrode inert-gas welding)* is considered. Here the operator can manipulate the heat source, the ionized inert gas, in a manner very like a gas torch flame.

Briefly, therefore, inert-gas welding produces a *fusion weld* where, because of the low heat input, there will be less danger from *heat treatment* and contraction stresses than with *electric arc fusion welding*. Because the weld is shielded from the atmosphere, any chemical effects are eliminated and this is the principal reason for use of the technique. With modern equipment the gas flow commences before the arc is struck, and continues for a few seconds after the arc is broken. It is also possible to have gas flowing in contact with the underside of the weld, or to fill a box with gas, and weld in this atmosphere.

There are now available techniques where argon–oxygen or argon–carbon dioxide gas mixtures are used. The gas ratios are varied to give improved ionization, thus better heat input, without causing excessive oxide formation. These are '*Argonshield*' and '*Argonox*', usable only on mild and low-alloy steel. The techniques are alternatives to *carbon dioxide welding*.

The terms *inert-gas welding*, *inert-gas shielded metal arc welding* and *inert-gas shielded welding*, as well as *MIG* and *TIG welding*, should be confined to uses with true inert gases. The use of these names in connection with welding using gas shrouds of carbon dioxide, nitrogen, etc. is technically incorrect and may lead to confusion, resulting in expensive faulty welding. The reader is advised to obtain further details on this method from a reputable supply house.

Metal Inert Gas Welding (MIG Welding)

A form of *inert-gas shielded metal arc welding* where the electrode is consumable and is fed at a constant speed through the shield of the torch.

Application range ALL METALS WHICH CAN BE WELDED

Across the metal electrode will flow the inert gas and, provided the speed of the electrode is controlled, a continuous welding process is achieved with the molten weld pool shrouded by the inert gas.

While this may be used as a hand technique, some skill is required to control the arc and the speed of the metal feed wire. It is, therefore, quite common that MIG welding is carried out as part of an automatic process where the welding gun is held on some form of tractor, or the component being welded is moved at a constant speed below the fixed arc. With this technique it is possible to control the length of the arc, and thus ensure that high-quality, consistent welds are produced at all times. It is now possible to electronically control the arc energy with some variation in arc length, thus simplifying the technique for hand control.

The *carbon dioxide shielded arc weld* with a continuously fed wire is commonly known as MIG welding. This is an abuse of the term as carbon dioxide is not an inert gas, and can cause confusion and serious technical problems if carbon dioxide is used on any metal other than steel.

Percussion Welding

A *forge weld* type characterized by impact force being applied. It is a form of *resistance welding*.

Application range COPPER, ALUMINIUM ETC. – HIGH CONDUCTIVITY METALS

With normal *resistance welding* the electrodes used to perform the weld on the metal may have comparable or even higher resistance than the material being welded. Thus, there is every chance that the electrodes will weld themselves to the component instead of, or as well as, welding the components to each other.

To overcome this problem, use is made of the condenser-discharge system, where the electrical energy is built up, and stored in a condenser which is triggered at preset intervals to discharge across the gap between the electrodes. This technique results in resistance heating of the component material, with less heating of the electrodes. When sufficient heat has been achieved, the electrodes are used to impact the weld slug in exactly the same manner as with normal *resistance welding*. The weld produced is identical to that with *spot, seam* or *stitch welding* made with normal resistance heating, that is it is a forge weld.

Obviously, it is theoretically possible to use this technique to continue heating to produce a *fusion weld*, but this is not normal use.

Plasma Welding

A method of welding using plasma as the energy source.

Application range ALL METALS WHICH CAN BE WELDED

With plasma, electrical energy is used to disturb the electron configuration of a gas over a distance. When the source of disturbance is stopped the energy is recovered as heat and very high levels can be achieved, in excess of 30 000°C. Any gas can be used but argon or an argon/air mixture are the most commonly used in welding. The heated gas will have high velocity in addition to heat energy and can be used as a very fine jet. This method of welding is expensive but can be useful where a large mass must be joined to a small component. It is also useful where distortion must be held to a minimum as the heat-affected zone can be very small because of the very small jet flame and the very high temperature. For the same reasons considerable care is necessary to prevent severe metallurgical problems caused by the rapid local cooling resulting from a large mass of cold metal adjacent to a small heated area.

Plasma welding can be an alternative to *electron beam welding* when it is necessary to minimize the areas affected by heat, but it does not require any vacuum equipment. The heat source is expensive but with advances in this technology plasma welding is finding increasing use as a specialist joining technique.

Pressure Welding

A general term given to any form of *forge welding* where the join is made by compression of the materials being joined.

Application range ALL METALS WHICH CAN BE FORGE WELDED

In general pressure welding is carried out at high temperatures, but no melting is involved and it can be carried out as low as room temperature with many metals, particularly those which do not work harden such as pure gold, lead, tin, aluminium, etc.

The process is described in more detail in this section under *forge welding*, but the term pressure welding is generally used for those techniques where the pressure cycle is well defined and can be seen to exist. Very often with *resistance welding* techniques the pressure cycle is not obvious, without some knowledge of the process details. Pressure welding is a term often used loosely to describe other forms of welding, thus some care is necessary in obtaining a more detailed definition of the process being discussed.

Projection Welding

A form of *electrical resistance welding* for sheet metal components.

Application range MILD STEEL, SOME ALUMINIUM AND COPPER BASE ALLOYS

The components are subjected to a pressing operation prior to welding, whereby one of the components to be welded has depressions, dimples or projections formed. This is then assembled with the mating part and electrical resistance heating applied across the dimple. With this, higher contact pressure and a higher electrical current is required than with *resistance welding*. This, then, is a form of *spot resistance welding* where the spots are preformed on the components to be welded.

Resistance Welding

A name given to the joining process using the electrical resistance of the metal involved, to supply heat for the weld process.

Application range ALL METALS WHICH CAN BE WELDED

Resistance welding is more economical and easier to control on sheet metal materials which have a high electrical resistance (low electrical conductivity). With these metals (mild steel is a typical example), using relatively small, high conductivity electrodes, generally in a copper alloy, a high concentration of electrical energy can be passed via the electrodes through the component being joined. The area between the electrodes will become hot, and when the desired heat is reached, a *forge weld* is produced. The heating will take the metal into the plastic range of the material, but should not produce any cast metal. The same electrodes are generally used to apply the compressive force which achieves the *forge weld*.

With higher conductivity metals, such as copper and aluminium, the difference in resistance between the electrode material and the components being welded is much less and considerably higher electrical current is necessary.

Unless care is taken, the resistance between the electrodes and the components will be as great as, or greater than, the resistance between the two components being welded, and thus there is some danger of welding the electrode to the component. This can be overcome by careful choice of electrodes, ensuring that they are clean and in good condition at all times, or by using condenser discharge of the electrical current. This very high electric-current discharge results in resistance heating of the components which are then subsequently *forge welded* as with mild steel. This is termed *percussion welding*.

It will be evident that the thickness of the two materials involved must be similar, and that the thicker the sheet metal the higher the energy required. With a thickness over 0.25 in (5 mm) approximately resistance welding is seldom economical and with even thicker materials technical problems will exist.

The inspection for resistance welding uses the *chisel test*. This can be applied to components but results in scrap, and more generally is carried out on test pieces of the same material and thickness as the components. These are welded under identical conditions to the components and are then prised apart with a chisel. A complete slug of one component must be left. Fracture along the weld line shows a faulty weld. Faults may result from the use of excessive heat causing some fusion or casting at the interface, or insufficient heat where the surfaces are not brought to the temperature required for forging. Investigation is necessary to identify the reason for the fault.

Resistance welding takes various forms, such as *spot, seam, stitch, roller, projection welding*, etc. all of which are described under their own headings in this section. There are very few obvious alternatives without design changes, but *plug welding* could be considered.

Roller Spot Welding
A form of *resistance welding*.

Application range ALL METALS WHICH CAN BE FORGE WELDED – CONFINED TO THIN SHEET METAL

The components pass between two rollers which act as the electrodes. To these electrodes is applied a series of pulses of electrical energy, which result in the resistance heating of the metal between the electrodes. The pressure applied to the electrodes or rollers is such that the heated material is then *forge welded*.

The rollers act as a method of feeding the component and do not, in fact, have a continuous roller movement, but pause while the electrical pulse is applied and pressure then applied. By adjusting the speed and pause-timing, it is possible to have overlapping *spot welds* or welds at any desired distance. The process is also known as *seam welding*.

The process has some advantage over *spot welding* where electrodes hold components at one spot and the welded assembly is then moved to the next area and positioned prior to repeating the welding cycle. It is obvious that only assemblies of reasonably homogeneous shape, such as cylinders or long flat components, are welded using this particular technique.

Seam Welding

A form of *resistance welding* where generally a lap-type forge weld is produced.

Application range ALL SHEET METALS WHICH CAN BE FORGE WELDED

In theory any material which can be *forge welded* may be seam welded, but the higher the electrical resistance of the material being welded the easier the process. The electrodes used are circular and revolve in steps gripping the two materials being welded.

The process consists first of a heating operation, then, when the material between the electrodes is at the correct temperature in the *forging* range, electrical resistance heating stops, and pressure is immediately applied by the electrodes to form a *forge weld*. Immediately on completion of each weld, the electrodes supply the necessary drive to move the components. It is obviously possible to produce a seam weld using normal electrodes, but this means that the assembly must be handled and indexed to produce the second spot. An alternative name for this is *roller spot welding*; further details of welding methods using electrical resistance heating are given under *Resistance Welding* in this section **Welding**.

It should also be noted that the term seam welding may be used to denote any weld join where a seam is produced. Resistance seam welds are tested using the *chisel test*.

Spot Welding

Resistance welding where an individual or 'spot' weld is produced.

Application range ALL SHEET METALS WHICH CAN BE FORGE WELDED

A comparatively wide range of materials can be used with the same restrictions applied as to other forms of *resistance welding*.

The term is sometimes used in place of *stitch* or *tack welding*.

With spot welding, individual electrodes are used to produce a local weld and the components being welded are then relocated for the next spot weld. Thus, the operator controls both welds and location, although the choice is often limited by the use of jigs and automatic or semi-automatic sequences. Spot welds are commonly used during the setting up of more intricate *resistance weld* assemblies where, when high-integrity components are involved, use might be made of what is termed *breaking spot* or *stitch welds*. With this, the weld produced is deliberately below standard so that if the components are not correctly located the weld can be parted or broken without scrapping the components involved.

Spot welds are tested using the *chisel test*, where the weld is parted and must leave a complete witness of one of the components. Fracture along the weld line shows a faulty weld. This would be expected only where a 'breaking' *spot* or *tack weld* had been produced, designed to fail at a low load. Further information on the spot welding process is given under *Resistance Welding* in this section **Welding**.

The term spot welding can be used as an alternative to tack welding when this has local individual spots of welding.

Stitch Welding

Generally a form of *resistance welding* where local 'spots', 'tacks' or 'stitches' are involved. The term can, however, be applied to other forms of welding such as *oxyacetylene* or *electric arc welding* where the weld is made locally, thus a 'stitch'.

Application range ALL METALS WHICH CAN BE WELDED

The purpose of stitch welding is generally to hold components in a temporary manner until full welding or some other form of joining can be accomplished. It can, however, refer to the procedure where interrupted welds are used along straight lengths for economical reasons, when low-strength welding can be accepted.

Stitch or *tack welding* of steel where areas of high local heat are surrounded by cold metal, can lead to serious brittle effects in the heat-affected zone if the steel being welded has any appreciable 'hardenability'. The extent of the problem will be a function of the heat input, carbon and alloy content of the steel, and the relative mass of cold metal adjacent to the weld. The higher the carbon or alloy content, the greater the 'hardenability', and thus the more critical the mass of cold metal required to cause hardness and brittleness.

It must also be realized that each stitch or tack can be a stress-raiser equivalent to the stop and start, or edge of continuous welds.

Stitch or tack welds should therefore never be allowed where there is any danger of impact loading, or where failure of the weld could result in any danger.

Where such local welds are essential on high-integrity components, then preheating is essential, and heat input should be controlled. They should, thus, be treated in a similar manner to full penetration welds under these circumstances.

Submerged Arc Welding

A specialized form of welding where the arc exists below a flux blanket.

Application range MILD AND LOW ALLOY STEEL

It is a mass-production process which necessitates the use of equipment requiring considerable control. It is generally confined to straight welding runs, but can make circumferential welds by rotating the component. With more specialized equipment, slightly more complicated shapes can be joined.

The procedure usually is that the prepared joint has a projection known as a 'starting-pad' where the arc is first struck between this projection and the wire used to make the weld. The wire will be of a specified size and type and will be fed at a controlled rate, depending on the speed of the tractor or the rate of movement used to traverse the weld and the weld current. Immediately after the arc is struck, powdered flux, which completely covers the arc, is fed into the weld pool, hence the name of the process. At this point the tractor commences its journey along the weld, or the material being welded is moved at a controlled speed under the arc. This latter procedure is commonly applied to circular components and results in continuous down-hand welding. The flux can produce a gas having high-energy ionization, thus increasing the energy available at the weld. The continuous weld wire can also take various forms to increase the rate of weld deposition.

The system is subject to considerable modification depending on the circumstances, and sophisticated equipment is available which loosens and sucks up the fused flux or slag. This material can on occasion be sieved and the unused flux returned for re-use. Subsequent weld runs can be applied on top of each other by reversing the direction, provided the flux and slag are correctly removed. In the case of circumferential welds this can be a continuous process, and full-penetration welds up to several inches thick can be produced. This form of welding achieves high-integrity joins at high production rates, provided the geometry of the components is relatively simple. All post-treatment techniques necessary for other forms of welding are required here, although the high heat input which is possible, and the thick plate material, achieve a considerable heat sink, thus reducing the necessity of post-heating with low-carbon steel. It should not, however, be assumed that submerged arc welding does not require pre- or post-heating. It will be appreciated that submerged arc welding can readily be applied only in the down-hand position, since it is not possible to feed the flux to submerge the arc in any other position. Using special asbestos belts, vertical welds in the '3 o'clock' position are possible. This is normally possible only with simple welds using special equipment.

The *electroslag* welding technique, although differing technically, results in vertical welds with characteristics similar to submerged arc welds.

Submerged arc welding is the normal electric arc welding which has been to some extent automated; it is confined to simple welds. Because of its high-quality, high-speed welding is possible and the operator need not be highly skilled. To some extent the use of *carbon dioxide welding* is now competing with submerged arc, but will not normally produce the good-quality high-speed welding possible with the submerged arc process. However, the former method is more flexible and can be used as a hand-welding technique.

Tack Welding

An alternative and more common name for *stitch welding*; the weld is local and often used to hold components together on a temporary basis.

Application range ALL METALS WHICH CAN BE WELDED

Considerable care is necessary when tack welding using the *electric arc process*, in that the heated area is local and surrounded by cold metal. If this has any appreciable hardenability, it can result in quench cracking or local embrittlement; this is discussed under *Stitch Welding* in this section **Welding**.

Thermit Welding

With this weld use is made of the energy, released as heat, from the chemical reaction when aluminium powder is allowed to react with iron oxide. Very high heat is obtained locally and, at the same time, the iron oxide is converted to molten iron.

Application range MILD OR LOW ALLOY STEEL

The filler metal is the molten iron while the aluminium is converted to aluminium oxide which, being lighter, floats to the surface and acts as a flux or slag.

This form of welding is a *casting* technique, where no *forging* is involved. The resultant weld depends to a large extent on the skill of the welder as the technique involves the production of a mould into which is placed the mixture of iron oxide and aluminium in powdered form. By some means, either electrical discharge or the use of chemicals such as gunpowder, this mixture is taken above a certain critical temperature when the reaction starts, and continues spontaneously and rapidly with the evolution of tremendous heat. This heat results in the melting of the adjacent interfaces, thus producing a true *fusion weld*.

It will be seen that the control obtainable once the reaction starts, is limited and, thus, the integrity of the finished weld depends almost exclusively on the design of the mould and the technique employed by the operator regarding quantity of chemicals used, degree of mixing and shape of the weld. At the least the finished product will probably contain some aluminium oxide, trapped within the weld, and it is unlikely that the weld will be completely free from iron oxide porosity. It should be appreciated that this technique precludes the use of any complicated geometry and the weld will require to be relatively simple. It is commonly used for joining rails.

Any control over the expansion and contraction stresses, and any pre- or post-heating necessary to control weld cracking because of hardenability problems can only be achieved using conventional heating techniques, and since thermit welding is generally used only when other techniques are not possible, it will be seen that it is unlikely that these precautions can be exercised.

In general thermit welding is confined to the production of fairly simple, relatively low-quality welds in circumstances where normal welding will be difficult because of access, or where very high local heat is necessary. It is probably the most portable welding method and, thus, can have some application where for any reason the weight of the equipment must be kept to a minimum. An alternative name is *Goldschmidt's process*, where this technique is used as a means of reducing certain metals from their oxides.

Tungsten Inert Gas Welding (TIG Welding)

A form of *inert-gas shielded metal arc welding* where the electrode used is tungsten, and thus is not present in the process after acting as the electrode for striking the arc; hence the name of the method.

Application range ALL METALS WHICH CAN BE WELDED

TIG welding is operated with a hand-held torch, the filler metal, also hand held, feeding the arc produced between the tungsten electrode and the component. When used in this manner, TIG welding has a considerable resemblance to *oxyacetylene welding* with the additional advantage that the argon gas shield is completely inert and, thus, the produced weld is free of oxide.

TIG welding can also be used where there is an integral filler as part of the component. In this case the welding arc is used to melt the filler, producing the necessary weld metal. Further information on the use of inert gases in welding is given under *Inert-gas Shielded Metal Arc Welding* in this section **Welding**.

The use of CO_2 gas is sometimes termed TIG welding but this is not correct as CO_2 is not an inert gas.

PART THREE

Tables for identification and comparison of processes

The following 23 tables are designed to help the reader in two ways:

(a) To identify the name of a treatment, or test, which might be of value or interest because of problems identified in service, or as an aid at the design or development stage.

(b) To identify the metal, or metals, to which the techniques might apply.

Thus a reader who has identified that corrosion is a problem can examine Table 7 which lists the techniques which will improve corrosion resistance. Knowing which metal is involved, the reader can then list the techniques which might be useful, and then refer to these in the book.

These tables are also considered useful in that they bring together, under one heading, all the known names for any technique, and are thus of some help in reminding readers of those techniques available.

The 23 tables are:

1. Adhesion testing processes
2. Alteration to properties – casting, forging etc.
3. Blasting processes
4. Brazing and soldering processes
5. Build-up or additive processes
6. Cleaning treatments
7. Corrosion resistance processes
8. Procedures for evaluating corrosion resistance
9. Decorative treatments
10. Fatigue resistance improvement processes
11. Flaw or crack detection
12. Friction characteristics improvement processes
13. Hardening treatments – increase in tensile strength
14. Heat treatments
15. Inspection processes
16. Joining techniques
17. Metal removal processes
18. Metal identification methods
19. Painting processes
20. Electroplating processes
21. Softening techniques – reduction in tensile strength
22. Strength testing methods
23. Welding processes

524 / TABLES

TABLE 1

Adhesion testing processes

The following tests are used to measure the properties of adhesion. They are therefore relevant to plating, painting, and similar processes. In addition to these processes there will be other similar types of inspection techniques where properties of coatings might be assessed.

Process	Metal	Process	Metal
Abrasive testing	All	Peel test	All
Adhesion testing	All	Pull-off test	All
Bend test	All	Repeated bend test	All
Blasting	All	Solderability testing	Copper, steel
Buffing	All	Taber Abraser	All
Burnishing	All	Wrapping test	All
Haworth test	All		

TABLE 2

Alteration to properties – casting, forging etc.

This section includes all the processes by which the basic properties of a metal can be altered. Thus casting, forging, pressing etc. are included but processes such as hardening or electroplating will not be found in this section.

Process	Metal	Process	Metal
Alloying	All	Hot working	All
Amalgamating	All	Imprest process	Aluminium
Bar drawing	Steel	Kayem process	Zinc
Casting	All	Lost wax process	All
Coining	All	Mechanical alloying	All
Cold drawing	All	Mechanical refining	All
Cold rolling	All	Mechanical working	All
Cold working	All	Mercast process	All
Deep-drawing	All	Plasma cutting	All
Die casting	All	Press forging	All
Dip moulding	All	Punching	All
Drawing	All	Scragging	Steel
Dry-drawing	All	Shell moulding	All
Electro-osmosis	All	Shrink fitting	All
Embossing	All	Sintering	All
Extrusion	Aluminium, copper, steel	Skin pass	Copper, nickel, steel
Flame cutting	All	Smithing	All
Forging	All	Spark erosion	All
Goldschmidt	Aluminium, magnesium	Spinning	Aluminium, copper, steel
Hard-drawing	All	Straightening	All

Process	Metal	Process	Metal
Strain or stress ageing	Aluminium, copper, steel	Stretching	All
		Swageing	All
Strain hardening	All	Temper rolling	Steel
Stretcher straining	All	Work hardening	All

TABLE 3

Blasting processes

This table includes all techniques where any form of blasting is involved.

Process	Metal	Process	Metal
Abrasive blasting	All	Grit blasting	All
Air blasting	All	Liquid honing	All
Airless blasting	All	Plumstone blasting	All
Almen test	All	Satin finish	All
Aquablast	All	Shot blasting	All
Cloud bursting	All	Sieve test	All
Dry blasting	All	Vapour blasting	All
Frosting	All	Walnut blasting	All
Glass bead blasting	All		

TABLE 4

Brazing and soldering processes

This table lists all processes where brazing or soldering are involved. There is no separate list of processes for soldering. All of these techniques are also listed under Table 16 on Joining.

Process	Metal	Process	Metal
Aluminium soldering	Aluminium	Tinning	Copper, steel
Bit soldering	Copper, steel	Torch braze	All
Brass weld	Steel	Ultrasonic soldering	Aluminium
Brazing	All	Vacuum brazing	Copper, nickel, steel
Bronze welding	Steel		
Silver soldering	Copper, nickel, iron, steel	Wave soldering	Copper
		Wiped joint	Lead, tin

TABLE 5

Build-up or additive processes

This table lists techniques used when for specific reasons an increase in size is required. Many of these processes will also appear under other headings such as Electroplating or Welding and there may be a number of processes under yet other headings which could also be used for increasing size.

Process	Metal	Process	Metal
Bronze plating	Steel	Linde plating	All
Buttering	Steel	Mellozing	Steel
Chromium plating	Copper, steel	Metal spraying	All
Dalic plating	All	Metallization	All
Electroforming	All	Metcolizing	Iron, steel
Electroless nickel	Copper, steel	Nichem	Copper, steel
Electroless plating	All	Nickel plating	Steel
Electroplating	All	Penybron plating	Steel
Enamelling	All	Plasma plating	Copper, nickel, steel, titanium
Endurance process	All		
Fescolizing	All	Roller tinning	Copper
Flame plating	All	Sherardizing	Steel
Hard chrome plating	Copper, steel	Silver plating	Copper, steel
Hard facing	All	Soft facing	All
Hard plating	Steel	Spra-bond	All
Hot dip coating	All	Sputtering	All
Impregnation	All	Tinning	All
Kanigen plating	Copper, steel	Tool weld	Steel
Lead–tin plating	Copper, steel, tin	Zinc coating	Steel

TABLE 6

Cleaning treatments

This table lists all of the processes which can be used to remove soil. The reader is advised to refer to the section on Cleaning to ascertain the definition and scope given to cleaning in this book.

Process	Metal	Process	Metal
Abrasive blasting	All	Liquid honing	All
Acid descaling	Steel	Micro-Chem	Steel
Alkaline descaling	Aluminium, steel	Needle descaling	Steel
Aquablast	All	Nitralizing	Steel
Barrelling	All	Pickling	Aluminium, copper, steel
Blasting	All		
Bullard Dunn Process	Steel	Plumstone blasting	Aluminium, steel
Drying	All	Polishing	All
Degreasing	All	Sand blasting	All
Descaling	All	Sawdust drying	All
Electrocleaning	All	Scouring	All
Emulsion cleaning	All	Scratch brushing	All
Flame cleaning	Steel	Shot blasting	All
Flame descaling	Steel	Sodium hydride	All
Flame scaling	Steel	Solvent cleaning	All
Fluxing	All	Solvent degreasing	All
Glass bead blasting	All	Trisec drying	All
Grit blasting	All	Ultrasonic cleaning	All
Hanson–Van Winkle Munning process	Steel	Vapour degreasing	All
		Wire brushing	All
Kolene K4	Steel		

528 / TABLES

TABLE 7

Corrosion resistance processes

This table lists the processes which can be used to improve the corrosion resistance of the metal stated. It will be appreciated that the variation in corrosion conditions is such that considerable additional information will almost invariably be required before there can be any certainty whether or not any of these treatments will be suitable for a specific metal or component.

Process	Metal	Process	Metal
Acid descaling	All	Dry film painting	All
Activation	All	Drying	All
Aldip proces	Steel	Durionizing	Steel
Al-fin process	Steel, iron	Electrodeposition	All
Alochrom	Aluminium, steel, zinc	Electrogalvanizing	Steel
		Electrogranodizing	Steel
Alodine	Zinc	Electroless nickel plate	Copper, steel
Alplate process	Steel	Electroless plating	All
Alrak process	Aluminium	Electrolytic polishing	Steel
Aludip process	Steel	Elphal	Steel
Aluminizing	Steel	Enamel plating	Aluminium
Alzak process	Aluminium	Evaporation	All
Angus Smith process	Iron, steel	Deep anodize	Aluminium
Anodic oxidation	Aluminium	Descaling	All
Anodic protection	Aluminium, steel	Eloxal	Aluminium
Anodizing	Aluminium, magnesium	Enamelling	All
		Endurance process	All
Anolok process	Aluminium	Etching	All
Antifouling	All	Feroxil test	Steel
Atrament process	Steel	Ferrostan	Steel
Banox	Steel	Fescolizing	Copper, steel
Barffing	Steel	Flame cleaning	Steel
Bethanizing	Steel	Flame descaling	Steel
Black anodize	Aluminium	Flame plating	All
Blackening	All	Flame scaling	Steel
Black nickel	Steel	Footner process	Steel
Blasting	All	Foslube	Steel
Blue anneal	Steel	Galvanic protection	Steel
Bonderize	Steel	Galvanizing	Steel
Bower barff	Steel	Galvannealing	Steel
Brytal	Aluminium	Gilding	Copper, steel
Buzzard process	Aluminium	Gold plating	All
Chromium plating	Copper, steel	Granodizing	Steel
Cladding	Steel, aluminium	Hard chrome plating	Copper, steel
Close annealing	Steel	Heat tinting	Copper, steel
Cobalt plating	All	Hot dip coating	All
Cold galvanizing	Steel	Ihrigizing	Iron
Colour anodize	Aluminium	Immersion coating	All
Cromodizing	Steel	Immersion plating	All
Dry film lubrication	All	Impressed-current	All

Process	Metal	Process	Metal
Iron plating	Copper	Sanding	Steel
Japanning	All	Satin finish	Steel
Kanigen plating	Copper, steel	Sawdust drying	Steel
Lacquering	All	Scratch brushing	Steel
Laxol process	Steel	Sealing	Aluminium, cadmium, copper, steel, zinc
Lead plating	Steel		
Mellozing	Steel		
Merilizing	Steel		
Metal-spraying	All	Selenious acid treat	Magnesium
Metallic paint	All	Sendzimir process	Steel
Metallization	All	Sensitizing	All
Metcolizing	Iron, steel	Sermetal	Steel
Micro-chem	Steel	Sermetriding	Steel
Molykote	Steel	Servarizing	Steel
Needle descaling	Steel	Shot blasting	Steel
Nickel plating	All	Siliconizing	Iron, steel
Onera process	Steel	Silver plating	Copper
Palladium plating	All	Sodium hydride	All
Passivation	Aluminium, copper, steel, zinc	Spra-bond	Steel
		Steam blueing	Steel
		Stop off	All
Peen plating	Copper, steel	Stove enamelling	All
Penybron plating	Steel	Temper blueing	Steel
Phosphating	Steel, zinc	Tin plating	Copper, steel
Plasma plating	Steel	Tin–nickel plating	Copper, steel
Polishing	All	Tin–zinc plating	Copper, steel
Porcelain enamelling	Aluminium, copper, steel	Tinning	Copper, steel
		Trisec drying	All
Powder painting	Steel	Vacuum deposition	All
Primer painting	Steel	Vacuum evaporation	All
Progrega	Steel	Vapour phase inhibition (VPI)	Steel
Pyro black	Steel		
Rhodanizing	Copper, gold, silver	Vitreous enamelling	Aluminium, copper, steel
Rhodium plating	Copper, gold, silver	Zinc coating	Steel
		Zinc phosphatizing	Zinc
Roller tinning	Copper	Zinc plating	Steel
Rust proofing	Steel	Zincing	Steel
Ruthenium plating	Copper, silver	Zincote	Steel
Sacrificial protection	Steel		

TABLE 8

Procedures for evaluating corrosion resistance

The processes listed are all involved in the testing of corrosion resistance. All are inspection techniques and as such will also appear in Table 15, Inspection processes.

Process	Metal	Process	Metal
Accelerated corrosion testing	All	Kerns test	All
		Porosity test	All
British non-ferrous test	All	Preece test	Steel
Cass test	Steel	Salt mist test	All
Corrodekote test	Steel	Salt spray test	All
Electrography	Copper, steel	Spark test	All
Ferrite test	Stainless steel	Strauss test	Stainless steel
Fingerprint test	Steel, zinc	Taber test	All
Holiday test	All	Thickness test	All
Huey test	Stainless steel	Weld decay test	Stainless steel
Humidity test	All		

Crack detection

Crack detection will always be a form of flaw detection and all the processes for identifying cracks will be found under Table 11, Flaw or crack detection. These processes will also appear in Table 15, Inspection processes.

TABLE 9

Decorative treatments

These are treatments whereby the visual appearance of a metal can be improved. Many such processes have the single purpose of improving the decorative value but the majority will probably also be used for some other reason, such as improvement of corrosion resistance or increase in surface finish. Thus many of these processes will also appear under other headings.

Process	Metal	Process	Metal
Alzak	Aluminium	Brassing	Steel
Amalgamating	All	Brass plating	Steel
Anti-fouling	Steel	Bright chrome	Steel
Barrelling	All	Bronzing	Steel
Black anodize	Aluminium	Browning process	Steel
Blackening	All	Brytal	Aluminium
Black nickel	Steel	Cladding	All
Black oxide	Steel	Colour anodize	Aluminium
Blasting	All	Dalic plating	All
Blue annealing	Steel	Damascening	Copper, steel
Brass colouring	Copper	Deep anodize	Aluminium

Process	Metal	Process	Metal
Dinanderie	Copper	Peen plating	Copper, steel
Dry blasting	All	Peening	Aluminium, copper, steel
Dry film painting	All		
Electrocolour process	Aluminium	Penetrol black	Steel
Electrodeposition	All	Penybron plating	Steel
Electroforming	All	Platinizing	Copper, steel
Electrogalvanize	Steel	Polishing	All
Electroless nickel plating	Copper, steel	Porcelain enamelling	Aluminium, copper, steel
Electroless plating	All	Powder painting	All
Electrolytic polishing	Aluminium, steel	Quicking	Copper
Electrophoresis/ electropainting	All	Red gilding	Copper
		Re-flowing	Copper, steel
Electroplating	All	Refrigerated anodize	Aluminium
Eloxal process	Aluminium	Repousse process	Copper, gold, lead, silver, tin
Elphal process	Steel		
Embossing	All	Rhodanizing	Copper, gold, silver
Enamelling	All		
Enamel plating	Aluminium	Rhodium plating	Copper, gold, silver
Etching	All		
Evaporation	All	Roller tinning	Copper
Fire gilt process	All	Rose gilding	All
Frosting	All	Sanding	All
Gilding	Copper, steel	Satin finish	All
Glass bead blasting	All	Satin nickel plating	Steel
Gold plating	Copper, steel	Scouring	All
Green gold	Copper, steel	Scratch brushing	All
Harperizing	All	Sermetel	Steel
Heat tinting	Copper, steel	Sermetriding	Steel
Hot dip coating	All	Silk screen printing	All
Immersion coating	All	Silver plate	Copper, steel
Immersion plating	All	Speculum plating	Copper, steel
Japanning	All	Sputtering	All
Kanigen plating	Copper, steel	Steam blueing	Steel
Kuftwork	Copper, steel	Stove enamelling	All
Lacquering	All	Temper blueing	Steel
Levelling	All	Tin plating	Steel
Liquid honing	All	Tin–nickel plating	Steel
Luminous plating	All	Vacuum deposition	All
Metal spraying	All	Vacuum evaporation	All
Metallic painting	All	Vitreous enamelling	Aluminium, copper, steel
Nickel plating	All		
Onera process	Steel	Zinc coating	Steel
Painting	All	Zinc plating	Steel
Palladium plating	All		

Electroplating processes see Table 20

TABLE 10

Fatigue resistance improvement processes

These processes have the characteristic that, by improving the standard of the surface, they improve the fatigue characteristics. Almost invariably these processes will also appear under other headings, generally Table 13, Hardening processes. Many processes in this table also appear in Table 12, Friction characteristics improvement processes, as there is a strong relationship between the physical surface improvement of friction and fatigue resistance.

Process	*Metal*	*Process*	*Metal*
Almen test	All	Peening	All
Aqua blast	All	Plasma nitride	Alloy steel
Auto frettage	Steel	Polishing	All
Banox	Steel	Sanding	All
Barrelling	All	Satin finish	All
Buffing	All	Shimer process	Steel
Burnishing	All	Shorter process	Steel
Carburizing	Steel	Shot peening	All
Deep anodize	Aluminium	Spin hardening	Steel
Electroless nickel plating	All	Sulphinuz	Steel
		Surface hardening	All
Flame hardening	Steel	Tufftride process	Steel
Hardening	All	Wet nitriding	Steel
Malcomizing	Steel		

TABLE 11

Flaw or crack detection

The headings in this table are all related to the identification of material defects. As such they are invariably inspection processes and will also be found in Table 15.

Process	Metal	Process	Metal
Acoustic test	All	Jet test	All
Air pressure test	All	Macroetch test	All
Anodize	Aluminium	Magna flux	Steel
Crack testing	All	Magnetic crack test	Steel
Demagnetizing	Steel	Magnetic particle inspection	Steel
Dye penetrant crack test	All	Paraffin test	All
Eddy current test	All	Pressure test	All
Electrolytic etch	Aluminium, steel	Radiography	All
Electrolytic polishing	All	Reflectogage testing	All
Endurance test	All	Sonic testing	All
Etching	Aluminium, copper, steel	Ultrasonic crack detection	All
Flaw detection	All	X-ray	All
Holiday test	All	Xeroradiography	All
Hydraulic test	All		

TABLE 12

Friction characteristics improvement processes

The friction characteristics are related to the surface only and can be a specific characteristic of material itself. The processes appearing here will invariably also appear under some other heading such as Fatigue resistance, Hardening or Welding.

Process	*Metal*	*Process*	*Metal*
Borax treatment	Steel	Lead–tin plating	Copper, steel, tin
Carburizing	Steel	Lime coating	Steel
Chromizing	Steel	Liquid honing	All
Chromium plating	Copper, steel	Malcomizing	Steel
Dry-film lubrication	All	Metal spraying	All
Durionizing	Steel	Metallization	All
Electroless plating	Copper, steel	Molykote	Steel
Electroless nickel plating	Copper, steel	Noskuff	Steel
		Onera process	Steel
Electrolytic polishing	Aluminium, copper, steel	Phosphating	Steel
		Plasma plating	Steel
De-burring	All	Polishing	All
Deep anodize	Aluminium	Porcelain enamelling	All
Descaling	All	Progrega	Steel
Dry blasting	All	Rhodanizing	Copper, gold, silver
Enamelling	All		
Endurance process	All	Rhodium plating	Copper, gold, silver
Fescolizing	Copper, steel		
Flame descaling	Steel	Sanding	All
Flame hardening	Steel	Satin finish	All
Flame plating	All	Scouring	All
Glass bead blasting	All	Scratch brushing	All
Graphitizing	Steel	Selective carburize	Steel
Hard chrome plating	Copper, steel	Selective hardening	Steel
Hard drawing	All	Sermetriding	Steel
Hardening	Aluminium, copper, steel	Shimer process	Steel
		Shorter process	Steel
Hard facing	All	Silver plating	Copper, nickel, steel
Hard plating	Aluminium		
Harperizing	All	Sintering	All
Ihrigizing	Iron	Soft facing	All
Indium plating	Lead, steel	Surface hardening	All
Induction hardening	Steel	Sulphinuz	Steel
Kanigen plating	Copper, steel	Tin plating	Copper, steel
Ion nitriding	Steel	Tufftride process	Steel
Lacquering	All	Wet nitriding	Steel
Lead plating	All		

Forging or forming processes
These are listed under the heading of Alteration to properties, Table 2. This book is on the treatment of metals and as such does not describe the forging or forming processes in any great detail. The treatments listed are there for reference only.

TABLE 13

Hardening treatments – increase in tensile strength
These processes are all related to an increase in tensile strength or surface hardness. Treatments such as Ageing and Hardening are presented here, whereas treatments such as Solution treat and Tempering, part of the hardening process, will not be found here but under the heading of Softening treatment, Table 21, as these processes improve the ductility. Almost all of these hardening treatments will also be found under some other heading.

Process	Metal	Process	Metal
Aerocase process	Steel	Eutectrol process	Steel
Age hardening	Aluminium, copper, nickel	Fescolizing	Copper, steel
		Flame hardening	Steel
Ageing	Magnesium, stainless steel, titanium	Flame plating	All
		Gas carburizing	Steel
		Graduated hardening	Steel
Air hardening	Steel	Grain refining	Steel
Alloying	All	Hard chrome plating	Copper, steel
Ammonia carburize	Steel	Hard drawing	All
Artificial ageing	(see Ageing)	Hardening	All
Austemper	Steel	Hard plating	All
Auto-frettage	Steel	Homocarb process	Steel
Bar drawing	Steel	Induction hardening	Steel
Black anodize	All	Interrupted ageing	Aluminium, copper, steel, titanium
Burnishing	All		
Cold rolling	All		
Cold working	All	Interrupted quench	Steel
Cyanide hardening	Steel	Malcomizing	Steel
Deep anodize	All	Maraging	Steel
Deep-drawing	All	Marquenching	Steel
Dew point control	Steel	Martempering	Steel
Dispersion hardening	Aluminium, copper, steel, titanium	McQuaid Ehn test	Steel
		Mechanical working	All
		Metal spraying	All
Double refining	Steel	Metallization	All
Dry cyaniding	Steel	Natural ageing	All
Dry-drawing	All	Ni-carbing	Steel
Electroless nickel plating	Copper, steel	Nichem	Copper, steel
		Nitrarding	Steel
Enamelling	All	Nitration	Steel
Endurance process	All	Nitriding	Steel *contd*

Process	Metal	Process	Metal
Nitrogen hardening	Steel	Selective carburize	Steel
Oil hardening	Steel	Selective hardening	Steel
Pack carburize	Steel	Self hardening	Steel
Patenting	Steel	Shimer process	Steel
Peening	All	Shorter process	Steel
Plasma plating	Aluminium, copper, steel, titanium	Shot Peening	All
		Siliconizing	Steel, iron
		Skin pass	Copper, nickel, steel
Porcelain enamelling	Aluminium, copper, iron	Snead process	Steel
Pot quench	Steel	Spin hardening	Steel
Precipitation hardening	Aluminium, copper, magnesium, nickel	Spinning	Aluminium, copper, steel
		Spra-bond	All
		Strain or stress ageing	Aluminium, copper
Progressive ageing	Aluminium, copper	Strain hardening	All
Quench hardening	Steel	Stretcher straining	All
Quench tempering	Steel	Sub-zero treatment	Steel
Recarburization	Steel	Sulphinuz	Steel
Refining	Steel	Surface hardening	All
Refrigerated anodize	All	Tufftride process	Steel
Sandberg	Steel	Vitreous enamelling	Aluminium, copper, iron
Schori process	All		
Scragging	Steel	Wet nitriding	Steel
Secondary hardening	Steel	Work hardening	All

TABLE 14

Heat treatments

This lists all of the heat treatment processes given in this book. Many of these treatments will appear under a separate table heading such as Hardening or Softening. Where a proprietary treatment exists which is a relatively little known name, this has not been included as it is felt that readers will use this section for general information, and not to obtain the trade name of a specific process.

Process	Metal	Process	Metal
Aerocase process	Steel	Interrupted ageing	Aluminium, copper, titanium
Age hardening	Aluminium, copper, magnesium		
		Interrupted quench	Steel
Ageing	Nickel, stainless steel, titanium	Inverse annealing	Steel
		Ion nitriding	Steel
		Isothermal annealing	All
Air hardening	Steel	Lead annealing	Steel
Ammonia carburize	Steel	Lead patenting	Steel
Annealing	All	Magnetic annealing	Steel
Artificial ageing	(see Ageing)	Malcomizing	Steel
Austempering	Steel	Malleabilizing	Iron, nickel
Balling	Steel	Maraging	Steel
Black annealing	Steel	Marquenching	Steel
Blue annealing	Steel	Martempering	Steel
Box annealing	All	McQuaid Ehn test	Steel
Bright annealing	All	Natural ageing	All
Close annealing	All	Negative harden	Steel
Cyanide hardening	Steel	Ni-carbing	Steel
Cyclic anneal	Steel	Nickel ball test	Steel
Differential heating	Steel	Nitriding	Steel
Double refining	Steel	Nitration	Steel
Dry cyaniding	Steel	Nitrogen hardening	Steel
Flame anneal	All	Noskuff process	Steel
Flame hardening	Steel	Oil hardening	Steel
Fluidized bed heating	All	Overageing	Aluminium, copper, nickel, stainless steel, titanium
Full anneal	Steel		
Galvannealing	Steel		
Gas carburizing	Steel		
Graduated hardening	Steel	Pack annealing	All
Grain refining	Steel	Pack carburizing	Steel
Graphitizing	Steel	Patenting	Steel
Hardening	All	Post heating	Steel
Homocarb process	Steel	Pot annealing	All
Homogenizing	All	Pot quenching	Steel
Quenching	Steel	Precipitation hardening	Aluminium, copper, magnesium, nickel, stainless steel, titanium
Hot working	All		
Hydrogen annealing	All		
Induction hardening	Steel		
Induction heating	All		

contd

Process	Metal	Process	Metal
Pre-heating	All	Solution treat	Aluminium, copper, magnesium, nickel
Process annealing	Steel		
Progressive ageing	Aluminium, copper, stainless steel		
		Spheroidize anneal	Steel
Quench hardening	Steel	Spin hardening	Steel
Quench tempering	Steel	Stabilize anneal	Steel
Recarburization	Steel	Stabilizing	Steel
Refining	All	Step anneal	Steel
Sandberg treatment	Steel	Strain or stress ageing	Aluminium, copper, steel
Seasoning	Aluminium, copper, iron, steel		
		Stress-equalization anneal	Nickel
Secondary harden	Steel		
Selective annealing	All	Stress-relieving/ stress-releasing	All
Selective carburizing	Steel		
Selective hardening	Steel	Subcritical anneal	Steel
Self-annealing	Steel	Subzero treatment	Steel
Self-hardening	Steel	Sulphinuz	Steel
Shallow hardening	Steel	Surface hardening	Aluminium, Steel
Shimer process	Steel	Temper blueing	Steel
Short-cycle annealing	Steel	Tempering	Steel
Shorter process	Steel	Thuriting	Steel
Siliconizing	Steel, iron	Tufftride process	Steel
Skin annealing	Nickel	Weathering	Aluminium, copper, iron, steel
Snead process	Steel		
Softening	All		
		Wet nitriding	Steel

TABLE 15

Inspection processes

Under this heading will be found all of the treatments involved in the inspection or quality control of metals and metal treatments. Many of these processes are destructive in nature and as such might not be truly classed as inspection but for the purpose of this table it was felt that any process used in quality control should appear.

Process	Metal	Process	Metal
Abrasion test	All	Anodizing	All
Accelerated corrosion testing	All	BNF test	All
		British non-ferrous jet test	All
Acoustic test	All		
Adhesion test	All	Cloud bursting	Steel
Air pressure test	All	Compression test	All
Almen test	All	Corrodekote test	Steel
Anodic etching	All	Creep test	All

Process	Metal	Process	Metal
Cupping test	All	Nickel fracture test	All
Demagnetization	Steel	Notch bar test	All
Dew-point control	Steel	Olsen test	All
Drifting test	All	Paraffin test	All
Drop test	All	Peel test	All
Dye-penetrant crack testing	All	Porosity testing	All
		Preece test	Steel
Eddy current sorting	All	Pressure test	All
Electrography	All	Pull-off test	All
Electrolytic etch	Aluminium, copper, steel	Radiography	All
		Reflectogage testing	All
Electrolytic polish	Aluminium, copper, nickel, steel	Repeated bend test	All
		Reverse bend test	All
		Rivet test	All
Elongation	All	Salt-mist testing	All
Endurance test	All	Salt-spray testing	All
Erichsen test	All	Sankey test	Steel
Etching	All	Scratch test	All
Falling weight test	All	Shape-strength test	All
Feroxil testing	Stainless steel	Shepherd test	Steel
Ferrite testing	Stainless steel	Shock test	All
File testing	Steel	Sieve test	All
Fingerprint testing	All	Silver ball test	Steel
Flattening test	All	Snarl test	Steel
Flaw detection	All	Solderability test	Copper, lead, steel, tin
Flex testing	All		
Fracture test	All	Sonic testing	All
Fuess test	All	Spark testing	All
Hardness testing	All	Spread test	Copper, lead, steel, tin
Holiday test	All		
Huey test	Steel	Spring-back test	All
Hull cell test	All	Steelascope testing	All
Humidity testing	All	Strain gauging	All
Hydraulic pressure test	All	Strauss test	Stainless steel
Jet test	All	Stress-rupture test	All
Jominy test	Steel	Stromeyer test	All
Keller's spark test	All	Sulphur printing	Steel
Kenmore process	Steel	Supersonic testing	All
Kern's test	Steel	Taber abraser	All
Macroetch test	All	Tensile test	All
Magna flux	Steel	Twisting test	All
Magnetic crack test	Steel	Ultrasonic crack detection	All
Magnetic particle inspection	Steel	Weld-decay test	Stainless steel
McQuaid Ehn test	Steel	Wohler test	All
Mechanical testing	All	Wrapping test	Steel
Metascope testing	All	X-ray	All
Nick break test	All	Xeroradiography	All
Nickel ball test	Steel		

TABLE 16

Joining techniques

Under this heading will be found the treatments which are used to join metallic materials together. Also in this section will be found specific treatments which are only applied when joining techniques are used.

Process	Metal	Process	Metal
Airomatic welding	All	Hard soldering	All
Al-fin process	Iron, steel	Heliarc welding	All
Arc welding	Steel	High-frequency induction welding	All
Argon arc welding	All		
Argonaut welding	Steel	Hydrogen brazing	All
Atomic arc welding	Steel	Hydrogen welding	All
Autogenous welding	Steel	Impregnation	All
Bit soldering	Copper, steel	Induction brazing	All
Braze welding	Copper, steel	Inert-atmosphere furnace brazing	All
Brazing	All		
Bronze welding	Copper, steel	Inert-gas shielded metal arc welding	All
Butt welding	All		
Controlled-atmosphere furnace brazing	Copper, nickel, steel	Inertia welding	All
		Integral welding	All
		Koldweld	All
Cladding	Aluminium	Lap welding	All
Die welding	Steel	Laser welding	All
Diffusion bonding	All	Loctite	All
Dip brazing	All	Metallic arc welding	All
Dip soldering	All	Metalock process	All
Dot welding	Steel	MIG welding	All
Dri-loc	All	Oven soldering	Copper, steel
Electric arc welding	Steel	Oxyacetylene welding	Aluminium, copper, steel
Electron beam welding	All		
Electropercussion welding	Aluminium, copper	Redux process	All
		Reflowing	Copper, steel, tin
		Resistance soldering	Copper, steel
Electroslag welding	Steel	Resistance welding	All
Explosive welding	All	Rivet welding	All
Explosive riveting	All	Roller spot welding	All
Flash butt welding	All	Shrink fitting	All
Flow soldering	Copper	Silver soldering	Copper, nickel, iron, steel
Forward welding	All		
Forge welding	All	Solder plating	Copper, nickel, iron, steel
Friction welding	All		
Furnace brazing	Aluminium, copper, nickel, steel	Soldering	All
		Spot welding	All
		Stitch welding	All
Fusion welding	All	Stud welding	All
Gas welding	Aluminium, copper, steel	Submerged arc welding	Steel
		Tack welding	All
Hammer welding	All	Thermit welding	Steel
Hand brazing	All	TIG welding	All

Process	Metal	Process	Metal
Tinning	Copper, steel	Upset welding	Steel
Torch brazing	All	Vacuum brazing	Aluminium, copper, nickel, steel
Ultrasonic soldering	All		
Unionmelt welding	Steel		
Unshielded metal arc welding	All	Wave soldering	Copper
		Wiped joint	Lead, tin

TABLE 17

Metal removal processes

These are treatments where the specific purpose of the process or its actual action results in removal of metal. Under most circumstances there will be more economical techniques, such as turning, grinding, milling etc. which should be considered before examining those listed.

Process	Metal	Process	Metal
Abrasive blasting	All	Fadgenizing	All
Acid descaling	All	Flame cutting	All
Air blasting	All	Grit blasting	All
Airless blasting	All	Harperizing	All
Alkaline descaling	Steel	Liquid honing	All
Aqua blasting	All	Macroetch test	All
Barrelling	All	Madsenall process	Steel
Blasting	All	Method X	Steel
Buffing	All	Polishing	All
Chemical machining	All	Plasma cutting	All
Deburring	All	Rumbling	All
Descaling	All	Sand blasting	All
Dry blasting	All	Sanding	All
Electrochemical machining	All	Scouring	All
		Scratch brushing	All
Electrolytic etch	All	Shot blasting	All
Electrolytic polishing	All	Spark erosion	All
Etching	All	Vapour blasting	All

Mechanical testing

All the mechanical testing processes appear under Table 22, Strength testing processes. Almost all of them will also be found under Table 15, Inspection processes.

TABLE 18

Metal identification methods

This lists the different techniques found in the book on methods of identifying metals.

Process	Metal	Process	Metal
Eddy current sorting	All	Macroetch test	All
Etching	Steel	Magnet test	Steel
File test	All	Metascope testing	All
Fuess testing	All	Steelascope testing	All
Hardness testing	All	Sulphur printing	Steel
Keller's spark test	All		

TABLE 19

Painting processes

All the painting techniques or methods covered in the book are listed here, together with other processes related to painting.

Process	Metal	Process	Metal
Cold galvanizing	Steel	Metallic painting	All
Dry-film lubrication	All	Molykote	Steel
Dry-film painting	All	Porosity testing	All
Electrophoresis/ electropainting	All	Powder painting	All
		Primer painting	All
Footner process	Steel	Progrega	All
Holiday test	All	Pull-off test	All
Hot-dip coating	All	Ransburg process	All
Japanning	All	Sermetal	Steel
Kern's test	Steel	Sermetriding	Steel
Lacquering	All	Silk-screen printing	All
Lime coating	Steel	Stop-off	All
Luminous painting	All	Stove enamelling	All

TABLE 20

Electroplating processes

These processes involve the electroplating of metals; all the metals which can be electroplated are listed here along with the associated techniques used directly in the plating process. No attempt has been made to table the proprietary names which are commonly found in electroplating as it is felt that the book will be used to find the definition of these proprietary processes, whereas the table will be used to identify a basic process.

Process	Metal	Process	Metal
Activation	All	Lead–tin plating	Copper, tin
Alloy plating	All	Levelling	All
Aluminium plating	Steel	Madsenell process	Steel
Back etching	Steel	Nickel plating	All
Barrel plating	All	Palladium plating	Copper, steel
Bethanizing	Steel	Pfanhausers Plating	Copper, steel
Black nickel	Steel	Pillet plating	Copper, steel
Brass plating	Steel	Platinum plating	Copper, gold, steel
Bright chrome plating	Copper, steel		
Bronze plating	Steel	Porosity testing	All
Cadmium plating	Steel	Red gilding	Copper
Chromium plating	Copper, steel	Rhodanizing	Copper, gold, silver
Cobalt plating	Steel		
Contact tin plating	Copper, steel	Rhodium plating	Copper, gold, silver
Copper plating	All		
Electrodeposition	All	Rosc gilding	Copper
Electroforming	All	Ruthenium plating	All
Electrogalvanize	Steel	Satin finish	Steel
Electro-osmosis	All	Satin nickel plating	Steel
Electroplating	All	Silver plating	Copper, nickel, steel
Fadgenizing	Zinc		
Ferrostan	Steel	Solder plating	Copper, steel
Dalic plating	All	Speculum plating	Copper, steel
Fescolizing	Copper, steel	Stop-off	All
Gold plating	Copper, steel	Thickness testing	All
Green gold	Copper, steel	Tin plating	All
Hard chrome plating	Copper, steel	Tin–copper plating	Copper, iron, nickel, steel
Hull cell test	All		
Hydrogen embrittlement	Steel	Tin–lead plating	Copper, iron, nickel, steel
Indium plating	Lead	Tin–nickel plating	Copper
Iron plating	Copper	Tin–zinc plating	Copper, steel
Lead plating	All	Zinc plating	Steel

Production of metals

The methods by which metals are produced are listed under Table 2, Alteration to properties. This book is on the treatment of metals and does not describe in detail the techniques used for production of metals.

TABLE 21

Softening treatments – reduction in tensile strength

The treatments listed in this table all result in a reduction in hardness or tensile strength. In most cases this will involve an increase in ductility. These processes also appear in Table 14, Heat treatment.

Process	Metal	Process	Metal
Annealing	All	Process annealing	Steel
Balling	Steel	Selective anneal	Steel
Black annealing	Steel	Self anneal	Steel
Box annealing	All	Short cycle annealing	Steel
Bright anneal	All	Skin annealing	Nickel
Close annealing	All	Soft facing	All
Cyclic anneal	Steel	Softening	All
Decarburizing	Steel	Solution treatment	Aluminium, copper, magnesium, nickel, stainless steel, titanium
Double temper	Steel		
Drawing	Steel		
Drawing-back process	Steel		
Flame annealing	All		
Full anneal	All	Spheroidize anneal	Steel
Galvannealing	Steel	Stabilize anneal	Steel
Homogenizing	Steel	Stabilizing	Steel
Hydrogen annealing	All	Step anneal	Steel
Isothermal annealing	Steel	Stop-off	Steel
Lead annealing	Steel	Stress-equalization anneal	Nickel
Lead patenting	Steel		
Magnetic anneal	Steel	Stress-relieving/stress releasing	All
Malleabilizing	Iron, nickel		
Negative hardening	Steel	Subcritical anneal	Steel
Overageing	Aluminium, copper, nickel, titanium	Temper blueing	Steel
		Tempering	Steel
		Thuriting	Steel
Pack anneal	All	Weathering	Steel
Pot annealing	All		

Surface build-up processes

These are listed in Table 5, Build-up or additive processes.

TABLE 22

Strength testing methods
This section lists all the different forms of mechanical test and, in addition, other techniques used to assess the strength or other physical properties of metals.

Process	Metal	Process	Metal
Air-pressure testing	All	Mesnager test	All
Bailey's creep test	Steel	Nick break test	All
Barr–Bardgett creep test	Steel	Nickel fracture test	All
		Notch bar test	All
Bend test	All	Olsen test	All
Compression test	All	Proof stress/proof strength test	All
Crack opening displacement	All	Repeated bend test	All
Creep test	All	Reverse bend test	All
Cupping test	All	Root bend test	All
Ductility measurements	All	Sankey test	Steel
Elasticity test	All	Schnadt test	All
Elongation	All	Shape-strength test	All
Endurance test	All	Shock test	All
Erichsen test	All	Snarl test	Steel
Fatigue testing	All	Spring-back test	All
Flattening test	All	Strain gauging	All
Flex testing	All	Stress-rupture test	All
Fracture test	All	Stromeyer test	All
Fremont impact test	All	Tensile test	All
Hardness testing	All	Twisting test	All
Impact tests	All	Wohler test	All
Izod test	All	Wrapping test	Steel
Jominy test	All	Young's modulus	All
Mechanical testing	All		

Strengthening
The techniques for strengthening metals are listed under Table 13, Hardening treatment.

TABLE 23

Welding processes
All the different welding techniques and processes are listed here. Welding is a major section of the book and the principal techniques of welding are brought together under that heading.

Process	Metal	Process	Metal
Arc welding	All	Inertia welding	All
Aluminothermic welding	Steel	Integral welding	All
		Koldweld	All
Arc welding	Steel	Lap welding	All
Argon arc welding	All	Laser welding	All
Argonaut welding	Steel	Metallic arc welding	All
Atomic arc welding	Steel	MIG welding	All
Autogenous welding	Steel	Nertalic process	All
Bronze welding	Copper, steel	Oxyacetylene welding	Copper, steel
Buttering	Steel	Percussion welding	Aluminium, copper
Butt welding	All		
Cold welding	All	Plug welding	Steel
Electric arc welding	Steel	Plasma welding	All
Electron beam welding	All	Postheating	Steel
Electropercussion welding	Aluminium, copper	Preheating	Steel
		Project welding	All
		Resistance welding	All
Electroslag welding	Steel	Rivet welding	All
Explosive welding	All	Roller spot welding	All
Die welding	All	Seam welding	All
Dot welding	All	Shielded arc welding	All
Endurance process	All	Skip welding	All
Flame cutting	Steel	Soft facing	All
Flash butt welding	All	Spot welding	All
Forward welding	All	Stitch welding	All
Forge welding	All	Stud welding	All
Friction welding	Steel	Submerged arc welding	Steel
Fusion welding	All	Tack welding	All
Gas welding	All	Techrotherm Rokos process	Steel
Goldschmidt process	Steel		
Hammer welding	All	Thermit welding	Steel
Hard facing	All	TIG welding	All
Heliarc welding	Steel	Unionmelt welding	Steel
High-frequency induction welding	All	Unshielded metal arc welding	All
Hydrogen embrittlement	Steel	Upset welding	Steel
		Weibel process	Aluminium, copper, steel
Hydrogen welding	Steel		
Inert-gas shielded metal arc welding	All	Wiped joint	Lead, tin
		Zerener process	Steel

PART FOUR

Appendices: useful information

APPENDIX 1

Elements with their symbols and atomic numbers

Element	Symbol	Atomic number	Element	Symbol	Atomic number
Actinium	Ac	89	Lead	Pb	82
Aluminium	Al	13	Lithium	Li	3
Americium	Am	95	Lutetium	Lu	71
Antimony	Sb	51	Magnesium	Mg	12
Argon	Ar	18	Manganese	Mn	25
Arsenic	As	33	Mendelevium	Md	101
Astatine	At	85	Mercury	Hg	80
Barium	Ba	56	Molybdenum	Mo	42
Berkelium	Bk	97	Neodymium	Nd	60
Beryllium	Be	4	Neon	Ne	10
Bismuth	Bi	83	Neptunium	Np	93
Boron	B	5	Nickel	Ni	28
Bromine	Br	35	Niobium	Nb	41
Cadmium	Cd	48	Nitrogen	N	7
Caesium	Cs	55	Nobelium	No	102
Calcium	Ca	20	Osmium	Os	76
Californium	Cf	98	Oxygen	O	8
Carbon	C	6	Palladium	Pd	46
Cerium	Ce	58	Phosphorus	P	15
Chlorine	Cl	17	Platinum	Pt	78
Chromium	Cr	24	Plutonium	Pu	94
Cobalt	Co	27	Polonium	Po	84
Copper	Cu	29	Potassium	K	19
Curium	Cm	96	Praseodymium	Pr	59
Dysprosium	Dy	66	Promethium	Pm	61
Einsteinium	Es	99	Protactinium	Pa	91
Erbium	Er	68	Radium	Ra	88
Europium	Eu	63	Radon	Rn	86
Fermium	Fm	100	Rhenium	Re	75
Fluorine	F	9	Rhodium	Rh	45
Francium	Fr	87	Rubidium	Rb	37
Gadolinium	Gd	64	Ruthenium	Ru	44
Gallium	Ga	31	Samarium	Sm	62
Germanium	Ge	32	Scandium	Sc	21
Gold	Au	79	Selenium	Se	34
Hafnium	Hf	72	Silicon	Si	14
Helium	He	2	Silver	Ag	47
Holmium	Ho	67	Sodium	Na	11
Hydrogen	H	1	Strontium	Sr	38
Indium	In	49	Sulphur	S	16
Iodine	I	53	Tantalum	Ta	73
Iridium	Ir	77	Technetium	Tc	43
Iron	Fe	26	Tellurium	Te	52
Krypton	Kr	36	Terbium	Tb	65
Lanthanum	La	57	Thallium	Tl	81
Lawrencium	Lw	103	Thorium	Th	90

contd

Element	Symbol	Atomic number	Element	Symbol	Atomic number
Thulium	Tm	69	Xenon	Xe	54
Tin	Sn	50	Ytterbium	Yb	76
Titanium	Ti	22	Yttrium	Y	39
Tungsten	W	74	Zinc	Zn	30
Uranium	U	92	Zirconium	Zr	40
Vanadium	V	23			

APPENDIX 2

Metallic elements with their melting points and specific gravities

Metal	Melting point °C	Specific gravity	Metal	Melting point °C	Specific gravity
Aluminium	658	2.7	Osmium	3050	22.5
Antimony	630.5	6.62	Palladium	1552	12.02
Barium	704	3.66	Platinum	1769	21.45
Beryllium	1285	1.844	Plutonium	640	19
Bismuth	271.3	9.8	Potassium	63.7	0.86
Cadmium	321	8.64	Radium	700	5.0
Caesium	28.4	1.89	Rhenium	31.67	21.2
Calcium	851	1.54	Rhodium	1960	12.4
Cerium	795	6.7	Rubidium	39	1.525
Chromium	1860	7.19	Ruthenium	2250	12.3
Cobalt	1493	8.8	Selenium	220	4.81
Copper	1083	8.96	Silver	960.8	10.5
Gallium	29.8	5.91	Sodium	97.8	0.971
Germanium	958	5.32	Strontium	770	2.6
Gold	1063	19.32	Tantalum	2950	16.65
Indium	156.4	7.31	Tellurium	452	6.24
Iridium	2443	22.65	Thallium	304	11.85
Iron	1535	7.9	Thorium	1800	11.3
Lead	327	11.34	Tin	231.84	7.2 (Beta tin)
Lithium	186	0.53	Titanium	1680	4.5
Magnesium	650	1.74	Tungsten	3370	19.3
Manganese	1244	7.44	Uranium	168	19.07
Mercury	38.87	13.55	Vanadium	1.735	6.11
Molybdenum	2620	10.3	Zinc	419.5	7.1
Nickel	1455	8.88	Zirconium	1850	6.53
Niobium	2468	8.6			

Note:
Specific gravity is the ratio of weight to volume. This is the same figure as the density expressed as grams per cubic centimetre.

APPENDIX 3

Temperature conversion table

To convert any temperature in Celsius or Fahrenheit degrees to the other, take the figure in the centre column, and if converting from Celsius to Fahrenheit, read off to the right-hand column; if converting from Fahrenheit to Celsius, read off to the left-hand column.

Example: 720°C = 1328°F
1200°F = 649°C

Formula for conversion:
Celsius to Fahrenheit = °C × 1.8 + 32 = °F
Fahrenheit to Celsius = °F − 32 ÷ 1.8 = °C

°C		°F	°C		°F	°C		°F
−128.9	−200	−328	−58.3	−73	−99.4	−37.8	−36	−32.8
−123.3	−190	−310	−57.8	−72	−97.6	−37.2	−35	−31
−117.8	−180	−292	−57.2	−71	−95.8	−36.7	−34	−29.2
−112.2	−170	−274	−56.7	−70	−94	−36.1	−33	−27.4
−106.7	−160	−256	−56.1	−69	−92.2	−35.6	−32	−25.6
−101.1	−150	−238	−55.6	−68	−90.4	−35	−31	−23.8
−95.6	−140	−220	−55	−67	−88.6	−34.4	−30	−22
−90.0	−130	−202	−54.4	−66	−86.8	−33.9	−29	−20.2
−84.4	−120	−184	−53.9	−65	−85	−33.3	−28	−18.4
−78.9	−110	−166	−53.3	−64	−83.2	−32.8	−27	−16.6
−73.3	−100	−148	−52.8	−63	−81.4	−32.2	−26	−14.8
−72.8	−99	−146.2	−52.2	−62	−79.6	−31.7	−25	−13
−72.2	−98	−144.4	−51.7	−61	−77.8	−31.1	−24	−11.2
−71.7	−97	−142.6	−51.1	−60	−76	−30.6	−23	−9.4
−71.1	−96	−140.8	−50.6	−59	−74.2	−30	−22	−7.6
−70.6	−95	−139	−50	−58	−72.4	−29.4	−21	−5.8
−70.0	−94	−137.2	−49.4	−57	−70.6	−28.9	−20	−4
−69.4	−93	−135.4	−48.9	−56	−68.8	−28.3	−19	−2.2
−68.9	−92	−133.6	−48.3	−55	−67	−27.8	−18	−0.4
−68.3	−91	−131.8	−47.8	−54	−65.2	−27.2	−17	1.4
−67.8	−90	−130	−47.2	−53	−63.4	−26.7	−16	3.2
−67.2	−89	−128.2	−46.7	−52	−61.6	−26.1	−15	5
−66.7	−88	−126.4	−46.1	−51	−59.8	−25.6	−14	6.8
−66.1	−87	−124.6	−45.5	−50	−58	−25	−13	8.6
−65.6	−86	−122.8	−45	−49	−56.2	−24.4	−12	10.4
−65.0	−85	−121	−44.4	−48	−54.4	−23.9	−11	12.2
−64.4	−84	−119.2	−43.9	−47	−52.6	−23.3	−10	14
−63.9	−83	−117.4	−43.3	−46	−50.8	−22.8	−9	15.8
−63.3	−82	−115.6	−42.8	−45	−49	−22.2	−8	17.6
−62.8	−81	−113.8	−42.2	−44	−47.2	−21.7	−7	19.4
−62.2	−80	−112	−41.7	−43	−45.4	−21.1	−6	21.2
−61.7	−79	−110.2	−41.1	−42	−43.6	−20.6	−5	23
−61.1	−78	−108.4	−40.6	−41	−41.8	−20	−4	24.8
−60.6	−77	−106.6	−40	−40	−40	−19.4	−3	26.6
−60.0	−76	−104.8	−39.4	−39	−38.2	−18.9	−2	28.4
−59.4	−75	−103	−38.9	−38	−36.4	−18.3	−1	30.2
−58.9	−74	−101.2	−38.3	−37	−34.6	−17.8	0	32

°C		°F	°C		°F	°C		°F
−17.2	1	33.8	11.1	52	125.6	54	130	266
−16.7	2	35.6	11.7	53	127.4	60	140	284
−16.1	3	37.4	12.2	54	129.2	66	150	302
−15.6	4	39.2	12.8	55	131.0	71	160	320
−15.0	5	41.0	13.3	56	132.8	77	170	338
−14.4	6	42.8	13.9	57	134.6	82	180	356
−13.9	7	44.6	14.4	58	136.4	88	190	374
−13.3	8	46.4	15.0	59	138.2	93	200	392
−12.8	9	48.2	15.6	60	140.0	99	210	410
−12.2	10	50.0	16.1	61	141.8	100	212	414
−11.7	11	51.8	16.7	62	143.6	104	220	428
−11.1	12	53.6	17.2	63	145.4	110	230	446
−10.6	13	55.4	17.8	64	147.2	116	240	464
−10.0	14	57.2	18.3	65	149.0	121	250	482
−9.44	15	59.0	18.9	66	150.8	127	260	500
−8.89	16	60.8	19.4	67	152.6	132	270	518
−8.33	17	62.6	20.0	68	154.4	138	280	536
−7.78	18	64.4	20.6	69	156.2	143	290	554
−7.22	19	66.2	21.1	70	158.0	149	300	572
−6.67	20	68.0	21.7	71	159.8	154	310	590
−6.11	21	69.8	22.2	72	161.6	160	320	608
−5.56	22	71.6	22.8	73	163.4	166	330	626
−5.00	23	73.4	23.3	74	165.2	171	340	644
−4.44	24	75.2	23.9	75	167.0	177	350	662
−3.89	25	77.0	24.4	76	168.8	182	360	680
−3.33	26	78.8	25.0	77	170.6	188	370	698
−2.78	27	80.6	25.6	78	172.4	193	380	716
−2.22	28	82.4	26.1	79	174.2	199	390	734
−1.67	29	84.2	26.7	80	176.0	204	400	752
−1.11	30	86.0	27.2	81	177.8	210	410	770
−0.56	31	87.8	27.8	82	179.6	216	420	788
0	32	89.6	28.3	83	181.4	221	430	806
0.56	33	91.4	28.9	84	183.2	227	440	824
1.11	34	93.2	29.4	85	185.0	232	450	842
1.67	35	95.0	30.0	86	186.8	238	460	860
2.22	36	96.8	30.6	87	188.6	243	470	878
2.78	37	98.6	31.1	88	190.4	249	480	896
3.33	38	100.4	31.7	89	192.2	254	490	914
3.89	39	102.2	32.2	90	194.0	260	500	932
4.44	40	104.0	32.8	91	195.8	266	510	950
5.00	41	105.8	33.3	92	197.6	271	520	968
5.56	42	107.6	33.9	93	199.4	277	530	986
6.11	43	109.4	34.4	94	201.2	282	540	1004
6.67	44	111.2	35.0	95	203.0	288	550	1022
7.22	45	113.0	35.6	96	204.8	293	560	1040
7.78	46	114.8	36.1	97	206.6	299	570	1058
8.33	47	116.6	36.7	98	208.4	304	580	1076
8.89	48	118.4	37.2	99	210.2	310	590	1094
9.44	49	120.2	38	100	212	316	600	1112
10.0	50	122.0	43	110	230	321	610	1130
10.6	51	123.8	49	120	248	327	620	1148

contd

°C		°F	°C		°F	°C		°F
332	630	1166	616	1140	2084	899	1650	3002
338	640	1184	621	1150	2102	904	1660	3020
343	650	1202	627	1160	2120	910	1670	3038
349	660	1220	632	1170	2138	916	1680	3056
354	670	1238	638	1180	2156	921	1690	3074
360	680	1256	643	1190	2174	927	1700	3092
366	690	1274	649	1200	2192	932	1710	3110
371	700	1292	654	1210	2210	938	1720	3128
377	710	1310	660	1220	2228	943	1730	3146
382	720	1328	666	1230	2246	949	1740	3164
388	730	1346	671	1240	2264	954	1750	3182
393	740	1364	677	1250	2282	960	1760	3200
399	750	1382	682	1260	2300	966	1770	3218
404	760	1400	688	1270	2318	971	1780	3236
410	770	1418	693	1280	2336	977	1790	3254
416	780	1436	699	1290	2354	982	1800	3272
421	790	1454	704	1300	2372	988	1810	3290
427	800	1472	710	1310	2390	993	1820	3308
432	810	1490	716	1320	2408	999	1830	3326
438	820	1508	721	1330	2426	1004	1840	3344
443	830	1526	727	1340	2444	1010	1850	3362
449	840	1544	732	1350	2462	1016	1860	3380
454	850	1562	738	1360	2480	1021	1870	3398
460	860	1580	743	1370	2498	1027	1880	3416
466	870	1598	749	1380	2516	1032	1890	3434
471	880	1616	754	1390	2534	1038	1900	3452
477	890	1634	760	1400	2552	1043	1910	3470
482	900	1652	766	1410	2570	1049	1920	3488
488	910	1670	771	1420	2588	1054	1930	3506
493	920	1688	777	1430	2606	1060	1940	3524
499	930	1706	782	1440	2624	1066	1950	3542
504	940	1724	788	1450	2642	1071	1960	3560
510	950	1742	793	1460	2660	1077	1970	3578
516	960	1760	799	1470	2678	1082	1980	3596
521	970	1778	804	1480	2696	1088	1990	3614
527	980	1796	810	1490	2714	1093	2000	3632
532	990	1814	816	1500	2732	1099	2010	3650
538	1000	1832	821	1510	2750	1104	2020	3668
543	1010	1850	827	1520	2768	1110	2030	3686
549	1020	1868	832	1530	2786	1116	2040	3704
554	1030	1886	838	1540	2804	1121	2050	3722
560	1040	1904	843	1550	2822	1127	2060	3740
566	1050	1922	849	1560	2840	1132	2070	3758
571	1060	1940	854	1570	2858	1138	2080	3776
577	1070	1958	860	1580	2876	1143	2090	3794
582	1080	1976	866	1590	2894	1149	2100	3812
588	1090	1994	871	1600	2912	1154	2110	3830
593	1100	2012	877	1610	2930	1160	2120	3848
599	1110	2030	882	1620	2948	1166	2130	3866
604	1120	2048	888	1630	2966	1171	2140	3884
610	1130	2066	893	1640	2984	1177	2150	3902

contd

°C	°F		°C	°F		°C	°F	
1182	2160	3920	1349	2460	4460	1516	2760	5000
1188	2170	3938	1354	2470	4478	1521	2770	5018
1193	2180	3956	1360	2480	4496	1527	2780	5036
1199	2190	3974	1366	2490	4514	1532	2790	5054
1204	2200	3992	1371	2500	4532	1538	2800	5072
1210	2210	4010	1377	2510	4550	1543	2810	5090
1216	2220	4028	1382	2520	4568	1549	2820	5108
1221	2230	4046	1388	2530	4586	1554	2830	5126
1227	2240	4064	1393	2540	4604	1560	2840	5144
1232	2250	4082	1399	2550	4622	1566	2850	5162
1238	2260	4100	1404	2560	4640	1571	2860	5180
1243	2270	4118	1410	2570	4658	1577	2870	5198
1249	2280	4136	1416	2580	4676	1582	2880	5216
1254	2290	4154	1421	2590	4694	1588	2890	5234
1260	2300	4172	1427	2600	4712	1593	2900	5252
1266	2310	4190	1432	2610	4730	1599	2910	5270
1271	2320	4208	1438	2620	4748	1604	2920	5288
1277	2330	4226	1443	2630	4766	1610	2930	5306
1282	2340	4244	1449	2640	4784	1616	2940	5324
1288	2350	4262	1454	2650	4802	1621	2950	5342
1293	2360	4280	1460	2660	4820	1627	2960	5360
1299	2370	4298	1466	2670	4838	1632	2970	5378
1304	2380	4316	1471	2680	4856	1638	2980	5396
1310	2390	4334	1477	2690	4874	1643	2990	5414
1316	2400	4352	1482	2700	4892	1649	3000	5432
1321	2410	4370	1488	2710	4910	1705	3100	5612
1327	2420	4388	1493	2720	4928	1760	3200	5792
1332	2430	4406	1499	2730	4946	1816	3300	5972
1338	2440	4424	1504	2740	4964	1871	3400	6152
1343	2450	4442	1510	2750	4982			

APPENDIX 4

Strength conversion table

N/mm^2	hbar	kgf/mm^2	$tonf/in^2$	lbf/in^2	N/mm^2	hbar	kgf/mm^2	$tonf/in^2$	lbf/in^2
10	1	1.02	0.647	1450	480	48	48.95	31.08	69,600
20	2	2.04	1.295	2900	490	49	49.97	31.73	71,050
30	3	3.06	1.942	4350	500	50	50.99	32.37	72,500
40	4	4.08	2.590	5800	510	51	52.00	33.02	73,950
50	5	5.10	3.237	7250	520	52	53.02	33.67	75,400
60	6	6.12	3.885	8700	530	53	54.04	34.32	76,850
70	7	7.14	4.532	10,150	540	54	55.06	34.96	78,300
80	8	8.16	5.180	11,600	550	55	56.08	35.61	79,750
90	9	9.18	5.827	13,050	560	56	57.10	36.26	81,200
100	10	10.20	6.475	14,500	570	57	58.12	36.91	82,650
110	11	11.22	7.122	15,950	580	58	59.14	37.55	84,100
120	12	12.24	7.770	17,400	590	59	60.16	38.20	85,550
130	13	13.26	8.417	18,850	600	60	61.18	38.85	87,000
140	14	14.28	9.065	20,300	610	61	62.20	39.50	88,450
150	15	15.30	9.712	21,750	620	62	63.22	40.14	89,900
160	16	16.32	10.36	23,200	630	63	64.24	40.79	91,350
170	17	17.33	11.01	24,650	640	64	65.26	41.44	92,800
180	18	18.35	11.65	26,100	650	65	66.28	42.09	94,250
190	19	19.37	12.30	27,550	660	66	67.30	42.74	95,700
200	20	20.39	12.95	29,000	670	65	68.32	43.38	97,150
210	21	21.41	13.60	30,450	680	68	69.34	44.02	98,600
220	22	22.43	14.24	31,900	690	69	70.36	44.68	100,050
230	23	23.45	14.89	33,350	700	70	71.38	45.32	101,500
240	24	24.47	15.54	34,800	710	71	72.40	45.97	103,000
250	25	25.49	16.19	36,250	720	72	73.42	46.62	104,400
260	26	26.51	16.83	37,700	730	73	74.44	47.27	105,900
270	27	27.53	17.48	39,150	740	74	75.46	47.91	107,300
280	28	28.55	18.13	40,600	750	75	76.48	48.56	108,800
290	29	29.57	18.78	42,050	760	76	77.50	49.21	110,200
300	30	30.59	19.42	43,500	770	77	78.52	49.86	111,700
310	31	31.61	20.07	44,950	780	78	79.54	50.50	113,100
320	32	32.63	20.72	46,400	790	79	80.56	51.15	114,600
330	33	33.65	21.37	47,850	800	80	81.58	51.80	116,000
340	34	34.67	22.01	49,300	810	81	82.60	52.45	117,500
350	35	36.69	22.66	50,750	820	82	83.62	53.09	118,900
360	36	36.71	23.31	52,200	830	83	84.64	53.74	120,400
370	37	37.73	23.96	53,650	840	84	85.65	54.39	121,800
380	38	38.75	24.60	55,100	850	85	86.67	55.04	123,300
390	39	39.77	25.25	56,550	860	86	87.69	55.68	124,700
400	40	40.79	25.90	58,000	870	87	88.71	56.33	126,200
410	41	41.81	26.55	59,450	880	88	89.73	56.98	127,600
420	42	42.83	27.19	60,900	890	89	90.75	57.63	129,100
430	43	43.85	27.84	62,350	900	90	91.77	58.27	130,500
440	44	44.87	28.49	63,800	910	91	92.79	58.92	132,000
450	45	45.89	29.14	65,250	920	92	93.81	59.57	133,400
460	46	46.91	29.78	66,700	930	93	94.83	60.22	134,900
470	47	47.93	30.43	68,150	940	94	95.85	60.86	136,300

N/mm^2	hbar	kgf/mm^2	$tonf/in^2$	lbf/in^2	N/mm^2	hbar	kgf/mm^2	$tonf/in^2$	lbf/in^2
950	95	96.87	61.51	137,800	1460	146	148.9	94.53	211,800
960	96	97.89	62.16	139,200	1470	147	149.9	95.18	213,200
970	97	98.91	62.80	140,700	1480	148	150.9	95.83	214,700
980	98	99.93	63.45	142,100	1490	149	151.9	96.48	216,100
990	99	101.0	64.10	143,600	1500	150	153.0	97.12	217,600
1000	100	102.0	64.75	145,000	1510	151	154.0	97.77	219,000
1010	101	103.0	65.37	146,500	1520	152	155.0	98.42	220,500
1020	102	104.0	66.04	147,900	1530	153	156.0	99.07	221,900
1030	103	105.0	66.69	149,400	1540	154	157.0	99.71	223,400
1040	104	106.0	67.34	150,800	1550	155	158.1	100.4	224,800
1050	105	107.1	67.99	152,300	1560	156	159.1	101.0	226,300
1060	106	108.1	68.63	153,700	1570	157	160.1	101.7	227,700
1070	107	109.1	69.28	155,200	1580	158	161.1	102.3	229,200
1080	108	110.1	69.93	156,600	1590	159	162.1	103.0	230,600
1090	109	111.1	70.58	158,100	1600	160	163.2	103.6	232,100
1100	110	112.2	71.22	159,500	1610	161	164.2	104.2	233,500
1110	111	113.2	71.87	161,000	1620	162	165.2	104.9	235,000
1120	112	114.2	72.52	162,400	1630	163	166.2	105.5	236,400
1130	113	115.2	73.17	163,900	1640	164	167.2	106.2	237,900
1140	114	116.2	73.81	165,300	1650	165	168.3	106.8	239,300
1150	115	117.3	74.46	166,800	1660	166	169.3	107.5	240,800
1160	116	118.3	75.11	168,200	1670	167	170.3	108.1	242,200
1170	117	119.3	75.76	169,700	1680	168	171.3	108.8	243,700
1180	118	120.3	76.40	171,100	1690	169	172.3	109.4	245,100
1190	119	121.3	77.05	172,600	1700	170	173.3	110.1	246,600
1200	120	122.4	77.70	174,000	1710	171	174.4	110.7	248,000
1210	121	123.4	78.35	175,500	1720	172	175.4	111.4	249,500
1220	122	124.4	78.99	176,900	1730	173	176.4	112.0	250,900
1230	123	125.4	79.64	178,400	1740	174	177.4	112.7	252,400
1240	124	126.4	80.29	179,800	1750	175	178.4	113.3	253,800
1250	125	127.5	80.93	181,300	1760	176	179.5	114.0	255,300
1260	126	128.5	81.58	182,800	1770	177	180.5	114.6	256,700
1270	127	129.5	82.23	184,200	1780	178	181.5	115.3	258,200
1280	128	130.5	82.88	185,700	1790	179	182.5	115.9	259,600
1290	129	131.5	83.53	187,100	1800	180	183.5	116.5	261,100
1300	130	132.6	84.17	186,600	1810	181	184.6	117.2	262,500
1310	131	133.6	84.82	190,000	1820	182	185.6	117.8	264,000
1320	132	134.6	85.47	191,500	1830	183	186.6	118.5	265,400
1330	133	135.6	86.12	192,900	1840	184	187.6	119.1	266,900
1340	134	136.6	86.76	194,400	1850	185	188.6	119.8	268,300
1350	135	137.7	87.41	195,800	1860	186	189.7	120.4	269,800
1360	136	138.7	88.06	197,300	1870	187	190.7	121.1	271,200
1370	137	139.7	88.71	198,700	1880	188	191.7	121.7	272,700
1380	138	140.7	89.35	200,200	1890	189	192.7	122.4	274,100
1390	139	141.7	90.00	201,600	1900	190	193.7	123.0	275,600
1400	140	142.8	90.65	203,100	1910	191	194.8	123.7	277,000
1410	141	143.8	91.30	204,500	1920	192	195.8	124.3	278,500
1420	142	144.8	91.94	206,000	1930	193	196.8	125.0	279,900
1430	143	145.8	92.59	207,200	1940	194	197.8	125.6	281,400
1440	144	146.8	93.24	208,900	1950	195	198.8	126.3	282,800
1450	145	147.9	93.89	210,300	1960	196	199.9	126.9	284,300

contd

N/mm^2	hbar	kgf/mm^2	$tonf/in^2$	lbf/in^2	N/mm^2	hbar	kgf/mm^2	$tonf/in^2$	lbf/in^2
1970	197	200.9	127.6	285,700	2190	219	223.3	141.8	317,600
1980	198	201.9	128.2	287,200	2200	220	224.3	142.4	319,100
1990	199	202.9	128.9	288,600	2210	221	225.4	143.1	320,500
2000	200	203.9	129.5	290,100	2220	222	226.4	143.7	322,000
2010	201	205.0	130.1	291,500	2230	223	227.4	144.4	323,400
2020	202	206.0	130.8	293,000	2240	224	228.4	145.0	324,900
2030	203	207.0	131.4	294,400	2250	225	229.4	145.7	326,300
2040	204	208.0	132.1	295,900	2260	226	230.5	146.3	327,800
2050	205	209.0	132.7	297,300	2270	227	231.5	147.0	329,200
2060	206	210.1	133.4	298,800	2280	228	232.5	147.6	330,700
2070	207	211.1	134.0	300,200	2290	229	233.5	148.3	332,100
2080	208	212.1	134.7	301,700	2300	230	234.5	148.9	333,600
2090	209	213.1	135.3	303,100	2310	231	235.6	149.6	335,000
2100	210	214.1	136.0	304,600	2320	232	236.6	150.2	336,500
2110	211	215.2	136.6	306,000	2330	233	237.6	150.9	337,900
2120	212	216.2	137.3	307,500	2340	234	238.6	151.5	339,400
2130	213	217.2	137.9	308,900	2350	235	239.6	152.2	340,800
2140	214	218.2	138.6	310,200	2360	236	240.6	152.8	342,300
2150	215	219.2	139.2	311,800	2370	237	241.7	153.5	343,700
2160	216	220.3	139.9	313,300	2380	238	242.7	154.1	345,200
2170	217	221.3	140.5	314,700	2390	239	243.7	154.8	346,600
2180	218	222.3	141.29	316,200	2400	240	244.7	155.4	348,100

APPENDIX 5

Hardness and tensile values approximate conversion table

Brinell			Vickers	Rockwell				Sclero-scope
10mm, 3000 kg		Approximate equivalent tensile strength for steel	Hardness number, HV, DPN or VPN	120° cone			$\frac{1}{16}$ in ball	
Dia. mm	Hardness No.	tonf/in² hbar		C scale 150 kg	D scale 100 kg	A scale 60kg	B scale 100kg	
2.00	945			1250	71	80	87	—	—
2.05	898			1150	70	79	87	—	—
2.10	856			1050	69	79	86	—	—
2.15	816			1000	68	78	86	—	—
2.20	781			975	67	78	85	—	106
2.25	745	163	251.7	950	66	77	85	—	100
2.30	712	155	239.4	910	65	76	84	—	95
2.35	683	150	231.7	850	64	75	84	—	91
2.40	653	143	220.9	790	62	73	83	—	87
2.45	627	137	211.6	750	61	72	82	—	84
2.50	601	132	203.9	715	59	71	81	—	81
2.55	578	127	196.1	671	57	69	80	—	78
2.60	555	122	188.4	633	56	68	79	—	75
2.65	534	117	180.7	599	54	67	78	—	72
2.70	514	112	173.0	572	52	65	77	—	70
2.75	495	108	166.8	547	50	64	76	—	67
2.80	477	105	162.2	523	49	63	75	—	65
2.85	461	101	156.0	501	48	62	75	—	63
2.90	444	98	151.4	479	47	61	74	—	61
2.95	429	95	146.7	459	45	60	73	—	59
3.00	415	92	142.1	441	44	59	73	—	57
3.05	401	88	135.9	424	42	58	72	—	55
3.10	388	85	131.3	409	41	57	71	—	54
3.15	375	82	126.6	395	40	56	71	—	52
3.20	363	80	123.6	382	39	55	70	—	51
3.25	352	77	118.9	369	37	53	69	—	49
3.30	341	75	115.8	358	36	52	68	—	48
3.35	331	73	112.7	344	34	51	67	—	46
3.40	321	71	109.7	332	33	50	67	—	45
3.45	311	68	105.0	321	32	50	67	—	44
3.50	302	66	101.9	310	31	49	66	—	43
3.55	293	64	98.8	299	30	49	66	—	42
3.60	285	63	97.3	290	29	48	65	—	41
3.65	277	61	94.2	282	27	46	64	—	40
3.70	269	59	91.1	274	26	45	64	—	39
3.75	262	58	89.6	267	25	45	63	—	38
3.80	255	56	86.5	260	24	44	63	—	37
3.85	248	55	84.9	253	23	43	62	—	36
3.90	241	53	81.9	246	22	42	62	—	35

contd

Brinell			Vickers	Rockwell				Sclero-scope
10mm, 3000 kg		Approximate equivalent tensile strength for steel	Hardness number, HV, DPN or VPN	120° cone			$\frac{1}{16}$ in ball	
Dia. mm	Hardness No.	tonf/in² hbar		C scale 150 kg	D scale 100 kg	A scale 60kg	B scale 100kg	
3.95	235	51	78.8	240	21	41	61	100	34
4.00	229	50	77.2	234	20	41	61	99	33
4.05	223	49	75.7	228	19	40	60	98	32
4.10	217	48	74.1	222	18	—	60	97	31
4.15	212	46	71.0	217	17	—	59	96	31
4.20	207	45	69.5	212	16	—	58	95	30
4.25	201	44	68.0	206	15	—	57	94	30
4.30	197	43	66.4	202	13	—	57	93	29
4.35	192	42	64.9	197	12	—	56	92	28
4.40	187	41	63.3	192	10	—	56	91	28
4.45	183	40	61.8	188	9	—	55	90	28
4.50	179	39	60.2	184	8	—	55	89	27
4.55	174	38	58.7	179	7	—	54	88	27
4.60	170	38	58.7	175	6	—	54	87	26
4.65	167	38	58.7	172	4	—	53	86	26
4.70	163	37	57.1	168	3	—	52	84	25
4.75	159	36	55.6	164	2	—	51	83	24
4.80	156	36	55.6	161	1	—	51	82	24
4.85	152	35	54.1	157	—	—	50	81	23
4.90	149	34	52.5	154	—	—	50	80	23
4.95	146	33	51.0	151	—	—	49	79	22
5.00	143	33	51.0	148	—	—	49	78	22
5.05	140	32	49.4	145	—	—	48	76	21
5.10	137	31	47.9	142	—	—	47	75	21
5.15	134	31	47.9	139	—	—	47	74	21
5.20	131	30	46.3	136	—	—	46	73	20
5.25	128	30	46.3	133	—	—	45	72	20
5.30	126	29	44.8	131	—	—	45	71	20
5.35	123	28	43.2	128	—	—	44	69	—
5.40	121	28	43.2	126	—	—	44	68	—
5.45	118	27	41.7	123	—	—	43	67	—
5.50	116	27	41.7	121	—	—	43	65	—
5.55	114	26	40.2	119	—	—	42	64	—
5.60	112	25	38.6	117	—	—	41	63	—
5.65	109	25	38.6	115	—	—	40	62	—
5.70	107	24	37.1	113	—	—	40	60	—
5.75	105	24	37.1	111	—	—	39	59	—
5.80	103	23	35.5	109	—	—	39	57	—
5.85	101	23	35.5	107	—	—	38	55	—
5.90	99	22	34.0	104	—	—	38	54	—
5.95	97	22	34.0	102	—	—	37	53	—
6.00	95	21	32.4	100	—	—	35	51	—
—	94			98	—	—	—	50.5	—

APPENDICES / 561

Brinell			Vickers	Rockwell				Sclero-scope
10mm, 3000 kg		Approximate equivalent tensile strength for steel tonf/in² hbar	Hardness number, HV, DPN or VPN	120° cone			$\frac{1}{16}$ in ball	
Dia. mm	Hardness No.			C scale 150 kg	D scale 100 kg	A scale 60 kg	B scale 100 kg	
—	93		96	—	—	—	50.0	—
—	92		94	—	—	—	49.0	—
—	91		92	—	—	—	48.0	—
—	90		90	—	—	—	47.5	—
—	88		88	—	—	—	46.0	—
—	86		86	—	—	—	44.0	—
—	84		84	—	—	—	42.0	—
—	82		82	—	—	—	40.0	—
—	80		80	—	—	—	37.5	—
—	78		78	—	—	—	35.0	—
—	76		76	—	—	—	32.5	—
—	74		74	—	—	—	30.0	—
—	72		72	—	—	—	27.5	—
—	70		70	—	—	—	24.5	—
—	68		68	—	—	—	21.5	—
—	66		66	—	—	—	18.5	—
—	64		64	—	—	—	15.5	—
—	62		62	—	—	—	12.5	—
—	60		60	—	—	—	—	—
—	58		58	—	—	—	—	—
—	56		56	—	—	—	—	—
—	54		54	—	—	—	—	—
—	50		50	—	—	—	—	—
—	49		49	—	—	—	—	—
—	48		48	—	—	—	—	—
—	47		47	—	—	—	—	—
—	46		46	—	—	—	—	—
—	45		45	—	—	—	—	—

Note:
Conversion of one hardness test figure to any other must only be an approximation.

Conversion of hardness test figures to ultimate tensile test is also an approximation and will also vary depending on the metal being tested. No hardness test, therefore, can be used to accurately define the strength of a metal.

APPENDIX 6

Impact data conversion table

joules	ft lb	kg/m	joules	ft lb	kg/m	joules	ft lb	kg/m
1	1	0.14	47	35	4.84	92	68	9.40
3	2	0.28	49	36	4.98	94	69	9.54
4	3	0.42	50	37	5.12	95	70	9.68
5	4	0.55	52	38	5.25	96	71	9.82
7	5	0.69	53	39	5.39	98	72	9.95
8	6	0.83	54	40	5.53	99	73	10.09
9	7	0.97	56	41	5.67	100	74	10.23
11	8	1.11	57	42	5.81	102	75	10.37
12	9	1.24	58	43	5.95	103	76	10.51
14	10	1.38	60	44	6.08	104	77	10.65
15	11	1.52	61	45	6.22	106	78	10.78
16	12	1.66	62	46	6.36	107	79	10.92
18	13	1.80	64	47	6.50	108	80	11.06
19	14	1.94	65	48	6.64	110	81	11.20
20	15	2.07	66	49	6.78	111	82	11.34
22	16	2.21	68	50	6.91	113	83	11.48
23	17	2.35	69	51	7.05	114	84	11.61
24	18	2.49	71	52	7.19	115	85	11.75
26	19	2.63	72	53	7.33	117	86	11.89
27	20	2.77	73	54	7.47	118	87	12.03
28	21	2.90	75	55	7.60	119	88	12.17
30	22	3.04	76	56	7.74	121	89	12.31
31	23	3.18	77	57	7.88	122	90	12.44
33	24	3.32	78	58	8.02	123	91	12.58
34	25	3.46	80	59	8.16	124	92	12.72
35	26	3.60	81	60	8.30	126	93	12.86
37	27	3.73	83	61	8.43	127	94	13.00
38	28	3.97	84	62	8.57	129	95	13.13
39	29	4.01	85	63	8.71	130	96	13.27
41	30	4.15	87	64	8.85	132	97	13.41
42	31	4.29	88	65	8.99	133	98	13.55
43	32	4.42	89	66	9.13	134	99	13.69
45	33	4.56	91	67	9.26	136	100	13.83
46	34	4.70						

Conversion: ft/lb to joules multiply by 1.356
joules to ft/lb multiply by 0.737
joules to kg/m multiply by 0.1
kg/m to joules multiply by 10
kg/m to ft/lb multiply by 7.37

APPENDIX 7

Hydrometer scales
Specific gravity conversion

Baumé	Twaddell	Specific gravity	Baumé	Twaddell	Specific gravity
0	0	1.000	19.3	31	1.155
0.7	1	1.005	19.8	32	1.160
1.0	1.4	1.007	20.0	32.4	1.162
1.4	2	1.010	20.3	33	1.165
2.0	2.8	1.014	20.9	34	1.170
2.1	3	1.015	21.0	34.2	1.171
2.7	4	1.020	21.4	35	1.175
3.0	4.4	1.022	22.0	36	1.180
3.4	5	1.025	22.5	37	1.185
4.0	5.8	1.029	23.0	38	1.190
4.1	6	1.030	23.5	39	1.195
4.7	7	1.035	24.0	40	1.200
5.0	7.4	1.037	24.5	41	1.205
5.4	8	1.040	25.0	42	1.210
6.0	9	1.045	25.5	43	1.215
6.7	10	1.050	26.0	44	1.220
7.0	10.2	1.052	26.4	45	1.225
7.4	11	1.055	26.9	46	1.230
8.0	12	1.060	27.0	46.2	1.231
8.7	13	1.065	27.4	47	1.235
9.0	13.4	1.067	27.9	48	1.240
9.4	14	1.070	28.0	48.2	1.241
10.0	15	1.075	28.4	49	1.245
10.6	16	1.080	28.8	50	1.250
11.0	16.6	1.083	29.0	50.4	1.252
11.2	17	1.085	29.3	51	1.255
11.9	18	1.090	29.7	52	1.260
12.0	18.2	1.091	30.0	52.6	1.263
12.4	19	1.095	30.2	53	1.265
13.0	20	1.100	30.6	54	1.270
13.6	21	1.105	31.0	54.8	1.274
14.0	21.6	1.108	31.1	55	1.275
14.2	22	1.110	31.5	56	1.280
14.9	23	1.115	32.0	57	1.285
15.0	23.2	1.116	32.4	58	1.290
15.4	24	1.120	32.8	59	1.295
16.0	25	1.125	33.0	59.4	1.297
16.5	26	1.130	33.3	60	1.300
17.0	26.8	1.134	33.7	61	1.305
17.1	27	1.135	34.0	61.6	1.308
17.7	28	1.140	34.2	62	1.310
18.0	28.4	1.142	34.6	63	1.315
18.3	29	1.145	35.0	64	1.320
18.8	30	1.150	35.4	65	1.325
19.0	30.4	1.152	35.8	66	1.330

contd

Baumé	Twaddell	Specific gravity	Baumé	Twaddell	Specific gravity
36.0	66.4	1.332	45.8	93	1.465
36.2	67	1.335	46.0	93.6	1.468
36.6	78	1.340	46.1	94	1.470
37.0	69	1.345	46.4	95	1.475
37.4	70	1.350	46.8	96	1.480
37.8	71	1.355	47.0	96.6	1.483
38.0	71.4	1.357	47.1	97	1.485
38.2	72	1.360	47.4	98	1.490
38.6	73	1.365	47.8	99	1.495
39.0	74	1.370	48.0	99.6	1.498
39.4	75	1.375	48.1	100	1.500
39.8	76	1.380	48.4	101	1.505
40.0	76.6	1.383	49.0	103	1.515
40.1	77	1.385	50.0	106	1.530
40.5	78	1.390	51.0	109.2	1.546
40.8	79	1.395	52.0	112.6	1.563
41.0	79.4	1.397	53.0	116	1.580
41.2	80	1.400	54.0	119.4	1.597
41.6	81	1.405	55.0	123	1.615
42.0	82	1.410	56.0	127	1.635
42.3	83	1.415	57.0	130	1.650
42.7	84	1.420	58.0	134.2	1.671
43.0	84.8	1.424	59.0	138.2	1.690
43.1	85	1.425	60.0	142	1.710
43.4	86	1.430	61.0	146.2	1.731
43.8	87	1.435	62.0	160.6	1.753
44.0	87.6	1.438	63.0	155	1.775
44.1	88	1.440	64.0	159	1.795
44.4	89	1.445	65.0	164	1.820
44.8	90	1.450	66.0	168.4	1.842
45.0	90.6	1.453	67.0	173	1.865
45.1	91	1.455	68.0	178.2	1.891
45.4	92	1.460	69.0	183.2	1.916

APPENDIX 8

Thickness conversion table

Millimetres (mm)	Microns (μm)	Inches (in)
1	1000	0.04
0.5	500	0.02
0.4	400	0.015
0.3	300	0.012
0.2	200	0.008
0.1	100	0.004
0.09	90	0.0035
0.08	80	0.0032
0.07	70	0.0028
0.06	60	0.0025
0.05	50	0.002
0.04	40	0.0015
0.03	30	0.0012
0.02	20	0.0008
0.01	10	0.0004
0.005	5	0.0002
0.004	4	0.00015
0.003	3	0.00012
0.002	2	0.00008
0.001	1	0.00004
0.0008	0.8	0.00003
0.0005	0.5	0.00002
0.0001	0.1	0.000004

Note:
In verbal discussion there can be some confusion in that millimetres (mm) are often referred to as 'mils', meaning one thousandth of a metre while thousandths of an inch are also on occasion referred to as 'mils', meaning one thousandth of an inch.

With inches 0.001 inch will be called 'one thou', and 0.0001 inch will be called 'one tenth'.

It is thus clear that considerable confusion can be caused unless the basic unit being discussed is defined.

In writing, the term 'mil' will almost always refer to one thousandth of an inch; this applies in particular to American literature.

APPENDIX 9

Companies and organizations involved in metal treatments and testing

Abrasive Developments Limited, Norman House, High Street, Henley-in-Arden B95 5AH.
Alcan Limited, Southam Road, Banbury, Oxon.
M. L. Alkan Limited, Stonefield Way, Victoria Road, Ruislip, Middlesex HA4 0JS.
Aluminium Federation, Broadway House, Cabthorpe Road, 5 Ways, Birmingham B15 1TN.
American Electroplaters' Society Incorporated, 56 Melmore Gardens, East Orange, New Jersey 07017.
American Society for Testing & Materials ASTM, 1916 Race Street, Philadelphia, USA.
Association of Metal Sprayers, Chamber of Commerce House, P.O. Box 360, 25 Harbarve Road, Birmingham B15 3DH.
British Aluminium Co Limited, Norfolk House, St James Square, London SW1.
British Steel Service Centres Ltd., Stourbridge Road, Lye, Stourbridge.
British Cast Iron Research Association, Alvechurch, Birmingham B48 7QB.
British Heat Treatments Limited (a division of Unochrome Limited), Strathclyde Division, 40 Milton Road, College Milton, East Kilbride, Glasgow G74 5DF.
British Metal Finishing Suppliers Association, 15 Tooks Court, London EC4.
British Non-Ferrous Metals Research Association, Grove Laboratories, Denchworth Road, Wantage, Oxfordshire OX12 9BJ.
British Oxygen Co Limited, Vigo Lane, Chester-le-Street, County Durham.
British Standards Institution, 2 Park Street, London W1.
British Steel Corporation, P.O. Box 403, 33 Grosvenor Place, London SW1X 73F.
Calorizing Corporation, Dumbuck Works, Dumbarton G82 1EP.
W. Canning & Co Limited, Great Hampton Street, Birmingham B18 6AS.
Cassel Heat Treatment, ICI, P.O. Box 216, Whitton, Birmingham B6 7BA.
J. J. Casting Investments (Heat Treatment) Limited, Wilson Industrial Estate, Van Road, Caerphilly CF8 3ED.
Ciba-Geigy (UK) Limited, 30 Buckingham Gate, London SW1E 6LH.
Copper Development Association, Orchard House, Mutton Lane, Potters Bar, Hertfordshire EN6 3AP.
Cobalt Information Centre, Chichester House, 278/282 High Holborn, London WC1.
Herbert Cotterill Limited, Unit 1A, Westfield Industrial Estate, Kirk Lane, Yeadon, Leeds LS19 7LX.
M. & T. Cruickshank Limited, Lydon House, 62 Hagley Road, Edgbaston, Birmingham 16.
Degussa Limited, Paul Ungerer House, Earl Road, Stanley Green, Handforth, Wilmslow SK9 3RL.
Deloro Stellite Limited, Stratton St Margaret, Swindon, Wiltshire SN3 4QA.
Dewrance & Co. Limited, Special Alloy Division, Great Dover Street, London SE1.
Diversay UK Limited, Cockfosters Road, Cockfosters, Barnet, Hertfordshire.
Dupont UK Limited, Dupont House, 18 Bream's Building, Fetter Lane, London EC4.
Elbar B.V. Industrieterrien Spikweien 5943 AD Lomm, Netherlands.
Engelhard Industries Limited, St Nicholas House, St Nicholas Road, Sutton, Surrey.
The English Abrasive Co Limited, Marsh Lane, Tottenham, London N17.
Exxon Chemical Limited, Arundel Tower, Portland Terrace, Southampton.

Fescol Limited, North Road, London N7 9DR.
Fothergill Engineered Surfaces Limited, Long Causeway, Leeds LS9 0NY.
Galvanizers Association, 34 Berkeley Square, London W1X 6AS.
General & Industrial Paints Limited, 28 Wadsworth Road, Perivale, Middlesex.
Haynes Stellite – Samuel Osborne & Co. Limited, P.O. Box 1, Clyde Steel Works, Sheffield 3.
Henry W. Peabody Limited, now Peabody Modernair Limited, 67 Wilson Street, Finsbury Square, London EC2.
Henry Wiggin Limited, Thames House, Millbank, London SW1.
Hockley Chemical Co. Limited, Hockley Hill, Birmingham 18.
ICI Limited
  Mond Division, P.O. Box 13, The Heath, Runcorn, Cheshire WA7 7QF.
  Paints Division, Wexham Road, Slough, Bucks SL2 5DS.
  Head Office: Nobel House, 2 Buckingham Gate, London SW1.
Imasa-Silvercrown, 188 Bath Road, Slough, Buckinghamshire.
Institution of Corrosion Technology, 14 Belgrave Square, London SW1X 8PS.
Institute of Metal Finishing, 178 Goswell Road, London EC1.
Institute of Metals, 1 Carlton House Terrace, London SW1Y 5DB.
Institute of Vitreous Enamellers, Head Office, Ripley, Near Derby DE3 3EB.
International Lead Zinc Research Organisation Incorporated, 292 Madison Avenue, New York NY 10017.
International Nickel Limited, Thames House, Millbank, London SW1.
Iron & Steel Institute, 4 Grosvenor Gardens, London SW1.
Johnson Matthey Metals Ltd, 100 High Street, Southgate, London N14 6ET.
Kolene Corporation, Detroit, Michigan, USA.
Korel Korrosionsschatz – Electronik, 4030 Ratingen 4, Lintorf, F.D.R.
Lea-Ronal (UK) Limited, Anderson Works, Tongue Lane, Fairfield, Buckston, Derbyshire SK17 7LG.
Lead Development Association, 34 Berkeley Square, London W1.
Loctite Company, Douglas Kane Group, Swallowfields, Welwyn Garden City.
M & T Chemicals, Scottish Sales Office & Depot, Wilson Place, Nerston Industrial Estate, East Kilbride, Glasgow G74 4QD.
Metachem Company, now known as Metadalic, 258b Ipswich Rd Trading Estate, Slouth SL1 4EP.
Metadalec Limited, Blackthorne Road, Doyl Estate, Colnbrook, Buckinghamshire.
Metal Finishing Association, 27 Frederic Street, Birmingham 1.
Metal Society, 1 Carlton House Terrace, London SW1Y 5DB.
Metalock Britain Limited, Metalock House, Crabtree Manorway, Belvedere DA17
Metco Limited (formerly Metalising Equipment Co. Limited), Chobham, Woking.
Molybond Laboratories, Placer Exploration, 34 Adelaide Street, Danderong, Victoria, Australia.
National Association of Corrosion Engineers, 2400 W. Loop South, Houston, Texas, USA.
Osro Limited, Tubro House, Mark Road, Hemel Hempstead, Hertfordshire.
Oxy Metal Industries (GB) Limited, Forsyth Road, Sheerwater, Woking, Surrey, GU21 5RZ.
Paintmakers Association of Great Britain Limited, Prudential House, Welsley Road, Croydon CR9 2ET.
Pennwalt Limited, Doman Road, Camberley, Surrey GU15 3DN.
Permanent Magnet Association, 301 Glossop Road, Sheffield 10.
Porcelain Enamel Institute, 1900 L. Street NW, Washington DC 20036, USA.
P.M.M., Unit G12, Wellheads Industrial Estate, Dyce, Aberdeen.

Pyrene Chemical Services Limited, Ridgeway, Iver, Buckinghamshire.
Rath Manufacturing Company, Jamesville, USA.
Sel Rex Limited (now Oxy Metal Industries – Precious Metals Division), Holyhead Road, Chirk, Near Wrexham, Wales.
Sermetal Division of Teleflex Incorporated, High Holborn Road, Cednor Gate Industrial Estate, Ripley, Derbyshire DE5 4QX.
Silvercrown (now Imasa-Silvercrown) 188 Bath Road, Slough, Buckinghamshire.
Steetley Berk Ltd, P.O. Box 3, Brampton Hill, Newcastle-under-Lyme ST5 0QU.
Sunbeam Anti-Corrosives Limited, Central Works, Central Avenue, West Molesey, Surrey.
Tin Research Institute, Fraser Road, Perivale, Greenford, Middlesex.
Union Carbide Co. Limited, Drake Way, Greenbridge Industrial Estate, Swindon, SN3 3HX.
Walterising Co. (UK) Limited, Waddon Marsh Way, Purley Way, Croydon, Surrey CR9 4HT.
Welding Institute, Abington Hall, Abington.
Wolfson Plasma Processing Unit, University of Liverpool.
Zinc Development Association, 34 Berkeley Square, London W1.